Marie Marvingt,
Fiancée of Danger

Marie Marvingt, Fiancée of Danger

First Female Bomber Pilot, World-Class Athlete and Inventor of the Air Ambulance

Rosalie Maggio

McFarland & Company, Inc., Publishers
Jefferson, North Carolina

Library of Congress Cataloguing-in-Publication Data

Names: Maggio, Rosalie, author.
Title: Marie Marvingt, fiancée of danger : first female bomber pilot, world-class athlete and inventor of the air ambulance / Rosalie Maggio.
Description: Jefferson, North Carolina : McFarland & Company, Inc., Publishers, 2019 | Includes bibliographical references and index.
Identifiers: LCCN 2019012361 | ISBN 9781476675503 (paperback : acid free paper) ♾
Subjects: LCSH: Marvingt, Marie, 1875–1963. | Women air pilots—France—Biography.
Classification: LCC TL540.M3672 M338 2019 | DDC 629.13092 [B] —dc23
LC record available at https://lccn.loc.gov/2019012361

British Library cataloguing data are available

ISBN (print) 978-1-4766-7550-3
ISBN (ebook) 978-1-4766-3407-4

Front cover: Marie Marvingt, 1906;
“French Monoplane,” 1909 (© 2019 PicturesNow)

Printed in the United States of America

McFarland & Company, Inc., Publishers
Box 611, Jefferson, North Carolina 28640
www.mcfarlandpub.com

DAVID
Liz, Anthony
Katie, Jason, Margot
Matt, Nora, Zoe, Evy

Acknowledgments

Marcel Cordier, award-winning French writer and poet, has done more than any other person to restore Marie Marvingt to her rightful place in history. He and I have worked together for many years, and he's a dear friend as well as a brilliant writer. In 1982, he founded the International Marie Marvingt Committee, whose first actions extended the lease on her burial plot and installed a plaque honoring her on the building where she had lived for seventy-five years. Since then, Marcel and the Committee have promoted Marie and publicized her name in hundreds of ways (his wife, prominent Lorraine artist Mido Cordier, designed the plaque).

Jeannie Jung-Pierron, Renaissance woman and also a dear friend, was the only publisher in France to take a chance on bringing out the first biography, *Marie Marvingt: La Femme d'un Siècle* by Rosalie Maggio and Marcel Cordier (Éditions Pierron, 1991).

The late Jacqueline Maire was a major inspiration, source of energy, and tireless gatherer of information. The late Jeanne and Louis Lhérault, connections of Marie's, were wonderfully helpful and generous as were so many individuals whose family names are known in Nancy:

Blanc, Coulet, Méchelle, Didier, Friry, Bridoux, Hulster, Thévenin, Bertrand, Knecht, Willotte, Plauche-Gillon, Besançon, Bouchet, Pessell, Rogez, Michel-Royer. Serge Laget, Philippe Bayart, and David Lam were fascinated by Marie and worked to illuminate her accomplishments. Several faithful friends were of tremendous support and assistance: Marie-José Lionel-Pellerin, Hélène Lapierre, Renée Schmit, Odette Friry, Nicole Diehl, Annie Becquer, Jacqueline Shernetzky.

Bonnie Zucker Goldsmith, gifted editor and dear friend, corrected infelicities and gave the manuscript its polish.

Many years ago the Dayton Hudson, General Mills, and Jerome Foundations awarded me a grant to visit France for my research; I'm still grateful.

Finally, this book relied on the countless nameless individuals and organizations that have so generously digitized old newspapers and magazines, making them available online for free. Information is priceless.

Table of Contents

Preface

Marie Marvingt (1875–1963), who is the most decorated woman in the world, set the world's first aviation records for women, was the first female bomber pilot, is the only person to be awarded the gold medal from the French Academy of Sports for all sports, invented the ambulance-airplane, fought in World War I disguised as a man, was a journalist for dozens of newspapers, gave thousands of conferences around the world on air rescue, and was aptly referred to throughout her life as "the fiancée of danger." She is also virtually unknown, even in her native France.

Marie Marvingt belongs in the canon of exceptional human beings, in the annals of sports, in any book of "firsts," in the history of the women's movement, as a role model in children's books, and as a name commonly recognized by most people. This book is a first step in introducing her to the English-speaking world. Her story is inspiring, entertaining, and educational. It needs to be told.

I have spent many of the years since 1980, when I first came across her name and marveled that no one appeared to have heard of her, researching her life by means of visits to Nancy, France, speaking with people who knew her, and spending uncounted months at the French equivalent of the Library of Congress, either in person or online.

The only published biographies are *Marie Marvingt: La Femme d'un Siècle* by Rosalie Maggio and Marcel Cordier (Éditions Pierron, 1991) and *Marie Marvingt: A l'Aventure du Sport* by Françoise Baron Boilley (L'Harmattan, 2013). The latter is oriented to a study of sports and gender. I've had articles on Marie published in *Women's Sports & Fitness*, *Cricket* (three-part series), *Jack and Jill*, and *Young American*. Not an impressive bibliography for "the most extraordinary woman since Joan of Arc."

This biography differs greatly from the French-language biography published in 1991. The earlier book was written in my second language but, most important, this biography benefits from digitized newspapers and magazines that provided a wealth of information on Marie not previously available. I enjoyed starting from scratch on this book, ignoring the earlier work entirely.

This biography differs slightly from the traditional in four ways: (1) the bibliography is especially inclusive in order to emphasize that this extraordinary woman's "unbelievable" accomplishments are corroborated by eyewitnesses; (2) the voice is intended to subtly reflect the manner in which her era's newspapers and individuals talked about and viewed Marie; (3) it is organized not chronologically but by subject matter—Marie engaged in, and achieved renown in, so many disparate domains, often simultaneously, that it would be otherwise difficult to tease out what she actually accomplished in each; (4) she is referred to throughout as "Marie" instead of as "Marvingt."

In her *New York Herald Tribune* obituary, Earl G. Talbott wrote, "To everyone, she was 'Mademoiselle Marie.'" The use of "Marie" not only seems truer to who she was, but I have been involved with her for nearly forty years, and we have been on a first-name basis for some time now.

I have included many of Marie's own words here. Short of meeting her, it is the best way to get to know her. I wouldn't be surprised if you finished the book feeling like an acquaintance of Marie's.

Introduction

"You aren't going to pretend that I'm a woman like any other, are you?"
—*Marie Marvingt*[1]

Near the French village of Machault, on an overcast December day in 1913, a farmer bundling sheaves heard an unfamiliar noise overhead. Almost before he grasped what he was seeing, a plane arrowed down through the clouds and plowed into a field some distance away.

He had never seen an aeroplane, but he read a newspaper when he could. He knew what it was.

Horrified, he watched the wheels stutter in the furrowed clumps of wet earth and flip the machine onto its back. When he finally reached the crumpled drift of wood and canvas, he cried, "Oh, the poor fellow! He is dead!"

From underneath, a voice.

"*She* is not dead. But whatever you do, don't smoke."

"I was, you see," explained Marie Marvingt later, "covered in gasoline."[2]

Airplane crashes, avalanches, volcano eruptions, earthquakes, enemy fire, and near-death experiences were a way of life for the woman known as the fiancée of danger.

Film buffs may be familiar with her avatar, introduced in the 1914 Pathé film *The Perils of Pauline*. The plucky Pauline represented the era's much-discussed New Woman. Determinedly single, her life was a series of close-calls, hair-raising predicaments, and danger—and modeled on Marie's.

Earl G. Talbott wrote in the *New York Herald Tribune*, "Death, with whom Marie Marvingt had flirted for eighty-eight years, finally came yesterday in a French nursing home, and ended the career of one of the most amazing women who ever lived. She was known as 'The Fiancée of Danger,' and scored so many firsts as a sportswoman, flier, and all-around daredevil that an accurate tally is well-nigh impossible."[3]

The model for "Pauline" deserved her own series. Instead, "one of the most outstanding feminine personalities of her generation,"[4] "the most extraordinary woman of the century,"[5] and a "ludicrously over-accomplished athlete who invented flying ambulances and won the only gold medal ever awarded by the French Academy of Sports for 'all sports'—yes, all of them,"[6] remains almost as unknown today as she was acclaimed during her lifetime. Who she was, what she did, and how she came to be forgotten is the memorable story of a woman who lived life out to the edges and engaged fully with her era. Long before "Be all you can be" was a cliché, Marie Marvingt lived it.

She was undoubtedly born with the ingredients for success: her father's predilection for sports, her mother's love of learning, her own innate gifts—fearlessness, physical coordination, a brilliant mind, personal charm, immoderate energy, and a matchless willpower that drove her to achieve every goal she set for herself.

Still, as Napoleon said, "Ability is nothing without opportunity." And Marie Marvingt was given opportunity, that one additional gift: she was born at precisely the right moment, the moment in history that allowed her to use her particular qualities of mind and body, her interests, character, personality, and heart. Never before and never after could she have been all that she was, or have done all that she did. Her life is inseparable from her era and the technological and human progress that marked it.

The eclectic and encyclopedically talented woman who "never missed a rendezvous with adventure" (*Le Petit Matin*) lived a life as full of headlines and fast-breaking stories as the newspapers she wrote for. Because Marie lived during the inventive, flourishing period before World War I—what the French call the heroic era—it was possible to become a hero, possible even for a woman. Marie didn't care if something had always been done by men. She only cared that she wanted to do it. Badly. And so she did it.

Having no role model for her life, she became one.

Because she lived on the cutting, deregulated edge of progress, rules could be broken. Could an unauthorized, non-military woman today fly a bombing mission because another pilot was sick? Could someone today decide to cross the North Sea in a balloon without notifying authorities, filing a flight plan, or telling anyone she was going, just because the wind was in the right direction? Neither these nor most of Marie's adventures could have happened in any other era.

Much was new then: telephones, electricity, sports competitions, bicycles, cars, planes, Tour de France, film, winter sports, high-circulation newspapers. A fertile, fabulous time to be alive, the age produced Cézanne, Monet, Matisse, Picasso, Rodin, Stravinsky, Debussy, Ravel, Proust, and scores like them.

Marie's imperatives reflected those of her era. Humanitarian ideals battled in her with a desire for individual adventure and achievement; her heart-sprung desire for competition, novelty, and even danger struggled with her head-rooted desire to serve others, to be useful and productive, to leave the world better than she found it. Only after she had satisfied the first imperative was she able to devote herself more fully to the second.

Her life easily lent itself to superlatives. "With her talent for sports, her taste for risk and adventure, and her multiple achievements in numerous domains, Mademoiselle Marvingt was one of the most extraordinary Frenchwomen of her generation."[7] Others went further, describing her as the most outstanding woman of the century,[8] the most incredible woman since Joan of Arc,[9] the universal sportswoman,[10] Leonardo da Vinci in a skirt,[11] and "this woman who has joined the ranks of the greatest men of her day."[12]

"If I had to choose just one adjective to describe the work and the personality of Marie Marvingt," wrote a journalist, "there is no better term than 'extraordinary.'"[13] Another said, "It's not a woman, it's a whirlwind."[14] Jean-Loup Nicolle told his young readers that he was about to relate to them the true story of a woman who became a legend in her own lifetime.[15] She was, they said, "as brilliant as she is unknown."[16]

It was only after Marie died that the legend became as insubstantial as smoke, so much so that those who encountered her story for the first time reflexively suspected it was a fabrication. However, with the digitization of hitherto unavailable vintage publications, what Marie achieved has become eminently clear. Thousands of articles detail her life minutely. It is possible to know where she was on almost any given day of the year.

Less obvious than what she did is who she was, how she accomplished so much, the sources and effects of her popularity and influence, her discreet love life, what became of her papers and medals and trophies, and how she could ever have come to be forgotten.

1

Early Years

"I was quite a tomboy, but they wanted a girl so badly they were glad to have me anyway!"—*Marie Marvingt*[1]

Marie Félicie Élisabeth Marvingt was born February 20, 1875, in Aurillac,[2] a small French town of 14,000 set among volcanic mountains and famous for its fine lace, wooden shoes, cheeses, and livestock.

Marie's mother, Élisabeth, was delighted to have a daughter while her father, Félix, soon found himself with the unexpected answer to his dreams. Marie was proudly named after both of them.

Félix-Constant Marvingt (1827–1916) and Élisabeth Brusquin Pallez (1840–1889) met in Metz, an industrial town in eastern France, where Élisabeth had grown up and where Félix was working for the Postes et Télégraphes. They were married July 6, 1861, at the Enclos-St-Laurent City Hall in Paris, but lived in Metz until it fell to Germany in 1870.[3] Not wishing to live under German rule, Félix immediately transferred to the P.T. in Aurillac.[4]

Félix Marvingt was a longtime postal employee, eventually becoming Director General of the Aurillac post office,[5] but his absorbing passion was sports. A swimming and billiards champion, he had looked forward to sharing his love of sports with his children. But the Marvingts' first three sons—Louis, Charles, and Eugène—died in infancy before Marie was born, and a fourth son born three years after Marie, "the second Eugène," was delicate and sickly.[6]

Félix was forty-eight years old when Marie was born, and he had given up hope of passing on his interests to the next generation. But little Marie brimmed with inexhaustible vitality of mind and body. Intelligent and athletic, she seemed to require constant and varied activities, the more demanding the better.

Her earliest years appeared to have been happy and carefree, sheltered in a loving family and a placid town where people's lives revolved around the weekly markets and the liturgical rhythms of the church. The deaths of babies born before her didn't seem to touch her, she didn't question brother Eugene's health, and for a long time she scarcely noticed that her mother was not as robust as she might have been. Besides, there were all the new things Papa was teaching her.

Nature and nurture collided in Marie with spectacular results. Having a sports-minded father who was eager to take his preschooler mountain climbing and swimming, who bought her one of the first bicycles, and who constantly urged her to go farther and faster would have been futile had Marie not been endowed with indomitable good health, uncanny physical talents, bottomless ambition and drive, and a fine mind.[7]

Marie said she learned how to swim when she was learning how to walk and by the time she was four, she could swim her age in kilometers (about three miles) in the Jordanne River. She remembered pitying friends older than she who simply waded a little ways into the water.[8]

"I could hear the river calling me," she said years later. "I still remember the shivers as my body hit the water, and how I loved frisking around in the mild currents. One day when I was four years old, I almost drowned. I remember the feeling as if it were yesterday, but even that didn't stop my love of swimming."[9]

The same year as that near-disaster, Félix was transferred from Aurillac to Lille,[10] but it must have been an unsatisfactory situation because in October of that year he retired[11] (according to Marie, he'd "had it" with the civil service) and the family moved back to Metz, possibly because Élisabeth's family lived there and she was not in good health.[12]

Despite the move, life didn't change much for Marie. It still revolved around her parents and Eugène, whom she appeared to love with all the fierceness of a strong older sister for a frail, sweet-natured boy who couldn't keep up with her games and activities.

Eugène was left behind with *Maman* while Marie and her father hiked and swam and climbed and biked. "As a little girl," Marie said, "my father turned me into his enthusiastic companion and the two of us spent his vacations mountain climbing together. I followed him with lighthearted eagerness wherever he agreed to take me."[13]

It was she who shared his fascination with sports, she in whom all the family's good health seemed to be concentrated.[14] Félix introduced her to one sport after another, making no concessions to her youth and inexperience. When he took her mountain climbing he expected her to keep up; he never looked back to see if she was still behind him.[15] When he taught her to play billiards, he competed against her as if he were in one of the regional championships he regularly won.[16] The first time she beat him, she experienced that incomparable sense of pride that comes from besting a champion—a feeling for which she would develop a strong taste.

Few girls of her era enjoyed such an upbringing.[17] And Félix must have been a markedly independent thinker to have raised Marie as a sportswoman.

Although he indulged his adorable, strong-willed daughter, Félix also demanded a great deal of her. No matter how hard she tried, what records she set, what prizes she won, Félix expected more: "You can do better, Marie."[18] Years later, after Marie won a balloon race from France to Wales, Félix said, "That wasn't bad, Marie, but you could have gone farther, on to Ireland."[19] But for Marie, outdoing herself and others meant more to her than any compliment. Just before her death, someone asked if her father had been proud of her. With a smile that must have gone back years, she said, "Oh, yes. Oh, yes, he was."[20]

(Félix Marvingt lived to the age of eighty-nine and saw Marie accomplish more than any other person of her day, more even than he had dreamed for her. Félix and Marie died at roughly the same ages; his lifespan was three months longer than hers.)

Uncommon willpower, self-discipline, and perseverance showed up early. When Marie was six, the woman who lived next door gave her a complicated needlepoint pattern. The neighbor disapproved of Marie's tomboy activities, although with a frail mother and an aggressive father, what could you expect? Her gift was her unsubtle push in the "right" direction. The woman's ultimate cleverness, however, was in telling Marie the project was probably too difficult for her.[21]

As much as she hated needlework, Marie would not be told something was beyond her. Off she went with her bag of colored embroidery threads, the stenciled material, and a needle. When it grew dark and she hadn't come in yet, her family went looking for her. They found her sitting under a tree determinedly pulling the needle through the canvas. She completed the work over the next several days, showed it to the officious neighbor, and never touched needlepoint again.[22]

To the consternation of Madame-Next-Door, Marie continued to swim, play billiards, climb mountains, and ride a newfangled bicycle. If Élisabeth had any reservations about all this, she was outnumbered by her strong-willed husband and daughter. Félix's joy in sports brilliantly played was exceeded only by his belief that Marie was more than ordinarily talented; Marie herself adored sports and was driven to perfect her skills. Who could stop them?

In Metz, Marie first attended a small school run by two older, unmarried sisters, the Desmoiselles Daurès.[23] She was later enrolled as a day student in a girls' school known for its well-turned-out young ladies, Sainte Chrétienne de St-Vincent.[24] The choice of school might point to a mother's hopes for a more a traditional daughter but, if so, Sainte-Chrétienne wasn't apparently the solution.

With her self-confidence, vitality, and good humor, Marie seemed to be popular with her classmates. Despite this, she had no close friends: no one else was doing the things she most liked to do.

By her own admission, Marie was a bit of a handful during her school years.[25] As quiet and obedient as a runaway circus horse wearing bells, she argued with the sisters in their winged headdresses and somber black habits. Still, she was an extraordinary student, learning easily and reading books with the speed of someone who fears they will be snatched from her.

Marie Marvingt, circa 1906, looking conventional enough.

Marie's classmates had been raised with strict ideas of what was ladylike. Although no one had ever specifically told them, "Young ladies do not ever take the controls of a clanging, smoking engine and drive it like a mad thing through Metz," the schoolgirls watching Marie from the side of the railroad tracks one sunny afternoon suspected that Marie was breaking rules they had yet to learn.[26] After that, they called her Daredevil Marie.

Many years later, at the 1955 reunion of the alumnae of Sainte-Chrétienne, Marie, by then a famous pilot and sportswoman, presided over the banquet—a small irony, given that she had been such a trial to the good sisters of Saint Vincent.[27]

And then there was Marie's other classroom.

"During the course of the year," recalled Marie, "we'd get one big circus or another in Nancy or in Metz—sometimes the Plège circus, sometimes the Rancy. I adored both of them.

"Each show time found me sitting in the first row, wide-eyed and clapping my hands fervently at the feats of the gymnasts and the horseback riders. Oh! The sight of those supple, light-footed girls, bounding up, almost flying off the backs of their large white or dappled horses, which then became living pedestals for their grace and their triumph. All night long, I dreamed of it."[28]

Gymnastics had already been adopted in Europe both as exercise and as a competitive sport. Since it was considered acceptable for women, Marie had been involved in gymnastics for years by the time she saw her first circus performance.

One morning, Marie begged her father to let her take lessons at the circus before show time. "I must have been very persuasive, or else my father spoiled me! He gave his permission, and each day after that saw me on the sawdust track, in the midst of all those empty bleachers, learning the secrets of trick riding and acrobatics."[29]

Despite his bureaucratic background, Félix Marvingt evidently had a streak of free spirit in him, or else he couldn't bear to deny Marie her birthright. How else to explain allowing a teenaged girl in that era to spend months frequenting the circus?

Circuses were extraordinarily popular in France during the nineteenth century. Antoine Plège and Théodore Rancy founded their famous circuses two years apart and were business rivals as well as close friends. The Rancy circus spent three months in the nearby city of Nancy on the Place Carnot every year, and it became an inspiration and proving ground for Marie.[30] She not only became expert in gymnastics, horseback riding, trapeze, tightrope, swaying slack-rope, trick-riding, juggling, and animal taming, but she learned discipline, endurance, and self-confidence, astonishing Félix, her circus friends, and possibly even herself with the speed with which she acquired a dizzying number of skills. One of her "firsts" developed directly from the circus years: she somersaulted on a galloping horse, the first person recorded doing so.[31]

In later years she never let people forget that "wild-animal tamer" was one of her skills.[32]

After the circus years, she continued lessons in dressage and *haute école* with the great equestrian Alphonse Rancy and in trick riding with Triault. It was unusual that she was allowed to learn *haute école* techniques, which had been strictly limited to cavalry officers until the last part of the nineteenth century. (Today only the Spanish riding school with the Lippizaners from Austria carry on this tradition.) She eventually became an all-around adept horsewoman, riding both sidesaddle and astride, and driving teams of two or four horses.[33] The rest of her life, she rode whenever she could, the newspapers rarely failing to mention the remarkable facts that she was wearing trousers and riding astride as men did.[34]

Engrossed as she was in circus activities, Marie only slowly realized that her mother's health was failing. Élisabeth Marvingt had never been strong. She became even frailer after giving birth to five children and seeing three of them die. Caring for Eugène also taxed her strength. Marie took on more and more of the housework, but even that didn't help.

Florence Nightingale's book on nursing had just been translated into French. Like many young Frenchwomen of that era, Marie idolized her as "the Lady with the Lamp" making her nightly rounds of wounded soldiers. Nightingale wrote that "every woman is a nurse," directing her words not just to professional nurses, but to anyone caring for a sick friend or family member.[35] Marie was comforted by being better equipped to care for Maman, little realizing her decades of work as a Red Cross nurse were taking root at the same time. The book's lessons of healing and service would stay with Marie for a lifetime. Not so the loved mother.

On March 19, 1889, when Marie had just turned fourteen, her mother died of what was then called a lung infection (perhaps an acute upper respiratory infection or pneumonia).[36]

Marie's life was abruptly cut in two: on one side, the carefree years of affection and security before her mother's death, on the other, all the long years after it. At a time when most of her friends were still playing games, she found herself in the important but thankless role of housekeeper for a grieving Félix and substitute mother for a bewildered Eugène. During her teenage years, when her head was filled with dreams of adventure and her hands were occupied with tedious chores, she stored up a dislike for housework that lasted all her life.[37]

Had her mother lived to guide her young daughter, she might have taught Marie, as her own mother had taught her, and as her mother's mother had done, to be a young woman of her time, readying her for marriage, motherhood, and passivity. Before her mother's death, Marie spent time in traditionally female activities, learning to cook, keep house, sew, and mend. Afterwards, she stoically met her household responsibilities, but with a ruthless and terrible efficiency that expressed her dissatisfaction.

Marie had a remarkable relationship with her father, who was delighted to buy her sports equipment and arrange for lessons. But the two of them shared activities and sports, not feelings and thoughts. Félix treated her the only way he knew how—as one strong-willed overachiever to another.

After Élisabeth's death, Félix moved his small family to the city of Nancy, some sixty-five miles south of Metz, where he rented one of the larger apartments in a beautiful old building on the Place de la Carrière (now a UNESCO historic site).[38] For seventy-four years, Marie would live at 8, Place de la Carrière, in an elegant Héré-designed building known as the Hôtel des Pages de Lorraine.[39] On the north, the building overlooked the crowded, colorful Rue des Écuries where years earlier the public stables ("écuries") had done their fragrant, noisy business. The other side of the building faced the popular and gracious Pepinière park.

In former times, the Hôtel's apartments were the epitome of luxury: soaring ceilings, heavy embroidered draperies, crystal chandeliers, huge fireplaces, and hand-carved woodwork. Its residents were cultured and well-to-do. Over time, the building lost much of its grace. Still, when a reporter interviewed Marie in her apartment years later, he was charmed by the children's voices floating in the window along with fresh breezes from the Pépinière gardens.[40]

Although Nancy had schools Marie could have attended, Félix felt it would be better for her and Eugène to continue their studies at Sainte Chrétienne de St-Vincent as boarders. On weekends and holidays, the three explored their new city. Nancy had been the capital of the dukedom of Lorraine for centuries until, under the terms of a 1735 treaty, the region was given to the ex-King of Poland, Stanislas I, who indulged his passion for architecture by creating a handsome, well-planned city, known today particularly for the spectacular Place Stanislas, a cobblestoned plaza surrounded by stately gray stone buildings, fountains, and statues highlighted with gold-edged wrought-iron gates, railings, balconies, and lacy lanterns. (The Place de la Carrière, where Marie lived, is separated from the Place Stanislas only by a large stone arch.)

After Stanislas died in 1766, the province of Lorraine became French. But because of its rich iron ore deposits (the greatest in Europe) and its heavy industry, textile

agne, c'est de m'avoir
lonné l'âpre désir de l'air,
'amour du voyage en plein
iel. C'est sur son conseil
que j'ai fait mon apprentis-
sage du ballon sphérique
avec ces maîtres aéronautes
qui s'appellent Blanchet,
Bachelard et Barbotte, et

que j'ai gagné le brevet de
pilote de l'Aéro-Club de
l'Est et de l'Aéro-Club de
France. En 1910, j'ai eu
la grande joie d'enlever le
premier prix du concours
de distance de l'Aéro-Club
de l'Est et d'aller de Nancy
à Neuf-Château, en Bel-

A page from Armand Rio's article about her multiple sports successes.

manufacturing, and fruit and wine production, the Lorraine region was often the booty of war. France lost the northern part of Lorraine to Germany in the 1871 Treaty of Frankfort. This history had real-world consequences for Marie. Schooled in Metz (part of Germany then), she studied German in school and spoke French at home. When World War I began, Marie wasn't alone in hoping France could heal the gaping wound that was the lost part of Lorraine.[41]

After she finished the secondary schooling she was fortunate to have (elementary education was guaranteed for all children but secondary education was still mostly reserved for the upper class), she began a period of intense training, indulging her pronounced taste for challenge and proficiency. Retired and over seventy when Marie began excelling in sports, Félix devoted himself to coaching and supporting her. As it always had, his life revolved around hers.

Marie swam nearly every day in the Moselle to improve her speed and endurance, made longer and longer bicycle trips, and entered a number of competitions. By the time she became visible to the larger world, she had mastered swimming, bicycling, horseback riding, archery, automobile driving, riflery, billiards, tennis, golf, gymnastics, and a broad range of track and field events.[42] Even after she began studying at the University of Nancy, she had a lively second life as a sportswoman, constantly training, taking lessons, or competing.[43]

And then, in 1897, gentle, affectionate Eugène died of a heart attack at the age of nineteen.[44] It was another devastating loss for Marie and her father. Eugène was buried in the Préville cemetery in Nancy, where the two remaining Marvingts would eventually join him. After his death, Marie threw herself into even more rigorous training and competing.

The world had seen nothing yet.

2

A Woman for All Seasons

"The sporting life of Mademoiselle Marvingt is of a most extraordinary kind. Swimming, cycling, mountain climbing, ballooning, flying, riding, gymnastics, athletics, fencing—there is not a single sport in which she does not shine. Where coolness, courage, and skill are required, in the aerodrome, on the mountains, in the sea, at the fencing school, she is always to be seen in the front rank."—"The World's Greatest Sportswoman," *Forest and Stream*, 1913[1]

Ernest Archdeacon, aviation pioneer, co-founder of the Air Club of France, and great philanthropist, wrote in 1910: "There is surely no other woman in the world with the athletic credentials of Marie Marvingt, and I wouldn't like to bet that there is even a man who has done as well."[2]

Marie began working toward her standing as a world-class athlete in her preschool days with swimming and mountain climbing. She added billiards and bicycle riding in elementary school, and from ages fifteen to twenty she began making her mark in canoeing, swimming, horseback riding, fencing, riflery, archery, martial arts, and gymnastics (vault, uneven bars, balance beam, rings), plus track, high jump, long jump, and weight lifting—all performed with "rare brio," according to Bernard Marck, French aviation historian.[3] She took up automobiling, tennis, golf, and juggling, and even learned to drive a locomotive and a steamboat.[4]

From then on, Marie was here, there, and everywhere, winning prizes in so many different sports that Armand Rio wrote a nine-page article for the popular weekly *Lecture Pour Tous* claiming that Marie more than earned her reputation as the world's greatest sportswoman. "There's nothing she hasn't tried," he said.[5] In 1913, *The Strand Magazin*e devoted nine pages to "the finest sportswoman in the world."[6] Newspapers around the world agreed.[7] In honor of the 2008 Olympics, a *Glamour* magazine panel chose the "10 Greatest Female Athletes Ever." Among the ten: "High-flyer Marie Marvingt, the 'fiancée of danger,' a noted cyclist, boxer, and skier."[8]

One of Marie's earliest loves, the bicycle, provided her with exercise, transportation, and an unremitting joy throughout her life.

"I've loved cycling for as long as I can remember," Marie told a journalist. "Do you remember those first bicycles? They had a huge wheel on which one was perched way up high. They were popular when I was just a toddler. From the first time I saw one of those in Nancy, tooling along and astonishing the open-mouthed passersby, I knew it was for me. I think I was probably one of the first Frenchwomen to climb on one. In cycling, I have—like everyone—several odd little wins to my credit."[9]

Those odd little wins included the Nancy-Bordeaux (1904), Nancy-Milan (1905), and Nancy-Toulouse (1906) races, which she won when she was in her twenties.[10]

At the time, everyone was mad about bicycles, but for women, the emphasis was on appearance, not function, speed, or physical fitness. Articles advised that ankle action was the chief factor in graceful, expert riding. Women were warned against spreading the elbows akimbo, turning out the toes or, worse, hanging over the handlebars while shuffling around on the bike seat. An 1890s photograph in *Vanity Fair* showed the cream of French society in the Bois de Boulogne, riding in a gentlewomanly manner that barely ruffled the ribbons on their chapeaux.

Pundits warned parents against letting their daughters play tennis or ride bikes: it could cause them to fall dangerously in love. Another, obviously repelled, wrote: "Do not speak to us of those androgynes in culottes who show all the world the same legs (usually unpleasantly spindly or thick as posts) they puritanically refuse to display even to their intimates at home. They sit embarrassingly forked on their machines like witches on broomsticks."[11]

Marie had been riding a bicycle for twenty years when that tirade appeared.

The first time Marie appeared on a bicycle to visit an uncle in Avranches, he refused to see her; a woman on a bicycle could only be of doubtful morality.[12]

After advancing through the series of improvements in the bicycle, from the tricycles and high-wheelers to the bicycle recognizable today (except that hers had the gear change on the pedal, not on the handle), Marie bought one and named it *Solidity*. She mentions "Solidity, my steel horse" in a 1905 article.[13]

She would eventually have four bicycles: *Solidity, Zéphirine, Désirée*, and *Reine*. *Zéphirine* was her favorite, perhaps because the two of them spent more than forty years together. Marie's "machines" (bicycles, balloons, cars, planes) were as important to her as other people's children were to them. With her trusty *Zéphyrine*, Marie began biking longer and longer distances, eventually competing in lengthy races. Being a practical Frenchwoman, and showing her usual inclination to look beyond the simple fun of a sport for its more useful purpose, she realized the bicycle was a reliable and inexpensive means of transportation.

Even when not training for biking events, Marie rode her *Zéphyrine* every day. Although her bicycle was described by less imaginative souls as irredeemably old-fashioned, and heavy as an oxcart, Marie herself affectionately referred to it as "the little queen."[14] Today, riding a bicycle is scarcely a daring exploit; it was exactly that in the early days.[15] Beginning in 1889, Marie left the streets of Nancy to bike her way around France and Europe.

Henri Desgrange, newspaper owner, former cycling champion, and future aviation enthusiast, launched the Tour de France in 1903 in an effort to boost circulation and thus outdo a rival sports newspaper. It worked. The first Tour de France drew huge crowds and doubled the circulation of his paper. For an entire month, the Tour de France dominated the news in France.

In the first years of the race, it was more a struggle for survival than a competition among entrants (referred to as the quick and the dead). The goal was simply to finish the race in one piece. The morbidly fascinated public could scarcely believe that a human being could endure such a grueling contest. It soon became the most popular sports event in France.

By 1908, Marie had won several important bike races. She had trained intensively, and felt ready for the Tour de France. She submitted her entry. And was refused. It was, they said, a race for men only.[16] At that time, the Tour de France was the most masculine thing imaginable; allowing a woman to ride was unthinkable, a challenge to male dominance.[17]

The course that year was 4,497 kilometers (2,789 miles) long, divided into fourteen stages. The 114 entrants left Paris en masse and proceeded to Roubaix, Metz, Belfort, Lyon, Grenoble, Nice, Nîmes, Toulouse, Bayonne, Bordeaux, Nantes, Brest, Caen, and back to Paris—effectively making a huge circle through France.

The course snaked over rivers, through forests, over eight steep hills, and through the mountainous Circle of Death, named for the vicious brown bears that roamed the area. (After riders had been threatened and even hurt by the bears, bikers and fans used to shout "Assassin!" at Desgrange when he appeared in public.)

Some of the mountain roads, like those over the Col de Porte and the Côte de Laffrey, were so steep that nearly everyone had to walk their bikes at times. In 1908, the route also took the bicyclists over the top of the Ballon d'Alsace, a rounded mountaintop; only one rider managed it without putting a foot on the ground.

Another rider was also noteworthy. Biking alone behind the field was "Daredevil Marie," who started each stage fifteen minutes after the men.[18] In the same way as the others, Marie spent nearly a month biking over some of the roughest terrain and highest hills in France, scarcely eating or sleeping. In the hot, humid July and August days, they faced dangers that ranged from mountain roads slippery with rain and without guard rails to devilish village children who liked to poke sticks through the bikers' wheels. And that didn't include biking accidents or breakdowns. Yet, even on unimproved roads, riding bicycles that were a far cry from today's efficient models, the contestants—and the one non-contestant—averaged an incredible 150 kilometers (93 miles) a day.

It seems likely that during that month, Marie dreamed more than once of a victorious entry into the racetrack at the Parc des Princes in Paris, the Tour de France's finish line. Custom dictated that the winner make an honorary tour of the track for all the enthusiastic fans. Marie knew she couldn't win because as far as the Tour de France was concerned, she didn't exist. That didn't stop her from riding as hard as everyone else.

Of the 114 Tour de France entrants in 1908, only 36 actually finished the course.[19] Thirty-seven, counting Marie.[20] There are reports that, given her time, she would have placed second that year. But her feat was unofficial, unrecorded, and unacknowledged. Even so, after that 1908 Tour de France, people began referring to her as *La Marie*, which signified the respect and affection in which she was held by her compatriots, and especially by sports enthusiasts. After describing Marie's feat, the authors of *The Big Book of Women's Sports* added, "History doesn't tell us if she was riding *Zéphirine*."[21]

Because of her unofficial status, it is impossible to verify Marie's Tour de France. Did she conform in all ways to the route, rules, and conditions of the official bikers? It can only be said that not once in all the years since 1908 has Marie's accomplishment been disputed. Today, some have doubts because it seems so improbable, but Marie did a number of improbable things in her life.

In 1909, a letter to *L'Auto* asked when there would be a women's Tour de France. The response: when the mountains of France have been flattened.[22] The writer evidently didn't know that the mountains had already begun to flatten.

Marie riding her *Zéphyrine* was a familiar figure to be pointed out to tourists in Nancy, Paris, and even London—where, in 1901, she was run over by a cab while biking in front of Westminster. (Fortunately, she said, her bike was not hurt.)[23]

"I just jump on my little queen here, whether I'm in Nancy or in Paris. Biking is an excellent sport and helps keep me in shape. Besides, in Paris, I can get through the worst traffic jams on my bike."[24]

Years later, Marie's *Zéphyrine* was stolen. She immediately applied for help from the neighborhood children.[25] That time they found it for her, but when it was again stolen, this time from in front of the Lotti hotel in Paris, friends chipped in to buy her a new bike, which she named *Désirée*. She had a hard time accepting the gift with grace. *Désirée* was certainly new, shiny, and well-made, but nothing could replace her dear old *Zéphyrine*, with whom she'd had so many adventures.[26]

Marie never stopped biking. On her seventy-fourth birthday she celebrated by riding 103 kilometers.[27] When she was seventy-six, she commemorated the anniversary of the death of her good friend Marshal Hubert Lyautey by biking 75 kilometers.[28] The Télé-Paris office recalled that Marie parked her bicycle out front to come in and visit them, "already in her eighties and still extremely alert."[29] Evidently.

At the age of eighty-five, she biked from Nancy to Paris (280 kilometers) carrying a forty-pound backpack. And at eighty-six, she was still biking 1,000 kilometers a year "just to keep in shape."[30]

The spiritual connection Marie found between bicycling and aviation may have been due simply to their invention a few years apart, or perhaps the fundamental attraction that people of her temperament had to speed and machines that pushed limits. The Wright brothers opened a bicycle shop in 1892, and four years later produced their own model. The shop income helped fund the development of their first workable plane. Oscar Lapize, winner of the 1910 Tour de France was a fighter pilot in World War I. Hélène Dutrieu, Marie's friend and the second woman in the world licensed to fly, was first a pioneering bicyclist, setting records in the 1890s. Glenn Curtiss, founder of the U.S. aviation industry, began his career as a bike racer. Even billionaire aviator and aerospace engineer Howard Hughes was in the Houston newspaper at age twelve, having

Marie passes her driver's exam: an 1890s photo on a country road outside Nancy.

built a "motorized" bicycle from parts of his father's steam engine. The link between bicycles and planes continues today: British inventors recently produced a flying bicycle.

Like bicycles and airplanes, automobiles are considered part of the scenery today, a workaday necessity. In the beginning, they were slightly sinister curiosities—again, especially for women. In 1904, a New York society physician advised against driving by "distracted women, which is most of them."[31] He felt strongly "that the art of driving a car will never become a profession for women because there is no career where the habituate feminine characteristics are more present, with more horrendous results. By its end, the twentieth century will see women take up medicine, law, teaching, writing, chemistry, but women drivers seem almost as inconceivable as women astronomers or engineers."

In a 1914 Paris newspaper, the Duchess d'Uzès disagreed. She said a woman could fulfill every requirement for driving an auto; she herself was the first woman to obtain a chauffeur's license. But, the paper said, "Mme. Lesueur, the authoress, is of the opposite opinion. She has ridden and driven horses, but says that 'between woman and horse there exists sympathy, comprehension or understanding, which, of course, cannot be created between women and machines.'"[32]

Daniel Cousin, writing the same year, admitted that there were women like Mademoiselle Marvingt and several others who were at home behind the wheel. "They are expert pilots so driving is child's play for them. But they are the exception. A woman," he said, "is generally too nervous to keep her cool for long. A woman driver might avoid one accident or several, but eventually, unnerved and tired, she'll let go of the wheel, and then everyone watch out! The moral of the lesson: Ladies, get yourselves driven. You're much better off in the back seat than behind the wheel. Believe me, that is the prudent choice—for you and for us!"[33]

Let go of the wheel?

It was a small leap for Marie to go from two wheels to four, and to be one of the first women to acquire a driver's license, probably in 1899.[34] To obtain what was then called a certificate of ability (the French driver's license didn't formally exist until 1921), a driver needed to demonstrate the ability to start, steer, and stop, and also show some know-how in case of a breakdown. Awarded by automakers, the certificate was technically available only to men over twenty-one.

A photo of Marie and an automobile on a country road outside Nancy was labeled "Marie Marvingt, Woman of Every Risk."[35]

Years later, Marie often said she didn't care for cars but in her early years, the automobile irresistibly presented her with fat wheels, shiny headlights, an unpredictable motor, the capability for going fast, and the chance to do something few others had done, perhaps even to set new records. In tune with her times, she embraced every novelty, especially if it involved risk and skill. How could she ignore the crazy excitement of the horseless carriage?

Dressed in the requisite heavy fur coat, Marie made several automobile trips and entered various races. In 1905, she drove from Bussang in northeastern France to her birthplace of Aurillac in south-central France (624 kilometers) in one day, a respectable accomplishment given that automobiles traveled only one-third as fast as today's cars and that early roads were primitive, wandering whimsically with the topography.[36]

When Marie pulled up near the statue of Pope Sylvester I in the center of Aurillac, she was hot, dusty, and tired, but exultant. The next day she held court near her begrimed automobile and chatted with people who remembered her and her family from years ago.[37] Automobiling lost much of its allure for Marie once she took to the air, but she was a lifelong member of the Automobile Salon and attended its yearly exhibition whenever she could.[38] She was, perhaps inevitably, drawn to the 1921 Grand Prix of Corsica, uniquely dedicated to sports cars and held in conjunction with the 100th anniversary of Napoleon's death. Marie drove the course before the race to report that "the roads were in fairly good shape, except for the climb from the Golo Valley."[39]

Marie rarely did one thing at a time, so along with her biking and automobiling, she was intermittently exploring and working on other proficiencies.

Near the end of the nineteenth century, fencing became a popular sport. The first regulated fencing competition was held in England in 1880, and fencing was first included in the Olympics in 1896. Modern fencing was thus coming of age when Marie was. She needed to see only one exhibition to appreciate fencing's grace, precision, and artistry. Her gymnastics and trapeze work combined with her innate coordination and quick mind allowed her to adapt to elaborately thought-out moves and countermoves and resulted in her skill with saber, foil, and epée, each of which has its own rules.[40]

The International Federation of Fencing wasn't founded until 1913, when Marie had been fencing for nearly twenty years, and there are no records of her fencing accomplishments. Photos of her in fencing gear are easily found, however, and newspapers always included her "known" expertise as a fencer. Beyond that, Marie appeared to have lingered for a while in the fencing arena, and then moved on to other sports.

In the same way, canoe racing was a minor piece of Marie's life, but it belongs to her history as a sportswoman. When she was fifteen, she raced alone (free of the chaperonage typical of her era) in a single-seater river canoe from Nancy to Coblenz, Germany, by way of the Meurthe and Moselle rivers (400 kilometers). She won first place.[41] For a fifteen-year-old girl in 1890 it was an exceptional achievement.

She apparently kept up her skills because in 1905, some thirteen years later, she took first place at Étretat in a race in which she had to stand to paddle her unstable, lightweight canoe.[42] As Étretat was famous for its hand-crafted artisan canoes, it was something of a center for canoe events. The 1905 race was well advertised, and Marie obviously wanted to be part of it.

A far cry from her races: This 1913 photo may show Marie during one of her few relaxed moments.

The idea of standing and paddling using an extended canoe paddle is far from new, but few women attempted this risky sport because of lack of opportunity enforced by social conventions. In the early years of the twentieth century, most women took to canoes in the most tentative (i.e., traditional) manner, enveloped in yards and yards of skirts, carrying parasols, and wearing large-brimmed hats. A woman's magazine wrote in 1905, "A most delightful end to a canoeing outing is tipping the canoe." Accompanying the article was a photograph of laughing, overdressed women standing in a foot of water.

Marie once nearly drowned during a canoe race at Le Havre.[43] Canoeing didn't have an international federation until 1924, and it has been an Olympic medal sport only since 1936. As she often did, Marie stood in the forefront of an emerging sport, curious, delighted, enthusiastic, and accomplished.

Marie held her first rifle at the age of fifteen, attracted by the degree of skill needed to become a sharpshooter. Seeing her natural talent for finding dead center of any target, Félix familiarized her with shotguns, carbines, and handguns. Marie was a champion at the new and popular pigeon shoot; in a competition at Reims, she brought down all fourteen of her clay pigeons, to the delight of the almost entirely male crowd.[44]

A few years later, at the 1906 National and International Rifle Competition, Marie competed with an entire army division to win the grand prize, firing at a target from 300 meters (the standard then for testing military rifles).[45] A few minutes later, she won another grand prize, using a Flobert carbine.[46] In the twenty-one days of trials, Marie won other awards, with her name in the paper nearly every day.[47] She was the only woman designated a first-class shot by the Ministry of War, and the Minister of War himself presented her with her award.[48] In interviews after the competition, she admitted she had not practiced much since her teenage years. However, she said, a true athlete rarely forgets a learned skill.

In 1913, Marie apparently still had her eye. Competing in Belgium in the first European women's rifle championship, Marie "distinguished herself," according to the papers.[49] With war on the horizon and several years of flying behind her, she took classes in shooting from a plane from General Jean Estienne, better known as the father of the tank.[50]

Another intense period in Marie's early life was devoted to the martial arts: boxing, wrestling, jiu-jitsu, and karate.[51] She worked out nearly every day with a punching ball, boxed in a ring, and held her own in the wrestling arena. In 1924, a note in the local paper announced that Mademoiselle Marvingt, who would be giving a series of lectures, had also challenged to a boxing match anyone who wished to accept.[52]

As a woman who often traveled alone, this training served her well. While still a teenager, she was walking along the quays at Le Havre when two thugs tried to grab her purse. She quickly sent them running with a couple of fast rabbit punches.[53] "In Brest one night when I was getting off a fishing boat, a mugger tried to attack me. But he had chosen his would-be victim poorly. I used my boxing and jiu-jitsu to dispatch him in the wink of an eye."[54] Years later, in her sixties, she was attacked by a pair of thieves. Again, her early boxing helped her knock one out and send the other away with a broken nose. Another time she used judo on a thief. She liked her skills to be useful.

Despite nearly drowning several times, nothing dampened Marie's enthusiasm for the water, especially for swimming. She associated it with her earliest days swimming with her father in the river. Early on, she recognized the health and fitness benefits of swimming, pointing out that the sport enlarges lung capacity, develops wind and muscle, and could also save a person's life.[55] Working against the water's resistance doubled its value, she said. She herself swam as often as she could, in rivers and the sea.

In 1905, the sports newspaper *L'Auto* sponsored a competition for professional swimmers to race in the Seine from one side of Paris to the other. Two men won first and second places, but third place went to Australian swimmer Annette Kellermann. Some 40,000 spectators lined the banks of the Seine to watch the race.[56]

The sponsors took note of the turnout and repeated the race the next year. Taking a leaf from *L'Auto*'s success, *Les Sports* co-sponsored a race two weeks later for amateurs set up exactly like the professionals' race.[57] On a beautiful, hot day in late July 1906, some 500,000 Parisians began arriving at 7:00 in the morning to watch the first international (seven nations had swimmers in the event) amateur swim competition across Paris by way of the Seine. The quays and bridges were packed with spectators. On the Seine itself, barges (also crammed with people), rafts, canoes, and police boats jostled for room.[58]When the race began, people followed it on the quays, by foot, bicycle, car, and carriage. The only unhappy faces that day belonged to those trying to fish; they were, according to the newspapers, scandalized and furious at this unexpected event chasing away their fish.[59]

The swimmers, Marie among them, dove off a pontoon at the Pont National in Bercy; one hundred feet ahead of each swimmer was a boat carrying a rower, a lifeguard, and the swimmer's trainer. When the racers arrived at the other end of the course, in Auteuil, an orchestra played the national anthem of the swimmers as each touched the marker—except for the Dutch swimmer. The orchestra didn't know that anthem.[60]

Marie, completely at home in the water, became the first Frenchwoman to finish that race, a distance of 12.5 kilometers.In doing so, she bettered Kellermann's record from the previous year by more than an hour.[61] Although two Swiss women completed the course in even better time, Marie was the first Frenchwoman to swim across Paris. Because the reappearance of the Olympics in 1894 had triggered intense feelings of patriotism among the French sporting public, Marie's win made her very popular.

Newspapers noted that her triumph came "despite an incident, noted by one of the officials, that retained Mademoiselle Marvingt for quite a long time at the Quai de la Rapée."[62] The inference seems to be that her time would have been better except for the incident, which remains unknown. In addition, a few of the competitors remarked afterwards that they "would have done better if there had been any water in the Seine."[63] For weeks prior to the race, several dams had been hermetically sealed to maintain more water in the city center, leaving the Seine "dead and exhausted."[64]

That same year, 1906, Marie carried off first prize in a swimming competition in the English Channel in which she defeated both men and women.[65] In 1907, she withdrew from the Paris race because of illness, unsure she could do well enough to beat her own record. It was the only known time she withdrew from a competition, or used illness as an excuse for anything.

Later that year she was one of fifty-four entrants in a swimming race across Toulouse in the Garonne. She came in third overall, but first in the women's category.[66] In an interview after this race with a journalist identified only as A.C., Marie admitted that she really ought to have made better time, "but you have to realize I hadn't really trained; that was the first time I've been in the water this year in a competition." Her interviewer asked if she hadn't been a little worried about Mademoiselle Monestié, a well-known swimmer. "Of course, I was. I knew I had better get going. The thing is, she's better trained than I was and besides she's younger than I."[67]

When a national society for the encouragement of swimming held a meeting at the Sorbonne for 3,000 attendees, a speaker remarked, "Among the most applauded, let us cite Mademoiselle Marvingt, the first Frenchwoman to swim the Seine."[68] At that time, competitive swimming was for the very few. Most women were still weighed down by long-sleeved jerseys and heavy woolen bloomers, stuck in the wading and dunking stage of water sports. Marie, however, had a special one-piece red bathing suit made for her, and has henceforth often referred to as the Red Amphibian.[69]

After listing several of her swimming victories for a reporter, she was pressed for more.

"Well, since you want to know everything," she said, "I also have to my credit the crossing of Lake Gérardmer and a night swim from Pallanza to the Borromean Islands."[70] (Pallanza is on the shore of Lake Maggiore, the second largest lake in Italy, and the Borromean Islands are in the middle of the lake.)

"This crossing was done at night and in rather unusual circumstances. The colonel of an Italian regiment garrisoned nearby in the town had gotten wind of my project, and he made arrangements for me to be accompanied by gondolas on which military music was being played. I have an unforgettable memory of crossing that enchanted lake with a perfect moon overhead, accompanied by the double music of the waves and the band!

"I was regaled by music of a very different sort in the Bay of Naples during the sixteen-kilometer crossing from to the Italian coast. There had been a storm for the preceding three days that kept boats from entering the Blue Grotto at Capri, and on the day I was to swim, the weather was still dreadful. All the tourists left, in fact. As for me, I was so exasperated at having to cool my heels that, in spite of the dire warnings of the Italian ferryman, I plunged into the waves and went to visit the beautiful Blue Grotto myself. Since boats couldn't get in, I decided to do it swimming!"[71]

Marie also swam from the Greek coast to the island of Samothrace, a distance of twenty-two kilometers.[72] In 1912, the Red Amphibian surfaced again. After an excursion boat accident, she had to stay afloat for twelve hours before being found by several startled men fishing before dawn.[73]

Although she was not a world-class swimmer, Marie was good enough to win a number of competitions. Her most important contribution to the sport, however, was using her celebrity, energy, and connections to make swimming obligatory in French schools, for both girls and boys.[74] She believed strongly that regular swimming could produce the world's fittest athletes, and that it served as a base for all athleticism. She devoted herself particularly to encouraging women to swim, pointing out that women floated more easily because they were generally smaller than men with lighter bones and more fat. Anyway, why shouldn't women get the benefits of swimming that men enjoyed?[75] Her example proved contagious.

"I've always been a fanatic of water sports," she explained. "They are wonderful for developing muscles using harmonious movements that allow the feminine body to keep its grace and elegance while contributing to its physical well-being and vigor."[76] When she told school children about the time she almost drowned at sea in a tempest, the lesson was that knowing how to swim can save your life.

Whether at sea level or high altitudes, Marie wanted to investigate every sensation. From 1906 to 1910, she was ranked number five among female mountain climbers throughout the world.[77] As early as 1903, she was being photographed on mountain peaks that had never known a woman's foot. That year she climbed the Trélaporte (the most westerly peak in the Mont Blanc massif) in culottes—and it's not clear whether it was the perilous exploit or the shocking breech of good taste that gained her more publicity.[78]

A much-reprinted photograph shows Marie at the summit of the Trélaporte, balancing on one foot, the other leg extended horizontally behind her, arms outstretched on either side. In the background is an unending chain of snowy mountains.

"I love the mountains passionately," she told journalist Armand Rio, "and I love them even more in the summer than I do in the winter. Then, you know, they are divine! One of the best guides in Chamonix, Camille Ravanel, taught me to appreciate them years ago in an excellent beginners' course. We climbed the Aiguille de l'M, which is located in the Mont Blanc massif. Oh, it was just a simple training course, but from that day forward I felt, in all its intensity, the extraordinary and pure joy that the mountains can offer their admirers. From the mountains I gained not only an unending obsession with their beauty—the white summits, where the sun, depending on the hour, illuminates them with fiery roses and purples under a vast sky—but also the imperious desire to climb, to climb still more across those vertiginous crevasses and those deep gullies, to the flank of the rocks, to the extreme heights where, at last, one is supreme. The mountains also taught me the thrill of conquering danger."[79]

She wrote a piece for the *Éclair de l'Est* about that first mountain-climbing experience, July 21, 1903.[80] Although mountain climbing was becoming popular, only a fraction of her readers knew what was involved. She was generous with details that allowed them to share her own joys. Describing a huge moulin, a vertical shaft created in mountain glaciers by water and debris, she wrote, "You couldn't even see the bottom of it. You could only hear the grumblings of an impetuous torrent. The edges of the shaft, formed like a funnel, were an intense blue. Whoever tried to get a closer look risked saying goodbye to this world. Only one hundred years later would the glacier give up its prey."

As they began to climb to the peak, "Camille the guide, Simond the porter, and I were advancing between high walls of sheer rock. Suddenly a hideous thunder broke the Alpine silence. Camille looked up and yelled, 'Get down on your right!' A huge rock avalanche came rolling toward us at a dizzying speed. I had only a second to throw myself into a little trench formed by the snow at the foot of a rock, and I ask you to believe that it was not only my imagination that made me feel the wings of the angel of death brushing past me! The infernal noise trailed off into the distance and finally disappeared. We crawled out of our shelters, not without continuing to tremble from the shock we'd had. As is traditional in the mountains, once the danger is over, we silently shook each other's hands."[81]

Avalanches develop when a mass of snow on a slope becomes so overloaded with new snow that its hold on the slope is loosened. The snow mass can be so unstable that the weight of a single skier can break the delicate bond, sending an entire mountainside hurtling below. As it gains speed, the downward-rushing mass breaks up into a river of flowing snow, generating a cloud of snow dust that may extend upwards for hundreds of meters. As the avalanche descends, it picks up more and more snow, which causes it to race even faster. At its peak, an avalanche can attain a weight of a million tons and a speed of 320 kilometers an hour.

Marie understood what they had survived. And to whom she owed her life. "After having escaped death by a wing's beat, I understood in a new way that your guide is more a friend and a comrade than a simple human compass. From that time forward, I've deeply admired these heroes who live daily in the heart of danger."[82] As a mark of this respect, Marie never failed to name every guide and porter in her group whenever she wrote of her adventures.

"Just before attempting the last part of the climb, we detached ourselves from the cord, put our Asti Spumante in the snow to chill, and emptied our sacks. Never had little toasts with foie gras and jam made by Simond's father accompanied by a good wine tasted so great. A blackbird approached us, waiting for our crumbs. Then Simond refastened the cord, and we headed up for the final climb to the peak."

Marie was almost poetic when describing her love of the mountains: "The sun sets every brilliant atom afire in the immaculate carpet of snow that we are about to deflower. Only a jackdaw soaring overhead animates the landscape's austere beauty. Once we gain the summit, I taste the new, inexplicable joy of arriving at the culminating point, the place where the mountain stops reaching, where the soul ceases to desire; it is almost a perfect form of natural satisfaction, perhaps of the kind experienced by philosophers who have finally understood, after long research, a truth in which their spirit finds contentment, and can at last rest."[83]

The final moments of Marie's first climb were sheer childlike fun. Faced with an expanse of icy snow, "Ravanel and Simond sat down, I did the same behind them, and we raced down the snowy slope, using our icepicks for brakes. Exhilarating!"[84]

In 1903 she was the first woman to climb the 4,013-meter Dent du Géant, a sharply pointed mountain "needle." From 1903 to 1905, she made a number of important climbs, including the Grépon, which had been considered unclimbable only two decades earlier.

Although a few Alpine peaks had been climbed before, modern mountain climbing really began in the Alps in 1850 when English sportsmen climbed there for pleasure. While the term mountaineering can be loosely applied to walking up low mountains that offer only moderate difficulties, it usually means climbing in areas where the terrain and the weather conditions make it hazardous and, for the untrained, deadly.

In the early days, only men climbed mountains. By the late nineteenth century, a few adventurous women were entering the field; Marie was one of the first of them. She is generally considered to be the first woman to climb the principal peaks in the French and Swiss Alps (from 1903 to 1907).

In 1913, Armand Rio wrote: "There is hardly a summit where Mademoiselle Marvingt has not planted her victorious ice-axe: the Dent du Géant and Monte Rosa, the Dent du Requin, and the Aiguilles Rouges, the Wetterhorn and the Aiguille du Moine, the Tacul, the Jungfrau, and how many others! Some of her mountain climbing, which would make the most experienced mountaineers hesitate, represented up to seventeen hours of climbing. She is the only woman who succeeded in climbing in one day the Grands Charmoz and the Grépon, August 22, 1905, with the Payot guides from Chamonix. Only two men had ever climbed the peaks before her!"[85]

A.F. Mummery climbed the Grépon in 1881, when the peak was considered unnavigable. It was Mummery who first explored the possibilities of rock climbing, which is done on the uppermost slopes of a high peak, after passing over its lower reaches of snow and ice.

Marie's climb of the Grépon was a sensation for three reasons: the Grépon was a difficult ascent—only twenty-four years earlier climbing it had been considered an impossibility; no woman had ever climbed it; she climbed another well-known and difficult peak the same day.[86] Articles titled "Girl Climber's Record" and "Alpine Bravery" began, "Mademoiselle Marie Marvingt, of Nancy, who is 22 years of age, may claim to be the best climber of her sex for this season. She has accomplished a brilliant series of ascents that any first class Alpinist might be proud of."[87]

She and the experienced guide brothers from Chamonix, Édouard and Gustave Payot, left Montanvert at 2:30 in the morning, reached Rognon towards 4:30, and the summit of the Charmoz by 9:00. Then, having crossed several peaks and descended the long couloir of the Grands Charmoz, the three of them entered the Mummery chimney at about 11:30.

That challenging climb, the crossing of several narrow ridges, and the dizzying rope descent of the Grand Gendarme were all done by 3:00 p.m.

Newspapers reported that the intrepid alpinist and her companions arrived at the top of the Grépon in a state of unparalleled delight. An approaching storm left them little time to celebrate, however. The air crackled with electricity, and they were obliged to leave their only ice axe behind because of the lightning. Over Chamonix, lightning arced continuously and thunder boomed. Moving rapidly through a hailstorm, the three climbers finally arrived back at the seracs (a pinnacle or ridge of ice on a glacier), which meant they had descended far enough to a known place where they would be safe from the storm.[88]

After this experience, Marie liked to warn climbers: "Never set off with fewer than three people on a cord."[89]

Marie and her guides returned to Montanvert at 8:30 that night, "not at all tired from their long day."[90] (Marie's considerable energy meant that she very often biked from Nancy to and from Chamonix for her mountain climbing and winter sports activities.[91]) Marie said later that she was repaid for the difficulty of the climb by a marvelous view of the surrounding mountain chain. This accomplishment earned her the descriptor of "greatest female mountain climber of our era."[92]

A 1905 article about the Alpine climbing season datelined Geneva noted: "Among the Alpinists of the fair sex, Mlle. Marie Marvingt of Nancy (France) and Miss Edith Lee-Baker, of Chicago, deserve special mention, as they have made a series of brilliant climbs, consisting of the highest and most difficult mountains in the Alps of which any first-class Alpinist of the sterner sex would be proud."[93]

In their September 1907 column on Alpine news, *La Montagne* reported: "Mademoiselle Marvingt, a medical student from Nancy, with Joseph Demarchi and Jacques Tisset, made several superb climbs: the Aiguille du Moine August 5; the crossing of the col de Passon and the col du Tour August 8; the crossing of the Fenêtre de Saleinaz and the col du Chardonnet August 9; an attempt to climb the Tour Noir August 10; and finally, on August 13, the ascension of the Dent du Requin."[94]

Mont Blanc is the deadliest mountain in Europe; to date, some 6,000–8,000 climbers have died there. In addition, crampons and carabiners hadn't yet been invented, nor had nylon. Ropes, to be of any use, had to be thick and heavy (and grew heavier when saturated with water).

From London to Tasmania, newspapers noted that "Mlle. Marie Marvingt, aged twenty three, of Nancy (France), spent a week among the 'aiguilles' of the Mount Blanc range, accomplishing a remarkable series of first-rate climbs. For a whole day her party was imprisoned in the hut on the Pavilion de Lognan by a violent snowstorm, and was nearly swept away by an avalanche on the Col de Saleinar, the mass of snow passing within a few yards of the climbers."[95] (One of the Australian newspapers ran this news under its "Gossip" rubric, undoubtedly because she was a woman.[96]) The public recognition meant as much to Marie as the medal she received in Rome for her mountaineering from the hands of the queen consort of Italy, Margherita of Savoy.[97] Marie particularly admired Queen Margherita for her unswerving support of the Red Cross and other charitable and cultural organizations.

A few years later, journalist Armand Rio said to Marie, "You can't have such an impressive list of mountain-climbing exploits without having some rather interesting anecdotes to tell."[98]

"Oh, as to that," responded Marie, "as many as you like! While climbing the Grépon, I missed by only a few centimeters being pulverized, annihilated, crushed by an enormous hunk of rock of about 12,000 kilos which, in the Mummery gully, had the fiendish idea of pitching itself upon our heads! I arrived at the summit only to be half electrocuted by the most terrible thunderstorm I've ever seen in the mountains, and God knows what a storm is like in the Alps."[99]

Scarcely stopping for breath, she continued, "During my climb of the Dent du Géant, my guides and I were surprised by a snowstorm. We wandered around for seven interminable hours on the glacier. To help us put this unpleasant memory behind us quickly, Fate in its fashion prepared another distraction for us. While we were scaling the needle, a climber on a nearby rope slipped, carrying his guides with him. They were hurtling toward an abyss when, happily, the hook on their supplementary rope caught them. We quickly went to their assistance, not without some personal danger. For several minutes during the rescue operation, we were all suspended from one iron clamp, our only hope. If it had given way, we would all have descended into Italy by the very fastest route."

Drawing used in the Jacques Boetsch article "A Balanced Life," apparently made from a 1903 photograph of her reaching the top of the Trélaport.

Marie's usual garb for mountain climbing was functional: flannel shirt, maroon culottes and matching vest of thick cloth, soft felt hat, and sturdy boots, tailored artistically for her by a local shoemaker.[100]

Once she hiked fifty-seven kilometers alone through the French Alps, a route that included two difficult rock climbs.[101] She scaled many Italian peaks with a group of Italian officers who were mountain-climbing fanatics.[102]

Charles Codron, a friend and admirer of Marie's, said, "Mountain climbing became a true passion for her; with a captivating verve, she used to relate her adventures naturally and unaffectedly, stories that made her listeners shiver."[103] He added that you would have to hear her professing her admiration for the Alps, their inhabitants, the guides, and all those with whom she shared these dangers to understand that "her heart beats with devotion and compassion for all humankind."

In 1911 the magazine *Femina* published a detailed article by H. Burlingham on female mountain climbers, with a two-page pull-out drawing of the major alpinists of the day ranked according to their reputations.[104]

"Mountain-climbing as a sport is, without argument, considered to be the most masculine of the sports. The hardiness, courage, and expertise needed belong, in the general opinion, to the stronger sex. It's generally believed that if women did try, it would only be as amateurs, dabbling just for fun.

"Not true. In fact, some of the most astonishing achievements in the mountains belong to women. One Frenchwoman whose mountain exploits deserve special mention is Mademoiselle Marvingt (of Nancy) who has a number of worthy climbs on her record."[105]

Ranked in first place was American Annie Smith Peck, who had climbed challenging peaks not only in Europe but in South America and in the Himalayas. Marie was ranked fifth.

Burlingham quoted Swiss travel writer Marc-Théodore Bourrit, who wrote in 1786, "Oh what a beautiful thing is a beautiful woman on a beautiful mountain!" and noted that a woman on a mountain was perceived by most people as a ridiculous idea given that even very few men ventured onto the "homicidal Alps."[106]

It was with some irony that, twenty-five years later, Marie flew over Mont Blanc in a plane, and saw the tiny funicular that ran up the mountain. She had reached the same heights, but with a great deal more effort and risk.[107]

Marie climbed a number of other mountains, including the Gabelhorn, the Breithorn, Col du Assoine, Tour Noire, Col d'Argentiche, and Aiguille du Tour, and she was also the first woman to ski up several mountains: Buet, col de Balme, and col de Voza.[108]

More than ten years after her peak mountain-climbing years, a 1921 article in *Le Figaro* still named Marie to a short list of the best pioneer mountain climbers.[109]

"As soon as winter arrives," wrote Armand Rio, "Mademoiselle Marvingt puts her bicycle in the garage and abandons her canoe and skiff to their hangars. She then inspects her ice skates and her skis, and takes off for the stunning kingdom of the snows to compete on the white trails of the Alps and the Vosges. There is no season that is not a sport season for this intrepid young woman who is constantly gripped by the fever to excel."[110]

Although the modern Olympic Games began in 1896, there were no official Winter Olympics until 1924 when they were held in Chamonix, a picturesque tourist resort in the French Alps. In the meantime, a series of winter sports championships were held in various mountain resorts in France.

"My first ski jump, in 1907."

La Grande Encyclopédie de La Montagne noted: "As of 1908 Chamonix became one of the best places for winter sports. In addition to some mostly English alpinists, the first clients came from the best of Parisian society; some of the first were the aviators Henri Farman and Marie Marvingt."[111]

Marie competed at Chamonix in 1908,[112] in Gérardmer in 1909,[113] and in Ballon d'Alsace in 1910.[114] In 1910 alone, she won twenty first prizes in skiing, ice skating, toboggan, and bobsled.[115]

In February 1907, not a single woman participated in the International Ski Week at Chamonix. The following year, nine enthusiastic sportswomen (including Marie) completed the three-kilometer ski competition.[116] By 1909 there were several ski contests open to women, including Marie.[117] In the beginning, however, the function of women skiers was primarily to attract tourists; the first women's ski runs were the same ones used for children. "Marie Marvingt ruptured and re-made this pattern."[118]

On January 27, 1909, Marie took two firsts, a second, and a fifth in skiing, ski jumping, and luge.[119] A month later, she won first place in women's skiing, first place in women's skating, and was singled out by newspapers for her exhibition ski jumps (she was classed among the best jumpers).[120] Again the next day, reported *L'Est Républicain*, "not to change her habits, Mademoiselle Marvingt takes first prize in women's skiing."[121]

She won first place in women's skiing in 1909, 1910, and on January 23, 1911, when she also took first place in women's ski jumping.[122] The following day, she took first prize or placed in six events.[123]

On February 11, 1911, she competed in the international military ski concourse in Lioran, placing first in the women's ski race.[124] To get to Lioran, she took the train to nearby Aurillac, where she had been born. Dressed in "a ski sweater and an elegant, warm fur coat," she was met at the station by the mayor and local officials, taken on a tour through the town, and generally fêted royally.[125] The local newspaper admitted that other women were doing well in sports, but they were loyally certain that "no one practices sports with the same degree of perfection that Mademoiselle Marvingt does."[126]

In 1909 George Casella wrote in *La Culture Physique*, "While most of the women at the winter resorts dressed traditionally, three sportswomen dressed like boys," but these three—Madame Frantz-Namur, Madame Pallier, and Mademoiselle Marvingt—were so successful at everything they did, he prophesied they would soon have many imitators. "When you're as good as they are," he wrote, "You can wear whatever you like!"[127] A photograph from the March 5, 1910, *Le Sport* of Marie in her ski costume was captioned, "Our colleague, Mademoiselle Marvingt (Myriel)

Marie and the Italian soldiers she skied with in 1911.

before the start of the women's ski competition in Gérardmer," and photos of her in the midst of ski-jumping competitions show her wearing the daring but practical culotte.[128]

Competitive skiing was relatively new for both sexes. The first big ski meet in the world was held in Oslo, Norway, in 1879 in front of His Majesty the King and 10,000 spectators. Marie was four years old. The best skiers and ski-teachers were thought to be Norwegian, and it was typical of Marie that she sought out Norwegian Harald Durban-Hansen when she wanted to learn this new sport. She was his first French student, and Durban-Hansen said he had no student he was prouder of than Marie.[129] In the end, she held fourteen ski championships.

Chamonix, January 19, 1913.

In 1911, Marie contributed six pages to a book on winter sports by Louis Magnus and R. de la Frégeolière: "Skiing is the healthiest, most invigorating sport there is, building and soothing muscles at the same time. And when it comes to skiing, forget those annoying comparisons between women and men, the silly claims that women aren't strong enough to accompany men down the slopes.... Skiing is more a question of skill, agility, and mental quickness than of brute force. Small, supple, even puny women often make better skiers than large, heavy, energetic men. ... The fact that you don't see more women skiing has less to do with their sex or their temperament and more to do with the contempt in which men hold sportswomen. Fortunately, we've arrived at a time when women can throw off the yoke."[130]

One sport seemed to lead to another for Marie, provided it was new and involved speed. It surprised no one when she took up ice-skating, and then won prizes in both racing and figure skating three years in a row.[131] (She was also well known as a roller skater.[132])

In an era of so much experimentation, of course, there were hits ... and misses. A 1910 photo shows Marie and several other women preparing for a skate-ski competition.[133] Each woman was wearing a ski on her right foot, a skate on her left foot. This sport apparently died aborning as no results of the race appeared in the newspapers, and there were no more photographs of skate-skiers. (Years later, she participated in another odd pairing: a balloon-automobile race.[134])

In another photo, also taken in Chamonix, Marie is the only woman in culottes among a group of athletes preparing to compete in the 1,500-meter speed-skating event.

She's wearing a knit hat, a dark skirt, and a pullover that comes down almost to her knees—foreshadowing a style of many decades later. The other women's skirts touch the tops of their ankles but Marie's culottes come to just below her knees, allowing her greater freedom of movement but exposing her to both cold and criticism.

On the back of the photo-turned-postcard, she wrote to Marthe Roumier: "I'm delighted with my stay here. The season has been superb, the parties very successful."[135]

Marie had little respect for bobsledding, which she said was "not a true sport and one that presents too many unnecessary dangers."[136] A mule was hitched to a log to trace a path, which was then watered to sheer ice. Since the weighty bobsleds reached over 100 kilometers an hour, with no safety provisions, accidents were often fatal.

Marie's sentiments were well known, which is why many were surprised for more than one reason when she won the first Léon Auscher Cup for luge on January 26, 1910.[137]

In an article bylined with her pen name, "Myriel," in the March 5, 1910, issue of *Le Sport*, Marie recounted how she happened to win the Léon Auscher international women's bobsled championship:

> The day before the big event, Monsieur de Rochette-Barron, president of the French Bobsleigh Club, came to me. "Are you participating in the Auscher cup?"
>
> "This is going to astonish you," I told him, "but I've never been on a bob."
>
> "That doesn't make any difference," said the president of the F.B.C. "You'll go down tomorrow and I'll bet you'll win. We're counting on you. We really want a win by a woman, to lessen the blow to the sport of bobsledding by the recent death of our most valiant and energetic friend Miss Hudson. Contrary to what's been reported, this young Englishwoman, who was at the brakes that day, was extremely competent in this violent sport—she'd been doing it for several seasons already. Yes, falls of that type are common but rarely, alas, fatal. From one false move, the *Fol-Bob* flew over one of the banked curves and fell heavily on the young captain, Miss Jeannion, and her group, who were badly injured. Miss Hudson was thrown several meters, and by some evil chance her head landed on a pointed rock sticking out of the snow. If this young woman had worn a helmet as we'd often advised her, we wouldn't be grieving today the loss of our charming friend."
>
> After this depressing and unencouraging recital of the dreadful accident—and I had witnessed several bad ones myself at the col de la Faucille and in Caux—Monsieur de Rochette-Barron spoke at such length and so eloquently of his favorite sport that I acquiesced, with certain conditions, to his foolhardy project.
>
> Later, I made my first descent on the *Ice Bird*, a bob belonging to Madame de Rochette-Barron. I tried to dissuade her but she insisted on climbing on behind me with two other riders. The iced trail was superb, and from the beginning, we flew at a great speed. But at the dangerous turn where, several days later during the Championship of France, seven of the ten bobs overturned, I didn't turn the wheel fast enough and the *Ice Bird* tossed us a good distance to swim through thick, wet snow where, happily, there were no rocks sticking up. Only Madame Rochette-Barron was trapped under the bob. Not without great emotion, we disengaged her from her position.

Marie (second from right) waits her turn for the speed-skating contest, 1911.

Like the rest of us, fortunately, she was scarcely hurt and didn't have too poor of a memory of my baptism on ice.

The next morning, the day of the championship, I practiced again with Messieurs Schmidt and Crochet, highly expert bobbers from Chamonix. Our speed wasn't as lively as the day before and I had the great pleasure, this time, of depositing my team, who had honored me with their confidence, safe and in one piece at the bottom of the trail.

Encouraged by this progress, I looked forward with delight to the race, which started at 2:30. I drew number 7. I refused to look at the times made by the other team captains going before me. Some of them had been bobbing for several seasons.

With my two new friends of the morning behind me, we listened for the three notes of the horn, followed by another series from the end of the course indicating that it was free. The timekeeper signaled to me.

Fat snowflakes were falling thickly and nearly blinded me. Inexpert as I was, I never thought of the black silk veil, used to such good effect several days later by la Frégeolière.

I took the first turn—the most difficult—nicely and I passed like an arrow between the two high white walls created recently by an avalanche that had obstructed the trail for a while. Under the stinging snow, our speed kept increasing.

My companions, bobbing together, were lying down completely on the one side during the second turn. We were practically flying, but I kept a firm hand on the wheel, with which I now felt completely at home. I scarcely saw the numerous spectators lined alongside the course, but I heard their shouts.

I arrived at the bottom at a dizzying speed in a whirlwind of snow. I'd done the course in one minute seven seconds. I won the Auscher cup.

A rather original parade concluded the course, which had been unmarred by a single accident. All the bobs, with their participants aboard, were pulled by horses across Chamonix, led by a band, and accompanied by visitors and inhabitants on skis. To the falling snowflakes were added roses and violets brought in from Nice and strewn over the victors.

Later, at the congratulatory glass of wine offered to us at the Hôtel des Alpes, Madame de Rochette-Barron told me she knew I'd be successful. I had to confess that I still hadn't changed my mind about the bobsled. However, without encouraging anyone else to take up the sport, I had to admit that the dizzying and intoxicating speeds were like nothing I'd ever experienced.

When it comes to sports, one shouldn't—in my opinion, at least—seek only the pleasure, the highs of the moment. You hope that in practicing this sport, you can translate it to something practical, useful, or scientific. But whatever could you find useful to do with a bobsled?[138]

The bobsled, described at the time as "a perfected luge, carrying three or four people," was little more than a child's sled except that it could go 60 to 80 kilometers an hour on a properly iced track. The bobsled required a great deal of cooperation among the riders, and serious injuries were common; a number of young people were seen walking around Chamonix at the time with bandages around their heads.[139] In *The Big Book of Women's Sports*, Serge Laget reported Marie's luge wins, and wondered about her style: Did she use the European method (seated, legs outstretched, heels in the snow) or American (lying flat, head facing forward, feet dragging on the trail)?[140] In both cases, the feet served as brakes.

Marie won or placed in a half-dozen other luge competitions from 1910 to 1913.[141] She must have kept up her skills, however, because on January 29, 1928, she placed in a race limited to luge aces.[142]

In the early 1900s, skiing was taught only in the military. It was a necessary skill for soldiers who sometimes needed to cross snow-covered mountains. In 1910 Marie opened her own ski school in the Alps, the first civilian ski school in France.[143] Based in Chamonix, she taught until 1914, taking her students on excursions so that they could sample skiing under varying conditions.[144] Two of them designed the school's logo: two winged red skis against a large embroidered "M."[145] The war obliged her to close the school.

In 1923, she founded another ski school, this one for her Berber friends in North Africa—except that they skied on sand.[146] It was in that same era that she experimented with and finally developed metal skis, testing them in the desert.[147] In *Skiing Heritage Journal*, Doug Pfeiffer writes about a pair of 1929 aluminum skis located in the Dauphinois Musem: "They were commissioned by Marie Marvingt, French adventurer, pioneer pilot, competitive swimmer and skier. She had two pairs hand-crafted—one pair for sliding across the sand dunes of the Sahara Desert, which reportedly she did for at least 50 miles on one of her odysseys (they didn't slide all that well), the other pair for her enjoyment of the abundant snows of the French Alps (same problem). The skis were an anomaly, and there was no manufacturing follow-through. But they are a part of the continuum of skiing's heritage."[148]

In 1909, Marie, writing as Myriel, introduced the readers of *Éclair de l'Est* to winter sports.[149] After apologizing for not having responded to readers' requests for information and advice, she outlined the joys and requirements of skiing and jumping, the girouette, the luge, the skeleton, the bobsled, and ski-joëring (skiers pulled by a galloping horse, with the first skier holding the reins). Her tips ranged from "Don't buy any equipment of inferior quality" to warning that while skiing down a hill might take two hours, it takes about twelve hours to ski up it first. She described the Telemark and Christiania stops, but admitted that beginners would still, despite themselves, make use of the Briançon stop, that is, sit down abruptly in front of the obstacle. In the end, her 3,000-word article provided a complete guide to winter sports by one who knew them all.

Life in the new winter resorts wasn't all work. Variable weather, low temperatures, snow conditions, and fewer winter daylight hours limited time spent on the ski runs, luge courses, and ice rinks. The off-hours offered excursions to other villages, sleigh rides, banquets, and dances that lasted until all hours for the high-spirited fans of vigorous activity. Marie reveled in the camaraderie, endless rounds of making plans for the next outing, and exchanges of the latest news and sports information. In between winning competitions during the day and socializing at night, Marie the journalist filed story after story, reporting back to Paris and London on the glittering resorts.[150]

On one of those beautiful frosty days, reporters stopped Marie dressed "in a short culotte and a balaclava that hid all of her hair and most of her face." Because of the way she was dressed, they saluted her: "Good day, sir!"[151]

She burst out laughing. "Don't joke," she said. "This is the best outfit for exercise. It's too bad the trails are so poorly maintained. Several sleds have been ruined, and the bobsleds have barely enough room to maneuver. If their drivers hadn't showed exceptional prudence, we'd have had some dreadful accidents. As for the women's trials, they appear to have chosen a spot where spills are inevitable."

"That happen to you?" they asked.

"Me?" she laughed with good humor. "I somersaulted with everyone else. And the photographer was totally pitiless. You'll see with what grace I executed that little move!"[152]

Newspapers had no trouble reporting men's sports in a factual manner, but when it came to women's activities, they couldn't seem to get beyond what the women wore or how beautiful and charming they were. Knowingly or not, newspapers trivialized women's participation in sports. For example, Marie proved herself an outstanding athlete at Chamonix in 1908, yet the photograph newspapers most often chose to run showed her at the wheel of her automobile-sled—a strange contraption with four skis underneath, a large-tread wheel in the center, two seats, and a steering wheel. This was certainly typical of Marie—driving the latest machine. And it was the sort of picture newspapers thought their readers would enjoy. Marie, however, would have preferred photographs of herself skiing, jumping, luging, or ice-skating.

Newspapers hesitated to give too much importance to women's participation in sports, always mindful of the slippery slope ("If women do this, next they'll be voting!"). Marie epitomized women's longing for a freer, more active life, and was thus a symbol of what some people most wanted for the future and what others most dreaded. However, skiing soon became both a means and a sign of women's increasing independence. Although there were women-only events, women could also often compete against men. When everyone was dressed for skiing, there was little difference between the sexes. Marie believed that women and men could meet as equals on skis, and that the conversations between them as they shared an activity were much more interesting, sincere, and personal than the vapid conversations of café society.[153]

In the chapter on skiing she contributed to the Magnus and de la Frégeolière book, Marie remarked that skiing was one of the sports "permitted" for women. She continued wryly, "Given men's limited tolerance for women's participation in sports, I know women will be suitably enthused to hear this."[154]

Armand Rio, one of Marie's admirers, pointed out that "only a few years ago, Frenchwomen were ridiculed by women from other countries who claimed that Frenchwomen were only good for playing a leisurely game of croquet with small boys. They made fun of Frenchwomen's nerves and vapors, their needlework and reading of romances in the *chaises longues* of convalescents. It was pure calumny, based however on a bit of truth. Frenchwomen devote far too little of their time to the outdoors and to sports. Today,

Painting by Vera de Landchevsky celebrating Marie's multiple sports (courtesy Odette Friry).

however, that's changed, and Frenchwomen are the respected and often victorious equals of sportswomen from England and the United States." He said that he could name outstanding Frenchwomen in most sports.

"However," he added, "to summarize all this, let us state immediately that Frenchwomen have the great honor of counting among themselves a woman who, more than any other—and by a great deal—has the right to claim the title of 'best sportswoman in the world,' Mademoiselle Marie Marvingt."[155]

Armand Rio was not the only person who thought so.

With the extraordinary eclecticism Marie showed as an athlete, it's perhaps not surprising that on March 15, 1910, the French Academy of Sports, "in a flattering exception"[156] awarded Marie Marvingt its gold medal for excellence in all sports.[157] To this day, she is the only individual of either sex to be awarded a multisports medal from the Academy.[158] A 1913 headline in the *Chicago Star Publications* summed it up: "Frenchwoman Has Made Remarkable Records in Everything That She Has Undertaken."[159]

Marie would certainly have wanted to compete in the Olympics had women been included. The first modern games were established primarily by Pierre de Coubertin in Athens in 1896. Marie was twenty-one at the time, as intrigued and passionate about them as a sports fanatic could be. Pierre de Coubertin, however, obstinately and categorically opposed any participation in the Olympics by women and men of color. He once said, "A female Olympiad would be impractical, boring, unaesthetic, and incorrect. The true Olympic hero is, in my eyes, the adult male individual. The Olympic games should be reserved for men, the role of women being to crown the winner."[160]

Only in 1928 at the games in Amsterdam were women allowed to compete (against Coubertin's advice). Marie was by then fifty-three years old and no longer interested. Today, the Pierre de Coubertin indoor arena in Paris pays tribute with a series of plaques to thirty outstanding athletes known as "the glories of sport." Marie's is engraved: "Marie Marvingt, omnisports, 1875–1963."[161]

Marie never met a sport she didn't like. She sailed land yachts (contraptions with sails and wheels that flitted across beaches)[162] and helmed sailboats, rode motorcycles, and was particularly fond of skiffs—long, narrow racing boats made for one person, usually fitted with a sliding seat and covered fore and aft with canvas. She learned to deep-sea dive and to explore caves (she spent sixteen hours spelunking in the Lacade Grottos, and explored the Daya Chicker in Morocco).[163]

Motorboats also came of age with Marie. They began to appear in the late 1880s, with the first international competition in 1903; France won it in 1905. Of course Marie was interested and intrigued. Noise! Speed! Possible danger! Once she had investigated its (to her) somewhat limited possibilities, she moved on.

Hunting for sport was widely accepted at the time, so Marie hunted tigers in Asia, seals and polar bears with harpoons in the Nordic countries.[164] She even participated at least once in "chasse à courre," or running hunt, a uniquely French sport dating to medieval times and involving hunters on horseback or on foot chasing a wild boar or stag with hounds.[165]

In one of her calmer activities, Marie is shown in a charming photograph serenely driving her horse-drawn buggy through the Bois de Boulogne.

She played baseball, hockey, polo, and water polo, and fished wherever the opportunity presented itself. Her involvement in certain sports was brief, but she made it a point to experience and achieve proficiency in every sport. Some of her passions became enduring patterns in her life, others simply provided a contrasting thread here and there.

"What's compelled me to devote half my life to sports?" she echoed a reporter's question. "But everything! Education, circumstances, and a pronounced taste—I was always very lively and demanding—for struggle and effort, for taking risks."[166]

To a reporter who asked which sports she was familiar with, Marie said, "Every one, without exception."

"But there must be an exception, especially for a woman. I bet you've never boxed."

"Wrong! I've done quite a bit of it, mon cher, but I never particularly distinguished myself in it."

"What about catch?" he asked.

"You mean as in 'catch as catch can'?" she asked. ("Catch as catch can" was a briefly popular no-holds-barred style of wrestling.)

"Hmmm, no, that one, I never thought of it. In my youth, that didn't exist. Is that even a sport?"

"Well, let's bet that you've never played football!"

"I'd win that one."[167]

In 1924, R.P., who obviously knew Marie, mused, "Marie Marvingt has always competed in sports with such fire and dash that the list of her wins soon becomes impressive. I wonder if she believed me when I said it would take a man much more talented physically than she to simply accomplish the half of what she's done. In order to relate everything about 'this woman like whom there is no man,' as Gounouilhon described her in the *Petite Gironde*, it would take several issues of *Le Cri*."[168]

"A strange need to spend herself drove her," wrote Georges Gygax. "A need to succeed better than anyone else, to triumph with joy over life's difficulties. This need made of her an outstanding athlete, one of a kind."[169] Other writers noted her acute need to surpass, not only others, but herself.[170] *Le Miroir des Sports* named her "the best sportswoman in the world"[171] while other publications revealed their affection, saying that she was, and remains, one of the most appealing figures in women's sports.[172]

Marie always maintained that sports were simply "a foundation for my intellectual activities, a guarantee of good health and equilibrium."[173] In practice, however, she seemed extraordinarily delighted by every sport that she engaged in.

She admitted, "It's the totality of the sports I've practiced that counts, and not really this or that performance, which in itself is insignificant."[174] Still, as Elizabeth Boselli said, "Mademoiselle Marvingt accumulates sports exploits that a good number of recordmen can envy."[175] And, in *La Vie au Grand Air*, Marcel Violette went so far as to describe her as "the creator of women's sports."[176]

He exaggerated. It takes many women to do that. On the other hand, the history of women in sports is nonlinear: a series of true and false beginnings, evolutions, innovations, leaps forward, brick walls, standing waters, cheerleaders, and critics. Marie ran her part of the relay well.

Marie welcomed, encouraged, and supported other women in multiple ways. Without saying a word, but by pursuing her own personal goals with willpower and talent, she provided a role model for thousands of girls and women.[177] In the early twentieth century, Chinese women's journals presented Marie as the model of a "new world trend" for women.[178]

In 1910, when Admiral Payssé's athletic club turned down his proposal for a women's auxiliary, he helped found Fémina-Sport, an omnisports women's club. He and others, including Marie, formed the French Federation of Women's Sports (FFSF), which hosted the first women's championships in 1917, a six-day affair with relay races, jumping events, field hockey matches, and demonstrations of gymnastics and boxing.[179]

In a 1919 article, George Rozet said that the negative reaction to women in sports is "a prejudice that belongs to yesterday" and he described the formation of a group called The Academy of Sports and Physical Education for Women, Young Women, and Infants (they eventually called it simply the Academy).[180] Marie was again one of the organizers.

Female sports pioneers like Marie competed in sports that were created, structured, organized, and run by and for men. It was not enough to join those sports as they stood. By creating their own forums and organizations, women could affect sports in useful, meaningful ways. As a journalist, Marie mentioned women in print hundreds of times, giving many women visibility in the press. She lobbied for girls' inclusion in swimming classes. She spoke to many all-girl groups to encourage sports, physical activities, and independence.

Evaluating Marie's place in the world of sports requires understanding the context in which she competed. She set numerous records simply by being the first or only woman to do something. Time and growing numbers of athletes mean that these titles are no longer available to competitors. Increasing specialization means there can be no more Marvingts, individuals who win championships in dozens of fields.

As one writer of her era said, Marie was a "brilliant exception" to all the rules.[181] Born at the perfect moment for her abilities, she still played the hand of winning cards she'd been dealt without missing a trick. She was fortunate in many ways, but what was exceptional in Marie was that she pushed her possibilities to the maximum: all that she could be, she was. A sports journalist put it this way: "Marie Marvingt is as much of an athlete as it is possible to be."[182]

3

The Shooting Star

> "Individuals of Marie Marvingt's caliber appear only several times a century."—J.G., *L'Est Républicain*[1]

Marie's love of the mountains led to her next passion, and one of her most renowned feats.

"I will always be grateful to the mountains," she said. "They gave me a keen desire for the skies, the love of being surrounded by pure air. My apprenticeship in balloons was inspired by my experiences in the mountains."[2]

She remembered the moment. "It was in 1903 and I was climbing the Giant's Tooth. I finally arrived at the 4,013th meter and set up camp on the tip of the summit. Looking about me, I couldn't help crying out, 'When will I have wings so that I can climb even higher and fly over this splendid panorama?'"[3]

A few years later, her question was dramatically answered by the development of the airplane. In the meantime, she was no stranger to the skies. She had already taken her first balloon ride,[4] and life hadn't been the same since.

The French unlocked the secret of aerial travel about one hundred years before Marie was born. Papermaker Joseph Montgolfier had the novel idea of launching a paper and linen bag into the air. He was having no luck with his creation until he saw his wife throw into the fireplace a paper bag, which then floated up the flue.

On June 4, 1783, Montgolfier and his brother filled a balloon with air and burned wool and straw beneath it to heat the air inside. The balloon rose 1,830 meters, traveled for ten minutes, and fell to earth two and a half kilometers away.

When aeronautics took off, it went quickly. A few months later, Jean Pilâtre de Rozier and the Marquis d'Arlandes became the first human beings to go up in a balloon. Watched by some 400,000 people, they floated nine kilometers over Paris in twenty-five minutes and landed unharmed.[5]

Nothing before so inflamed and captured the public imagination. Aeronautical designs decorated everything from wallpaper to jewels. Songs, poems, and clothing all glorified balloons.

Women were part of aeronautics almost from the beginning; a woman went aloft just a few months after the first men did. Ballooning became so common that on May 20, 1784, four women from the court of Louis XVI went up in a balloon with Pilâtre de Rozier himself, and it grew truly fashionable around the turn of the twentieth century. In a captive balloon, anchored to the ground by stout ropes, passengers could enjoy a balloon ascension and the view from above without floating wherever the wind took them. They were not entirely safe, however, as sudden gusts of wind could dash them to the ground.

Free balloons left the earth behind and drifted with the wind. They rose higher when the air inside was heated or when more gas was introduced. They descended by releasing gas from a hydrogen balloon or by letting the air cool in a hot-air balloon. Dumping out ballast (bags of sand or other heavy objects) made the balloon lighter, allowing it to rise.

Marie's aerial career began in 1900 under the tutelage of "the three B's," Bachelard, Barbotte, and Blanchet.[6] Georges Blanchet took Marie up the first time.[7] As they gained altitude, she let out a whoop of exhilaration, alarming friends on the ground until they saw her waving happily. When Blanchet asked how she liked it, Marie exclaimed, "Glorious, absolutely glorious!" Her passion for aeronautics inspired her to say, "My greatest adventure, my biggest achievement, will come in a balloon."[8]

Immediately after her first flight, she enrolled in the Eastern Air Club, a section of the French Air Club that served the eastern provinces. The camaraderie, skills, and practical assistance of this close-knit group would be important to her all her life, and provide some of her fondest memories. Many euphoric balloon trips followed, with Georges Blanchet at the helm of the *Archimède* or with her other mentors, Edouard Bachelard and Ernest Barbotte.[9]

For her first trip as a pilot, Marie entered the 4th International Aeronautics Congress race held September 24, 1909, in a new 1144 Astra, which she named the *Eagle*.[10] Her passenger was Pierson de Brabois, who would be competing against her in another race the next year.[11]

The other balloons in the race were piloted by good-humored competitors who knew each other well: Marie's former teachers Ernest Barbotte and Georges Blanchet; her good friend, the painter and founder of the Eastern Air Club, Emile Friant[12]; and Monsieur Ferry, the local pharmacist.[13]

Just before the always fateful "Let go!" May 15, 1910.

As the balloons rose, Marie noticed spectators at all the windows and on many roofs, watching and cheering. At one point during the trip, a tree snagged their drag rope and refused to let go. Marie and de Brabois hauled in the heavy rope, many feet of it, but just as de Brabois touched a knife to the rope to cut it, they jerked free.[14] (Trees were not always friendly; Marie once ended up hanging upside down, caught by her skirt and with one foot wedged between two branches. Unable to free herself, she had to await rescue. On another occasion, she was caught by her belt.[15])

Interviewed after the race "in her elegant salon on the Place de la Carrière," she said she was utterly enchanted with the splendid trip, the views over the Black Forest, and the warm welcome she and Monsieur de Brabois received when they landed in Grabennendort. At the train station (after an inspection of their balloon because they were in Germany), officials gave "un petit speech" to which she responded in German.[16]

"We returned to Nancy on the same train as the other pilots. We were, all of us, dazzled and delighted with our trips."[17]

The same day Marie's interview was published, the newspaper reported on the disastrous wreck of the dirigible *La Rèpublique*, in which all four aboard were killed.[18] Lighter-than-air machines came with few guarantees of safety.

Later, Marie wrote her own article about the trip for the *Éclair de l'Est*:

"At Aulnois, where we crossed the frontier, people yelled to us from below: 'Stop! Come down! You're heading into Prussia.' We thanked them courteously, and continued. As we came into Sarrebruck, we were so low that we almost ran into a factory chimney, but we threw out a little ballast and passed overhead. We flew over numerous factories. All those great ovens, spitting out jets of flames, with their fairy lights, could be worrying for aeronauts wanting too close a look. The streets of Sarrebruck appeared as long luminous strips cutting through the thick black veil of the moonless night—heavy, tepid, hinting at a storm to come. At a quarter to eleven, seeing a light in a window, we called out to the human shadow inside, who informed us we were in Assveiler. We heard the clacking of a windmill, then the fresh song of a waterfall. ... While my companion slept a little, I listened to the silence, my head leaning against the anchor, analyzing my impressions, and I experienced the exquisite joy of floating through the dark, in the most absolute isolation I've ever known."[19]

She joined the Stella, a balloon club just for women organized that year by Marie Surcouf, the first woman to earn a balloon license.[20] It sponsored several splendid balloon rallies over the next few years, featuring balloons decorated with roses, violets, hortensias, peonies, and cornflowers.

As a member, Marie wasn't a good fit. She didn't much care what her balloon was decorated with as long as it went higher and farther than everyone else's. Most of the women in Stella had wealthy husbands underwriting their interest in aeronautics, and the two honorary presidents in 1909 were men. Even at Stella-sponsored balloon rallies, as few as three balloons might actually be piloted by women.[21] For most Stella members, being a balloon passenger and part of an elite social group were the persuasive factors.

By 1911, the Stella had 200 members, but only a few women were pilots.[22] An active and visible organization, the Stella gave women a footing in the world of aeronautics, even if it emphasized independence and skill less than Marie would have liked.

On October 10, 1909, Marie piloted a balloon with a passenger, landing near Vitry-le-François; the newspaper reported they returned to Nancy by the 3:00 a.m. express.[23]

October 21, she took up three passengers, including the director of the glassworks in Portieux.[24] "Enormous crowds" were reported by the newspapers, which helped by announcing well ahead of time any balloon ascensions in the area as well as Marie's participation.[25]

And then came October 26, 1909.

Marie and her team arrived at 3:00 a.m. to begin inflating her Astra "Continental" balloon, *The Shooting Star*.[26] The sun was almost overhead by the time all 1,200 cubic meters were full. Marie then detached the gas pipe from the neck of the balloon. Pinching her nose and holding her breath, she stuck her head up inside the balloon to make sure there were no holes and that the red cord and the white cord were not entangled. Satisfied, she withdrew her head and took a deep breath. Because hydrogen was expensive, Marie used coal gas, which contained hydrogen but was toxic. A negligent aeronaut could die from carbon monoxide while making this last pre-flight examination.

Marie ushered her passenger, Major Garnier, into the gondola ahead of her, latching the door behind them with thick leather straps. It was time for the "weighing off." Marie tossed out bags of sand until the balloon just barely stayed on the ground. At the end, she was throwing out the sand handful by handful to arrive at the delicate balance between ground and air.

She turned to her passenger and asked a question with her eyebrows. It was his first trip in a balloon, but he smiled: he was ready. Turning toward her ground crew, Marie gave the order to let go of the ropes that held her down: "All hands off!"[27]

The next day, the newspaper reported on the "great event" (they used English to emphasize the marvelousness of it all). "Mademoiselle Marvingt took off with Monsieur Garnier. The weather was cool, a kindly breeze was blowing, and the two aeronauts seemed headed toward Metz. They probably landed last night somewhere."[28]

They did not.

A telegram arrived in Nancy from London the next day announcing that the pair had crossed the North Sea and landed on the south coast of England, something that had never been done before.[29]

Because the predominant winds over the North Sea blow east, the crossing is easiest from England to France, which is what the Frenchman Jean-Pierre Blanchard and the American John Jeffries did January 7, 1785. Like Marie and Major Garnier, they barely made it, throwing out every bit of available ballast and even most of their clothing. But when Jean Pilâtre de Rozier and P.A. Romain tried to cross the English Channel in the other direction, from France to England, their hydrogen caught fire, and both men plunged to their deaths. Forty years after Marie's crossing, Paulette Weber, an experienced balloonist with 345 ascensions behind her, also died in the attempt.[30]

In addition to being "dangerous and more than risky,"[31] Marie's crossing took place in a violent gale.[32]

The news flew around the world. The Melbourne *Australasian* wrote, "One could hardly ask for a more striking display of self-reliance by a woman than was exhibited this week by Mlle. Marvingt, a French balloonist. … Mlle. Marvingt, who had charge of the balloon, had laughingly asked her companion if he felt inclined for a trip to England."[33]

The local newspaper was eager to interview Marie. "We went to 8, Place de la Carrière, hoping to get an account of her adventure.[34] 'Mademoiselle isn't home,' said her maid. 'However, she sent me a telegram from England with two words: Coming home.'

"'Do you think she will return directly to Nancy?' we asked.

"'Well, she might stop in Paris to see some friends. But I think she'll need to rest before taking off again so she might be home Friday night. And then she has that balloon to drag behind her. It's a cumbersome bit of luggage.'"[35]

Two days later, as soon as they heard she was home, "we showed up at her door." A journalist known as G.M. interviewed her for *L'Est Républicain*: "The intrepid aeronaut to whom fatigue is apparently unknown, said she was just leaving to see to repairs on her balloon and then visit some friends, but she could give us a few minutes on her way. So we accompanied her, asking questions as fast as we could."[36]

They inquired about her health because one of the bulletins from Paris mentioned an injury. "'False. Absolutely false,' she said with a little vexed air that suited her very well, 'How on earth can the newspapers say such a thing. Would I be walking like this if I'd been injured?'"[37]

October 26, 1909: Marie and Monsieur Garnier heading for England, although they didn't know it then.

She declared the trip to be "Magnificent! Marvelous! Splendid!"

"Weren't you ever frightened?" they asked.

"'No! I was only awed by the magnificent and imposing spectacle before me. And, anyway,' she added in a more somber tone, 'I'm protected, you know. Besides, it was not the moment to lose one's cool!'"

What about Monsieur Garnier?

"Admirable!" said Marie. "Not for a moment did he show anything less than a laudable attitude. On the other hand, for a 'baptism' it certainly was a salty one!"

After a little thought, she admitted that "the situation wasn't exactly brilliant" and "I wasn't too displeased to see the lighthouse at Southwold."

Then they asked a sensitive question. "So, you didn't leap out of the gondola once you were over land, as some are saying?"

"False! Worse than false! No aeronaut worth the name would do such a thing. And would I, I the pilot, responsible for my passenger, do anything so crazy? It's unthinkable."

"Well, then, what happened?"

"We were letting the gas out of the balloon and everything was fine when we ran into a tree so abruptly the basket rocked violently and I was thrown out headfirst into a bush. By the time I got untangled, the balloon was floating away. I wasn't too worried because it was beginning to deflate so it couldn't go far. Nobody wants to jump out of a balloon, believe me. Actually the same thing happened to Monsieur Garnier, when the drag rope tangled in a poplar and pitched him out."[38]

G.M.'s article included details of the trip that Marie gave them in the interview. But being a journalist herself, she used the log she kept during the flight to describe her historic, near-fatal trip in a front-page story in *Le Petit Journal* and for the January 1, 1910, issue of *L'Aérophile*, a French aviation magazine that devoted an unprecedented four pages to her account. Some excerpts:

> Since our ascent hadn't been publicized, just a few friends were there to help inflate *The Shooting Star*. It was a gorgeous morning with a strong wind from the North, which made me think we might have an interesting trip—and I was happy that I could finally avoid the routine flight over Château-Salins.
>
> Louis de Brabois was in charge of the delicate weighing operation, and Monsieur Mercier, the uncle of my passenger (who was taking his first balloon trip), photographed our departure.
>
> At precisely 11:07 a.m. came the 'let go!' and we rose with alacrity.
>
> I stabilized us at 900 meters and we passed over Laxou, Champigneulles, Bellefontaine, and Frouard. Over Custines, we glided across an ocean of fat, downy clouds that sheeped off into the distance as far as we could see. Above our golden balloon, the sky was a pure, captivating blue.
>
> While I was checking my instruments, my passenger studied the map. Through two great holes in the clouds, like giant eyeglasses, appeared Pont-à-Mousson and Mousson. Toward noon, we crossed the river Seille, just above the Cheminot. Under us, we admired the shadow of our balloon with its iridescent circle, so dear to balloonists.
>
> At 12:35 p.m., we approached the broad ribbon of the Moselle river and burned past the arches of Jouy. As far as we could see, the river snaked away into the distance, silvered by the sun. Ars, with its huge foundries, groaned, pounded, smoked. To the right, in the great desert plain of Frescaty we could see the nest of that great bird, the Zeppelin.
>
> We passed to the west of Metz, and the clouds shredded enough so that we could admire the old cathedral; it was with deep joy that I contemplated, from the balloon, the city where I'd lived for six years.
>
> We were over the battlefields of Gravelotte at Saint-Privat when the first condensation appeared and cost me some ballast. By 1:30 we passed Briey at about 700 meters of altitude. Our speed wasn't that fast; we were averaging about 24 kilometers an hour. At 2:45, over Rédange, we had new condensation, golden dust in the wind, and we stabilized at 1,000 meters.
>
> At 3:30, no more visibility. For twenty minutes we scudded through milky clouds, then finally caught sight of Dipparck. Below us, picturesque Luxembourg caught and kindled our admiration: chateaus, dungeons, autumn woods in tones of rusty reds.
>
> At a quarter to five, we were treated to the unforgettable panorama of Diekirch. A fortress, seeming no larger than a child's toy, overlooked the town, which was beginning to light up in the dusk. A great ribbon of train tracks ran along the Sure river, both bordering the elegant city on its west.
>
> In the falling twilight, the tall chimneys and gray slate roofs took on bluish hues and reflected the street lamps, lending a magical aspect to the city. I find Diekirch to be one of the most artistic and pleasing cities that I've ever seen from above.
>
> We're still heading north. Up ahead, in the distance, I recognize the somber pyramid, lit by several lights, that indicates Vianden and its picturesque ruins.

At 5:30 we pass over two rather large groups of city lights, which must be Spa and Verviers. Night has arrived, and our speed increases. In a few minutes, at 400 meters of altitude, we are over the immense city of Aix-la-Chapelle, passing the Cathedral on our right. After we light our electric lamp, we hear rising from the shadows of the great city cries of "Balloon! Balloon!"

We had our dinner while we crossed the Rhine. After half an hour of scudding along very rapidly, we saw an enormous factory, which we recognized by its outsized chimneys for the famous Krupp cannon foundry in Essen.

As we crossed the Lippe river, the wind picked up significantly. We were at 300 meters and fled through the landscape at an insane speed.

About 7:00, we exchanged a few words with people on the ground; they answered us in Dutch and told us we were near Enschede.

At the very moment that I was considering a forced and fairly immediate landing, we changed direction completely, the compass showing true west.

I laid out the situation for Monsieur Garnier and said to him, "Let's cross the North Sea!"

Emile Garnier, who crossed the North Sea in a balloon with Marie.

Having done a lot of sailing in that area, I knew that when you have a strong wind from the east, you are almost always carried straight to England, and that you rarely get a change of wind. Our speed by now was dizzying.

We had barely resolved to attempt the crossing when we arrived at a huge expanse of water that must have been the south part of the Zuyderzee. Waves furled wildly up over the dikes. The sprawling twinkling city on our left had to be Amsterdam.

It is 7:50 p.m. The night is dark but clear, we have five sacks of ballast left, and it is with complete confidence of our success that I decide to undertake the crossing.

Leaving the Zuyderzee, we pass over the immense gray splotches of the polders, where we hear the cries of seabirds. Then we arrive at a second, smaller body of water that must be the Y canal. Twenty minutes later we're over the lighthouses. Matter-of-factly, we watch their circling, multicolored lights disappear behind us. It's 8:15.

We've barely arrived at the North Sea when condensation stars form again, and our drag rope feathers the surface of the water for a moment.

The sea is running strong. I toss some ballast over the side and we ascend again. We are in complete darkness. Only the electric lamp from *The Shooting Star* remains. We had so hoped for a moon, but there is nothing.

The balloon is maintaining altitude, so after the initial dispensing of ballast at the coast, I don't use any more.

Alas, at 9:00 when we are out in the middle of nowhere, everything changes. The temperature drops suddenly. Almost immediately the barometer glass is covered with ice and the cords of the balloon become stiff and frozen. Despite our fur coats, we are freezing. At one moment, I was shivering so badly that Monsieur Garnier had to take over the drag rope. At 1,600 meters of altitude, slowly, as though we were in the midst of winter, the snow begins falling in large, sticky flakes. By the light of our lamp, I even see needles of ice in the air, a warning sign that a storm is approaching.

Our fur coats are soon white, and the balloon begins to descend from its 1,600 meters. Even a sack of ballast fails to stop the downward chute. Covered with heavy snow, the balloon has no lift and we end up about twenty meters above the sea where our drag rope hits the water with a raucous smack.

Enormous waves, furious waves slap against our wicker basket. The drag rope, pulled this way and that by the water, tilts us and spins us. Are we going into the sea?

I throw out some more sand and we rise again to 1,700 meters, but the snow continues to fall, and the statoscope warns of a new and precipitous descent. A third time, our hemp drag rope hits the sea with a resounding impact.

The snowflakes fall thick and fast, obliging us to imagine a moment not far off when we have no more ballast. The tempest continues. I empty our last sack of ballast. The needle of the statoscope turns slowly to the right, while the barometer stops at 2,200 meters. It is 10:00.

When we are at a decent altitude, thanks to the last of our ballast, I begin to consider the seaworthiness of our basket. I remove our canvas covers from their bags. Wet, they can weigh us down and destabilize us. I get out the anchor and the anchor chain. We drop vertiginously. Tossing out the tarps helps very little in the end.

And then, torrential rains set in. We are soaked. Some tiny, far-off light shines from the horizon, but otherwise all is dark.

Suddenly, the basket is shaken violently. We drag rope again. Only a few meters away, the sea booms, froths, lies in wait for us.

The tempest is superb. The magnificent spectacle of unchained nature fills us with awe and admiration, and does not frighten us. Nevertheless, all our rigging is soaked, and all we have left in the way of ballast is the anchor. Our five sacks would have been enough if it hadn't snowed. But we are still in a maritime desert of emptiness. I remember the nickname an old friend bestowed on me—the fiancée of danger. Which of us would win this round? Having met this fiancé many times before, I am not about to lose confidence at this point.

I have Monsieur Garnier toss our last ballast over the side: from now on, our anchor will dwell forever under the waves. The drag rope pulls out of the waves and we take off for our last ascension. The statoscope, which I tried to keep protected, isn't working any more. The needle of the barometer goes straight up and stops at 2,600 meters.

At this altitude, the terrible gnashing and groaning of the sea grows fainter, and Phoebe, whom we have so desired, appears to us through thick layers of cotton batting. I jot down a few words in my log book, but our respite doesn't last long, and we plunge as quickly as we'd previously ascended.

We appear to have finished the aerial portion of our trip. We are destined for the sea, but we will do battle. We have decided to defend ourselves with all our might.

The drag rope pulls through the sea, which seems to be growing calmer. It is 11:30. We check the compass again and see that we're still headed in the right direction, due west. If *The Shooting Star* will fight with us, we will surely reach the English coast. We're in the hands of God. There's nothing left to do but await that decision.

We will not give the cold waves the joy of extinguishing our light, although we are sinking closer and closer to the sea that hungers for us. Behind us, our drag rope traces a long silver trail in the rolling swells.

In just a few moments we are going to know the cold embrace of the sea. I give Monsieur Garnier a few indispensable instructions for the landing in case I alone should be torn from our basket. Despite the gravity of the moment, we are stoic, plastered to the floor of the basket, clinging to the cords.

Watch out! An enormous wave, at least ten meters high, breaks over us, drenching us clear to the bone and cutting off our breath. The first caress of the sea is brutal. Violently, it tears off our coats and hats. Under the water flowing back into the sea, sparks leap up from the wicker as it squeaks and ratchets. We throw ourselves backwards to tilt the basket upright, the drag rope reestablishes a little equilibrium, and the balloon lifts several meters from the salty waves. For just a moment we breathe, but scarcely two minutes later we are once again submerged. The impact of the sea breaks the glass on the barometer, and it falls into the water behind us.

Suddenly, off to our left, a distant red light appears—a fishing boat fleeing the tempest—but in the dark they probably don't see us. Even while doing everything we can to stay upright, we continue to admire the splendor of these powerful, phosphorescent mountains of water.

After half an hour of struggle, we become almost accustomed to parrying the blows. Every two minutes, we disappear under a wave and, until the next one comes, we rise a little. But the least inattention could be fatal. Without clinging tenaciously to the cords, we could be carried away by the waves.

For more than an hour, the waves—still immense—whip us, bruise us, shake us, bang our heads violently against the basket. But we remain determined. My companion is admirable. Despite our critical situation, he doesn't lose his *sangfroid* and his calm for one second.

With the violent struggle, we are working hard, and despite our wet garments, we're not too cold.

Toward half-past midnight, about fifteen meters from us, a colossal shadow passes by, a phantom vessel with the wings of a giant. We hear screams of fear and horror as they see our golden sphere whip past them, but we are going at such a breakneck speed that their rear light is soon nothing but a tiny distant star. Then, nothing. What must those sailors have thought? I don't know which was worse for them—the fear that we were something not quite of this world or the fear that we would collide with them.

Three huge lights break through the dark. Could that be the coast? Because we're moving fast, we can soon see, between the assault of two waves, some buoys that probably indicate a reef. Then, far off in the same direction, we see the long luminous trail of a transatlantic ship. Like the sailboat, it disappears into the shadows.

Suddenly a gold star embraces the horizon! This time, it is surely a lighthouse, and we are heading directly for it. Saved! Very quickly, the star grows bigger and we can see its turning projections. For about three-quarters of an hour we see it from afar. We redouble our energy to battle the tireless tempest.

Toward 1:30 in the morning, I cry: Land! England! At first doubtful, Monsieur Garnier realizes that, this time, we are close to victory. But our speed is dizzying—we're going at least 80 kilometers an hour—and straight ahead of us is a cliff. Having escaped the sea, are we now to dash ourselves against the earth?

As soon as we reach the beach and the drag rope leaves the water, we rise 100 meters and easily get over the cliff. In any case, I still had our soaked fur coats, which I could have tossed overboard to give us some lift. The area looks suitable, so I begin thinking about our landing.

I warn Monsieur Garnier that the shock of landing is bound to be violent, and that he must stay in the middle. I pull the valve to release the gas, but it doesn't seem to be working. Our speed is hair-raising but as we're very close to the ground, I pull vigorously on the cord to open the panel and release gas. Impossible to budge it. Saturated with water, it must have formed a snarl at some point.

I'm still pulling on the cord—in vain—when the basket is abruptly snagged by a tree and turned upside down. I dive headlong into a bush. As I am falling, I feel the worst sort of turmoil watching the balloon take off again and worrying about the fate of my passenger (the press unjustly accused me afterward of having abandoned him on purpose). Happily, as deflated as it was, the aerostat couldn't go far.

Not without some difficulty, I manage to disengage myself from the brambles and thorns. Hair soaked and full of twigs and pine needles, shivering and chilled to the bone, in the dark, under a torrential rain, I head in the direction the balloon took.

The ground is swampy and I make little headway. After about two hundred meters, I come to a creek that the storm has turned into a torrent. Trying to swim across it, given the current, would be insane.

Regretfully, I turn around and return to the place where I fell out. I tie my scarf to the tree we ran into so that I can find it again if I need to.

The lighthouse that we'd seen from the sea was nearby, and certainly civilization couldn't be far away. With the rain and the wind, which nearly knocks me over, I follow a road, ten centimeters deep in water, edged by trees and pasture land. The countryside seems deserted but after about a kilometer, I see houses. A number of little one-story English cottages, bordered by gardens, are lined up in the shadows.

In one window, I see a light. At the same time, a bicyclist rides by, so I call out to him. His fright is apparent when, by the light of his lantern, he makes out a feminine silhouette, but what a silhouette … ! In a few words of English I tell him about my trip, which quickly explains to him the state of my attire.

He is astonished and immediately must tell me that the transatlantic ships take eleven and a half hours to come from Holland. How could it have taken us only five hours to cross two hundred kilometers of sea? He also tells me he hasn't seen such a violent tempest in years. He eventually takes me to a nearby house, where he tells three charming Englishwomen my story, and they hasten to make me comfortable.

But first, I'm concerned about Monsieur Garnier. Still shaking and unable to go out and look for him, I give several villagers as much information as I can, including the scarf tied to the tree.

Then I get into dry clothes and warm myself by a huge fire that quickly erases the memory of my recent chills.

It's a quarter past two. I spend interminable minutes of worry and anguish: what is happening out there over by the cliff? For me, this is the most painful moment of our adventure.

Finally, toward four o'clock in the morning, a police officer returns carrying my scarf, but he is alone. Seeing my anxiety, he reassures me immediately that my traveler is safe. What joy! In addition, my balloon is tied down and is being guarded by another police officer.

Toward 4:45 in the morning, a happy Monsieur Garnier rejoins me. He had been distraught about my abrupt and involuntary departure from the basket. He tells me about his own header into an oak tree a little distance from my landing. The panel cord had continued to stick, so he fell from branch to branch, and to the ground, incurring only several light contusions.

Our kind hosts offered Monsieur Garnier a room and some dry clothes. Over a meal, the two of us relived with awe and delight that unique night, filled with impressions neither of us will ever forget.[39]

The day after Marie and Major Garnier landed on the English coast, they traveled by train to the Liverpool Street station in London (where "considerable interest was aroused by the arrival of the intrepid balloonists"[40]) to make arrangements to return home. "We were treated like heroes," remembered Marie.[41] Journalists and crowds carrying French and British flags gathered at Marie's London hotel to take a look at the woman who had beaten almost impossible odds to become the first person to cross from France to England in a balloon. The story made the first page of most major newspapers, and the *London Daily News* was so exhilarated it ran the story with four stacked headlines:

Balloon Adventure
Occupants Drift Across the North Sea
Lady's Thrilling Story
Exciting Descent Near Southwold[42]

London papers described Marie as a handsome, dark-haired woman wearing a serge costume, a cloak, and a motor-cap. Monsieur Garnier "is an electrical engineer. He has a shaggy brown beard and wore a gray Norfolk suit. Both travellers were somewhat worn, but were perfectly cheerful."[43]

Marie told reporters, "We ran short of food as we had only taken one bottle of champagne and some sandwiches with us, but we had an appliance for making coffee which stood us in very good stead."[44]

Marie described her 1,200-cubic-meter balloon, *The Shooting Star*, as "the very last word in balloons."[45] It may have been, but at take-off, a rope connected to the ground was overlooked, tilting the gas-ring and releasing pounds of precious hydrogen, which would have given her needed lift during the crossing.[46]

Marie was later awarded a medal from the Eastern Air Club for her historic crossing. In the 1909 French scientific and industrial almanac, editor Émile Gautier wrote that a number of rather banal balloon ascensions had taken place here and there during the summer and early autumn, "but this monotony was broken by the trip as unexpected as it was incredible of Mademoiselle Marvingt and Monsieur Garnier."[47]

The popular weekly, *La Vie au Grand Air*, accompanied their announcement of the historic crossing with photos of Marie, Garnier, the balloon on the ground in England, and a map of their route: "An aeronaut, Mademoiselle Marvingt, very well known to the readers of *La Vie au Grand Air*, has just accomplished a feat as dangerous as it was daring.... The wind that night, if you remember, turned into a tempest."[48] Except for referring to the imprudent audacity of deciding to cross the North Sea in the first place, the article was highly complimentary.

The *Aérophile* devoted a full page to her: "Although a number of flyers have recently earned the right to be included in our gallery of famous aeronauts, it seemed to us that today they must give way to a female aeronaut, Mademoiselle Marie Marvingt."[49] They emphasized that, in the daring crossing, she was the pilot, the skipper of that ship.

And so it seemed that Marie's greatest adventure, her greatest achievement had indeed come in a balloon. But there were other adventures, other achievements yet to come.

Leading up to the May 15, 1910, Grand Aeronautics Fête in Nancy, the local paper interviewed Marie by stopping her on her bike downtown.

"Where do you think you'll end up this time, Mademoiselle Marvingt? Will you try the North Sea crossing again?"

"The wind is capricious. Yesterday it wanted us to go to Central Europe. Today, well, I think it wants us to head toward England. Who knows where it will want to go tomorrow?"

"So, it's all the same to you?"

"Perfectly the same, as long as the weather is decent. I'm hoping for some sun for our Pentecost ascension."[50]

The reporter assured her that the barometer was looking good.

"So much the better!" said Marie, jumping back on her bike and pedaling off.[51]

Only six balloons took part in the first competition sponsored by the Eastern Air Club. In addition to Marie were two of her best friends, Emile Friant and Georges Blanchet, and the local pharmacist, Monsieur Ferry. Pierson de Brabois, with whom she'd had a great trip the previous September, was piloting the fifth balloon, and a Monsieur Rebuffet had the sixth. Still, the crowds began arriving for the race in mid-afternoon, and the train station reported selling 60,000 tickets to people arriving for the fête.[52]

Preparing a balloon for departure told Marie who her friends were. They were the ones, like the Bergeret family, who showed up in the dark to help her start preparations.[53] One longtime friend, Madame Voge, often assisted Marie, and later, at the age of ninety, was still smiling at the memory of a particularly colorful ascension from the Place Stanislas.

Marie's passenger that day was Tonin Bergeret, son of the balloon enthusiasts who were so loyal to Marie; his mother had asked her to take him up.[54] Tonin, his brother Charles, and his father Albert were among Marie's most helpful assistants. They would be there to lend a hand a month later, on June 16, 1910, when she obtained her balloon pilot's license, number 145, the second woman in the world to do so.[55]

When the start was given for that May 1910 fête, the six balloons rose one after another like a colorful bouquet of flowers over the heads of a spellbound multitude. Fifteen hours after leaving the ground, Marie and Tonin landed near Neufchâteau in Belgium, winning the first-place gold medal of the Ligue Nationale Aérienne.[56] Equally delightful to Marie was the joy of out-ballooning Georges Blanchet, three-time winner of the French Air Club's balloon event.[57]

A few days later, Tonin Bergeret described their trip for *Le Sport*. "Despite the weak wind, Mademoiselle Marvingt had ambitious plans. 'If the wind doesn't change direction,' she said to me, 'I'd like to land in Brest. Otherwise, if it blows us toward the Channel, we'll cross it.'

"Coming from another pilot, this language would have astonished me. But, finding myself facing Mademoiselle Marvingt, I wasn't surprised."[58] They kept in mind the possibility of crossing the Channel, but the inconstant winds and, finally, a thunderstorm, indicated that landing in Belgium would be prudent.

Tonin wrote, "She decided to set the balloon down right next to the train station so we wouldn't have to walk so far carrying the balloon to catch a train home. With her customary expertise, she managed to set that monster down on the ground, less than a meter from the spot she'd selected."[59]

As they were folding up the balloon, he asked her, "Are you pleased with this ascension, Mademoiselle?"

She said simply, "One more, one less."

He thought to himself, "It may be all the same to her, but I think this trip might count." He finished the article: "And, in fact, when we got off the train in Nancy, I was delighted to hear that Mademoiselle had taken first place!"[60]

Two days later, she took another little trip. In reporting a "New Ascension of Mademoiselle Marvingt," the newspaper said that Tuesday she took off by herself in a 600-cubic-meter balloon a little after noon. About 5:30 in the afternoon, the paper learned that Marie had landed in Saulxures-lès-Nancy and would return to Nancy by auto at 6:00.[61]

"One of our reporters caught up to her on his bike and managed a conversation with her, bike-to-automobile.

"'Where are you coming from, Mademoiselle?'

"'From Saulxures.'

"'Oh?'

"'Yes.'

"'So what were you doing there?'

"'Buying a dozen eggs.'

"'Well, that's an original way of doing it. I guess your balloon will take you anywhere.'

"Laughing delightedly, Mademoiselle Marvingt holds up a basket of fresh eggs.

"Some balloonists bring back impressions of their flights; but Mademoiselle Marvingt, she brings back an omelette!"[62]

A month later, the French Air Club held its 6th Grand Prix distance competition with sixteen entrants.[63] The Paris *Journal* said, "For the first time, a female pilot will be competing, Mademoiselle Marvingt, champion of the Eastern Air Club." They added that the wind, which was the terror of airplane pilots but the joy of balloon pilots, was going to be ferocious. "There will be some fantastic flights this afternoon where great courage and energy will be needed."[64]

They were correct. After three hours of rain, the balloons began taking off, but the violent northwest wind meant that as soon as each balloon reached a certain altitude, it oscillated violently pendulum-fashion. Marie was piloting a brand-new balloon made by the Carton company and carrying a passenger, Monsieur Chagnot, which was remarked upon: "Certain pilots carried female passengers. Mademoiselle Marvingt also had a passenger, but in that case, it was the weaker sex who was in charge."[65]

Marie landed at 3:15 in the morning in Rondefontaine, a village of thirty-one inhabitants, near the Swiss frontier.[66] She won the prize for being aloft the longest, but Georges Blanchet won for distance as he landed in a lake in Bavaria—his third win in this competition.[67]

By 1910, the airplane was siphoning off public interest in balloons (although the Paris *Journal* insisted there was still no sport as chic as ballooning[68]). In her usual intense, energetic way, Marie took to the air in the new planes, but until the outbreak of World

War I, she continued to pilot balloons whenever she could.

She won a medal from the Eastern Air Club for a balloon race in 1911, and in 1912, she took fourteen trips.[69] That year, she reported, "I also gave rides to thirty-six people taking their first balloon trip. I love to see the excitement and delight on people's faces when they experience that ineffable feeling for the first time!"[70]

Marie (at right) was the only woman competing in the Grand Prix of the Aeroclub of France, June 26, 1910. Next to her is Mademoiselle Carton, whose family's company constructed Marie's balloon.

On a sunny Thursday in June 1913, the newspapers reported that Marie was "in the clouds" again, piloting her *Myriel*.[71] She took two passengers on a three-hour sightseeing tour over Paris, passing within a few meters of the Eiffel Tower. The next month, this time flying the *Lorraine*, she was the only woman competing in the nineteen-balloon race held in Lille.[72] Another dozen balloons went up that day just for fun, painting a colorful sky for the immense crowds: photographers, cinematographers, mounted police, balloon crews, passengers waiting for their turn. Prizes were given for the first five winners; Marie came in sixth. In August, at the Châlons-sur-Saône balloon festival, she piloted her own aerostat, *The Shooting Star*, taking off to the shouts and waves of an admiring and enthusiastic crowd.[73]

Marie didn't always land her balloon as well dressed as she had been upon ascending. Once when she and a friend were ballooning across the Moselle River at one of its widest points, she had trouble gaining altitude even after unloading every bit of ballast. She and her male passenger scrutinized every item in the gondola and pitched out anything that wasn't tied down, including their picnic lunch and bottle of wine. When the basket was stripped bare, they looked at the river—which was still much too close for comfort—and made their choice. They peeled off their heavy fur coats (typical balloon apparel because of the cold air at higher altitudes) and tossed them overboard. Next came their shoes—heavy, utilitarian things whose loss nonetheless only slightly affected the balloon's altitude. Heavy woolen sweaters followed the shoes, along with thick scarves and gloves. Their situation worsening, they tossed out hats, ties, jackets, vests, socks, pants, and a long skirt. Alas, their efforts were fruitless. The next time the gondola touched the water, it was sucked under, and the pair had to swim to shore dressed only in their "small clothes." It was probably just as well; had they fallen into the river fully dressed, their cumbersome garments might have sunk them like stones. Only a few people actually saw the embarrassed aeronauts' arrival on land, but Marie enjoyed the joke on herself and told the story often.[74]

Marie was one of twenty-three competitors for the 10th Grand Prix of the French Air Club held July 18, 1914.[75] An enormous crowd had invaded the Tuileries gardens in Paris to enjoy the colorful air show. The paths were full of jostling spectators, the terraces of the Orangerie and the Jeu de Paume overflowed with families and picnic baskets, and the Place de la Concorde was saturated with people. Political and sports figures watched from the reserved section. While a military band played, 2,500 passenger pigeons were released and soared upwards with a great flapping of wings.[76]

Marie's many successful balloon trips cloud the reality that ballooning was always risky, and often fatal. In the midst of the festivities, the race's start was tragically spoiled when the *Toto*, piloted by Marie's friend Georges Blanchet, went up seven or eight meters heading toward the Champs Elysées only to be slammed earthward by a gust of wind from off the Seine. Blanchet had only a sprained jump, but his passenger, Henri Duval, suffered a skull fracture and internal injuries. Blanchet was using a balloon previously owned by Paul Leprince who, thirteen months earlier, had died when his balloon exploded because of the imprudence of a spectator's lit cigarette. Sixty people were also injured.[77]

Twenty-two balloons left the Tuileries that day. Although the racers planned to go from Paris to the Irish Sea, some balloons landed in France due to weather conditions. "Others, among whom, a woman, Mademoiselle Marvingt, the well-known aviatrice, crossed the English Channel. It's the first time that a woman, piloting a balloon, has crossed the English Channel."[78]

Both the English Channel and parts of the North Sea lie between France and England—but at different points, with varying distances and weather. Marie had already crossed from France to England by way of the North Sea in 1909. Now she had made the crossing over the English Channel.

In 1906, the Englishwoman Beatrice Swanston Brewer piloted a balloon from England to France across the Channel.[79] Because of the prevailing winds, this was the easier crossing. But no one before Marie had piloted a balloon from the continent to England, either across the North Sea or the English Channel. Now she had done both.

Marie was piloting her 1,200-cubic-meter *The Shooting Star* in that competition. Some entrants were flying 1,600-cubic-meter balloons, which gave them a certain advantage. With her was Quénardel de Warcy, a good friend from Reims and a devoted fan.[80] Marie wrote later that the decision to continue across the Channel was made by two identical votes: "Hip! Hip! Hurrah! A la conquête de l'Angleterre."[81]

After a calm crossing and a smooth passing over England and most of Wales, they landed in Aberystwich in Wales near the Saint-Georges Canal, one of the most difficult landings Marie had ever experienced.[82]

They found themselves in a beet field, and began deflating and rolling up the balloon. "We had been working for an hour and a half when another balloon landed near us. It was Dubois. Then, another, one of the German entrants. Was it possible we were all going to end up in this little lost corner of Wales?"[83]

Her companion was as pleased as she was with their trip, but he was a bit anxious:

Had they gone the longest distance? Would they be the winners? If not, would they at least be ranked well?

Marie remembered, "His young heart fussed at these thoughts; my indifference annoyed him."[84] As it turned out, they placed seventh out of twenty-two.[85]

Ever the journalist, Marie wrote an article that appeared July 22 in the *Éclair de l'Est* describing "the picturesque beauty of the French countryside's tangled mountain chains and hillsides" from a vantage point of 2,000 meters. She spoke of "the countryside fleeing beneath our feet" and "the foamy fringes of the waves." She piloted her balloon, she said, in a manner that would neither "tempt God, nor forget prudence."[86]

Marie evidently didn't mind placing seventh, or even her father's remark that she could have gone farther, on to Ireland. She concluded the article by saying none too shyly, "One thing I'm absolutely certain of at this point is that I can be ranked among the top balloon pilots in the world. I am modestly content with that."[87]

Despite numerous other interests, Marie continued to enjoy the balloon whenever she could. The only problem with balloons—a fairly major one—was the pilot's lack of control. "After that fatal 'Let go!'" Marie once said, "human willpower loses nearly all its sway."[88]

If anything was important to Marie, it was willpower. Balloons offered splendid adventures, but they also presented immoderate perils because they could be controlled only vertically, and sometimes not even very well that way. Marie was accustomed to taking charge of her destiny, and she wasn't happy being at the mercy of the winds.

In spite of this drawback, balloons gave Marie something that nothing else had—feelings of being at one with creation. If piloting a balloon meant sacrificing a certain control, she accepted it. Perhaps she survived the North Sea crossing—the same one that killed other talented aeronauts—because she knew how to respect and ride the elements instead of fighting them.

Marie was thus ready to embrace the next stage in the development of human flight. Once again, she was in the right place at exactly the right time.

4

Queen of the Air

"We must fly and fall, fly and fall, until we can fly without falling."—Otto Lilienthal, 1896[1]

Marshal Ferdinand Foch, supreme commander of the Allied forces on the Western Front at the end of War I, was working in his Paris office at the Invalides when General Weygand, his longtime aide and second-in-command, announced that Marie Marvingt had stopped by to say hello.

Weygand added, "Unfortunately, sir, you won't be able to see her. You have an appointment now with …" (he named one of the crowned heads of Europe).

Marshal Foch had known Marie since 1913 when he'd been posted to Nancy as head of the XX Corps, and they'd developed an enduring friendship.

"Show her in, General," he said. "A queen of the air always takes precedence over a king of the earth."[2]

At the dawn of the twentieth century, Alberto Santos-Dumont, heir to a coffee fortune, constructed a single-seat, gasoline-powered dirigible that was the most advanced flying machine of the day. At the same time, Henri Deutsche de la Meurthe, wealthy sportsman and co-founder of the French Air Club, offered 100,000 francs to the first person who could fly from the Air Club headquarters in St. Cloud to the Eiffel Tower and back again. The offer received a great deal of publicity, partly because of the intense interest in aviation and partly because of the new and controversial Eiffel Tower, built in 1889 and, until 1930, the tallest structure in the world. Coincidentally, Alexandre Gustave Eiffel was also attracted to aviation, spending the last years of his life studying the effects of air motion on airplanes.

And so it was that in 1901 the astonishing Eiffel Tower, the developing field of aviation, and the daring and inventive Santos-Dumont all came together to fan the flames of French interest in heavier-than-air flight.

Parisians dining in the Eiffel Tower's upper restaurant that day were astonished to see Santos-Dumont guide his unwieldy machine around the tower. He flew so close that diners could look down and see him holding the controls in the small wicker basket of his dirigible. He received an award from the Brazilian government as well as the 100,000 francs prize, which he divided between his employees and the poor of Paris.

Two years later, on December 17, 1903, Orville and Wilbur Wright succeeded in keeping a plane in the air for twelve seconds. That flight and the three others that took place that day were the first powered, sustained, and controlled airplane flights in history. Twelve seconds doesn't sound impressive today, but it represented a great leap forward in humanity's conquest of the skies, and it was enough to fire the imaginations of people around the globe.

The French, already keenly interested in flight, threw themselves, their money, and their passions onto the altar of aviation. They were the first to buy rights to the Wright brothers' patents. Years before the Wright brothers, the French Air Club had been founded to encourage aerial locomotion. Members met at Maxim's of Paris to discuss the latest news and rumors about who was doing what in aviation circles. It was also the French who, early in the history of aviation, founded the still-existing International Aeronautic Federation to evaluate both pilot and airplane performance.

Marie's journey to becoming one of the most well-known airplane pilots of the day began at the same time aviation did. While Santos-Dumont was circling the Eiffel Tower in his dirigible, Marie was aloft in a balloon. As a journalist working for several newspapers, she began covering a group called the Flying Fools. Those dreamers and daredevils, whose names have since become legendary—Blériot, Farman, Voisin, Latham—were considered hotheads and a little crazy besides.

On October 7, 1909, Louis Blériot was issued the first French license to fly airplanes, Glen Curtiss from the United States received the second French license (and the first American license). Because aviators are superstitious, there is no license #13, but #14 and #15 went to Orville and Wilbur Wright. (Since diplomas had already been earned by experienced pilots, the first licenses were given in alphabetical order to avoid hurting anyone's feelings, which is why #1 is Bleriot, and #15 is Wright.[3])

Aviation was moving faster than any new field in memory. Understanding its enormous appeal, Marie, writing as Myriel, took her readers behind the scenes, reporting detailed anecdotes, interviewing the heroes of the day, and introducing readers in an intimate way to this new world of risk, glory, and violent emotion. She toured the Voisin works, watched the debuts of new planes at Issy-les-Moulineaux, interviewed Alberto Santos-Dumont ("Santos" to her) in his workshop, and wrote article after article about

Marie at the controls (such as they were) in 1912.

the planes and the pilots. On August 4, 1909, Myriel wrote, "The astounding exploits of these bird-men haunt many minds, evoking numerous fantasies in certain imaginations. Sometimes these dreams become reality."[4]

Later that month, she was outside Reims on the plains of Bétheny, covering the first international air meeting.[5] Part of the annual champagne festivities, the event drew such celebrated pilots as Henry Farman, Glenn Curtiss, Louis Blériot, Hubert Latham, and Wilbur Wright, as well as some 250,000 spectators who jammed the town and flocked to the airfield.[6]

The early airshows were unforgettable blends of tragedy and comedy, fear and courage, the commonplace and the extraordinary. Having adopted the airplane with unprecedented fervor, the French idolized the foolhardy pilots who dared defy gravity. Despite torrential rains and strong winds, spectators watched thirty-six monoplanes and biplanes roll through thick, greasy mud to line up on the field.

In an imaginatively written 1939 piece for the *Tribune*, aviation writer George E. Pelletier returned to the time when Marie "watched the immortals at Reims that hot summer day in 1909 and drew therefrom the inspiration and nerve for which has been hers ever since. ... As Mlle. Marvingt inspected the thirty-six ships 'in the line,' watched Latham nonchalantly roll his own cigarette in flight, or giggled as Bunau-Varilla politely doffed his hat to the ladies each time he flew past the crudely constructed, flag-decked grandstand, the urge to fly quickened within her. 'Oh, to pilot one of those ships of the sky,' she said to herself over and over again."[7]

Marie complained of the boring morning: "The birds were all sleeping. Because of the winds, they wouldn't come out of their cages."[8] "Because of the weather, nobody could fly all day long. When a pilot finally appeared near 7:00 in the evening, the crowd applauded him like a conqueror!"[9] The altitude contests began late in the day and Hubert Latham went up in his Antoinette despite the wind, despite the rain, and set a new world record.

From the first, Marie adopted Latham as her hero. The story she filed said, "Latham, as always, was the king of the birds; he outflies the eagle."[10] In a photo taken at that meet, Latham wears an elegant beige Burberry, a cigarette hangs from his lip, and his hat's visor faces backwards. He was "cool" before "cool" was cool. Marie added, "As I write these lines, Latham has taken off again and is silhouetted against the moon. At this time of evening, in the faint twilight colors, the effect is magical: we must be dreaming."[11]

Unfortunately for Latham (and his admirer), the next day Henri Farman beat Latham's altitude record—and the crowd went wild. Railings were torn down, police were scattered, and screaming masses of enthusiastic spectators swept across the field to tow the biplane back to its hangar.[12]

French president Fallières and future English prime minister Lloyd George were there, seated in a box right over an ominous sign with an arrow saying "ambulance." Two first-aid stations, with several physicians, 150 stretcher-carriers, and two ambulances, were kept busy. Many of the frail planes crashed, although no one was seriously injured or killed.[13]

After seeing record after record broken, government and military officials from several countries left that first international air meeting convinced the world was on the brink of a transportation revolution. The Reims air show proved that the airplane was there to stay and that people would pay for the privilege of sitting in grandstands to watch the human birds in action. It also stimulated aircraft manufacture across France and Europe.

Marie wrote for the *Eclair de l'Est*, "It's official: Aviation Week is a success, and enthusiasm for flight is at its height. Our brave fliers have demonstrated to the public that 'the conquest of air' is no empty phrase, and that before long, aviation's popularity will soar. Many fine minds are obsessed by this new sport."[14]

One of the fine, obsessed minds was, of course, hers.

Less than two weeks later, early on a beautiful Friday morning, Roger Sommer, cloth manufacturer and record-setting pilot, headed out to the airfield in Jarville. The weather was perfect for flying, with virtually no wind. Sommer flew up about a dozen meters, circled the airfield, flew a little higher over some trees, and landed. Seeing his sister-in-law among the spectators, he invited her to take a spin.

At the airfield to interview several pilots, Marie watched Sommer's sister-in-law eagerly climb in. As soon as the plane landed, Marie ran to the first-time passenger to congratulate her and ask what it was like.

"Delightful! Absolutely delightful!"

Marie appeared to have more questions but, sensing a little jealousy in the air, Sommer asked, "Would you care to accept a tour, Mademoiselle?"[15]

A fortune falling down the chimney of a poor family could not have produced a more stunning effect. According to one newspaper, "It wasn't even happiness, it was delirium, something other-worldly, that appeared on the face of the well-known sportswoman Mademoiselle Marvingt."[16] She threw her photography equipment and her coat on the ground, replacing her hat with a helmet, while crying, "I am so happy! Oh, I am so happy!" And she climbed into the plane.[17]

Little more than a big kite with two seats for passengers and a broom handle for the pilot, the machine was generally known as Sommer's chicken coop. As "firsts" go, Marie's flight wasn't much. At five or six meters above the ground, they flew to the end of the field, turned around, and came back to the hangar. But Marie declared she was "in heaven!"[18]

Newspapers carried photos of Sommer assisting her into the plane, and of the two of them seated in what passed for a cockpit. That's how, on September 10, 1909, Marie Marvingt left the ground in a fixed-wing aircraft for the first time.

By then, she had seen all the planes and most of the pilots in action. She knew just what she wanted. Within weeks she had signed up to take lessons at the newly created Latham School for Pilots in Mourmelon.[19] She was Latham's first student, the school's only female graduate—and the only one of his seventeen students to survive for any length of time.[20]

The insouciant Hubert Latham, 1910.

Arthur Charles Hubert Latham, born into a wealthy family, was the first aviator to attempt to cross the English Channel in an aeroplane. His Antoinette IV suffered engine failure and he had to ditch in the Channel, thereby becoming the first person to land an aircraft on water. The undamaged fuselage remained afloat, so Latham lit a cigarette and awaited rescue by the French torpedo-destroyer Harpon that was following him. He liked to say he preferred wild weather to calm: "A few months ago, I was flying with Mademoiselle Marvingt. The weather was absolutely ideal. Suddenly, I was grabbed by a whirlwind. It was all I could do to avoid crashing. I'm much more comfortable in a storm."[21]

Younger than Marie, he must have seemed like the much-loved younger brother she'd lost. Latham was the pilot with the sensitive touch. It was said that he could sense a stray gust of wind before it arrived.[22] Marie liked to tell the story of how, when he left Los Angeles after an air meeting, the clocks of the city rang out to mark his departure.[23] Speaking with noticeable emotion the year after his death, she called Latham her "brilliant advisor and faithful friend."

In the meantime, sadness still in the future, Marie trained joyously with Latham on one of the Antoinette monoplanes—the first woman to fly one—and eventually passed ("brilliantly") her pilot's license tests on it.[24] The Antoinette was sleek and graceful, the most beautiful and expensive of the first planes, but it was unstable, lacked power, had an ungainly control system, and was generally unable to bank effectively in turns—it effectively merited its reputation for being dangerous (although the same could be said of most of the early planes).[25] It was also "the worst possible machine for a beginner."[26] Hubert Latham was the only pilot who had been successful with it, and Marie loved it despite its faults.[27] The year she began flying it, 1909, the plane had been standardized after much experimentation and modification.

Even so, the Antoinette didn't have many fans. Ernest Archdeacon, who adored Marie, wrote late in 1909, "Mademoiselle Marvingt, who has taken a dozen lessons with Latham, would already no doubt be flying if the Antoinette planes only flew for longer periods of time, like the Voisin planes."[28] Two Antoinettes were available for flight lessons; Latham's twenty student-pilots, who had to wait days for obliging weather, needed a great deal of patience. Lessons cost 4,000 francs for three months of instruction. Flight training involved classes in general aviation, aviation mechanics, and light motors and machines. Hands-on learning included exercises in the studio, practice flights, and factory visits. Written tests in mathematics, industrial design, and logarithmic calculations, and oral exams in math, elementary chemistry, mechanics, and general physics preceded the practical in-the-air tests.[29]

The only woman, Marie was highly respected—partly because of her accomplishments in sports and aeronautics and partly because she was also a Red Cross nurse and a journalist, both of which added to her prestige. "But," reported Jules Chancel, "at the Mourmelon camp, she wanted to be treated as a comrade, so that's the way the other pilots regarded her, with perhaps a bit of extra courtesy and deference thrown in. In effect, this young woman with the friendly handshake and the look that was both sweet and energetic, was the perfect comrade. She carried thread and needle in the pocket of her culottes to sew on buttons for those who needed it, and she watched through the night over those who were sick."[30]

Marie took thirteen months to obtain her pilot's license. Although she might stay three months or even five months at a time in Mourmelon, she had other commitments: she made numerous balloon flights during that time, competed in winter sports at Chamonix, and worked as a journalist. In addition, she was learning on the Antoinette, which more than one flier deemed the worst possible machine for a beginner.

After she obtained her license, Marie the journalist told the readers of *Le Sport* what was involved if they too wanted to learn to fly. In a wonderfully evocative description of the debut of aviation, she related her experiences as a pilot in training:

A 1910 photo of Hubert Latham and Marie Marvingt.

> Upon arriving in Mourmelon-le-Petit, you take an ancient bus a very long half-hour to the ultra-modern village of the aviators. Then, depending on whether you're a biplaner or a monoplaner, the driver will drop you at the Hôtel de l'Europe, where the Farman and Voisin people stay, or at the Hôtel Marillier, chosen by the Antoinette school. At the latter, you'll be kindly received by Madame Martinat and her son Richard. One of the pilots—René Labouchère, the Count de Robillards, or Louis Chatain, according to their habitual kindness—will offer you a place in their auto so you can get out to the airfield, which is about four kilometers from Mourmelon-le-Grand. There you'll be received by Monsieur Robert Gastambide, director of the school, who will take you around to the workshops and to the numerous hangars where thirty to forty large birds will be sleeping.
>
> Next you'll be introduced to the brilliant and amiable master pilot Laffont, who will explain the monoplane to you in every detail. After long and clear theories of the controls and the motor, you'll be invited to get onto the practice equipment, an ingenious invention of Lieutenant Clavenad that trains beginners' reflexes. Two assistants have programmed movements identical to the Antoinette's so that this apparatus imitates the effect of wind gusts and turbulence that students will encounter later during their actual flights.
>
> When your professors consider you ready, one of them—Laffont, Gobé, or Cure—will take you up for your first trip in the air. This will last only ten minutes at about 40 or 50 meters, and you'll simply be a passenger, free to appreciate the exhilaration of flight.
>
> The second time you go up, the professor will be seated directly behind you, prepared to correct any mistakes you might make as you use the control for turning right and the control for going down. Once you've mastered that, he'll let you use the left control, then both at the same time, and finally, the direction bar, worked by the feet. Lastly, you'll be taught how to take off and to land, both very tricky operations.
>
> Normally, you'll go up 25 to 30 times accompanied before you're able to solo. However, you can pass the tests for your French Air Club pilot's license after only five or six flights. For that, you fly three times around a closed circuit of five kilometers, following a route, and landing in a space of 150 meters.
>
> It's impossible to be precise about how long an apprenticeship might take. It depends on the student's aptitude and on the weather, since lessons are generally given only between 4:00 or 5:00 p.m. and 5:00 or 6:00 in the morning. Between lessons, you spend a certain amount of time studying the motors and the planes or listening to the professors on the theories of flight.
>
> When there's a great deal of wind or fog, students engage in sports or games: riflery, football, the boomerang, boules, etc. A few have begun hunting wild geese from planes, with the professor at the controls, and the student shooting game.
>
> Water battles are a great favorite with students. Some days you'll see students with buckets of water running across the roofs of the hangars. Latham is expert with the lasso, and once he sights someone, he never misses. There seems to be an infinite variety of entertainments. Every day there's something new. The students like to pretend to be famous fliers, and are delighted when spectators ask for their autograph.[31]

The high-spirited fun didn't stop at the training camps. In one of the first big air meetings, Bielovucic was grounded by "a stupid incident." He and Garros were blowing projectiles through a tube and one hit Bielovucic in the eye.[32]

Marie continued, "When someone goes up in the air for the first time, or someone earns their pilot's license, there are great ritual celebrations. You eat out somewhere, in Sept-Saulx or in Reims. And, by the way, you never bring a melancholic face to a dinner table full of aviators.

"The Rose Room at the Hotel Marillier, always reserved for the Antoinette school, has been the scene of many a joyous evening. One of Latham's little party tricks, at the end of the dinner, consists of cutting a flying apple in half with a knife when someone throws it to him from the other side of the room.

"The long months that must be devoted to aviation in the Mourmelian desert pass relatively quickly, thanks to the friendship and fun that's always to be found among the students. The days you spend at Mourmelon will never be forgotten."[33]

A training plane often had no cockpit, only a seat for the instructor and a perilous perch behind for the pupil. The pupil could reach over the instructor's shoulder to use the control stick but there was no way for the pupil to touch the rudder bar. However, since the pilot often sat on a wicker seat placed directly atop the fuel tank, that situation wasn't enviable either. ("Fire," pilots used to say, "was our third passenger.")

In the early days of aviation, those daring young things in their flying machines sought excitement, adventure, and glory in the skies. Most of them also found death.[34] Attracted by rich prizes, people were hurriedly building hopeless contraptions, using unsuitable materials and construction that would not have been tolerated in a third-rate bicycle shop. Quality wing spars were being attached to the fuselage with tiny bolts and flimsy brackets that couldn't take sudden strain. Some 87 percent of all the early fliers died in aviation accidents.[35] *L'Aerophile* regularly featured a prominent pilot in its column, "Contemporary Aviators." The next issue of the biweekly would often use the same photo for his obituary.

Marie knew the risks she faced. She passed her pilot's test in the Antoinette, which had been involved in a number of deaths before her flights in it. Because the French aviation world was a tightly knit group in those golden days, the tragic death of one bright young flier after another affected all of them. Marie watched her friends disappear—people with whom she had flown, shared meals, discussed planes far into the night. Mingling with the band of daring young men and women was an unseen but faithful member, Death.

Marie (second from left) and the other student pilots at the Antoinette school; she was the only one of Latham's pupils to survive after a few years.

The same year Marie learned to fly, aviation fatalities darkened the headlines scores of times. Léon Delagrange, one of the top-ranked pilots in the world, was killed while flying at Bordeaux. Charles Wachter, considered one of the most skilled of the Antoinette pilots, died while making an altitude flight. He was diving down from 300 meters when both wings suddenly folded back. His wife and two young daughters were watching.[36] Brothers Daniel and Nicholas Kinet were killed in plane accidents a month apart. After them, five more died during August and September, including Georges Chavez, a 23-year-old Peruvian running champion who had swept through the aviation world like a human comet and was the toast of the Parisian boulevards. He died trying to fly over the Alps, possibly becoming numbed by the cold.

In December, *Le Matin* announced that the death of two more pilots had cast a pall over the airfield. "One wants to think it's a false rumor or at least an exaggeration. Alas, it's only too true."[37] All the fliers at Mourmelon were friends of the deceased Laffont, even though he'd only been there six months. "Frank and open, he loved to laugh, even at himself, because he tended to stutter."[38]

Marie told the reporter that she'd taken lessons from him. "Laffont was prudent and attentive to minutia, perhaps even excessively so. He never flew without checking all his lines. He never tempted fate. I even heard him say once that if he thought he was going to break so much as one finger, he'd quit flying."[39]

When daring young pilot Pierre-Marie Bournique crashed in May 1911, his good friend Émile Aubrun was teary-eyed as he told *Le Matin*, "Imprudence, foolish imprudence has taken an excellent aviator from us. He had been warned, you know. Even yesterday I said to him, 'Don't go up. You're going to kill yourself.' This morning another pilot told him the same thing. When Sommer watched him fly, he had to cover his eyes. A while ago, Mademoiselle Marvingt watched him fly. I was with her later when she spied his name in the newspaper. She marked it with a tiny cross and looked sadly at me. She knew, too."[40]

In January of 1911, a syndicated article ran a photo of Marie, Raymonde de Laroche, Hélène Dutrieu, Valentinto Duels, and Mademoiselle Steir as "the most brilliant French 'aviatresses'" and said, "French women have taken up the perilous sport of aviation with surprising daring and success. Every day one of them flies higher and farther. Many have been upset and received serious injuries and it cannot be long before one will be killed. There is an agitation to forbid women engaging in their dangerous pastime, but they will submit to no such regulation."[41] (Between 1909 and 1914, fifteen women obtained pilots' licenses while 1,600 men did.[42])

Not everyone was so pessimistic: "Whenever an aviatrix climbs into the seat of her aeroplane, and starts out for a flight the men, spectators and aviators alike, usually worry over the possibilities of an accident. For the popular notion makes women sensitive and panicky, and the air is full of unknown dangers, which must be met with coolness and prompt action. But, so far, the women of the air have shown that they are as good as men in handling the aerial ponies."[43]

Given the conventions of the time, much was made of the deaths of female pilots. However, the women themselves did not believe that the death of one of them was somehow "worse" or "sadder" than that of a male pilot.

Amelia Earhart said, "Feminine fliers have never subscribed to the super-sentimental valuation placed upon their necks. I am sure if they fall they can endure their share of misfortune, whatever it be, as quietly as men."[44]

After her disappearance, Amelia's husband found a note she had left for him: "Women must try everything in the same way men do. When they fail, their failures should serve as a challenge to others."[45]

The legendary Hélène Boucher, said to be the fastest woman in the world, passed like a meteor in the skies of aviation. From her first flight at the age of twenty-two, she was obsessed by one idea: to pilot her own plane. Less than four years later she died in an air crash. Her casket was displayed near Napoleon's tomb, in the *Invalides*. Never before had a woman received such an honor.[46]

The talented and vibrant Raymonde de Laroche, a friend of Marie's, was the first woman in the world to earn a pilot's license; she qualified for the 36th license issued by the French Air Club on March 8, 1910. Eleven years younger than Marie, she took up flying lessons in 1909 at Mourmelon, in the same place and just several months earlier than Marie did. Within a week of her first lesson, she broke an arm in a plane accident. Two months later, she suffered another accident while taking her pilot's license test. Five months after that, at the Reims air show, she sustained seventeen fractures and numerous wounds and contusions. After eight months of misery, she was released from the hospital and headed straight for the airfield.[47] In 1912, she was in an automobile accident in which Charles Voisin, the famous airplane maker, was killed, and in the following years she survived a number of other air accidents. Ironically, it was as a passenger that she met her death in an airplane at the age of thirty-two.

Harriett Quimby, the first licensed female pilot in the United States, was famous for her 1912 crossing of the English Channel in a monoplane without instruments or parachute. Less than two months later, she flew her Blériot monoplane in a Boston air show. The Blériot was one of the best performers of the day, but very unstable. Balance had to be maintained with a bag of sand placed at a particular point in the plane when there was no passenger. When there was a passenger, he or she had to sit absolutely still. A shift of this center of gravity meant disaster. A Mr. Willard, who was in charge of the meet, flew with Harriett. Toward the end of the flight, he apparently moved, causing the monoplane to spiral out of control. Willard's body flew into the air followed a few seconds later by Harriett's.

Marie lost many comrades and friends, attended scores of funerals, wrote dozens of obituaries. Yet she never hesitated to fly. Had no one been prepared to assume those lopsided risks, aviation as we know it would not exist.

The esprit de corps among those young fliers was an inebriating elixir of belonging and daring and accomplishment. A family of sorts, they nick-

Mademoiselle Marie MARVINGT, pilote aéronaute-aviatrice, a obtenu le premier brevet féminin sur monoplan Antoinette. Première détentrice de la Coupe Fémina, 27 novembre 1910.

A 1910 postcard celebrating her new pilot's license.

named each other, kept in close touch with postcards, breathed a rarer, hipper air than their non-flying peers. One of Marie's postcards read, "Thanks my dear Fiv, for the chicken cage."[48] Impossible to know now which chicken cage or who precisely "Fiv" was, but she hoped to see him soon in "Mourmeluche," their slangy nickname for the airfield at Mourmelon. With the specter of death hovering over the airfields, early pilots shared feelings that could truly be known only by one another.

Early steps to leave the earth behind were almost laughably tiny. The Wright brothers' first flights were properly speaking more hops than flights, and success was measured in feet flown and seconds in the air. Marie used to say that the first airplanes rose such a small distance that "you had to get down on all four paws to see if it had actually left the ground."[49]

Planes flew so low, you could hear pilots talking to each other. They sometimes received directions from someone on the ground, pointing. Armed with only the crudest maps, pilots often simply leaned over the side to see where they were going. The machines themselves were unstable, liable to tip over on the ground or burst into flames without warning. Wings were designed to fold up so that planes could be housed in an ordinary garage, but they sometimes folded in mid-air.

The wooden-framed planes themselves were so light that they could be lifted and moved around by the tail, although this lightness meant that they were also prone to blowing over on the ground unless firmly anchored. To propel oneself skyward on such a frail assembly of wood and canvas, while sitting sling-like suspended over the void, at the mercy of rain, hail, snow, wind, and dust, took an indescribable audacity.

Jules Chancel told the readers of *Lecture Pour Tous* what it took for Marie to get a pilot's license: "I've left for last the story of the only woman, Mlle. Marvingt, who lives bravely in the midst of all these young men and who will soon have her license, earned by means of courage, *sangfroid*, and perseverance. During long months, she has lived the hard life of the camp, getting up at 4:00 in the morning, patiently waiting her turn in the cold, in the mud. She has acquired the indispensable strength and endurance required to maneuver a plane in high winds. The process is extremely painful, and it's certainly not within the reach of all women, but Mlle. Marvingt comes to aviation from a particularly adventurous and challenging life."[50]

Aviators Kimmerling, Aubrun, unidentified woman, and Marvingt, waiting, perhaps for the winds to die down, at Champirol, 1911 (courtesy Mireille Bertrand).

From October 1909 until she obtained her license in November 1910, Marie studied principally under Latham, but also worked with Alexandre Laffont (head pilot at the Antoinette school), René Labouchère (Latham's cousin, who helped develop the Antoinette engines), and Charles Wachter.[51] The newspapers said she was one of the hardest-working students there. To broaden her skills, she took instructions from Deperdussin's military aviation school and she flew a Hanriot, which she found easier to maneuver than the Antoinette.[52] Newspapers followed her progress, commenting on her "making beautiful flights," "landing magnificently," or "piloting remarkably well."[53]

The flying camps were of unending fascination to readers; according to *Le Matin*, "everyone" was at Mourmelon: master-pilots, student-pilots, army officers, foreigners, plane builders, simple sportsmen, "and ... even a woman."[54]

Marie said her most stirring flight "was beyond all dispute the first that I made alone in my Antoinette, on the 4th of September 1910. For quite a while I'd had complete control of the machine while flying with my last teacher, Laffont. And I had as much confidence as a learner can have. Despite that, everything felt completely different when experiencing the unequaled sensation of being alone up in the air. There was a little wind and I quickly rose to 60 meters. The monoplane acted far better than the one I usually flew. The first turn caused me real uneasiness but, once completed, was transformed into joy unalloyed. I was nervous about coming down to earth, but my landing was quite unexceptional. It was done: I had flown!"[55]

By the time she took her test for her pilot's license on November 8, 1910, she was more than ready. Perched among the struts and wires of her fragile monoplane, she waved to onlookers, started the engine, and ... flew the course without incident.[56]

Although this would be unsurprising today—in fact, it would be required—it was unusual enough for the public to wreathe Marie with superlatives. The Antoinette was as difficult to fly as it was delightful to look at, and every other pilot who obtained a license on the Antoinette had suffered damage or injury in the process. Marie was the only one to land the plane intact.[57]

Louis Charvet wrote in his history of aviation, "In the beginning all you had to do was be able to take off and stay in the air a while."[58] To obtain a license, anything short of the death of the pilot was acceptable. Many pilots crawled out of a wrecked plane to receive their licenses on the strength of having survived the experience.[59] Because of this, Marie's flight was a sensation.

According to the newspapers, "This morning Mademoiselle Marvingt brilliantly passed the three aerial tests needed to gain her pilot's license. She flew to a height of sixty

Marie on a Deperdussin at the Fêtes d'Aviation, April 7-8, 1912.

meters and, with great regularity, made the circuit of the airdrome. Later, despite some turbulence, she flew over the countryside for fifteen minutes and then landed in a superb planed descent."[60]

Marie said later she'd never forget that morning: "During the second test, a biplane took off just sixty meters from me. I altered my course to avoid its slip-stream, and I went up to twenty-four meters. But during the whole circuit, I couldn't see it. I could only think of the catastrophe that happened to Dickson and Thomas. Colleagues on the ground told me later that the biplane and I were only twenty meters apart. With the new oblique wings, this will no longer be a problem."[61] (Dickson and Thomas, flying in Milan, were unable to see each other and collided; both planes were telescoped, and Dickson was killed.)

Marie was awarded the French Air Club pilot's license number 281, the third woman in the world licensed to fly fixed-wing aircraft.[62] Referring to Marie's place as third woman to be licensed, aviation historian Edmond Petit wrote, "If you knew her, you'd be astonished that she wasn't first."[63]

A lengthy piece in the December 1910 *Washington Post* by John Trevor Custis was subheaded "Granting of Pilot's License to Three in Europe Indicates That They Mean Business, and Are Not Satisfied to Confine Themselves to Being Mere Passengers."[64] Custis wrote, "When women first invaded the field of sport by playing championship tennis and golf many of the old-timers shook their heads in disapproval. But it was only a beginning, for in these latter days, when myriad members of the tender sex are breaking into every active field of man's domain, from dog-catching to banking, their sporting blood is growing hotter."

Letting his fancy fly free, he supposed that "it will come to pass, ere long, that when Mr. Joy Bird seeks to escape the ennui of commercial life by means of an aeroplane trimmed with hot fowls and cold bottles, he will fly near some other machine to borrow the 'makings,' only to discover Mrs. Joy Bird and the little Birds in the act of enjoying an aerial picnic of their own.

"In noting the advent of women to the world of aviation, I am not considering the dilettantes who are mere passengers in men's aeroplanes. Give almost any up-to-date woman a well-tailored costume labeled an aviation suit, together with a becoming veil and a smart pair of high-topped shoes, and she will gladly take a chance on making her husband a widower by acting as ballast for a flying machine. Not to have had a ride in an aeroplane will soon be as horrid a thing to confess as never to have been to Paris. It is not this sign of the times that makes the strong men of aviation weep and fear for their laurels."[65]

What makes them weep? "It is the granting of pilot's licenses to women that indicates they mean business. Three women in Europe have already received such licenses after complying with the conditions demanded of men and women alike. And it is now recorded in the newspapers that they are breaking each other's records for distance flights—breaking them, apparently, with a nonchalance that would indicate that they consider such things as merely a part of the day's work, sandwiched in between a pink luncheon and an afternoon tea. A specimen item in a French paper relates the fact that Mlle. Marvingt beat Mme. de la Roche's record for long-distance flight in an aeroplane.... Mlle. Marvingt does not use a ladylike type of machine like the Demoiselle. No, indeed. She pilots an Antoinette, which is one of the heaviest and most difficult monoplanes to manage, a high-powered monster of the air. This is the type of machine that was brought to this country and used in a number of excellent flights by Hubert Latham, and Mlle. Marvingt is a pupil of Latham's. As the weather was cold when she made her flight, she wore a fur sack invented by Latham for his own use, which made her look like a feminine arctic explorer. (Fashion papers will kindly copy.)"[66]

In 1911, the Sydney *Sunday Times* wrote, "Though many aviators assure us that flying is less dangerous than it appears, it might well be imagined that the new science or sport would offer few attractions to women."[67] They then showed this was not the case by reprinting from the American magazine *Collier's* a piece in which five female pilots gave accounts of "some thrilling personal experiences of their flying career."[68]

In Marie's contribution, "The Intoxication of Flight," she wrote, "Having experienced severe turbulence and having barely avoided an in-air collision as a student, I was much less surprised when I experienced them while on board alone."[69]

She finished with lines much quoted afterward: "This new sport is comparable to no other. It is, in my opinion, one of the most intoxicating forms of sport, and will, I am sure, become one of the most popular. Many of us will perish before then, but that prospect will not dismay the braver spirits. In devoting themselves to the new cause, those who have the true aviator's soul will find in their struggle with the atmosphere a rich compensation for the risks they run. It is delicious to fly like a bird!"[70]

Aviation historian Bernard Marck quotes her as saying, "What a wonderful trampoline towards the sky and the unknown."[71] Marie generally erred on the emotionless side, but once she revealed the three ineffable "emotions" she experienced while flying: speed, wind, and enhanced perception.[72]

The same year that Raymonde de Laroche, Hélène Dutrieu, and Marie received their pilot's licenses, a Parisian magazine wrote, "If there is one sport that is not at all suitable for women, in our opinion, it is aviation! Why should a woman expose herself to continual and imminent danger when she doesn't have to? She risks a life that could be better spent as the joy and guiding light of some loving family."[73]

Aviation meeting in St. Etienne, 1911.

Writing about "the conquerors of the air," aviation journalist Robert Marchand said that "the most astonishing, the most curious of all the female pilots remains, and will probably always remain, Mademoiselle Marie Marvingt, licensed in 1910 in an era where one could count female pilots on the five fingers of one hand."[74] (Those five were, in order: Raymonde de Laroche, Marthe Niel, Marie Marvingt, Héléne Dutrieu, and Jane Hervieu.)

In *Kingdom of the Air*, de Saint-Fégor wrote, "In France, in 1910, we have seen alongside the men-birds a good number of gracious women-birds." He included Marie, "the well-known sportswoman. ... With their courage and fortitude, they cede nothing to the men."[75]

From the Perth *Daily News*: "Mademoiselle Marvingt ... is now a duly qualified aeronaut. She has received a pilot's certificate, and is an actual sky pilot. ... She seeks to demonstrate that in the air, as on the earth, a woman's as good a man as anything in peg-tops. ... The sex aspires to the higher planes, and these are all aeroplanes nowadays, so prepare for the airwoman. Soon the man who attempts to shoot Folly as she flies will in all probability bring down Polly."[76]

Nineteen days after obtaining her license, on a freezing-cold November 27, 1910, Marie set the first world records for women in time aloft and distance flown.[77] It was the first time a woman's flight had been officially timed, measured, and verified.[78] Marie had insisted her flight be recorded because she was attempting to win a Cup recently offered by Pierre Lafitte, owner of the women's magazine *Femina*.[79] The winner would be the Frenchwoman, duly licensed by the French Air Club, who by the end of 1910, had flown the longest distance in a nonstop flight. The competition inspired much excitement among women who had not, thus far, been offered a front seat in the aviation world. At last they would have their own competitions and records.

Marie made numerous training flights for the Cup. Nearly a month earlier, she'd made a nonstop flight of fifty minutes, with a stiff wind and heavy fog, reaching 1,000 meters of altitude. She must have expected to do even better with weeks ahead of her to perfect her time and distance.[80]

On that Sunday morning, Marie prepared for the cold, wearing woolen leggings, a skirt of stout cloth, a woolen jersey, a sheepskin mantle, a fur cap covering the top of her head, and fur-lined gloves. Later, seated in the plane, she slipped into the soft fur sack invented by Latham that covered practically her whole body.

She couldn't take off as scheduled because mechanics had to warm up the congealed oil. Despite a violent wind and glacial temperatures, she decided to fly.

Marie and her Antoinette stayed in the air for fifty-three minutes, covering forty-five kilometers.[81] When she descended from her plane, she was carried off the field in triumph.[82]

A little later at the St. Etienne meeting: Marie lands in an acacia tree.

The news went around the globe: "The world of sport is bowing at the feet of a new idol. This time it is a daring woman. Miss Marvingt is her name and she has broken the record for long-distance flights for women in an aeroplane."[83] As a French aviation journal put it, "Many pilots with mustaches couldn't have done as well."[84]

Marie described her flight a week later for *Le Sport*: "I had to constantly adjust the altitude and the steering because of the gusts of wind. Blown off-course in the turns, I had to take them wide. To distract myself from feeling like a bird in a cage (my circuit was only three kilometers around), I watched the officials below. … The cold had frozen my face, but I continued to fight the wind's buffets, which increased as the sun rose. Sometimes the wind was so strong I scarcely moved forward. And then, alas, on my fourteenth round, my motor sputtered and I had everything I could do not to touch down. On the next lap, the motor was scarcely working and no matter what I tried, I was forced to land. I was heartbroken as I wasn't tired and the cold was tolerable. I saw then that the gas throttle, which I'd been told never to touch, was off; with gas inadequately reaching the engine, no wonder the latter failed. … Now that I've made a little trial distance flight under rather difficult conditions, I'm ready to beat my own record and those of any competitors."[85]

Once Marie set the first records, others prepared to better them. Particularly able and eager was the Belgian pilot Hélène Dutrieu, whose life paralleled Marie's in many ways[86]: they were born two years apart (Hélène was younger) and they died two years apart (Hélène died earlier); Marie was the third woman in the world to be licensed to pilot a plane; Hélène was the fourth; both were named to the Legion of Honor; Marie was an all-around sportswoman and the creator of aeromedical aviation, and Hélène was a cycling champion, automobile racer/stunt driver, wartime ambulance driver, and director of a military hospital; both had close relationships with their fathers (Hélène's was an army officer and she never flew without her good-luck charm, a pair of his gaiters). The two women proved to be exceptions to the tragic rule of death in the skies. But they couldn't have known that in advance. They had to take their chances like everyone else.

Newspapers liked to frame the two as rivals,[87] but they considered themselves friends and colleagues. Friends or not, Hélène wanted the Femina Cup too.[88] But flying in the winter was doubly challenging. An article published in sunny California in the *Los Angeles Herald* expressed astonishment that anyone would fly in the cold but added, "Women as well as men are proving that aviation is a winter sport. … And two young French women, Mlles. Marvingt and Dutrieu, are making daily flights to win a prize offered by a woman's journal."[89]

The *Tacoma Times* of Washington mentioned other competitors: "Four Little Women, All in a Row, Soaring For Prizes, Hi-lee, Oh, Hi-low": "A desperate battle in the clouds is being waged by four women of France, with aeroplanes as steeds, and the end will not be until midnight of Dec. 31. The object is the winning of the 'Coupe Femina.' The air women are Mlles. Hélène Dutrieu and Marie Marvingt, Mmes. Niel and Franck."[90]

On December 21, 1910, Hélène Dutrieu did it. She flew for one hour and nine minutes, covering almost sixty-one kilometers. Unless another woman could top that before midnight, December 31, the Femina Cup was hers.

Marie wrote, "Having flown in poor conditions, I'm going to try again. When this record is beaten, I want to be the first to try."[91] She made two more attempts to regain the Cup, but on the first try, "after a splendid departure," she was interrupted by severe rain. The second time she took off with enough gasoline to fly for three hours, but after forty-five minutes, she was obliged to land because of problems with the motor.[92]

Four days later, an article appeared in the *L'Est Républicain* in which Marie appeared chagrined at having to explain a failure that she felt was due to circumstances beyond her control:

> The first and principal reason I was unable to win back the Femina was that my new plane wasn't ready until 10:30 a.m. on December 30. I'd been waiting impatiently at the Air Club since the 27th. To be certain it was ready, it had to be flown several times, with the mechanics checking everything after each flight. Gobé, the master-pilot, made two trial runs and the motor seemed to be fine. Thus, despite a rather strong wind, I decided to make a new attempt. It was the first time I'd flown a plane with a 60 HP motor and with such a big gas tank (110 liters). Despite the extra weight, takeoff was good. From the beginning I felt balanced, and took the corners wide. My motor was throbbing regularly, and I felt confident of a good flight.
>
> On the eighth round, I was annoyed by the rain lashing my face and fogging up my glasses. Large, menacing clouds made me fear it was going to continue but it was only a shower. Then the wind picked up and I ran into rather powerful turbulence. I still thought I could stay up. It was work, but I was able to maintain control of my machine.
>
> After forty minutes of eventful and spirited flight, my motor suddenly worried me: big black puffs were coming out of the distributor and the power of my 60 HP diminished. I tried to give it more gas, but it grew weaker and weaker. Something abnormal was going on. To my profound regret, I decided to land.[93]

The problems with the motor were serious. "Since repairs would take more than a day, I had to give up the idea of winning back the Femina Cup."[94]

Marie Marvingt and Hélène Dutrieu were featured on the cover of *Aviation Illustrée* for having established the first (Marie) and the second (Hélène) world aviation records for women.[95] But it was Hélène who won the fancy Art Nouveau Femina Cup, which was pictured with Marie on one side of it, Hélène on the other.[96]

Marie did not, as is sometimes reported, win the Femina Cup the next year in Turin.[97]

In the next few years, Marie flew the Antoinette, Deperdussin, and Voisin planes. She put in hours of flight time at the Reims and Mourmelon airfields, and trained dozens of new pilots.[98] One of the reasons she was so popular in military circles all her life was that she had given so many aviators their first lessons.

One of the best of the early air shows was held on the prairie-turned-improvised-airfield of Champirol July 29 to August 8, 1911. More than 150,000 spectators gathered to watch some of the finest aviators in the world demonstrate the novel and dangerous sport of flying. Marie flew her famous monoplane, *Antoinette*.[99] As always, she was the crowd's darling, their love for her mixed with fears for her safety, along with a dash of that half-pious, half-morbid thought: What if they actually had to, god forbid, witness her tragic death? At that thought they always cheered a little more loudly.

Because air shows were so new, people hardly knew how to behave. At this one, overexcited spectators frustrated the pilots by blocking the ramps reserved for their planes. A troublemaking local politico argued with the police about removing a barrier (the crowd backed him). Then he argued about keeping spectators out of the hangars (the crowd disagreed). After a while, the crowd began to police itself and to boo its more obnoxious members.

The reserved and box seats were occupied by the rich and elegant from all over France. Sitting in the inexpensive seats, available for ten or twenty sous, was a veritable anthill of humanity. On July 29, the phenomenal crowds meant that by 4:00 in the afternoon there was no point in trying to catch a trolley out to the airfield from nearby St. Étienne. The trolley company complained that it had to triple and quadruple services.

According to the *Tribune*, "Taxis were reserved ahead of time. Déronzy, owner of the luxurious taxi-autos, could not keep up with the demand. A wealthy ribbon manufacturer offered him twenty francs to drive him to Champirol. But Déronzy had to refuse because he had already been retained by Garros, Bielovucic, and the gracious aviatrix Marie Marvingt!"[100]

A few days later, Marie and Marcel Loridan hired a taxi to get out to the field, but ended up smashed against a lamp pole when their driver tried to avoid a careless bicyclist. The driver's arm was broken, Loridan had chest injuries, and Marie was bruised.[101] This would be the first of three accidents in less than two weeks for Marie; the other two were in planes.

A local sportsman and mountain-climber wrote an article for the newspaper giving his impressions of the air show.[102] He mentioned the crowd of pedestrians, cyclists, automobiles, and trucks heading toward the airfield, the enthusiastic reception given the aviators by the fans, and the various airplanes and aviators he could identify:

> Nearby I spot a familiar figure, Mademoiselle Marvingt, the aviatrix. Although I haven't had the honor of knowing her as an aviatrix, I've met her many times, and I've been singularly impressed with her talents as a skater, skier, and mountaineer. She's certainly the most complete sportswoman we've ever had. Her exploits are legendary. All the guides in the French Alps know her and love to tell about her vertiginous mountain climbs. She's a past master at skiing and, lashed to her great runners of ash, she's made winter crossings of the French Alps, the Vosges, and the Massif Central. I can't resist the pleasure of presenting my respects to her. Her charming welcome makes me forget the trouble I had getting here today.
>
> "So," she says, "What brings you here?"' I explain my presence and we chat a little. Talking to her brings back all the good memories of our adventures. I ask, "By the way, are you abandoning mountain-climbing? Mountain-climbing and aviation don't seem to go together."

Marie sets the first women's aviation records and is carried off the field in triumph.

"Don't go together?" she exclaims. "I don't agree! Listen, I've just come from Turin, where I was at an air show and, together with a group of Italian officers, I climbed the Concarena, the Badil, and the Corne del Dante."[103]

Marie received more publicity than usual during the Champirol air show because of what the newspapers called her "uncommon little adventure." Scheduled to give a demonstration flight on her fragile and rather unstable Antoinette, Marie took off at 6:30 on the evening of August 6, despite an approaching storm. She knew immediately that something wasn't right: she was unable to gain altitude. Henry Beaubois later wrote that "her considerable expertise allowed her to avoid a tragic accident when her motor went out over the heads of a dense crowd who, perfectly unconscious of their danger, were applauding the big white bird."[104] In addition, "Mlle. Marvingt had, for choice of landing places, the crowded café itself, or the little adjoining skittle-alley. With extraordinary dexterity, she succeeded in landing, with no damage to herself, and with but little to the machine, in the confined space of the alley, which was surrounded by houses on three sides and a stone wall on the fourth."[105]

Marie disappeared from view, and the horrified crowd suspected the worst.

But she came to rest, "gentle as a flower,"[106] in an acacia tree that stood in the courtyard of the Café de la Terrasse. Below her a group of men from the village were in the middle of a game of *boules*.[107] Marie said later she didn't know who was more surprised—she or the men playing *boules*.[108]

In the meantime, according to the *Echo de Sanflorain*, "Her accident caused a considerable panic in the stands. Night had fallen; more than 100,000 people, who didn't know Marie had landed, ran into the fields around the aerodrome, searching for the spot where she might have crashed."[109] As the Paris *Journal* explained, "When someone crashed or disappeared from the airfield, the crowds usually ran after them, trying to be the first to find the disaster."[110]

Along with the crowd and police on horseback, Kimmerling and Aubrun looked for her, questioning farmers in the area. Marie was fortunately unhurt, and her plane suffered only slight damage to the left wing—not because of the accident but because of greedy souvenir-hunters.[111] A photo of a smiling Marie sitting calmly at the wheel of her Antoinette as it perched in the tree accompanied most accounts of the Champirol air show.

In those days, aviators spoke of "breaking wood." Planes were preposterously fragile and it didn't take much to damage them, so skilled pilots were complimented when they managed to land without any breakage.[112] Marie was proud of having made over 900 landings without "breaking wood," a record unequaled at that time.[113] She was teased about the incident at Champirol because, although her plane had not been hurt in the accident, the acacia tree in which she landed was the worse for wear; in that sense she had "broken wood."

Four days later, she crashed again, this time on the ground but, again, suffered no injuries.[114] Her first serious airplane accident didn't happen for another two years, and it wounded her pride more than her body, although her injuries took weeks to mend.

During a routine flight from Reims, she had to land in a barley field and ended up trapped beneath her flipped-over plane.

From the clinic, she wrote to her good journalist friend, Frantz Reichel at *Le Figaro*, asking him to reassure and thank the many people who had sent her get-well wishes since she wasn't up to it yet.[115] In doing so, he gave readers details of the crash—on page one.

Renewed Activity Everywhere

AERO
AMERICA'S
AVIATION WEEKLY

Vol. II
No. 2

April 15,
1911

TEN CENTS

MARIE MARVINGT ON HER ANTOINETTE

Marie on the cover of a U.S. aviation weekly, 1911.

Marie thought she had chosen a soft spot for her emergency landing, but she hadn't counted on recent rains turning the earth into a prehensile mud. Once she and her plane had reversed positions, it being on top of her, "my helmet was completely driven into the mud, blood bathed my face, and I had to carve a little trough in the ground with my left hand so I could breathe."

Unable to move, she waited "interminable minutes" until she heard the farmer's voice: "

"Oh, the poor fellow! He is dead!"

"From my little carved-out trough, I shouted, '*She* is not dead. But whatever you do, don't smoke.' I was, you see, covered in gasoline."[116]

In another interview, she assured readers that her career was far from over. "I plan to take up my activities again as soon as I get rid of these bandages."[117] A few weeks later, *Le Figaro* summed it up: "Aerial locomotion has given women the opportunity—for them particularly appealing—of affirming their taste for adventure and their disdain for danger."[118]

The newspapers, which had made of aviation a symbol of technology, innovation, and social progress,[119] followed Marie regularly in the pre-war years, reporting almost every time she took to the air, whether in Reims, Turin, London, or Toul.[120] A report on a meet in August 1912 noted that the crowds were not disappointed in the show put on by "a little bit of a woman, Mademoiselle Marvingt, at the commands of a Deperdussin monoplane."[121] But the next year, when Marie was both flying and writing about a meet in Reims, she realized that the crowds were not nearly what they had been in previous years. Progress—in the form of longer and higher flights—was the downfall of air meets. Planes were going so high they couldn't be seen. Pilots trying for endurance records flew so long that spectators lost interest. Ironically, developing aviation science left spectators with little to do.[122]

When the Antoinette factory closed, Marie switched to a Deperdussin, then the fastest machines in the air and much easier to pilot than the Antoinette. She wanted to get her military pilot's license, which required a more stringent test than the French Air Club's. She trained hard for about a year,[123] but sometime in 1913 she decided to devote her time and energy to her ambulance-airplane project. She may also have realized the futility of expecting French authorities to allow a woman in their air force.

By the time the war began, Marie had hundreds of flights to her credit, more than many male pilots. Those pilots who earned their licenses from the French Air Club before August 2, 1914, the day France and Germany declared war on each other, were called Vieilles Tiges ("old twigs"). These were the legendary fliers from the so-called heroic age of aviation.[124] As Vieille Tige #281, Marie received all the admiration and respect always accorded to these pioneer pilots.[125]

She and her air sisters would have been useful in the war effort, but women were never officially allowed to fly for military purposes during World War I. Marie ("the impossible is not Marvingt") managed it, first by replacing an injured pilot needed for a vital mission, then by taking advantage of her many military friendships. Her Legion of Honor citation and her *Croix de Guerre* medal state that she piloted a bomber on at least two occasions during attacks on German-occupied Frescaty (near Metz) in 1915.[126] She is acknowledged by most reputable sources to be the first woman to do aerial combat.[127] Marking the anniversary of Marie's service, the Australian Women of Aviation Worldwide Week's theme in 2015 was: "Serving with honour: 100 years of female pilots in combat."[128]

In the beginning, aviation was considered a sport, and was covered on the sports pages. By 1910, aviation news appeared under aeronautics (which referred to ballooning). Only later did aviation become a serious field of its own. (The new vocabulary shifted as well: women were "bird women,"[129] "aviatrices,"[130] "aviatresses,"[131] "aeroines,"[132] and "fly-erettes."[133]) Marie had always seen aviation's wider uses: "I'm convinced that aviation is going to change and become very practical. For now, it's the most marvelous school you could dream of for people to learn courage and perseverance."[134]

Marie spent the inter-war and after-war years lecturing on air rescue, training personnel, and working relentlessly to get air ambulances established in every country's first-aid armamentarium. Although her years of setting aviation records were over, she never lost her interest in all things aviation.

In North Africa she flew numerous air reconnaissance missions. On two tours across the United States in the 1930s, she spoke to numerous audiences about aviation, visited airplane factories and aviation schools, and met with military aviation officials. On October 4, 1935, she was invited to sign a pair of copper wings with her name and the date to add to the famous fliers' wall in the St. Francis Chapel of the Mission Inn in Riverside, California.[135]

Marie was the only woman in the world to hold four pilot's licenses simultaneously: balloon, seaplane, airplane, and helicopter.[136] She also flew dirigibles and gliders, and was licensed as the first air-rescue pilot, doubly qualified as a pilot and as an emergency medical technician.

Journalist Georges Gygax said succinctly, "She piloted anything that would fly."[137]

Gliding was part of the early efforts of the Wright brothers and others to fly a heavier-than-air vehicle, but eventually it gained a place of its own among sports enthusiasts. Marie liked to say gliding was useful to her for studying meteorology and air currents, but it's likely that she also enjoyed its singular exhilarations. Once the engine-driven plane caught on, gliding drifted into the background, but Marie retained a certain fondness for it.[138]

She never much cared for dirigibles, referring to them disdainfully as simple engines.[139] On the other hand, she learned to pilot them and wrote articles about them.[140]

She remained enthusiastic about flying machines for the rest of her life. The day she turned eighty, a U.S. army officer came to visit her in Nancy. With a smile, he asked, "How would you like to take a little ride in my superjet tomorrow?"[141]

Marie jumped up and cried, "Tomorrow! What's wrong with today?"

She demanded to be taken to the base immediately, where she passed the medical exam required for supersonic flights. Declared in good physical condition by the military doctor, Marie climbed into the narrow cabin of a McDonnell F-100 Super Sabre supersonic jet fighter. Soon she was flying over her beloved Nancy at 1,200 kilometers per hour. Taking over the controls from her officer friend was the best birthday gift she could have received.[142]

American fliers, who weren't allow to fly supersonic jets after the age of thirty, expressed their admiration to her when she returned to the base.

"Bah!" said Marie. "I didn't even go 'Bang!'"[143]

Several weeks later she was at the airfield at Issy-les-Moulineaux, where she learned to fly the Djinn 1221, a new "cold-jet" helicopter.[144]

Gabriel Bichet said that when news flew through the newspaper offices that an eighty-year-old was flying a helicopter, sceptical journalists who had, as he puts it, seen a hundred revolutions thought, Sure, we've got some grandpa standing in the doorway of a plane, with someone holding him up just long enough for us to take a picture.[145] The reality was so different that they all made good stories of it.

Marie later took and passed the test to obtain her helicopter pilot's license on a Djinn, at that time the only operational jet-engined helicopter in the world. The story of the helicopter-licensed eighty-year-old woman was picked up by newspapers around the world.[146] Some included photos of her at the controls, others showed her with physician-pilot Valérie André, the "Helicopter Angel" of the Indo-Chinese War. At the time, Gabriel Bichet described Marie as "an extravagant octogenarian with the look in her eyes of a young woman in love," and added, "The news of her latest activity is both highly unlikely and rigorously authentic."[147]

In 1960, when Marie was eighty-five, she was still up in the air. At the controls of another jet-engined helicopter, she soloed, showing she still had a lively eye and a quick hand.[148] The picture that appeared in the papers featured her smiling broadly while pulling the plexiglass cover of the cockpit over her head. As one newspaper put it, "She seemed as comfortable today in the cockpit of a helicopter as she was half a century earlier in one of the first airplanes."[149]

"The newspapers said I'm going to get my license to fly jet-engined helicopters," Marie commented, "but I don't think so. Such a fantasy is too expensive and my resources are modest. In any case, I don't think such a license would make me any more useful than I am now."

At the age of eighty-six, Marie renewed her pilot's license for another year.[150] She underwent a medical exam at the military hospital in Nancy, which she passed with the French equivalent of "flying colors" ("haut la main"). Pilots who invited her to fly with them at that time knew she loved nothing better than co-piloting; they delighted in accommodating her.[151]

During her later years, an American journalist asked Marie about her future projects. "I am the fourth woman signed up for a trip to the moon," she replied.[152] The interviewer didn't take her seriously, but a few years later when he was working in the archives of a U.S. space research service, he came across a list of those requesting a place on the first spaceship to the moon. Fourth woman on the list: a certain Marie Marvingt.[153]

When she died, a reporter from Nancy recalled her saying she wanted to celebrate her ninetieth birthday on the moon. "Marie Marvingt never realized this last wish," he wrote, "She has taken off, it is true, but to a much higher place."[154]

The life of the young balloonist of 1900 and the helicopter pilot of 1955 framed and chronicled the stupefying advances of aviation from its infancy to maturity. And she loved flying until her dying breath. A friend who was with her at the end said Marie talked as though she were still flying her plane.[155]

5

Masquerade in Uniform

> "How do you tell the history of a war in which more than nine million combatants and nearly seven million civilians across the world died by bullet, fire, hunger, and disease? How do you describe the experiences of a war that ignited two revolutions, brought down four monarchies, scarred a generation and culminated in major political and territorial changes that cast shadows to this day?"—*Intimate Voices From the First World War*[1]

At first, World War I was just a rumble on the horizon that nobody wanted to hear. When hostilities finally broke out, the war was expected to last four months, not four years. Soldiers went to war imagining that one decisive engagement could turn the clock back to the good old days. Instead, some eight million French soldiers (out of a population then of thirty-nine million) fought endless, futile, costly battles, and many found themselves on the five-hundred-mile-long combination slaughterhouse and cesspool known as "the front." Life would never be the same.

Just before the war began, Mildred Aldrich wrote, "It will be the bloodiest affair the world has ever seen."[2] Although dreading the thought of war, she expressed the feelings of many in western France, including Marie whose family roots were there: "I have felt as if I could bear another one if only it gave Alsace and Lorraine back to us."[3]

To the French, the soil of Alsace-Lorraine, which had been ceded to the Germans after the war of 1870, was sacred, a symbol of earlier French glory and home to many who still considered themselves French. Its return to France would be the only positive outcome for the French in the debacle to follow.

Of all combatants (65 million soldiers), nearly 9 million died, 21 million were wounded, and some 8 million were captured or missing. Civilian deaths amounted to nearly 7 million.[4] On July 1, 1916, on one single day, the British Army suffered nearly 60,000 dead and wounded.[5] Some 116,000 Americans died in the nineteen months of U.S. participation—more than twice as many as in Vietnam, nearly twenty times as many as in Iraq and Afghanistan.[6] During the first two weeks, French casualties totaled more than 300,000 men killed or captured, with little or no ground gained.[7]

France would eventually mourn the death of an entire generation.

World War I was rife with insanity. Nineteenth-century armies fought with twentieth-century weapons: tanks, machine guns, howitzers, aircraft, submarines, poisonous gas. When the war began, most people, including Mason Patrick, who would later lead the U.S. Air Service in Europe, had never seen a plane in flight.[8]

Making the conflict and the death counts more visible and horrifying was the development of cheap printing, which allowed newspapers to proliferate, along with compulsory primary education, which made newspapers readable by more people. "No war before 1914 had been so written about; no war after 1918 would be so uniquely captured by a single medium."[9]

Marie's wholehearted support of the war effort was deeply rooted. Her father's family produced a small army of military ancestors: officers, generals, colonels, and commanders.[10] Marie wrote to a cousin in August of 1914: "I'll never forget the sight of Nancy preparing for war. We are calm, resolute, and enthusiastic, sure of victory. But alas! Such sacrifices between now and then. We already have a great number of wounded. My father's health is excellent. He has only one regret: that he is no longer able to fight the Prussians. I've asked to be put on the roster of volunteer pilots. In the meantime, I'm working in a surgical clinic."[11]

On the day after Christmas in 1914, a zeppelin dropped eleven bombs on Nancy, along with an envelope weighted with a French bullet containing photos of German officers and these dedications: "Merry Christmas from Kaiser William II" and "From German pilots, on behalf of the Kaiser."[12]

Marie, a major with the Red Cross, wrote: "I was on night duty when the bombs hit at 5:30 in the morning. I heard their engines twenty minutes before the first bomb. They were flying only 500 meters high and they stayed a long time right over the hospital. I went outside and was trying to see through the thick morning fog when the first bomb hit with the most appalling racket. Flying debris was all around me; it fell on the garden like rain. They were aiming at the hospital, but they hit the place next door, fortunately damaging only the building. As soon as they flew past the roof of my section, I went inside to reassure my patients that they were no longer in danger, but they were calm and not in the least panicked."[13]

She described the bombing for *Samedi Soir:* "Every three minutes there is a new explosion. The clouds change color, following no doubt the path of the bombs. Another shell, falling perhaps on chalky ground, sends up a milky cloud. At 7:50, the eleventh and final explosion creates a brick-red, luminous cloud but not as high as the previous ones. Otherwise, it's a beautiful day here with spring-like sunshine contrasting with the gray cotton floating over Nancy. There was only one victim, lightly wounded in the leg, and that's a miracle because debris weighing over a kilo was found more than 1,000 meters from the explosions. During this fourth attack on Nancy, which had no discernible military goal, they did the same old thing except that they came a little earlier, forcing our brave old people to head for the basement without breakfast. Meanwhile, twenty minutes after the last explosion, some fifty youngsters, armed with shovels, are searching the many shell-holes for interesting debris. A great joy seizes the group when one of them, after much work, discovers the intact shell case of a 380. It's a 'toy' of at least 80 kilos, much too heavy for childish arms. But this clever little tribe goes looking for a stretcher and triumphantly carries off its prize."[14]

Earlier, as soon as war became fact, Marie had signed up as a Red Cross nurse.[15] She also attempted to enlist as a volunteer pilot, planning to either fly a bomber or join a squadron as a seaplane pilot.[16] Her request was turned down, however, leaving her bitterly disappointed. Popular notions about "a woman's place" defeated her, at least at first. Worse still, the new aeronautics department of the army requisitioned Marie's plane.[17] No, you cannot fly but, yes, we'll have your airplane.

At the time, there was little interest in military aviation. Even Field Marshal Foch, who became a friend of Marie's, said, "Airplanes are interesting toys, but of no military value."[18] And General Weygand, who was Field Marshal Foch's right arm in 1914 and an even better friend of Marie's, apparently believed that victory depended on controlling the terrain and planes couldn't control the terrain.

Jacques Mortane (real name Jacques Romanet), a pilot and journalist who later wrote about Marie in his 1937 *Women in Sports and Aviation*, was thirty-one years old when war broke out. He enlisted, requesting a spot in aviation, but ended up unloading drums of petrol and caustic soda. A fellow pilot told him, "When I mentioned my pilot's license to the recruiting officer, I might as well have been talking about my high school certificate. 'Oh,' he said, 'It's you, the pilot. Aviators are lunatics. You'll have to change, lad! Can't be doing with mavericks in wartime.'"[19]

If flying had been dangerous before the war, military aviation was even more unhealthy. The average pilot sent to the front had barely three to six weeks to live; 77 percent of French pilots died.[20] The airplane was younger than the pilots, and many of the planes still didn't have brakes.[21]

For the first time, airplanes turned the skies into battlefields, and both Germany and France dropped bombs from planes. But in the beginning, planes were used only on scouting missions, to observe ground troops. A German report in 1914 said "the duty of the aviator is to see, and not to fight."[22] In those early days, when enemy fliers passed each other in the air, they waved, their kinship as pilots outweighing other considerations.

The military air force was in its infancy compared to the army and the navy, and its pilots were essentially considered infantry; flying at low altitudes, they received directions from officers on the ground. But planes made a great deal of noise and there was no telecommunications system yet, so the orders from the ground often consisted of futile shouts, frantic gestures, even written notes, which were carried away on the wind. Early hostilities were equally primitive. One French pilot returning with a hole in his plane complained that a German pilot had thrown a brick at him.[23]

In 1915, the London *Daily Mirror* ran a piece called "Joan of Arc of the Clouds." It noted that the "Maid of Orleans went to battle on a charger. Mademoiselle Marvingt, another gallant young Frenchwoman, means to go one better, and dreams of offering the Germans battle in the clouds. For several years she has been well known in aviation circles in France. She has frequently distinguished herself by her skill and daring, not only flying, but in other sports. And now she has come to Paris to ask the Minister of War if he will accept her as an aerial scout."[24]

L'Est Républicain told the story, too. Almost lost in the huge crowd outside the enlistment offices on the Place des Pyramides in Paris, a small woman in a dark, tailored suit was recognized by several men as Marie Marvingt, who was known to want to enroll in the air corps.[25]

She needed documents. "Admit," she said to the officer, "if I fall behind German lines, I'm just a civilian, and I'll be shot like any other."[26] She pointed out that in England and in Russia, the air service was accepting women and hadn't had any problems. The article finished, "Mademoiselle Marvingt is using her gifts for eloquence, along with her vivacious words and her energetic chin. Will she succeed?"[27]

The French government officially refused to mobilize or militarize women.[28] Doing so might imply that women were full citizens, and thus entitled to women's suffrage and civic equality, both bitterly opposed by French leaders. Unofficially, however, there were possibilities.[29]

An AP story released around the world on August 14, 1914, said, "For the first time in history there are going to be women aviators acting as scouts during the war. Three women, all of them experienced fliers, have volunteered their services. They are Mme. Pallier, Mlle. Marvingt, and Helen du Plagino."[30]

Newspapers endlessly debated whether women should serve as military aviators. A few days after the AP story, the *Boston Sunday Post* headlined: "France May Send Five Into Fighting Zone—Poincaré Believes They Are Well Equipped for Scout Duty and Bomb Throwing": "At least five of the fair sex, all expert aviatrices, have volunteered their services and expressed the hope that they may be allowed to get into the thick of the fighting. And so the question arises: 'What chance have the Germans now?'"[31]

Under the subhead, "Known for Her Beauty," the *Post* article introduced Marie: "Long before she took up flying, Mlle. Marvingt was noted for her rare beauty, and many times attracted attention in the smart set of Paris. Almost as soon as airships became fairly common in President Poincaré's country she entered into the sport, first for the fun there was in it and then as a professional. In the American ex-pat colony she is well known, having taken many Americans up in her machine."[32]

The article posited that the reason for including women as pilots was simple: "Women really make better airbirds than men. By experiments at Harvard and in Paris it has been proven that woman's arc of sight is twice that of man's and is better adapted to flying. It has been shown, the French authorities say, that diffused attention is the opposite of concentration. Woman has the first, man the second. By diffused attention is meant the quality of mind that can take in a number of details at once."[33]

The "science" that followed gave other reasons for woman's superiority as an aviator. She already had to multitask, having responsibilities for children, household chores, and "pleasing her lord and master"; her skeleton more closely approximated that of the bird, her bones being lighter and more hollow than man's, her specific gravity lower; her craving for excitement and her nervous endurance were greater than his; her intuition was more powerful as was her ability to rapidly perceive a number of causes and instantaneously synthesize them. And, finally, high altitude and strong winds created a tendency to sneeze, and "it is well known that women sneeze more mildly than men and in this respect, therefore, they have another natural qualification for aviation, which men lack."[34]

Two months into the war in Europe, U.S. newspapers asked, "Women Aviators for the Allied Army?" The running head read: "A serious but startling plan to organize a fighting corps from the many charming, darling, and skilled flying women of England and France."[35]

Saying that the "idea has been received with enthusiasm by many women, who reject the ancient theory of feminine weakness and declare that their sex is fully as fit to face peril as the man," the article asked: "Why should not women serve as military aviators? They have the courage and the skill. They have the patriotism. If they are inferior in weight and muscle to men, these qualities are not of the first importance to the aviator. In steadiness and nerve, which are the essential qualities in the air, the modern girl claims to be fully the equal of her brother. Should women be exposed to danger? They've been employed in all sorts of occupations that shorten and endanger life. Women have taken part in every war, some in the last Balkan war. Women risk their lives with Red Cross forces and as nurses."[36]

The piece mentioned Marie among the brilliant French airwomen: "Mademoiselle Marvingt began her training with two men, Wachter and Laffont, who were subsequently killed. This did not daunt her. … A fearful prospect is thus sketched for the ordinary foot soldier tracked by a hawk-like woman in the air. Primitive, nerveless, quadrupedal in structure, ignorant of danger and contemptuous of the other sex, she planes above him, ready to annihilate him, calmly and coldly, with high explosives or blazing gasoline!"[37]

French military authorities apparently weren't reading the Boston and Spokane papers. But if the front door isn't open, try the back door. The U.S. periodical *Automobile Journal* reported in 1915 that members of the Women's Automobile Club of France were petitioning the government to allow women to fill vacant spots as ambulance drivers.[38] Among the 200 members were group president Jeanne Pallier, the Duchess d'Uzès, Hélène Dutrieu, and Marie Marvingt, the same women arguing for female pilots.[39] Marie was also a member of the Patriotic Union of Aviatrices, which was invested in promoting women's contributions to the war.[40]

A small irony of the official refusal to let Marie fly was that she continued to train young pilots. Marie was godmother and teacher to scores of the young men who flew in World War I.[41]

Soon dozens of newspapers in France and the United States reported that Marie and Hélène Dutrieu were flying scouting missions.[42] Because scouts only looked and didn't fight, this was considered quasi-acceptable. The official stance was still "no" to women pilots, but unofficial exceptions were made.

Most other areas of the French economy evinced no opposition to women moving into business and positions of influence to replace the men gone to war. The template of women being "good enough" to do "men's work" in time of war is repeated throughout the centuries. The pendulum would swing again, but in the meantime, women had a taste of power and responsibility in a world beyond the home.

Marie was determined to be a real soldier of the air, to fly military missions. But bureaucracies move slowly, and they positively creep when they are also unwilling. Marie and her colleagues began collecting signatures for a petition, but before they reached the required number, Marie had already flown two bombing missions.[43]

One day, while giving emergency aid to incoming wounded, Marie discovered that one of her patients was the injured pilot of a plane urgently needed to take part in a raid over Germany. There was no other pilot available.[44]

No other pilot available? How could there be no other pilot available when Marie Marvingt, pioneer aviator *extraordinaire*, was at hand?

Adding charm and logic to the urgency of the situation, Marie convinced the base commander to let her pilot the plane. On March 25, 1915, she flew to the rendezvous and then continued to the German air base of Frescaty (near Metz) where she and five other French pilots bombarded the aerodrome.[45] For her actions there, and for a second bombing sortie, she received official commendations including the *Croix de Guerre 1914–18* with palms and the Legion of Honor (the citation on the latter reads: "Authorized by M. Millerand, Minister of War, Mlle. Marvingt took part in two bombing raids on Frescaty aerodrome").[46]

Thus, in 1915, Marie became the first woman to pilot a fighter plane in combat and the first bomber pilot.[47] (This information was made public only in 1932.[48])

After that, she made several solo reconnaissance flights over the Italian front, but her activity as a war pilot was irregular, unrecorded, and usually extemporaneous. If anyone inquired, Marie was a volunteer nurse.[49]

In the mess hall in Metz, Marie overheard a group of infantry soldiers at the next table scorning the "easy life" of pilots in the war.[50] Although pilots risked their lives in newly developed and not always reliable planes, they didn't have to face death at close quarters, slog through miles of exhausting terrain, and fight and live in filthy, wet, unsanitary trenches. The action, the hardships, the horrors of war took place in the trenches. That's where the war would be won or lost. They concluded that one of those fancy pilots wouldn't last a day in the trenches.

Marie was certain she could be as tough and effective in the front lines as any foot soldier, and she decided to prove it.[51] The fact that women weren't allowed in the military, let alone in the front lines, didn't deter her. (She was not the first woman to be there unofficially. Clara Barton had written earlier, "But later, it chanced—just how, no one knew—/ That the lines slipped a bit, and some 'gan to crowd through; / And they went,—where did they go?—Ah! where did they not?"[52])

For many years, Marie refused to reveal names or details about how she obtained the required papers and managed to enlist, but she hinted that a "highly placed military chief" was complicit.[53] Only in a 1956 interview did she admit that it was with Marshal Foch's authorization and assistance that she became "Private Second-Class Beaulieu," a foot soldier in the 42nd Battalion.[54] An old friend, a lieutenant in the infantry, agreed to help her.[55]

With her hair hidden under the round, tight-fitting helmet and sporting the infantry uniform, Marie blended in well enough. Although she was an excellent actor and could change her voice at will, she generally avoided speaking and, when possible, stayed near her officer friend so he could run interference for her. Soon after her arrival, the 42nd Battalion was sent to the front. One of Marie's duties was to take a turn at the listening post in a front-line trench to locate and report enemy gunfire.[56]

In a 1920 piece in *Le Miroir des Sports*, a journalist known only as R.G. wrote that the Germans knew Marie had flown those missions and when she later gave conferences to French prisoners in Switzerland, she understood that being arrested would lead to her execution.[57]

Fighting in the trenches was hellish: bodies decomposed in shallow graves, adding to the stench of overflowing cesspits, sulfur-smelling stagnant mud, and the body odor of men un-washed for weeks or months. Ubiquitous rats gorged on human remains and grew fat as cats. Lice, trench fever, pneumonia, and fungal infections flourished unchecked.

World War I: after Lieutenant Beaulieu is discovered to be Marie (top left).

Twenty years later, in the United States, Marie was asked to talk about the time she spent in the trenches. *The Washington Post* noted: "Her experiences in this capacity are some she has never related. 'I will tell some of them when I talk later at the

Franklin Institute in Philadelphia,' she said, 'But not now.'"[58] In fact, she didn't tell the story in Philadelphia or anywhere else. Marie had experienced the unspeakable and she never described those days to anyone.

Once the 42nd held a front-line trench for forty hours straight, a grueling, soul-searing experience. A crack shot, Marie fired and was fired upon.[59] Between exchanges of gunfire, there was little time for anything but the demands of warfare. However, Marie made a few friends among soldiers she could trust with her secret. She was thirty-six at the time, enough older than most of the young men in the infantry to impress them with her audacity. They admired her and were delighted to be taken into her confidence.

A journalist identified only as R.P. wrote: "Marie Marvingt fought the war. Yes, I see people smiling and I can hear them saying, 'Oh, sure, we know that—she was a nurse in a some snug, comfortable little hospital.' It's true that Marie Marvingt in her white uniform rendered important services to the war effort, but the trenches and the listening posts recognized her under her helmet and soldier get-up, and twice the aviator did the Germans the honor of bombing them.

"A story for you. One day, while in the army, she was told to deliver a message to a colonel in a neighboring regiment, who was actually her cousin. After crossing through an attack zone that would make the bravest think twice, undistinguished in her uniform, she presented herself to the colonel and handed him the note. Without recognizing her, the colonel offered her a cigar and a little money 'to drink to my health.'

"Marie said coldly, 'We don't behave like that in the family.' Stupefied, the colonel assumed he was faced with a drunk soldier until Marie pulled off her helmet. The colonel was mortified to find one of his relatives in such ambiguous circumstances."[60]

René Métrot, who wrote as "Micromegas," said he heard the story with an additional bit: Just before Marie removed her helmet, with her cousin still baffled by the strange soldier, Marie clapped him on the shoulder and shouted, "Arrest this man!" Métrot added that after explanations and laughter, the colonel and Marie embraced, sharing a tender moment in the midst of the ugliness.[61]

"Beaulieu" lasted forty-seven days before incurring an injury that needed treatment.[62] Exiting from the first-aid tent, Marie carried her helmet in her hand, revealing her long thick hair. Nearby 42nd Battalion soldiers gaped at the sight. Marie told friends later, "You should have seen their love-starved faces when they realized they'd been sharing their quarters with a woman!"[63]

Marie was sent home, where newspapers reported that a woman had reached the front lines and fired on the enemy. Marie herself was very clear: "Yes, I fought in the front lines. Yes, I fired on the enemy."[64] Although she was later awarded another palm on her *Croix de Guerre* for her bravery in action, she said that what she enjoyed most about the masquerade was running into members of the old 42nd Battalion years afterwards and reminiscing about their experiences.[65]

Georges Clemenceau ("War is too important to leave to the generals") had become the French prime minister and General Foch the new commander-in-chief of the army. As Marie's friends and admirers, as well as men who understood her unique contributions to the war effort, Clemenceau and Foch are thought to have intervened to save her from interrogations and reprimands. Possibly to avoid future incidents, one or both arranged for her to join the Italian 3rd Regiment of Alpine Troops in the Italian Dolomites where, although technically classified as a nurse, she was part of their operations to ski out wounded soldiers on stretchers, and to ski in needed supplies the same way.[66] (She was, of course, the only woman in the regiment.) Skis were being used for the first time in military efforts along the Austrian-Italian front running through the Alps. Marie's skills as nurse, soldier, skier, and mountain climber were all crucial as beleaguered troops could be reached only by a combination of hiking, climbing, and skiing. Her knowledge of avalanches was particularly useful.

Marie sent her cousin Marthe a postcard-photo of herself using ropes to descend a wounded soldier from a high mountain pass. She wore the Alpine infantry uniform, with a plumed hat, white collar, high boots, and a cape that came down to her knees. She wrote: "Here's a glimpse of the war in the mountains. You can probably guess that I am in my element!"[67] Despite her cheery message dated December 23, 1916, some 200,000 tons of snow and ice had buried 500 soldiers a week earlier, and on one day alone, December 13, 1916, known as White Friday, some 10,000 soldiers perished in avalanches.[68] Overall, avalanches accounted for more than half the casualties inflicted in the Alpine fighting in that war.

Marie spent six (some reports say ten) months helping with the exhausting, dangerous, but vitally necessary work of provisioning and evacuation.[69] Like everyone else, she endured cold that dropped to forty below zero, snow that was sometimes more than four meters deep. She saw men die of pneumonia, and she herself had to be evacuated to the rear after having spent an hour under an avalanche.[70]

A comment made by General François Mazaud probably meant as much to Marie as her many honors: "I salute you, Mademoiselle, as one of the finest soldiers in the service of France that I have ever known."[71]

Another general once described Marie as "the perfect comrade, a good pal."[72] (And a desired dance partner. She liked to say that Italian officers did a mean tango.[73]) She was admired in military circles for her courage, optimism, and charisma. She kept her military friends all her life, as well as a souvenir from the war years: her military bearing—proud, stoic, dignified.[74]

War wears two faces: the one that kills and maims, and one that binds up wounds. Before, during, and after her activities as pilot and soldier, Marie donned her Red Cross nurse's uniform and applied herself to her love of serving others. She cared for hundreds of wounded soldiers, many of whom wrote her letters of thanks or remembered her at Christmas for years afterward. Because she was so skilled a surgical assistant, much of her work was done in operating rooms, either the hastily rigged ones in field hospitals, or in the repurposed hospitals behind the front.[75]

Marie in the uniform of the 3rd Regiment of Chasseurs Alpins in World War I.

Marie's war duties were interrupted by another kind of nursing. Her father, who had been vigorously healthy into his late eighties, was seriously ill. They had lived together since the death of her younger brother, although Marie was more often away than at home. As he got older, the two of them could be seen riding around Nancy, Félix in a sidecar attached to Marie's bicycle.[76]

Postcards sent from Nancy often mentioned him. Ten months before he died, she wrote Marthe: "My father is most appreciative of your kindness and he asks me to thank you for the superb asparagus. All is well here and we're happy to know that the same is true for you. … I've just returned from the front in Alsace and the Vosges, where my patients have been evacuated. My congratulations to you on your *sangfroid* during the recent bombing."[77]

On March 10, 1916, at the age of eighty-nine, Félix Marvingt died with Marie by his side.[78] For the first time in her life, Marie was entirely alone. There was now no one to whom she could say, "Look, Papa! No hands!" Félix had been demanding, pushing her to do more and go farther, but she was also the very heart of his life, and she knew that.

A one-sentence announcement of his death appeared in the paper.[79] The funeral was small and private. This was wartime, and death was becoming commonplace.

The rest of her life, whenever she was in Nancy, she biked out to visit her father's and brother's graves. On her way to the cemetery, Marie stopped so often to visit with a boy selling textiles on the rue St. Jean that they became friends. In the same way, she and the daughter of the florist from whom she bought floral wreaths for the graves became lifelong friends.[80]

But it was 1916 and Marie was needed elsewhere. She settled her father's affairs, made a last visit to the cemetery, and returned to war.

She spent the next six months on the Italian front as a war correspondent, often getting around on a bicycle.[81] She was the only woman to obtain from the French authorities a safe-conduct pass for the entire Italian front.[82] Marie was working in a hospital in Lyon at the time, so she applied for a safe-conduct pass from the Lyon Prefect, who sent a telegram marked "very urgent" to the Prefect of Nancy:

"Please let me know as soon as possible if you see any problem with my renewing the visa you issued last July 24 (number 133) to Mademoiselle Marie Félicie Élisabeth Marvingt, war correspondent and pilot currently residing in Lyon. The Consul General of Italy in Lyon has told me confidentially that last September he received an order to deny an Italian visa to Mademoiselle Marvingt."[83]

He received an immediate response: "From a security standpoint, nothing unfavorable is known against her. Although her private conduct is open to criticism, there is no valid reason to refuse her a visa for Italy."[84]

Despite the positive, if lukewarm response, or perhaps because of it, the Prefect in Lyon contacted a higher authority, who responded with Marie's birth and residence information, and added: "Mademoiselle Marvingt is very well known in this area for her sports accomplishments. This young woman requested in March 1914 permission to fly over the forbidden French zones. Because of the war, no permission was granted." The report went on to say that "Mademoiselle Marvingt has rather liberal morals, and although no precise facts have been established, her private conduct leaves something to be desired. Despite that, her patriotism has never been in question. The petitioner is absent from Nancy for many months at a time, traveling either in France or abroad, where she gives a number of conferences, notably in Switzerland, where she was in 1916. Given the above, I see no problem in granting her a visa for Italy."[85]

Women who failed to follow traditional rules of female behavior were often assumed to be guilty of other infractions. It seems unfair that in the absence of any "precise facts," French officials could drop such gratuitous hints about Marie's "liberal morals." Fortunately, the French love a good exhibition or a display of courage and daring. Because Marie brought glory to France with her accomplishments and because she was so visible during the Great War, she was never maligned publicly and she escaped some of the bitter criticism leveled at other role-breaking, rule-breaking women.

In December of 1916, she wrote her cousin Marthe that she'd spent an extraordinary ten weeks on the front, having finally gotten her visa.[86] A newspaper reported in August 1917 that Marie had just returned from passing "long months on the Italian front. She brings back from her stay over the winter in the middle of our allies a collection of extremely interesting photographs. Mademoiselle Marvingt, who is in touch with the chiefs of the Italian army and who went everywhere, even to the extreme points of the front, is preparing a series of lectures with photographic projections that will give an idea of the unheard-of difficulties of the war."[87]

She must have returned to the front after that. In November 1917, she wrote from her Parisian hotel, "I had the misfortune of being part of the sad retreat of our allies, and was almost taken prisoner. I left Udine in the middle of the night on my bicycle several hours before the Germans arrived. I took part in a splendid charge of the Italian cavalry, which greatly slowed the enemy's advance."[88]

From 1914 to 1918, Marie sometimes wore a soldier's helmet, sometimes a nurse's cap, and sometimes her fur-lined pilot hat. Her constant traveling during wartime, her multiple roles, the celebrity that gave her a laissez-passer almost anywhere, and her connections with numerous highly placed military leaders gave rise to a confidence among those who knew her that she was involved in information-gathering.

Her dressmaker's niece said everyone always assumed that Marie was in the Secret Service. "She had her 'entrées' at the Prefecture, you know," she said.[89] More important, she had the much-desired franking privilege, a concession by the government to people who had rendered the state some great service, allowing them to mail letters and packages without paying postage.

"Not everyone gets the franking privilege," said the niece. "And besides, Marie used to disappear from Nancy for eight, nine, ten months at a time."[90]

Where did she go?

"Yes, you see? That's what I mean," said the woman suggestively. "Where did she go?"[91]

The inference was clear, at least to her.

"It wouldn't surprise me at all if she had been in Intelligence during the war," said a close friend. "I always got the impression from Marie that she was."[92]

In 1918, American reinforcements arrived in Europe. Clemenceau and Foch took up the offensive, the Germans began retreating, and on November 11, 1918, they asked for an armistice. At eleven on that Monday morning came the order everyone had been praying for: "Cease fire!"

Although the Russians bore the heaviest burden in terms of casualties, the French were not only close behind in numbers of killed and wounded, but much of the war was fought in French cities, towns, and countryside. The north of France was devastated, the state coffers were empty, and nearly one of every six French citizens was a war casualty. (United States casualties were one of every 263 Americans.)

With such a high percentage of war dead or wounded in France, it seems odd that Marie—who bombed a German air base twice, nursed hundreds of wounded soldiers, many in field hospitals, filed reports as a war correspondent on the Italian front, provided important assistance to the Alpine troops, and fought in the front lines—escaped the slightest injury. The French could only say, "Ah, but she is the Fiancée of Danger."

A few years later, during French pacification" efforts in North Africa, Marie accompanied French, Italian, and Spanish troops as surgical assistant, nurse, war correspondent, and pilot.[93]

Marie's visas show that she spent time in 1921 and 1922 in German cities such as Bonn, Cologne, Mayence, and Coblentz.[94] In December 1926, she requested a safe-conduct pass to visit occupied countries; it allowed her to make "several voyages, on foot, on horseback, in car, and by train."[95] As an internationally known pilot, she would certainly have visited airfields, aviation centers, and military groups. In the 1920s, she warned her audiences of the dangers posed by a Germany that appeared to be preparing its aviation for a war of revenge.[96] She described the aerial fleet of the Reich as "powerful."[97]

"France doesn't want war," she said then. "We must do the impossible to avoid it."[98]

In Paris in 1931, she responded to an admirer's request for a few words by writing: "World peace? I desire it with all my heart. But I do not believe in it. For that, one would have to change the German brain. One need not resort to prescience in order to predict the future. The German attitude should have opened the eyes of France much sooner."[99]

When the Second World War began, Marie put on her Red Cross nurse's uniform again, this time with the rank of colonel.[100] By now, she was known as the inventor of the ambulance-airplane, the moving spirit behind air rescue, and the founder of a program to train flying nurses. She continued this work—giving conferences to raise funds and promote air rescue, helping organize new groups and run others, training personnel—while devoting much of her energy to surgical nursing wherever she was needed. In 1939, she opened a home for wounded aviators in the Dordogne and spent several of the war years running it.[101]

Like the whispers during World War I, rumors of Marie's work in the Resistance during World War II have always been part of her story but only recently, according to Marcel Cordier, president of the International Marie Marvingt Committee, have documents surfaced corroborating it.[102] Several people have maintained that her conferences gave her the perfect cover for moving around France and other European countries, and that her military connections were unusually solid, considering her status as a civilian and as a woman.[103]

By the time World War II began, Marie had been lecturing around the world for over twenty years. Pictures of her and excerpts from her speeches appeared in the papers, and she was a familiar figure to millions. Even so, before giving conferences in war-torn Europe, she had to sign an agreement to speak only of neutral topics.[104] She kept to her agreement, but she was not forbidden to socialize or see people outside her conference hours.

The first public mention of her Resistance work appeared in 2016 in the *Sud Ouest*: "Marie Marvingt, Resistance Fighter, Is Honored."[105] The Dordogne village of Sainte-Alvère, where Marie founded and managed her rest home for wounded pilots, recognized her connection with the Maquis of Durestal by placing a plaque ("Colonel Marie Marvingt lived here") on the home she inhabited for several years. The Maquis was the name for the core of the French Resistance and Durestal was the most important Resistance camp in the Dordogne area. Well-hidden in a thick oak forest, it was invisible to planes overhead and impossible to approach on the ground without being seen. From the spring of 1943 until the end of the war, sometimes as many as 600 Resistance fighters lived there. Despite the newspaper mention, there is little information about Marie's connection with the Maquis. As the mayor's office pointed out, those involved "have always been very discreet about it."

Coincidentally, a member of the Resistance, whose mother was Marie's dentist in Nancy, took refuge near her soldiers' home. They formed an immediate friendship, based on their shared hatred for the occupier. He remembers the evening conferences and daytime school talks Marie gave all over the region, and he conceived a lasting admiration for her, always referring to her as "an ardent patriot" or "a great patriot," his way of letting it be understood that she was as involved in the Resistance as he was. As a former parachutist, he and Marie also shared their love of the skies. He attended her funeral.[106]

And so, the evidence was there, subtle but unprovable, until the spring of 2018, when Colonel Pierre-Marie Paoletti discovered the medal the French government awarded her for her work in the Resistance. Her medal was given *avec rosette* (with a little rose), indicating that she was among the 10 percent of the Resistance heroes who had contributed the most and faced the greatest dangers on behalf of their country. A heretofore unknown medal has thus been added to Marie's already impressive collection. And she can add one more title to her legend: Resistance Hero.

6

The Wings of Mercy

> "Men have made of the airplane a machine of sport, commerce, and war. Mademoiselle Marvingt has undertaken the most magnificent of crusades—that of making it serve the cause of humanity; medical aviation is called to render the most precious services."—Marshal Philippe Pétain (1931)[1]
>
> "How much the Army owes to medical aviation, and especially to its godmother, Marie Marvingt!"—Marshal Franchet d'Espèrey (1929)[2]

The invention of the ambulance-airplane and the establishment of air rescue throughout the world grew organically from Marie's life: she was a highly qualified and experienced pilot, and she was a highly qualified and experienced nurse.[3] In combining these two roles she became the originator[4] and the "fairy godmother"[5] of air rescue.

When Marie said, "I have always been drawn to medicine,"[6] it is reasonable to suppose that her interest in the healing arts emerged from her childhood: the loss of three infant brothers, her mother's ill health and affecting fragility, and the death of her younger brother, Eugène, who had been sickly most of his life.

Since adolescence Marie had admired Florence Nightingale, who had elevated nurses to the status of competent, professional, and disciplined—in short, respectable—members of the medical profession. The public had previously perceived nurses as camp followers, prostituted women, or at least women unconcerned with their virtue. In Florence Nightingale's wake, single women of probity embraced the skills of nursing and did so with dignity and devotion.

Intending to become a physician, Marie began her medical studies at the Faculty of Medicine of the University of Nancy. (Photos from that era show the faculty of medicine as entirely male.) She sought the best professors to study under, attended their clinics, and assisted when they were operating.[7]

A sponge for knowledge, Marie acquired a solid foundation in medicine and surgery, which served her well even after her medical career was interrupted by her sports activities and, more definitively, by World War I.

When the war broke out, Marie was already a licensed nurse, so she signed up with the Red Cross, which rendered service in ways that suited Marie's interests, talents, and goals. She enjoyed the camaraderie, the pulling together for a common purpose, and the lifelong Red Cross friends she made. Red Cross workers often helped people anonymously and rarely received thanks for their demanding and, in wartime, often dangerous work.[8]

An old photograph shows Marie and another Red Cross nurse, each pedaling a bicycle, with a stretcher carrying a patient strung between them. This contraption filled a need for convenient transport when ambulances were in short supply. Marie's biking skills were put to good use.[9]

Marie (right) and a Red Cross friend transport a patient.

In 1914, as numerous auxiliary hospitals opened, Marie began working under several of her old medical faculty friends as a surgical assistant. She returned to this work off and on during the war and was proud of a personal best: she and René Gauthier de Luxeuil, with whom she worked for two years during the later war years, performed twenty-seven operations over two nights and one day at the Hôpital du Panthéon in the Latin Quarter, without leaving the operating room for more than a few minutes at a time.[10] It was during that period that Marie invented a new surgical suture to minimize scarring.[11]

The government mobilized more than 20,000 Red Cross-trained nurses in 1914, but they were quickly overwhelmed with casualties and had to seek volunteers; more than 60,000 women eventually served as nurses.[12] Marie did much of her wartime nursing in field hospitals, often hastily constructed sheds or tents, overflowing with the wounded where four operations and ten amputations might be done every hour in the same crowded space. The work was grueling, gruesome, and distressing. At the war's outset, 80 percent of the wounded soldiers had head injuries because protection didn't arrive until 1915 when the army produced the Bourguignotte helmet. Medical personnel worked amid the stench of gangrenous wounds, the sounds of men choking to death on the fluid in their gassed lungs, and the sight of bodies mangled beyond repair.

Along with her companions, Marie worked with aching feet, chapped hands, back pain, sleep deprivation, exhausted muscles, an unremitting fatigue, and the sounds of death in the air and on the ground. Shells whined. Guns pounded. Airplanes droned. Motor convoys rumbled. Feet tramped. Provisions wagons creaked.

In World War I, 10 percent of all nurses in field hospitals were killed on the front. In all, the Red Cross reported 600 nurses killed in service and 2,500 wounded.[13]

Even as Marie worked to treat injuries and restore health, she knew she was an unwilling accessory to a military system that would send her patients back to the front as soon as she had completed her work. This may have been the most difficult part of war nursing.

Despite regulations against female chauffeurs, Marie occasionally drove an ambulance. As a driver, she retrieved the injured from first-aid stations in the trenches—where wounds were patched and limbs amputated—and transported them back to hospitals.

Again, when World War II was declared, Marie put on her Red Cross nursing uniform and spent 1939 and 1940 working as a surgical assistant.[14] During this period, she also went back and forth to oversee her refuge for aviators in Sainte Alvère.[15] She often said, "Pilots are my brothers," and more than one flyer who arrived there was astonished to be greeted at the door by the legendary Marie Marvingt herself.[16]

Until the end of her life, Marie retained her connections with the Red Cross, attending meetings, gatherings, and retreats, and volunteering when needed. At the age of eighty-three, while being interviewed about a lecture she was giving that evening, she suddenly glanced at the no-nonsense watch on her wrist and said, "I must run. There's a patient waiting for me at the hospital. We're going to perform a rather delicate operation."[17]

Marie's life was studded with successes in so many fields that it's difficult to isolate a theme, one single activity that might have been more significant for her than all the others. But when asked, she always pointed to her career as a nurse: "Ah! That means a lot to me. My greatest joy has been relieving pain, binding up wounds, caring for the injured."[18]

Indeed, Marie identified so closely with her role as a nurse that wherever she went, she was always, at some level, Marie the nurse. The journalist J.-J. Leblond wrote, "She's the first everywhere! Indefatigable and never satisfied, Marie Marvingt travels all the main roads. She hunts the seal in the polar waters. Yes, but while she's there, she nurses the Eskimo peoples. By ten hours she beats Madame Citroën as they race their cars across the Sahara, but at the end she teaches and nurses nomad tribes. The sponge divers of the Red Sea enjoyed her company, but it was so that she could nurse them."[19] (Someone reportedly has Marie's old diving watch.[20])

Marie never entirely left the world of medicine, keeping abreast of current innovations, enlarging her knowledge, attending conferences. When she was eighty, she wrote an article highlighting several dramatic changes in medicine: "Trepanation From 1914 to 1918 and Neurosurgery From 1950 to 1955."[21] At the same time she was developing another paper—a little less rigorously scientific—on "the temperament and character of actors according to their palms, their handwriting, and the best role of their life."[22]

When interviewed in 1958, the subject of new surgical techniques arose. Marie said that as early as 1939 she had assisted at heart operations, and her patients were still alive and well. "I remember one cardiac intervention that almost ended disastrously. The patient was saved by massaging the heart, which we hear a lot about today. But that method isn't new. We were doing it twenty years ago. In that particular instance, the surgeon and I took turns for sixteen hours before we persuaded that obstinate heart to regain its normal functions."[23]

A registered nurse, Marie served decades with the Red Cross.

In the early years of studying medicine and working as a nurse, Marie was also flying, setting records, training young pilots, and becoming more and more skilled in the air. In Marie's view, every activity, every new knowledge must have some practical end. Her sense of high purpose and idealism made her a true descendant of the humanitarians, who believed that art must not only please the senses, but be socially useful as well.

"At first I was into sports," she explained. "But all that was merely a training period which went nowhere. What really interested me, what I truly wanted, was to be *useful*, to serve. About then I became involved in aviation—first balloons, then heavier-than-air machines. In the meantime, I had always been involved in medicine and had assisted at dozens of operations. I knew the importance of rapid intervention. It came to me then: why not transport the wounded by airplane?"[24]

The answer appears so obvious today that it is difficult to understand the determined objections Marie faced from military, government, and civil authorities who seemed unable to grasp the benefits of flying medical aid to the injured, and flying the injured to medical centers.[25]

Marie made her first efforts in this direction in 1910, less than seven years after Kitty Hawk.[26] Aviation had scarcely any credibility with most people, and those who were interested in it regarded it as a sport. It was inconceivable to connect something so risky, so clearly still in its infancy, and a sport besides, with a serious purpose. She might as well have tried to convince people that tennis or golf might somehow save lives. But Marie saw that heavier-than-air machines were developing at a fantastic pace. What might not be possible now would surely be possible in a few years.

"You cannot imagine the opposition I encountered," Marie said. "However, I refused to let skepticism, setbacks, or criticism change my vision."[27]

She developed a workable design for what she called her ambulance-airplane, and by 1911, she was approaching air force and military personnel with her idea, urging them to create a medical aviation branch equipped with rescue planes carrying nurses, stretchers, and surgical supplies.[28] She spelled out air rescue's two clearly defined roles: flying the injured to medical centers, and flying doctors, surgeons, assistants, and medical supplies to those unable to be transported or in inaccessible areas. She stressed that air rescue was important not only during wartime, but for such peacetime catastrophes as earthquakes, volcanic eruptions, cyclones, floods, tidal waves, epidemics, and maritime emergencies. But the air-ambulance would be most useful, she maintained, in serving distant, isolated communities without medical facilities. (The rescue plane she first conceived in 1910 bears a strong resemblance to the helicopters used in Vietnam to carry the wounded on stretchers in the pontoons under their fuselages.)

In June 1912, she was still meeting with officials, still fruitlessly. She asked General Hirschauer if she could share his booth at the first Aviation Salon that year. Visitors to the booth seemed intrigued by a drawing of her ambulance-airplane, and they enjoyed chatting with her, but in the end, they moved on to the next exhibit.[29]

After encountering shut doors and blank faces everywhere she went, Marie remembered the one door that was always open to her: the nearest newspaper office. She had always been good copy, and she didn't disappoint then.

With her typical precision, she explained to a reporter, "My plane would be a Deperdussin monoplane with three seats run by a Gnome 100 HP motor, and furnished with a telegraph. It wouldn't be used at first so much to transport the wounded as to find them, notify physicians of their whereabouts, and bring medical supplies to the first-aid stations. I've been looking at a sort of flying stretcher, which could be adapted to my plane. I'm planning to name my first air-rescue plane after my late friend, Captain Écheman. In order to get the money to build it, I plan to lecture all over France. Bit by bit I'll gather what I need to realize my greatest dream as a Frenchwoman—to give the French military its first ambulance-airplane."[30]

Marie's concept was rendered visible and memorable in a much-reproduced 1914 drawing by Émile Friant, painter, sculptor, balloonist, founder and president of the Eastern Air Club, and a good friend of Marie's.[31] In the drawing, Marie crouches at the head of an injured soldier lying on the ground. An open wicker basket on her right holds medical supplies. Marie applies chloroform to a wad of cotton while the military physician, sleeves rolled up, attends to the wounded man. In the background, lightly sketched, is Marie's dream, an air-ambulance named the *Captain Écheman*. The only color in the otherwise black and white drawing, red, is used for the Red Cross symbol on top of the medal Marie is wearing, on her armband, the doctor's armband, and the two underwings and back rudder of the plane.[32]

Later, posters made from the drawing were sold for 10 francs each to raise money for the ambulance-airplane (printers in Nancy donated their work); postcards were also made and sold for the same purpose.[33] Friant's drawing inspired the sculptor Jules Déchin (the Goff Memorial in Washington, D.C., is his work) to re-create the scene in a bas-relief bronze work cast in a Swiss foundry.[34]

In 1931, Marie established the Captain Écheman Challenge, a prize to be given annually to the aircraft company producing the best example of any plane that could be quickly converted to an air-ambulance.[35] By that time, Marie was no longer asking for dedicated ambulance-airplanes, but had been encouraging military and civilian groups to use planes

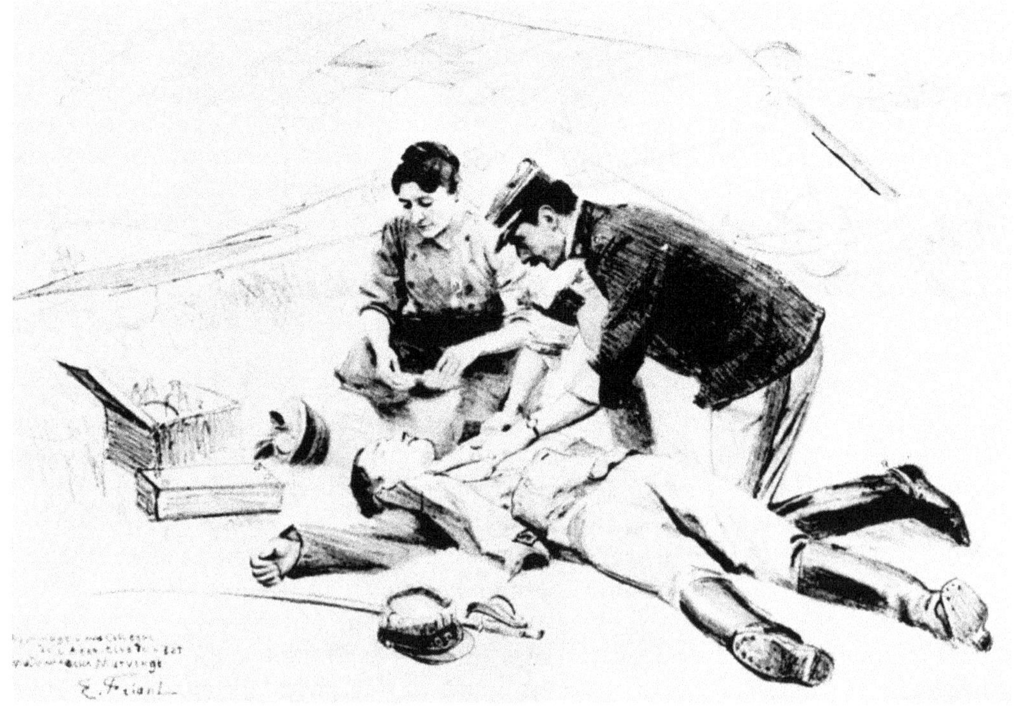

Emile Friant's much-reproduced drawing of the first ambulance-airplane.

they already owned and to adapt them to air-rescue use by, for example, turning seats into stretchers. The purpose of her Challenge was to stimulate design of the convertible plane. The prize was a copy of Déchin's sculpture, and the jury consisted of board members of the Friends of Air Rescue. To keep the Challenge, the company needed to win it three years in a row.

In highly publicized and well-attended events, Marie awarded the Challenge to such aircraft as the *Potez 42* and the *Bréguet 28-T* (in 1931), the *Potez 29* and the *Ford Trimotor* (1933).[36] In 1932, Marie donated Déchin's original model to the air museum at Chalais-Meudon, which was later moved to what is now the Musée de l'Air et de l'Espace.[37]

By the time World War I broke out, Marie had persuaded the French army to use a few planes for medical emergencies. Practice drills were held to test the planes and personnel.[38]

Journalist Jacques Noetinger related stories his surgeon uncle had told him about flying air rescue missions with Marie, landing on the front lines to care for injured soldiers, occasionally operating *in situ* on a nontransportable patient.[39] Ambulance-airplanes so speeded up the process of getting the wounded from the firing line to the hospitals that they could have inspired the apocryphal World War I soldier who said, "You hear a hell of a noise, and then the nurse says, 'Sit up and drink this broth.'"[40]

Along with explaining her idea to the newspapers around the world, Marie embarked on a series of fundraising conferences,[41] telling audiences across France, "With the aeroplane in use by the medical department of an army, more wounded men on the battlefield can be cared for. Think how much time could be saved in getting aid to a wounded soldier. In the aeroplane a physician could be transported quickly from place to place, supplies could be carried high in the air, scores of hidden wounded could be discovered, and the hospital efficiency of any army corps could be greatly increased by the use of the aeroplane."[42]

She needed 36,000 francs to build her plane, which meant giving hundreds and hundreds of conferences.[43] She would eventually give thousands of conferences in her lifetime, including some radio talks.[44] When she had finally collected enough money, she took her design (at that point, an amphibian monoplane with two fixed seats, convertible seats, air mattress, mica windows, removable stretcher under the fuselage) to the well-known aviation engineer Louis Béchereau (he was the genius behind the Spads, considered the best French war planes in World War I) at the Deperdussin factory. After Béchereau added his own engineering brilliance to the project, Marie finally, in 1912, placed an order for the first air-rescue plane.[45]

She already knew that converting her idea to reality was not going to be easy. Still, the next obstacle was unanticipated. Before her plane could be completed, owner-manager Armand Deperdussin embezzled the operating funds and the company went bankrupt in mid–August 1913.[46] Apparently unfazed, Marie set about organizing a new round of conferences to finance a second attempt.[47]

In France and all over Europe, in Africa, Asia, even in the United States, a familiar sight emerged: Marie Marvingt standing behind a lectern with a pitcher of water and a glass at her elbow. Traveling thousands of kilometers by air, train, car, truck, and even camel, to say nothing of bike and her own two feet, she spoke in person as well as on radio and television to a range of groups: military, government officials and bureaucrats, medical personnel, high school students, school children, and aviation fans. Everywhere she went, the halls were full.[48] One morning in Le Mans, she spoke to a gathering of

1,000 school children.[49] Oftentimes, she had to add an extra talk later that week for those who couldn't get in.[50]

Over a period of forty years, she gave more than 6,000 conferences[51]—about one talk every two-and-a-half days. But she made use of her considerable gifts for efficiency. She sent a postcard from Sarreguemines in 1931: "I just beat my record. Four school talks in one day!"[52] In Algeria, the newspapers reported that she gave "twenty-five conferences in nineteen days, which is, one must admit, a remarkable feat."[53] At the age of seventy-eight, she wrote a friend, "This is one of the best tours I've had. I've given fifty-one lectures in twenty-seven days … and that's only in Algeria! They are so delighted they've asked me to do the same tour again."[54] Marie was proud of raising all the money for her plane herself, without being sponsored or funded by others.[55]

What did she talk about 6,000 times?

In aviation's early years, people needed to be persuaded that flying was possible, safe, and desirable, that it was much more than a sport for the wealthy and foolish. Readers subjected to article after article about plane crashes tended to think air disasters were far more common than they actually were. Fear and ignorance made for an airplane-averse public.

People also wanted to understand how flying worked, what good it might do them and their families personally, and why, as a nation, France should support the military, tourist and commercial possibilities of aviation. It took a great deal more persuasion than anyone, including Marie, thought it should. But in each talk she gave, it was as if she were describing the miracle of flight for the first time. Rochefoucauld claimed that "enthusiasm is the most convincing orator," and Marie was a most enthusiastic fan of aviation.

She also persuaded many of her listeners to enroll in various air-related groups, such as the French Aeronautics League, thus building a popular base for aviation, which needed both moral and financial support.[56] Then, when her audiences had accepted that aviation might be all right, Marie turned to the heart of the matter: air-rescue operations. She was profoundly convinced that flying the injured to hospitals and flying medical help to the injured, particularly those in inaccessible parts of the world, was critically important. This was her real mission.

Once she helped her audiences appreciate the benefits of aeromedical evacuation in both war and peace, she asked for contributions for the *Captain Écheman* ambulance-airplane that she intended to donate to the French military services.[57]

Her talks—although their purpose was fundraising—were immensely popular for several reasons.[58]

Full of fire and verve, Marie was phenomenally entertaining. She knew how to tell a story, how to make a point succinctly, how to evoke both sympathy and laughter. She'd had diction lessons, she'd spent time on the stage, and she possessed an intriguing collection of unusual slides and videos. According to one writer, Marie "never used oratorical gimmicks, but instead, with a heartfelt passion, evoked in her audiences almost the same feelings that she had herself experienced during her adventures, feelings that still vibrated within her."[59] Success breeds success: she was an often-requested speaker on the lecture circuit, which produced even more requests for her appearances.[60]

Marie was a celebrity, known to her audiences before she walked onto the stage. She had thrilling adventures of her own to relate, and she was on a first-name basis with other newsmakers of the day. Celebrity sells.

Few women spoke before audiences, especially not as the sole speaker for an entire afternoon or evening. People came to hear her out of curiosity, or to show their daughters what might be possible for them. A few came prepared to dislike her and all she stood for. She was a phenomenon.

She left her audiences much more knowledgeable than they'd been. She introduced dates, facts, figures, history. She told people things they hadn't known, and in ways they wouldn't forget.

Although her talks were promotions for aviation and air rescue and were fundraisers for her *Captain Écheman* award, they were grounded in patriotism, which meant a great deal at that time.[61] Marie had worked vigorously for French interests, had fought in two wars, nursed and sheltered soldiers, bombed enemy air bases, and invented the concept of the life-saving ambulance-airplane. The patriotic French applauded her passionately.[62]

In the United States, *The Washington Post* described her "rapid-fire flow of words" and commented, "Probably few women in the world have a wider range of interests than this fast-talking, enthusiastic Frenchwoman."[63] Her magnetism as a speaker worked for a variety of audiences; one newspaper marveled at her ability to captivate an auditorium full of girls and boys, all under eight, who maintained a "religious silence" throughout Marie's talk.[64]

One measure of her success was that wherever she spoke, she was, from that day on, welcome to come back. Word of delighted audiences spread throughout France, so that when she was due to speak, she was given advance raves from the local paper and readers were warned that seats usually sold out quickly. Many articles included photographs, interviews with Marie, or excerpts from newspapers in other cities with post-speech compliments.

Little wonder she drew standing-room-only crowds. She cleverly (but in all sincerity) first sold herself and her mission to the newspapers so that they could, in turn, sell her to the public.

Typical of news reports of her many conferences is one in 1932 that began with a description of the glitterati in the box seats: politicians, foreign consuls, military officers, physicians, representatives of the Geographical Society. "And then there were her friends—'they were legion'—among the overflowing hall, all there to hear the traveler, pioneer, and 'the intrepid one,' attired in evening gown, who would elicit many 'oooh!'s of admiration, of approval, and even of stupefaction!" Columns and columns of newsprint followed, encapsulating the highlights, the anecdotes, the laughs, the information conveyed, the spirit of "la Demoiselle Sans Peur" ("the Young Woman Without Fear").[65]

"Mademoiselle Marvingt has mastered the art of seduction. You would have to hear her to appreciate her energy, her willpower, and, with everything else, her soul that remains supremely feminine. She is, without a doubt, the most audacious heroine of our era."[66]

She handwrote invitations to her talks to be reproduced locally.[67] She sent invitations to individuals and to the newspapers, or posted them on kiosks. She tailored each talk to the area, to current news, and to her particular goals at the moment. She generally invited people to come hear a patriotic talk on aviation and ballooning called "Two Hours in the Air."[68]

"Thanks to cinematography," she wrote on the invitations, "you will taste the joys of flying over the countryside, then I will have you fly over the Mediterranean. I'll tell you the story of my crossing of the North Sea, and many other stories. Come, be entertained, and give generously to the Wings of Mercy." She signed it with her name followed by a handful of her titles and honors.[69] At one point she was charging "10, 8, 5, and 3 francs" for seats, but she always offered discounts for students and the military.[70]

Marie often said that the Wings of Mercy were but a small consolation for what she called the criminal wings of war. In 1933 in *Les Ailes*, Marie wrote, "For some time now, 'flying to the aid of the wounded' is no longer a metaphor."[71] She pointed out that during the last battle in Tafilalet, more than 5,000 wounded owed their lives to ambulance airplanes. "I've always wanted ambulance airplanes to fly alongside combat planes."[72]

She encouraged and enjoyed questions, comments, and objections from her audiences. One particular question that came up in the early days was about the risks to the patient while landing. Marie explained that statistics put patient deaths on landing at 8 percent, but since the patients needed to be operated within 48 hours to survive, the chance had to be taken. Audiences nodded.[73]

Her school talks were always overbooked as students loved her colorful stories and her animated way of telling them. Whenever she asked who would like to come flying with her, nearly all hands waved wildly. One of her most popular moves was to leave behind a few coupons good for a free ride in an airplane.[74]

She always finished by signing up new members of the Aeronautical League, asking graciously if the teachers would collect the enrollment forms and the 3 francs per child for her to pick up later. She helped establish aviation clubs in all the schools she visited, naming each one after an outstanding figure of aviation such as Captain Écheman or Marshal Lyautey. One named itself the Marie Marvingt Club .[75]

Students wrote her letters and stopped her in the street. Once, in Tunisia, a group of young fans commissioned a bust of Marie by Rachel Hautot, a renowned French sculptor who lived most of her life in Tunisia.[76] The whereabouts of this bust are unknown, since it disappeared with the rest of Marie's souvenirs and medals after her death.

In a useful gesture in 1952, the grateful Ministry of Aviation recognized the uncommon good Marie had done with her many conferences, and placed at her disposal the ambulance-airplane *The Cardinal Luçon*, along with two pilots, so she could give even more conferences with fewer logistical arrangements.[77]

A newspaper article commented on her first use of the plane to fly between Bordeaux and Marseille: "It is well within Mademoiselle Marvingt's character that she would wish to pilot the plane herself. The first day she took the controls was also the 52nd anniversary of her first balloon ascension, and the 43rd anniversary of her first flight on the Antoinette."[78]

Marie's conferences energized her; French media repeatedly referred to her as "indefatigable."[79] Unlike introverts who head back to the hotel after speaking and pull the covers over their head, Marie spent few evenings of her lecture tours alone in a hotel room. According to sightings by those who knew her, newspaper reports, and postcards sent to friends and relatives, Marie enjoyed her evenings at one reception or another, at dinners in her honor, or at a theater performance or movie.

In a 2003 article in *Aviation, Space, and Environmental Medicine*, physician and aeromedical aviation specialist David M. Lam remarked, "The development of air ambulances was one of the most important advances in military medicine in the twentieth century. It is often forgotten today how difficult a task it was to achieve military, governmental, popular, and medical support in the early years of the century for this then-heretical concept. While many individuals were involved in this development, one of the most influential and effective proponents was Mademoiselle Marie Marvingt, of France. In the area of air evacuation, Marie Marvingt was a true visionary, ordering the construction of an air ambulance in 1912, and devoting the remainder of her long life to gaining its full acceptance in the medical armamentarium."[80]

In 1914, the *Indianapolis Star* ran a lengthy article, "The Woman With War Wings," in which they singled out Marie and her plans for aeromedical operations, calling her "the nerviest of all flying women."[81]

While promoting the cause of air rescue and endlessly fundraising for ambulance airplanes, Marie was also planning ahead: trained personnel would be needed to staff the planes. She envisioned, and established in 1934, three types of instruction: courses for nurses who already had their diplomas to learn about the particular needs of air-rescue nursing, courses for physicians who needed to learn about aviation in order to make proper use of ambulance airplanes, and basic medical courses for pilots who needed to have some understanding of emergency medical procedures. Marie insisted there be no age limit for applicants; the only requirement besides professional qualifications was passing a health exam.[82]

These courses, given by military or civil professors on various airfields, were also typed and sent to regional physicians who were unable to attend the classes. To complete the program, each applicant had to spend a minimum of 12 hours in the air. Marie was emphatic about this. Medical personnel needed air experience so they would be calm in the face of, for example, air turbulence. During their training, Marie stipulated that, if necessary, pilots were to shake or rock the plane to simulate the experience.[83]

She was delighted when "after thousands of conferences, of radio programs around the world, articles, and taking all sorts of steps, after two international conferences on air rescue, my nurse-pilot project is finally going to be realized: Doctor-General Cadiot, Director of Health Services, has given me the go-ahead for these classes. Before long, I hope, together with the relevant ministries, the National Aeronautics Federation, Red Cross societies, air clubs and their auxiliaries, the Friends of Air Rescue, these certified flight nurses will be able to fly in military ambulance airplanes in times of peace or war, in France or in the colonies."[84]

As part of establishing air rescue in Morocco, Marie had already organized courses for nurses, physicians, and pilots so that each understood the intersecting requirements of aviation and medicine. Much earlier, during World War I, she had originated the idea of flight nurses, giving classes in aviation medicine. Single-handedly she created "an admirable legion of air nurses" from a group of young nurses and aviators.[85]

Twenty years later, in the early 1930s, she helped set up classes to train nurses and pilots for aerial medical work.[86] Under the auspices of the French Air Club, a flight nurse corps was begun in 1932 under the leadership of the Marquise de Noailles and Madame de Vendeuvre. Marie assisted in developing the training program, including practical and theoretical flight nursing, as well as flight physiology.[87] A number of influential Frenchwomen, as well as Red Cross groups, and the Aeronautics League supported the training program. Diplomas proclaimed the new specialization: nurse-pilot-air rescue personnel, eventually shortened to IPSA (Infirmière Pilote Secouriste de l'Air, or Paramedic Pilot).[88] Marie received the first diploma.[89]

According to a textbook on aerial medicine, "The first recorded flight nurse and probably the true founder of flight nursing was Marie Marvingt."[90]

By the late 1930s, air-rescue personnel included more than three hundred licensed flight nurses. About forty of them also had their pilots' licenses, another fifteen were licensed as navigators, four were licensed for public transport, and about fifteen had passed their parachuting tests.[91] Air-rescue personnel were not required to parachute to the aid of the wounded, but they were trained in parachuting for their own safety. Marie said firmly, "Air nurses need their parachutes just as naval nurses need their life belts."[92]

A 1939 article in the *Oakland Tribune* by aviation writer George E. Pelletier was subheaded, "France's 'Angel of Mercy,' World's Third Woman to Fly and Veteran of the World War, Is Now Recruiting Her Countrywomen Into a Flying Ambulance Corps to Serve in the Next War." The editor's note said Pelletier interviewed Marie by phone and letter to reveal "a little-publicized phase of European aviation and war preparation."[93]

"Flying ambulances manned by women pilots and staffed by trained nurses will be commonplace in the next war.... Fast planes will swoop down over the latest battleground before the din has died or the smoke cleared away and will head for well-marked ground stations, where there will be gathered those felled in the last attack. Properly emblazoned to distinguish them from combat aircraft and exempted by international concordat from ground attack, the planes will come to a gentle landing on broad terrain prepared for them just to the rear of the battle lines. Crews of nurses and stretcher-bearers will assist the wounded aboard. Then the hospital planes will roar off toward some far-off base hospital, where more effective medical care can be given to the casualties of the fray. Other planes, also bearing the mark of mercy upon their sides and wings, will drop food or medical supplies or even medical personnel by parachute to groups of wounded where landing is impossible.

"The world organization of an international flying hospital corps, similar to the Red Cross, is an objective toward which Mlle. Marie Marvingt, France's flying nurse, has been working for nearly thirty years. 'But'—and she reverted to her native French to express herself fully with a phrase and a sigh—'*ce sera un travail de très longue haleine*.' That is, it's going to take a long while and a lot of work yet to get it to the point of efficiency which she envisions.

"It's a far cry indeed from the first International Air Meet at Reims just three decades ago to battalions of multi-motored aircraft, equipped with cots, stretchers, first-aid supplies and trained nursing personnel. But Mlle. Marvingt's double career as a trained nurse and pioneer airwoman of the world has linked them both."[94]

And from the other side of the globe that same year, a Sydney, Australia, newspaper announced, "'La Belle France' is at war, and Frenchwomen are playing their part with an efficiency which matches that of their highly trained men. Nurses no longer find themselves in the comparative safety of base hospitals. Many of them are attached to the Flying Ambulance Corps.... With typical French foresight for her country's needs, Mademoiselle Marie Marvingt, France's flying nurse and pioneer airwoman, has been working out a scheme to establish a flying hospital for nearly thirty years.

"Through all these years Mlle. Marvingt has harried officials in an attempt to convince them of the need for flying hospitals and for aircraft of a type required for the exigencies of war. Her persistence has won the day, and she won the right to organize 'L'Aviation Sanitaire.' Scores of women pilots and nurses have been recruited by her, and as part of her campaign to make it an international organization she has made several visits to America to confer with Government officials in that country. In her own country she had the co-operation of such authorities as the late Marshals Foch and Joffre and of more recent military leaders. That her scheme has caught the imagination of the young women of her country is evident from the rush of recruits since the war began, and France has more than five hundred nurses with at least ten hours' flying experience."[95]

Meanwhile, in North Africa, there had been 6,548 military aerial evacuations from 1920 to 1934 (with only three deaths). General Albert-Ernest Spick, Director General of Health Services in Morocco, saw that the air-ambulance could be equally valuable for civilian efforts, and he knew how to get the job done quickly.

On January 12, 1934, Spick wrote to Marie: "After you created air rescue, you helped us establish military medical evacuation. Now it's time to set up civilian air ambulance operations."[96]

On January 15, 1934, in Rabat, Marie enlisted the considerable assistance of Dr. Jules Colombani, Director of Public Health, to form an action committee to popularize, develop, and encourage civilian air rescue in Morocco. With him as honorary president, a group of medical, aviation, and political officials met on January 22 to hear Marie describe the project: converting ordinary planes into air ambulances as needed for peacetime disasters as well as for wartime injuries and to medically serve people living in areas with little access to health care. In addition to suitable aircraft, trained personnel were needed: pilots who were also doctors and nurses, and doctors and nurses specially trained in the requirements of air rescue. On February 24, 1934, ASCAM (*Aviation Sanitaire Civile au Maroc*) was officially launched, with Marie as founding president. Six weeks from start to finish: Spick had chosen well.[97]

The list of honorary presidents, officers, board members, and others involved in ASCAM appeared to include everyone over the age of five in Morocco, but actually simply

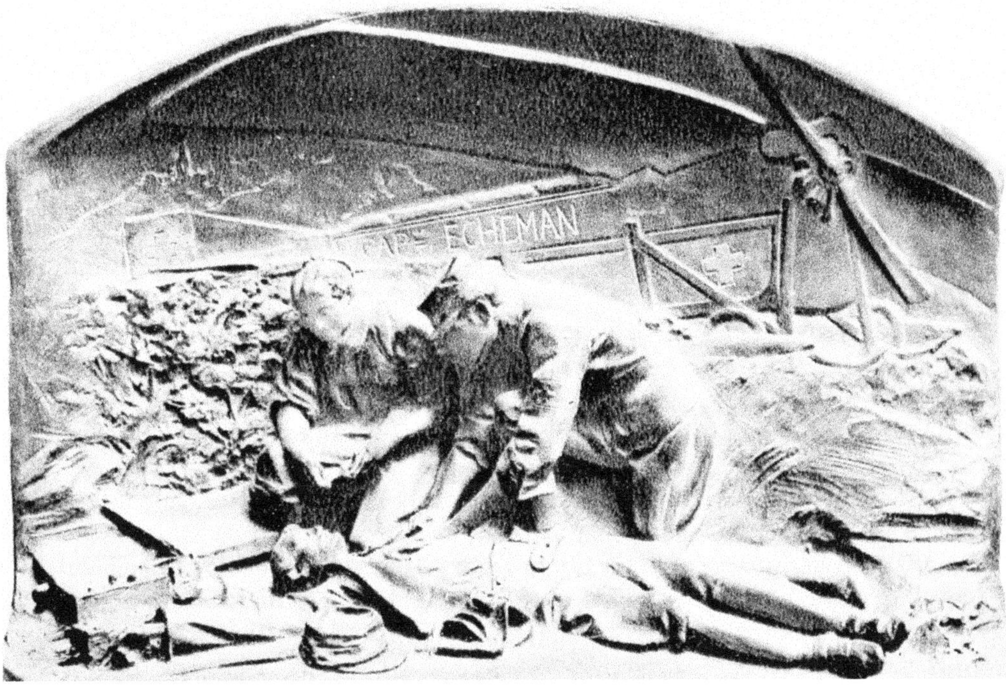

The Echeman Challenge, which was given to airplane manufactures that made the best convertible-to-ambulance planes.

drew upon all possible organization leaders (military, Red Cross, aeronautical federations, flying clubs, health organizations, emergency groups, medical, pharmaceutical, and political entities) to ensure broad support for the project. "Broad support" was a hallmark of Marie's: "It is vital that we coordinate our efforts in order to attain a single goal: to heal or diminish suffering."[98]

Although Marie moved quickly once she had the go-ahead from Spick, preparation for this moment had been long in the works. She'd established air rescue in France, acquiring knowledge and credibility; she spent ten years traveling in North Africa introducing and popularizing the concept of air rescue; she had searched out potential airfields and effective hubs of operation; and she had encouraged Moroccan aircraft manufacturers and engineers to produce planes easily converted to ambulance use (and perhaps win her Captain Écheman Challenge).[99]

The establishment of civilian air rescue operations in Morocco earned Marie the country's Medal of Peace.[100] Her model for air-rescue operations would spread globally over the decades that followed. For nearly fifty years, Marie promoted the concept of air-medical transport and the value of ambulance-airplanes. Known on several continents as the originator of the "Wings of Mercy," she worked tirelessly to encourage each nation to set up its own air-rescue programs.[101]

Unlike many prophets, Marie lived to see the fruits of her efforts. During World War I, although ambulance-airplanes weren't regularly used, they flew some 7,000 wounded soldiers to safety.[102] Marie was particularly touched by the appreciation of survivors; one American mother sent her a package every year to thank her for saving her son's life.[103]

In the 1920s, more than 5,000 wounded were transported by air ambulance in North Africa during the Riff conflicts.[104]

In World War II, aeromedical evacuation was used by nearly every warring nation for the transportation of millions of wounded. General Eisenhower classed it among the three most important medical advances of the war, the other two being blood transfusions and penicillin.[105] World War II air-rescue operations proved to Marie that "the aerial route is the most rapid, the most gentle, and the most hygienic of all and our duty as nurses is the same in the air as it is on land or sea: To try to save lives and to sometimes succeed, but always, always, to reduce human suffering."[106]

In the January-February 1940 issue of the *Journal of the Scientific and Historical Society of Aerial Documentation*, Marie wrote that her "timid little girl of yesterday—Air Rescue—that so many people snubbed has become a strong, robust woman with the eyes of a lynx, and a heart that beats more rapidly than the wind. She has already saved more than 8,000 wounded persons, in peacetime as well as in wartime."[107]

In 1954, ambulance-airplanes once again saved numerous lives following the Orléansville earthquake.

Marie believed audiences reacted most strongly to stories of individual lives that had been saved by her "robust woman with the eyes of a lynx," so she spoke of General Giraud, whom she'd seen in the hospital in Taza. He told Marie he had been wounded at 7:00 in the morning in Djebel Alpha, and, two hours later, needing surgery urgently, he was on the operating table, thanks to an air-ambulance.[108] She liked to list other well-known and heroic officers whose lives had been saved by quick transport in air ambulances.[109] She talked about her cousin, Sister Marie Andréa, who was evacuated from Algeria to Paris for a fractured femur,[110] and of another cousin, flown from Le Havre to

Bern, Switzerland, for surgery.[111] And the pilot Gorlacher, who transported a woman to Tiznit in three-and-a-half hours, instead of the eighteen hours her husband would have needed to get her to the hospital.[112] And she mentioned Dr. Georges Blanc, Director of the Pasteur Institute in Casablanca, who often flew critical vaccines to isolated areas.[113] Audiences applauded and contributed enthusiastically.[114]

The First International Congress on Medical Aviation was held in Paris at the Oceanographic Institute in 1929 under the patronage of the President of the French Republic and the Queen of Belgium, drawing representatives from forty-one nations.[115] Marie always credited Robert Charlet for pulling this Congress together, although most reports named her along with Charlet and others as organizers and facilitators. *Le Monde*, however, gave rather more influence to Marie, saying the Congress was the result of her "tenacity."[116]

That same year, 1929, Marie and Robert Charlet co-founded the Friends of Medical Aviation, headquartered in Paris, to support the development of aeromedical services.[117] Marie served as first president, and then as vice-president throughout its existence. For many years, the organization published an internationally distributed illustrated monthly, *L'Avion Sanitaire*.[118] She and Charlet worked well together until 1936, when he died in an airplane crash. She attended yet another friend's funeral.[119]

The Second International Congress on Medical Aviation was held in Madrid in 1933.[120]

The Third International Congress on Medical Aviation gathered in Brussels in June 1935, with twenty-one countries represented.[121] To attend, Marie flew back from North Africa after having spent nineteen months there; she wouldn't have missed this Congress for anything.[122] Her principled determination to involve as many supporters and interested groups as possible was bringing results. The vice-presidents at that time included a professor from Italy, a doctor-general from Poland, a doctor-general and baron from Sweden, a doctor from Spain, a prince from Romania, a count and doctor-major from Belgium, and a doctor-senator from France.[123]

At the Congress, she expressed two fervent wishes: first, that properly certified air nurses be authorized to fly on military as well as civilian missions in times of peace or war and, second, that the 300 women who already had pilot licenses should be authorized to fly air-ambulances in emergencies.[124] (Marie had suggested the second at the 1929 Congress, but had been turned down because, as she later wrote, "When 'votes opposed' was called, a single voice was raised—that of an eminent French physician who had been in an airplane only once!"[125])

Marie called the Fourth International Congress, held in France in 1935, "unforgettable, possibly because there was an exhibition of seventy airplane-ambulances, both civil and military, and they transported 190 immobilized patients."[126] Subsequent congresses or air-rescue gatherings were held in Dijon, Nice, Oleron, Budapest, Esch sur Alzette, Orly, Reims, and Lille.[127]

In alternate years, the Friends of Air Rescue met as part of other congresses. In 1931, it was the National Aeronautic Colonial Congress,[128] where Marie was introduced to the assembly as "the fairy godmother of aeromedical aviation."[129]

L'Aérophile solicited her opinions after that congress. She was most complimentary about the work being done in air-rescue operations, but admitted she would have liked to have heard more first-person stories from pilots and injured passengers. Apparently many scientific papers were read, one after the other. Marie said, "Perhaps you'll tell me that these papers were so well written that there was no need for discussion. But I'm a big believer in the saying, 'From discussion is born light.'"[130]

Later, she mentioned that pilot visibility was often poor. She said, "The pilot can feel the ground coming up, but can't see it. Why not add a window to the fuselage?"[131]

In 1947, Marie was invited to speak at the Third National Congress on French Aviation, where she told her audience, "In the near future, it is likely that the Red Cross in *all* countries will possess medical planes in the same way that they now own ambulances."[132] At that assembly, she asked for international agreements on medical aerial transport so that air-ambulances could be granted priority in take-off and landing in the world's airports. She wanted medical pilots to be permitted to fly over forbidden zones to reduce evacuation time. She also asked that rescue pilots not have to go through customs formalities, and that an international emblem be designed for all medical crews. The emblem, which would appear on the wings and backs of rescue planes or on commercial or military planes temporarily converted for rescue work, would quickly identify them. She asked for international cooperation so that the injured citizen of any country in the world could obtain fast, adequate medical care.[133]

She looked at certain other countries with longing. "In Sweden, Norway, Finland, Siam, and England, the governments have allied themselves with Red Cross organizations to supply ambulance-airplanes. And in Poland, the Red Cross actually owns air-ambulances, and all a transport costs is the price of a second-class train ticket! In cases of public calamities or for those without funds, the service is free. Alas, France isn't there yet."[134]

She repeated her demand that female pilots be allowed to pilot ambulance-airplanes; she failed to see why qualified pilots of either sex weren't treated equally. In England, she pointed out, female pilots had been flying passenger planes since 1927. She told of letters written to her by women, asking why there wasn't more progress in this respect. She spoke admiringly of women in Turkey and Greece, with whom she stayed in contact, who were organizing their own air-rescue services; one group already had its own ambulance-hydroplane.

She waxed indignant about Kerkeunah, an archipelago of small islands off Tunisia. Its 20,000 inhabitants had no doctor and no pharmacy. Two or three times a week a little speedboat brought supplies to "these disinherited islands. I studied them and their resources and their needs. You could land almost anywhere on the flat, sandy expanses. While I was there, three people died unnecessarily because they lacked simple but immediate medical assistance. The director of the schools said to me, 'What security, especially for our dear children, if we only had access to an ambulance-airplane that could fly people to the hospital in Sfax.'"[135]

Marie finished with, "Alas, there are too many places like Kerkeunah."

She made a heartfelt plea for international cooperation—among nations, among Red Cross organizations, among aviation, health, and government groups. Having seen and grieved the destructive power of planes in two world wars, she felt more strongly than ever that "after sowing terror and death, wings must once again take up their roles as saviors."[136] Countries needed to weld together the chain of humanitarian aerial networks that were being created or that already existed.

Marie was endlessly creative in presenting air rescue to audiences. Through her conferences, she built large networks of interested supporters, while also indulging her love of travel.[137] However, she could reach more people through the media and when she could interest reporters in telling the story of the ambulance-airplane, so much the better. When she couldn't, she wrote numerous articles for as many publications as she could inveigle to run them.

Always one of the first to notice and embrace the newest technology, Marie immediately saw film as a way of broadening her reach. In 1934, she produced, appeared in, and directed part of the documentary film, *The Wings of Mercy* (later, she would do the same with another documentary, *Saved by the Dove*) to "popularize the idea of flying to the aid of the wounded."[138]

The first part of the film shows the Swedish Red Cross using air-ambulances in Lapland. The second half was filmed entirely in Morocco, with the support and resources of the military and with the cooperation of the airfield in Meknès. Against the picturesque landscapes of Fez, Marrakech, Meknès, Djéma-el-Fana, and Ouarzazat, the dramatically portrayed history of air rescue is followed by suspenseful footage of air-ambulance rescues in France and North Africa; the use of planes to transport medical personnel and vaccines to inhabitants of isolated villages; centers set up to prevent trachoma, a serious problem in North Africa; wartime incidences of doctors flying to the injured on battlefields; and the transport of post-surgical patients from emergency field operating rooms back to sophisticated medical centers. In one scene, Marie, playing herself as the first aeromedical pilot, gives an injection of camphor oil to a gravely injured man while flying over the Grand Atlas mountains.[139]

Cinematographer Gaston Chelle was associated with a form of cinema popular at the time, "mountain cinema"; he was the cinematographer of the classic *Crossing of the Grépon*. It would have been astonishing had Marie found anyone else to work with. Voice-overs were done by Alexandre Némirovsky and Robert Charlet, both prominent in air-rescue work.[140] Since she knew Morocco so well, Marie served as location scout, selecting backgrounds for ground and air scenes. During the filming, Marie flew in one plane while her pilot communicated her directions to the pilot of Chelle's plane.[141]

Given the landscape, the scope of the film, difficult weather, great distances to cover, and the nineteen hours of footage, it wasn't easy getting their 2,000 meters of film, but, as one newspaper put it, "What Woman wants, God does too."[142]

The film toured European cinemas and thrilled audiences from the Pyrenees to the Rhine with vivid stories of air-ambulances, making an excellent case for their widespread adoption. Marie showed the film at aeronautic events and at aeromedical evacuation congresses. Both a documentary and a tourist film, it was designed to illustrate the importance of air-ambulances as well as to draw visitors to North Africa (which is why Marie chose the most picturesque sites as backdrops).

An untouched copy of *The Wings of Mercy* was found in French army archives in 1993. *Saved by the Dove* remains undiscovered.[143]

Once again, Marie pulled in as many individuals and organizations as possible: she had the widespread support of the Friends of Air Rescue, the Red Cross, the Ministry of War, the Ministry of Air, and French officials in Morocco.[144] Marie was in North Africa when *The Wings of Mercy* premiered in Paris on November 23, 1934.[145] She wrote her cousin: "They tell me it is magnificent. Perhaps will you see it?"[146] Robert Charlet did see the film when it opened, and wrote to Marie:

"I've just this morning seen the film with its sound track, of which the part in Morocco was filmed under your talented and competent direction. I want to tell you immediately what a magnificent success it is, and to compliment you on the manner in which you directed the eye of the cinematographer. I can't imagine a production more faithful, evocative, or beautifully brought to life as the way the film covered the subject and is such a pleasure to the eye."[147]

For such an independent, outspoken woman, Marie played well with others, bringing disparate groups together to work toward a common goal. She was loyal and scrupulously honest about giving credit where credit was due. In the case of these films, she said that the general plan for the first film was laid out by Némirovsky and Charlet, and that she simply added a few scenes. In the same way, numerous individuals and organizations contributed to the many-tentacled, decades-long process of bringing aeromedical evacuation to its present state, and Marie mentioned every one of them in her scores of articles and interviews over the years.

The French government honored Marie for her work in aviation medicine on November 5, 1937, when the Ministry of Social Services made her a Knight of the Public Health—an honor reserved for the most outstanding public servants.[148] The Nancy newspaper, *L'Est Républicain*, grumbled about how long it took the government to recognize Marie's efforts. "But, as it is never too late to repair a regrettable omission, there will be many citizens of Nancy, of Lorraine, or simply of France who will rejoice at this honor and address once more their warmest congratulations to Mademoiselle Marie Marvingt. When noble words and beautiful actions are allied, a medal is not, after all, but a modest recompense."[149]

In January 1955 the National Aeronautics Federation of France and Abroad presented her with the grand prize *Deutsch de la Meurthe* (awarded to one person annually) in recognition of her development of medical aviation against so much opposition. At the time, Marie said, "If we have given wings to the world, we have the obligation to ensure that they are the wings of the dove of peace."[150]

On February 20, 1958 (which happened to be her birthday), Marie was awarded two more medals in recognition of her air-rescue work, one from the Department of Public Health and one from the Service of Aviation Health.

David M. Lam summed up her contributions to aeromedical evacuation after years of studying her work: "Marie was one of the first to publicly espouse the use of aircraft for medical purposes, and spent much of the remainder of her long life in proselytizing, arguing, and convincing the medical, military, and civil communities of the need for, and the safety of, airplane ambulances. (She was truly farsighted in this regard. It must be remembered that at this period the airplane was still considered more dangerous than useful by the vast majority of the general public.)"[151]

Of all Marie's titles, she herself preferred to be known as the "Inventor of the Wings of Mercy." When she had to identify herself, she used some variant of that descriptor. She would be happy to know that her tomb in Nancy reads, "Marie Marvingt, Founder of Medical Aviation 1910–1946."

7

The Fiancée of Danger

"You want to make funeral arrangements ... for yourself?"—Clerk in French funerary establishment[1]

"There are some women who seem to be born without fear, just as there are people who are born without the ability to feel pain. The painless ones go around putting their hands on hot stoves, freezing their feet to the point of gangrene, scalding the lining of their throats with boiling coffee, because there is no warning anguish. Evolution does not favor them. So too perhaps with the fearless women, because there aren't very many of them.... Providence appears to protect such women, maybe out of astonishment." —Margaret Atwood[2]

Marie Marvingt was first described as the "fiancée of danger," a title never associated with any other woman, in 1903. She had just climbed the almost inaccessible Dent du Géant in the Alps when a certain Monsieur Château-Thierry de Beaumanoir exclaimed that she could be none other than the Fiancée of Danger.[3] The name caught on, and afterward newspapers on both sides of the Atlantic used the descriptor in most articles about her.

Women at the time were either affianced to man or to God (if to neither, they had the good sense to remain quietly at home where they were not an embarrassment to society and their families). The public seemed to like knowing that Marie had a man in her life, even if he was rather unusual. Marie herself accepted the sobriquet tongue-in-cheek. She was a journalist; she knew what sold papers.

Someone who neither expresses nor displays fear may still harbor fearful feelings invisible to others. Of Marie Marvingt, it can only be said that she appeared fearless.[4] Witnesses report that even in the most dangerous situations, her attention focused entirely on the threat, while perhaps also admiring a magnificent storm, being delivered of a stunning new insight, or expressing her joy at the intensity of the moment.[5] She was not an interior sort of person. Extroverted, energetic, drawn to action, Marie's orientation was almost wholly outward, rarely inward.

One credible witness to Marie's attitude in life-threatening situations was her companion on the balloon crossing of the North Sea, Colonel Émile Garnier. They fought for hours to outwit the almost overpowering elements, and openly discussed the possibility of their deaths. Although Garnier's courage and *sangfroid* were equal to Marie's, the experience must have felt doubly precarious to him: this was his first balloon ascension, and he was obliged to depend on his companion's expertise. The experience left him with an unflagging admiration for Marie, and in the years that followed he spoke often of her tenacity, grace under pressure, and fearlessness in the direst of circumstances.[6]

Referring to her title, Marie said, "It's a fiancé that I don't fear, you see. He's a childhood friend and I've seen his face many times."[7] Land, sea, and air—Marie escaped death dozens of times. She experienced wars, shipwrecks, volcanoes, plane crashes, earthquakes, muggings, serious diseases, a burst appendix, a disappearing island.[8] She reassured friends who worried about the risks she took: "I'm not as stupid as it might seem sometimes."[9] (And she used to caution readers, "You young people who are reading this: Always keep your cool in times of danger, and always let prudence be more important to you than daring."[10])

A journalist said Marie never missed a rendezvous with danger, describing her "adventures following adventures, so unbelievably numerous that they seem to be born of the conjoined imaginations of Jules Verne and Blaise Cendrars."[11] Others characterized her as the victim or witness to some twenty catastrophes.[12]

It was perhaps inevitable that Marie's dramatic life inspired the iconic popular film series, *The Perils of Pauline*. When the series was conceived, Marie had been known for over a decade as "Daredevil Marie" and "the world's greatest athlete." In 1909, she had defied death by crossing the North Sea in a balloon. In the first episode of *The Perils of Pauline*, Pauline too faced death in a balloon. Like Marie, Pauline was a "plucky girl reporter" who often recounted her exploits in articles and interviews where the major attraction was the description of the various risky situations she had managed to survive. Marie was an outstanding fencer, tennis player, and pilot. Onscreen, Pauline also fenced, played tennis, and flew planes. Both the real and the fictional heroines became involved in the war. Pauline too carried a gun, fought tough enemies, and even became a Red Cross nurse. Pauline refused to marry because she was unwilling to renounce her "adventurous spirit" and her "desire to live and realize the greatest thrills."[13] Pearl White, who played Pauline, said that each episode was a woman-always-in-danger film.[14] Danger was central to the series' themes, plots, and title.

Pauline was an American export, but the Pathé company was French, and its U.S. branch was managed by Louis Gasnier, a French director. In *Exporting Perilous Pauline: Pearl White and the Serial Film Craze* (2013), Italian academic Monica Dall'Asta points out "the obvious resemblance between the Pauline character and the amazing French athlete, mountaineer, and aviator Marie Marvingt," and wonders whether "the scenarists of *The Perils of Pauline* had chosen the name of Pauline Marvin as an explicit homage to this incredible woman ... but there is no doubt that Marie Marvingt's exploits could be an ideal source of inspiration for a feminine-oriented action serial. In fact, the cinematic qualities of her balloon adventure can be seen in the episode in which Pauline experiences a similar situation."[15]

In *Exporting Perilous Pauline*, author Marina Dahlquist says that few film genres have been as historically significant as the spectacular American action serials of the 1910s featuring powerful female heroines: "The serial queens represented a new and independent female type of protagonist within sensational, action-packed story lines. ... the genre offered young women a novel template for negotiating gender stereotypes. ... The versatility and bravado of Pauline and her serial 'sisters' in performing tasks traditionally associated with masculine brawn—battling in fistfights, handling pistols, and demonstrating agility in stunts—were the genre's most striking, and gender-bending, negotiations."[16]

Dall'Asta says that "for the American audience, Pauline simply represented a hyperbolic transfiguration of an already familiar, but still quite novel, feminine type that was emerging in both society and popular media. ... In any event, the all-sport record-holding woman and pioneer aviator, Marie Marvingt—also known as 'the fiancée of danger,' 'the most important woman in France since Joan of Arc,' and 'the universal sportswoman'—had already become familiar to the American audience when *The Perils of Pauline* was released in 1914."[17] In France, Pearl's character was even more familiar to audiences; they'd been encountering her prototype every day in their newspapers with stories by and about Marie.

Marie had, in fact, been doing everything (and more) that "Pauline" did, but it took "Pauline" to carry the empowering message farther than Marie ever could have.

An eerie resemblance existed between Marie Marvingt and Pearl White, the actor who played Pauline Marvin. Both were daredevils at heart. In her 1919 autobiography, White says that her earliest dream was to be a bareback rider in the circus; she eventually toured with a circus one summer and earned enough money to pay for school for her sister and herself. "I started out to be a daredevil and dare I must," she wrote. Marie's early fascination with the circus and horseback riding and her training with the Rancy circus were springboards for her, too. To prepare for her role, Pearl had to learn to play tennis, swim, drive a car through water, fire, and sand, go to sea in a yacht, survive a car smashup, and ascend in a balloon. "I've even learned to fly an aeroplane," she said, "a feat that took me many months."[18]

White's indifference to fear was acquired rather than innate, as it seemed to have been with Marie. As a child, White's heavy-handed father beat her if she ever showed what he called cowardice or if her opponent didn't end up looking worse than she did.[19] (Pearl's mother died shortly after giving birth to Pearl[20]; Marie's mother died when Marie was 14.) As an adult, she learned to live with fear, ending up in much the same emotional place as Marie did: it was simply part of what they did.

"I have actually gotten to like fear," she wrote," and to like the sensation of taking some very dangerous chances that frighten me. My old heart beats a ragtime, and I face the music feeling more thrilled than I would be doing something in which I knew there was no risk."[21] Her contract for *The Perils of Pauline* stated the studio was not responsible if she died or was injured. In fact, during filming she incurred cuts, bruises, and sprains, several with lifelong consequences.

Just as one of Marie's greatest achievements was the balloon crossing of the North Sea, so one of White's greatest challenges was her ballooning scenes: "There are some things that can't be learned and you just have to take a chance. Ballooning is one of them. This merry old sport nearly cheated me out of fame and fortune." She says that before filming the balloon sequences, "I remember I borrowed a pencil from the camera man, and wrote out my will. I did this to be funny, but it wasn't long until I felt pretty certain the joke was on me."[22]

Like Marie, White was awed by her balloon experience in the midst of a storm. It was such "a wonderful sight watching the old storm carry on between us and the earth that I got all over being afraid for the moment."[23] The balloon scene in the movie was so similar to Marie's experiences that its inspiration is undeniable.

White bought herself a funeral plot in Paris and arranged for her own funeral, much as Marie had done in the early days of aviation. Having lived with danger all her life, White had a practical perspective on death. (Pearl White died of cirrhosis, due to the alcohol and drugs she used to medicate a back injury suffered while filming *Perils*.) The two women thus shared a certain pragmatism, disdain for death, daredevil natures, motherlessness, physical beauty, athletic abilities, and an uncommon fearlessness.

Marie took risks sometimes because it was the only way to accomplish her goals (flying in the rickety early planes, for example). Other times she had no choice; danger sought her out (in 1912 an excursion boat on which she was riding capsized, and she had to swim twelve hours before being rescued).[24] And sometimes she embraced danger for the pure thrill of it (on a dare, she dived from a dirigible into the Grand Canal of Venice).[25] An old friend of Marie's said, "Wherever there was danger, that's where she went."[26] The numerous risks she took, or encountered, earned her a place on a 2011 list of "The Top 10 Female Risk-Takers."[27]

Marie had both a predilection for danger, and a contempt for it.

"The taste for taking risks," said a friend of Marie's from Nancy, "often accompanies people of truly great spirit. Certain ones, like Marie, do it like a game … with infinite grace." She added, "Marie was richly self-sufficient. People like that follow the dictates of their consciences in responding to the appeals of their own inner selves with unquenchable courage and a many-faceted vocation to 'service' in the highest sense of the word."[28]

There are no images of Marie walking with her fiancé, but this early photo of her indicates her willingness to dive into almost anything.

Marie repeatedly, perhaps defensively, emphasized that in everything she had done, every risk she had taken, every activity she'd engaged in, she had searched for a higher goal to give her involvement meaning, to make her expertise and joy useful and beneficial to humankind.[29]

She transformed her love of flying into the worldwide air-rescue movement and the invention of the ambulance airplane. Her love of swimming led to her campaign for new laws that mandated swimming for school children because of its health benefits. Her enchantment with Esperanto led her to teach people from different cultures to communicate with each other. Her multiple athletic activities led her to urge young people to take up a sport. Her skiing led to the wartime rescue of soldiers in the Dolomites.

"Nothing I have done," she said, "has been done solely for the thrill involved or for the publicity, but for what it can teach me about nature, about humankind, and about myself. When people ask me why I have done these things—taken these risks and accepted these challenges—I have to give them the same answer the mountain climber gave when someone asked him why he kept trying to climb Everest: 'Because it is there.' And that's why I became an adventurer, if you will ... because there are things to be conquered, things to be done."[30]

One journalist added, "But without her ever precisely saying so, one gets the impression that she enjoyed herself a great deal at the same time."[31] General Barthélemy, who knew Marie well, commented on her adeptness in balancing deeply ingrained, but opposing and normally irreconcilable, tendencies.[32] She managed to indulge her love of risk, adventure, and novelty as well as her desire to be constructive, useful and benevolent.

Marie never took senseless chances, and she never made the same mistake twice. An elevator operator in Paris told about seeing Marie in the lobby of his building. Hoping to give her a lift and perhaps get her autograph, he asked, "What floor, Mademoiselle?" But she only smiled and walked up six flights of stairs. She was eighty-five years old at the time.[33]

Later, he learned that Marie had once almost been killed when she was riding in an elevator that broke loose from its connections and plunged all the way to the basement.[34] Marie did not need another elevator ride to confirm her initial impression that elevators were foolish and untrustworthy. Besides, she wasn't in control in an elevator, and she rarely did anything she couldn't bend to her own will.

A lover of mountaintops and open skies, Marie would not have been fond of elevators even without her accident. Mildly claustrophobic, she hated being "boxed in."[35] The Paris subway was one of her pet peeves. "There's no air in there," she said.[36]

Marie's idea of danger often differed from most people's. The same woman who avoided elevators and subways thought nothing of taking up flying at a time when 87 percent of all pilots died in aerial accidents.[37] These were people with whom she'd flown, dined, talked aviation, and made plans. She attended scores of funerals. In the periodicals and newspapers of the day, death and disaster were constants. In one aviation column containing an item about a success of Marie's, the other four items were: "To the Victims of Aviation," "An Accident," "A Plane Crash," and "An Airplane on Fire."[38] Still, she went up again.

Despite the hundreds of flights she made, Marie's record as a pilot was amazingly accident-free. Once the propeller fell off her plane, but she managed to land unharmed.[39] Another time, "in August of 1912, I was flying at 1,000 meters over Château-Thierry when a part of the fuselage burned up. It's a miracle that the control stick didn't malfunction. At 1,000 meters, you'd better believe that's sure death!"[40]

Landing in the tree at St. Étienne, crashing again a week later,[41] digging herself out of the mud in a farmer's field after a plane crash—all were sources of amusing stories to Marie rather than brushes with death. (One reporter referred to her as the Mistress of Death, but the press and the public never picked it up.[42] She remained the fiancée of danger.)

The dangers she confronted in two world wars, in the early days of aviation, and in her historic crossing of the North Sea were serious ones. Lesser (to her, at any rate) encounters with tricky situations and tight corners included one that underlined her nickname, "The Red Amphibian."

"When visiting Le Havre," Marie explained, "I used to go aboard my friends' boat at dusk to set the nets." (Charles Codron, a neighbor from Chamonix, says Marie occasionally signed on as a deckhand on fishing boats.[43]) "One night when everyone was sleeping, I decided to take a little swim. But the weather worsened unexpectedly and before I knew it, the waves had carried me so far from the boat that I had no hope of regaining it.

"I swam all night long, sure that my hour had come. By dawn the current had carried me, with the good Lord's help, near a fishing boat. When the fishers saw the splash of red from my suit and my wildly waving arm, they approached and threw me a line. They were so astonished to see me out in the middle of nowhere that they thought at first they had caught a mermaid!"[44] They also had questions as to how she managed to survive a storm that had raged all night, whipping up huge waves and battering her with rain and winds.

While scuba-diving for bioluminescent sessile sponges in the Mediterranean off Tripoli, she was sixty meters underwater when she came face to face with a shark. Later, to a group of schoolchildren, she confided, "I will never forget its eyes. *Mesdemoiselles*, if you ever run into a shark, remove yourself from it as fast as you can!"[45]

She recalled another minor incident: "In 1900 in London, where I was riding my bike, a cab actually ran over my back, right in front of Westminster."[46] Over fifty years later, journalist Jacques Noetinger nearly ran Marie down in Paris. Again, she was on her bike, but she stopped to excoriate him as a "clumsy driver" and worse. Once she realized who he was, however, she smiled broadly, and the two of them chatted for a while.[47]

Fascinated for years by volcanoes, Marie set off immediately for Italy in April of 1906 when she heard that Vesuvius was erupting. While thousands were fleeing the area by boat, train, and foot, Marie pedaled her bike from Nancy to Naples, and from there on to Vesuvius. Always a participant rather than an observer, Marie borrowed equipment from the director of a nearby Italian conservatory so she could collect materials for analysis.[48] She described how her horse's nostrils became singed from the smoke as she collected samples of rock and lava; "I was half roasted climbing around on Vesuvius in the aftermath of that eruption."[49]

In Greece on a lecture tour, she decided to visit the tiny uninhabited island of Gouliani, which lay off the coast. She spent the day watching the enterprising island rabbits, looking for wildflowers, and enjoying the sunshine.

That evening she returned to her hotel on the mainland. But the next morning she wanted to take a picture of the charming island where she had spent such a pleasant day. At first, she thought she was looking in the wrong direction. She could find no trace of Gouliani. Residents of the coastal city informed her that during the night the sea had swallowed up the island. Marie said she broke out in a cold sweat when she realized what had happened.[50]

In 1952, during the North African campaigns, Marie was in Tunisia traveling from Gafsa to Gabès with friends of hers—a father, mother, and their eighteen-year-old daughter. About forty kilometers from Gafsa, a violent explosion shook the car. The young woman immediately fell over on the seat, bleeding profusely from wounds in her arm and chest. Her terrified screams startled her father into pulling over to the side of the road. But Marie insisted that he keep driving as fast as he could while she tried to comfort his daughter and control the bleeding.[51]

Another round of fire went off and a bullet ricocheted off the driver's hand, hit his wife in the leg, and lodged in the unfortunate daughter's abdomen. Driving at top speed, the car arrived in Gabès where the wounded were cared for at a military hospital. All three family members survived.[52]

Le Monde reported the incident, saying that Mlle. Marvingt, "today seventy-two years old, who forty years ago was the most famous of the French aviatrices, was also in the automobile, but was not hit." Included in the article were reports of a bomb blowing up a store in Tunis, another placed in a letterbox in Sfax, and that evening, an explosion in aTunisian pharmacy.[53] It was not a safe time to be in Tunisia.

Marie's quick thinking was mentioned at the inquest, along with the comment that at least one of the many bullets that struck the car ought, by rights, to have hit her. Incidents such as these enhanced the Marvingt legend. Marie herself attributed her good luck in that instance to the holy medal given to her by Pius X, which she always carried with her.[54]

After being fired on by rebel troops while testing her metal skis in Morocco, she had a similar close call in another area of Tunisia. She wrote her cousins, "I am still the fiancée of danger, and I have rubbed shoulders with him again today."[55] She told friends that she benefited from "baraka"—in Islam, a beneficent force from God that flows through the physical and spiritual spheres.[56]

Marie admitted, "I have lived dangerously, but with serene trust in Providence."[57] She once listed several chapter titles from her memoirs (since lost): "What It Feels Like to Drown," "Thrown Into the Air Like a Cannon Shell," and "Buried Alive."[58]

Marie's hatmaker recalled mentioning appendicitis during one of their visits. Marie interjected that she had undergone emergency surgery to have her appendix removed in the Sahara.[59] This incident didn't merit a mention in any of her newspaper interviews or postcards. It must have seemed insignificant to her. She may even have deliberately omitted it: any ordinary person could have appendicitis.

Only one other time did she mention an illness. Following the 1932 earthquake in Ierissos, she was stranded in Stavros. The simple rail line leading out of the city had been destroyed by the earthquake, a storm prevented boats from coming in, and she was sharing a small room with a number of other wayfarers ... and a high fever. After three days of treating her fever as best she could, she was elated to be told someone downstairs was asking to see the Frenchwoman. "Finally, in this forsaken corner of the world, a Greek who spoke French! And how lucky was I! He was a doctor and, what's more, he knew me! He had been at a conference in Salonica three weeks earlier that I had given for 600 physicians."[60]

Those who nursed Marie through her long, near-fatal microbial illness in North Africa feared several times for her life. Her recovery from the illness, with its several relapses and hospitalizations, added to the myth that nothing could kill Marie Marvingt.[61]

All her life, Marie liked to say she had never known a genuine feeling of fear.[62] Approaching her eighty-eighth birthday, she still maintained that she feared nothing.[63] Many years earlier, she said, "Someone once predicted that I would die in an accident. I know I'm marked for death, but I manage to live very well with that thought, I assure you. For example, when I participate in a sport in which I risk my life, I go into it with my eyes open, calculating all possible eventualities. I take every precaution, and I make all necessary arrangements before I start.

"I always have to laugh when I remember the stupefied face of a funeral director to whom I made my usual interested visit upon arriving in a city that's one of the major centers of aviation. I must have been the first aviator ever to inquire about my funeral ahead of time! When I do something dangerous, I leave my affairs in order."[64]

Then she added, with a smile in the dark eyes, "When one has a deep faith, one doesn't fear death. I am convinced that beyond this life there is another, so what is there to fear?"[65]

A daily newspaper in Tunis once announced her death, "in capital letters," said Marie. "I've kept the article, which is surrounded by a wide black border. It tells how a blood clot carried me off after a landing in Philippeville. My family was notified and, for a few days, prayers were said for my soul."[66]

Marie took the deaths of others very seriously, as shown by her attendance at so many funerals and her visits to so many bereaved, but she spoke of her own death with an insouciant matter-of-factness. Only once did she appear awed by a narrow escape.

In India for a series of talks, she made a plane reservation in Calcutta for the next city on her itinerary. Her bicycle was already in the plane when she received a message from local aviation officials asking her to stay over one more night and talk to a group that had been hoping to hear her. Passengers were starting to board the plane when Marie canceled her seat. "I apologized, you know," said Marie, "but I decided not to fly."

That plane crashed, and 140 people died.[67]

8

Even Cats

"I will never marry. I couldn't bear the ties of marriage, and I can't imagine any man putting up with me for long. Climbing mountains is a lot more interesting to me than washing dishes."—Marie Marvingt[1]

People have always been curious about the love lives of the famous and accomplished. Lively speculation about Marie's private life has been ubiquitous ever since her earliest appearances on the world stage.

In her 1905 book, Matilda Betham-Edwards wrote, "A foreigner suddenly plunged into French society and quitting it without any chance of modifying first impressions would affirm that there were no single women in France."[2] This was an exaggeration, of course, but not by much. Being single at that time was not the norm. In the case of an attractive celebrity, it raised questions and eyebrows.

Several factors contribute to the cloud of unknowing that surrounds Marie and her affections. First, she was discreet. Second, journalists respected her private life to a degree unknown then or now; many of them were her employers, colleagues, or friends. Third, much of her life was spent out of public view—on mountaintops, in the desert, in other countries, in war, in hospitals.

And so, did she have a love life? Apparently.

From the first moment she conceived it, the ambulance-airplane she designed, raised money for, had built, and donated to the military was named "Captain Écheman." The prize she awarded to airplane manufacturers for the best transformable ambulance airplanes was called the Écheman Challenge.

The logical name, given Marie's history to that point, would have been "Hubert Latham." Her first flight teacher and great friend, dead too young, was the debonair, charming, and adored pilot Latham. Marie often brought up his name, mourned his loss for years, and publicly protested the lack of credit he received for his multiple attempts to fly across the English Channel.[3] He was obviously important to her, and everyone would have understood the significance of her naming a plane after such a heroic aviation pioneer. But he was a friend, only a friend.

In the larger world, no one knew much about Captain Écheman.

In his own way, Paul Maurice Écheman was a male incarnation of most of Marie's traits: profoundly patriotic, an outstanding sportsman, a pioneer pilot, endowed with heroic impulses, and a humanitarian in the making. Born January 28, 1877, in Angers (where a street is named after him), Écheman studied engineering for four years, then joined the 35th artillery of the French military in 1899. Two years later he was awarded a medal of honor for saving the lives of soldiers in his battalion during a fire.[4]

He attended the famous French cavalry school St-Cyr at Saumur in 1903 and 1904, and joined the 2nd Cavalry Division in Lunéville, where he stayed for four years. In 1909, he was named to the 24th artillery division in Tarbes as an instructor; he was made captain the next year. In 1911, he served as head of the airfield at Angers-Avrill.[5]

He enrolled at the Blériot aviation school February 1911, receiving his pilot's license in March (#466; Marie's was #281) and military pilot's license #10 in May, one of the first pilots so licensed. He was then posted to the military aviation group in Vincennes. Referred to as one of "the great birds at the dawn of aviation" and as "one of our best aviators," his name appeared in the newspapers for flights from Pau to Paris, for example, for his military training flights, for his "aerial adventures."

In between his studies, military service, and training at the aviation camps, Écheman could be found on the ski slopes, on horseback, on a bicycle. After they met at Mourmelon, he and Marie participated in many of the same aviation meetings and sporting events in the pre-war years.

In the early years, flying lessons were given and planes were put through their paces on any accessible flat field in rural areas, whence the term "airfield." The newly licensed pilots and the eager would-be pilots at these aviation camps were a lively group defined largely by youth, wealth, leisure, and a passion for aviation. (Among others, the Prince of Orléans was there when Marie was.[6]) Because weather conditions had to be just right before the temperamental early planes could take to the air, groups of energetic young people found themselves out in the countryside waiting. And waiting. As Marie said, "We watched the anemometer very closely in those days. If there was a wind of so much as five kilometers we dared not chance it. Sometimes we were as long as two weeks between flights."[7]

During the day, the students biked or raced their cars on gravel roads or found a swimming hole. At night, they gathered in their hotels for protracted, jovial dinners. Long days and long nights had to be gotten through. No one dared leave in case the hoped-for calm appeared.

Marie and Paul both spent a great deal of time at these hastily constructed, crude airfields, part of a unique band of daredevils who shared experiences and feelings that outsiders would never understand. After boring,

Cover photo of the February 17, 1912, *La Vie au Grand Air*.

or sometimes exhilarating, days, the two would return to their hotel. With so much in common, Marie and Paul talked of sports, the newest planes, and plans for the future. During the long waits for good flying weather, there were undoubtedly opportunities to disappear quietly and anonymously from the energetic mix of people and activities.

The ambiance of the early aviation camps resembled many of the situations in which Marie spent her life. She regularly worked with men, often under unconventional conditions of great risk—in aviation, war, sports—when, knowing death was never far off, life and choices acquired a sense of urgency.

Écheman's picture graced the cover of the February 28, 1911, issue of *Armée et Marine*, leading a pack of skiers at the competition held in Lioran. Inside that issue was a photo of Marie, winning the women's ski.[8] They were two of a kind, winners and leaders, but with compassion and generosity hard-wired into both their natures.

Tall and fit, with chiseled features, a mustache, and dimples, Écheman's personality and genuine kindness outshone even his good looks. A photo of him and Marie appeared on the cover of the February 17, 1912, issue of *La Vie au Grand Air* with the caption: "In Chamonix last week, these two pilots seem to have made a date to rest up from their aerial fatigue. Thus the famous birdwoman and eclectic sportswoman, Mademoiselle Marvingt, and the captain-pilot Écheman devote themselves to the pleasures of ski and bobsled."[9] Marie had postcards made of the photo, and wrote across the front "Souvenir of Chamonix." Another photo, in *Le Miroir des Sports*, shows them figure-skating as a couple.[10]

When Paul Écheman died, at 35, in an airplane accident at Étampes on May 14, 1912, all the papers carried the news. The weekly publication *L'Illustration* ran substantive articles.[11] After announcing that military aviation had lost yet another pilot, the piece went on to say, "The sudden and disconcerting catastrophe at Étampes is a particularly cruel sorrow for *l'Illustration* because Captain Écheman died while flying a plane that we had planned to offer to the Army. Like other patriots wishing to support their country, we located a plane that seemed to be good and bought it to donate to the military."

The plane was named the *Henri Lavedan*, after French writer Henri Léon Émile Lavedan, still living at that time. "This type of plane had already been responsible for one death, so the engineers and mechanics were careful to refine it painstakingly and double-check everything. In its refurbished state, it had been flown fifteen times by M. Perreyon, one of the best instructors at the Blériot aviation school. Écheman was chosen to make the final trials. Perreyon took it up first, and then Écheman climbed into the cockpit.

"He was chatting with us with perfect serenity as they removed the blocks from in front of his wheels. He pulled a leather vest on over his sweater, changed his military cap for a helmet, adjusted his sunglasses, and took off as tranquilly as if he were on horseback or riding a bicycle.

"Once he was up, the flight was so smooth that we were no longer even watching him. Suddenly the machine dropped. At about thirty meters he seemed to regain the horizontal for a half-second perhaps, then the plane arrowed straight down to the ground. It took only about two seconds. Officers and firefighters ran to him. Both his legs were broken, his head was covered with blood, and the cross he wore was twisted and encrusted into his chest. He was taken to the hospital in Étampes but he died two hours later without regaining consciousness. We are all still trying to understand exactly what happened. In addition, Captain Écheman—tall, elegant, confident, elite sportsman that he was—was considered by other aviators to be an experienced pilot with a rare and sturdy *sangfroid*, the top student of that incomparable pilot, Captain Bellenger. Écheman was particularly outstanding at the last maneuvers and he had been awarded the Legion of Honor in 1911 for his accomplishments as a pilot.

"The only possible conclusion is that the machine was in perfect shape, and that Écheman did nothing wrong. In the end, such accidents are unfortunately inevitable at the current stage of development of aviation. The effects of heat and air currents are as yet imperfectly understood."[12]

His plane crashed at 6:00 p.m. He died at 8:00 p.m., "after two hours of agony," the *Figaro* reported. The *Figaro* also differed from *L'Illustration* in their description of the accident: "After a superb takeoff, Captain Écheman rose to about 1,000 meters and after a few minutes began to descend. Imperceptibly, he let himself speed up, and his machine began to race and take off. At 100 meters from the ground, he saw his error and tried to raise the nose of the plane, but he was too late! Those watching from the ground saw him make, at 90 meters' altitude, a last and desperate attempt to pull up, but in vain. At a dizzying speed the monoplane crashed and pulverized."

Two photos accompanying the article were graphically tragic for anyone who loved Paul Écheman: one showed him smiling at takeoff; the other showed him lying on the ground next to his ruined plane. In the two weeks surrounding Écheman's death, nine pilots lost their lives.

A ceremony, touching in its simplicity, was held in the hospital chapel at Étampes on the morning of May 18, 1912. Gathered around were family, close friends, municipal officials, all the civil and military pilots attached to the various airfields in Étampes, and, of course, Marie. After a few prayers, Commandant Félix, director of the military aviation school at Étampes, said a few words of farewell "to the valiant friend who died almost in front of me."

Écheman's body was transported to the little church in Montreuil-sous-Bois where a funeral Mass was celebrated.[13] Marie was there, along with the governor of Paris, the former Inspector General of Aeronautics, and many officers, pilots, and civic figures. Écheman was eulogized for being part of a movement to make France a nation of the physically fit, for his sailing skills, his skiing, his piloting, the medals he was awarded, his considerable charm and lightness of spirit. Condolences poured in from every city he'd ever lived in.

Marie and Paul figure-skating at Chamonix, 1911

Later, at the Montreuil cemetery, Lieutenant Colonel Estienne spoke with military precision of all the services Écheman had rendered his country. In October of that year, a bust of Écheman done by Maillard for *L'Illustration* was dedicated in a simple ceremony attended by a few military leaders and pilots. And Marie.[14]

Shortly before his death, Écheman had written a letter to his younger sister, who was a religious in the order of Notre-Dame-de-la-Retraite. The two had always been close and she had begged him to give up his dangerous "hobby." He responded:

I want to fly, and I mean to fly, perhaps for a very long time, now more than ever because huge changes are in the making, perhaps even war. Those who can handle this difficult life and this unique role must do so: lives are at stake. Because of so many things, I want to fly. I love this life, I love it with savage joy, with passion. I could, of course, from a military standpoint, do something other than fly—return to a careful life and do thousands of little nothings. … And, my dear, without wishing to hurt your feelings, may I ask you not to speak so lightly of this? Call boating or skiing or mountain climbing a hobby, if you wish. But flying, that's something else. It's almost a vocation. And, yes, it's dangerous. On the ground, you can watch out for attacks, but in the air, there are sneaky gusts of wind that can do you in. We're not afraid of death up there but we think of it just enough to understand that we're doing something serious. We don't have to fear the big words like "death." We must call things what they are. If you like, you can think of pilots as loudmouthed seekers of glory, or as acrobats trying to invent some original moves, or as sweet-natured maniacs chasing the elusive chimera. But we, we military pilots, we serious ones, are nothing like that.

We choose our planes carefully, ones that have been proven over time—although in the case of aviation, 'over time' is laughably short. These are not toys that we climb into with confidence. We know where the danger is, what we must avoid, where we can reasonably go, and what we can reasonably do. Is that to say that we are absolutely certain not to crash? Of course not, and it's from that uncertainty that is born a certain legitimate pride that we take in our work. In this lamentable inaction in which we are currently stagnating, it seems to some of us that aviation reawakens our national energy, and that it is good for our poor France to know that some of us are willing to give ourselves without second thoughts.

I've always noticed a certain sense of well-being among sailors. Yes, their lives may be routine, but there is always the possibility of battling the elements or even, occasionally, battling an actual enemy. There is always that possibility, and they feel superior to land people who have only their routines.

In the same way, the sailors of the air have their routine moments, but you never know when you may have to battle an element even more perfidious and dangerous than the "big green," the sea. You know, when I'm up there, and nothing is happening and there's not much to do but look at the huge mute map before me, I sing. I like to sing as loudly as I can, right into the wind coming off my propeller, the invocation to St. Georges that we used to sing when we were at Saint-Cyr: "And then, when the day comes and our destinies have been worked out, at the hour when the black angel smiles and signals to us, grant us, O Lord, the gift of, if you please, not dying in our beds!"

And there you are, little sister, the reason that your word "hobby" made me feel bad, but I don't mind now that I've told you what it's all about. Love, Paul."[15]

In later years, Marie mentioned Paul Écheman's name often in her speeches when referring to the ambulance airplane. After she became emotional a few times, she began saying simply, "My ambulance airplane is named for a friend of mine who was one of the heroes of the early days of aviation, one of the first and glorious victims fallen in the service of their country and for the cause of aviation."[16]

Sometimes she asked someone else to talk about him, as when M. Charles-Louis Julliot introduced her and her Challenge, named after Captain Écheman "who was, and who will remain, one of the great figures of early aviation, who died May 14, 1912, in an air accident that even today remains inexplicable. He was an outstanding soldier, an outstanding pilot who believed in the future of aviation."[17]

And that is all that is known about Paul Écheman. When those who knew Marie were asked whether she had ever had a lover, the response varied from an abrupt, "No, of course not" to "Oh, but certainly."

Marie was a Roman Catholic and attended Mass in Nancy whenever she was at home. This fact alone determined for some people what she might or might not have done. The local newspaper said, "She was a practicing Catholic. She would never have done anything against her religion."[18] And an older military friend and his wife said, "She told us everything, and she never said anything about affairs."[19] But then Marie would not have.

Good Catholic or not, she wore on the same silver chain both a medal blessed by the Pope ... and a skull. She told fortunes, looked into spiritualism, missed Mass often. There were cafeteria Catholics even then.

Marie let a few close friends understand that she had had "adventures."[20] One old friend said, with pauses, "According to conversations we had, yes, there were several men. Oh, not many. ... When she was in the desert, you know.... When she was ... well, anyway, several men. She had such a forceful character, you know ... nothing stopped her."[21]

One of Marie's godsons, A. Lorac Gerbaud (his mother was a childhood friend of hers), said she often stayed with his family when she was in Paris, and he got to know her and hear about her "love affairs in the military and sports worlds."[22] She wouldn't have discussed matters of the heart with an older military couple from Nancy, but it was different with a young godson in Paris.

Newspapers referred to her as "the intimate friend of Marshal Lyautey whom she accompanied on a number of campaigns,"[23] but to the two of them, it was a joke, something to keep their biographers busy. Rumors were undoubtedly fueled by frequent photos of them; for example, at a reception for Air Minister Laurent-Eynec, they were sitting together.[24] They were close, but it was largely a matter of shared interests and years of working together.

Her name was associated with other public figures: Clemenceau, Millerand, and a number of the early aviators. Some people liked to point out that with a few men she used the familiar "you" ("tu") rather than the formal "you" ("vous").[25]

Marie's dressmaker's niece spoke of Marie's magnificent eyes, adding that Marie had admitted to her that she had "known" (in the biblical sense) several great Frenchmen.[26] A male friend said that Marie was rather flirtatious "in the good sense of the term," and had a great deal of success with men.[27]

Her general attitude toward men—attested to by many—was one of camaraderie. As soldier, nurse, journalist, and sportswoman, she was in close contact with many men, whom she treated as brothers and friends, and they reciprocated in kind. An aviation general's wife said that according to her husband, Marie "was a perfect friend, not very flirtatious but kind, with a great deal of spunk and gaiety. She had an extraordinary energy."[28]

Pretty, flirtatious at times, talkative, and charming, Marie practically invented the notion that men and women could be friends and co-workers, an odd idea at the time.[29] She obviously valued her professional relationships, and would have been unlikely to mix work and pleasure very often.

A man who knew her around 1905 described her admiringly: "Although she could do sports as well as any man, what made this young woman such a remarkable personality was that she also knew how to retain the feminine graces, which made her so likeable and interesting—and makes a person regret that we don't have more like her in France, women you can admire profoundly for their talents, who are truly independent but worthy of the name of 'woman.'"[30]

According to Marcel Cordier, Marie once said, "I practically live like a Carmelite nun." But she was eighty-one years old at the time.[31]

Gaby Curral Couttet, writer and daughter of a hotel keeper in Chamonix where Marie often stayed, knew Marie well, and said she watched Marie fascinate even a worldly explorer: "She used to spend hours exchanging ideas with Lucien Tignol, Legion of Honor, one of our friends. I had nicknamed him 'Plum Pudding,' which made him furious, but which suited him, because he was maroon from head to toe—suit, sweater, puttees, shoes, hat, and bronzed skin! Handsome as a god, charming, he made plenty of hearts beat faster. Even certain high society women became jealous of each other if he spoke to one of them. He was an explorer, traveled the world in dangerous and forbidden places, climbed mountains, feared neither deserts nor nomads, and had risked his life numerous times. A remarkable speaker, he had a seductive voice and kept his audiences breathless. And, with all that, he was completely blown away by Marie."[32]

That a few of her male friends were more than friends was where Marie showed discretion. She left a few words, a few clues, about her love life; that must suffice.

A reporter asked her, "Why have you led the life you have? Didn't you want to marry, have children, and become the good fairy of your own household?"[33]

Marie replied, "I wasn't made for that life. Action, danger, and adventure have always attracted me, and always will. Of *course* I could have gotten married! Heavens! I received many offers and declarations of love. I could write a book about it! But, no, destiny didn't want it that way, and I must admit I helped destiny along a little." Those who knew Marie were nearly unanimous in saying that she had a lot of suitors.

Another time she explained, "If I had been married, I could certainly never have lived the life I've lived, done the things I've done. I've not been just a sportswoman. I hope to have done some good in my life—I invented the ambulance-airplane, I fought at the front, I've tried to be useful in all circumstances. Everything I've done has been for humanitarian or scientific reasons. I could never have done that as a married woman."[34]

Marie was so frequently asked why she didn't marry that she became impatient with the question, frustrated that reporters were more interested in her single status than in her accomplishments. When the question of marriage came up once too often, Marie snapped, "Look here, anybody can get married. And anyone can have babies. Even cats!"

Years later, at the Toul-Rosières base, eighty-four-year-old Marie was giving a talk when an American woman asked why she had never married and had children. Marie sighed a little and gave her "even cats!" response one more time.[35]

One newspaper reported, "She has never married—lack of time, she assures us."[36]

As perfectly independent as she was (after her father's death, there was no one in the world she needed to answer to), Marie would not have liked the legal and social limitations of marriage. Her choice to remain single was as unusual at that time as her pursuit of sports, devotion to aviation, solo travels, invention of the ambulance-airplane, and everything else she did. Marie didn't live to discover her life; she lived to create it.

She never claimed to be much of a housekeeper (and rightly so, according to visitors to her apartment in Nancy),[37] but whenever Marie recited the long list of things she'd done, she always wrapped up triumphantly with, "And don't forget I got a First Prize in Cordon Bleu."[38] She liked to mention that her paternal grandfather was a chef and that she was as good at cooking as she was at sports. When reading over press clippings, she would add corrections and additions in the margins, always heavily underlining "first prize in cooking."[39] This obviously meant something to her, possibly a reaction to perceived insinuations about how "womanly" she was. But it was also her nature to want to be good at everything—and to be seen to be good at everything; her cooking prize was the only proof of her talent in the domestic arts.

Surveying the crowded salon in her apartment, a journalist remarked, "Diplomas, medals, trophies, photos of all that you've accomplished. All the same, there must be something you haven't done."

"Children!"[40]

The issue of children must have touched Marie both personally and as a patriot. The German annexation of Alsace and Lorraine in 1870 deprived France overnight of 1.5 million citizens. From 1870 to 1939, the government exhorted the French to have more children (and even today offers substantial benefits to new parents in an effort to stimulate the birth rate). The German population was nearly double that of France's, and the two countries had clashed three times in the space of sixty-six years, so the population question was significant.

If she ever regretted not having children (she did not appear to), she made up for it by surrounding herself with them whenever possible. Marie spoke to thousands of schoolchildren.[41] She hoped not only to attract young people to aviation and sports so they would be physically fit, but to encourage them to attain moral fitness as well. When she had free tickets to the circus, she took her neighbor's children.[42] She used to visit a family with eleven children three times a week, to entertain them and give their mother a break.[43] And a fond father insisted she truly loved his children.[44]

On one of her school visits, a young fan brought Marie a tiny model plane. The other pupils took note and the following week a flotilla of model planes were delivered, child by child, to her hotel.[45] A school principal noticed that "Mademoiselle Marvingt not only enjoys talking with children, but she possesses the art of getting them to talk."[46]

At a school in Nancy, Marie impressed sixteen-year-old Nicole with her colorful personality and her strength of mind. Nicole says she has remembered Marie's words all her life: "When you really want to do something, let nothing stop you. Go after your goal even when it seems impossible."[47]

At the end of her talk, Marie asked all those who wanted to be pilots to raise their hands. Nicole was the only student to do so. Marie carefully examined Nicole's ears and told her she had a talent for aviation. "I never knew what she was looking for or what she could have possibly seen in my ears!"

Later, Marie sent Nicole a postcard: "As I mentioned during my little talk, our schools are asleep, and one must wait to become a pilot. But perhaps you could start by taking gliding lessons when the School of Pont St. Vincent begins taking students. With my best wishes to you for success, Marie Marvingt."[48]

Today Nicole shrugs and says, "My dream of flying remained a dream. I was only sixteen then, and with no money, what could I do? But I never forgot the amazing Mademoiselle Marvingt."

Marie especially enjoyed the youngsters in her neighborhood in Nancy, inviting them into her apartment and watching them grow up over the years. She used to tell them, "You're my only family."[49] As for them, they often climbed the stairs to her apartment without invitation. She was half actor, half storyteller, and utterly fascinating to them.

During World War II, when her bicycle *Zéphyrine* was stolen from the courtyard she asked the neighborhood children to help her find it. The delighted children embraced the search as a sacred mission. They spent the afternoon scouring Nancy, finally locating the abandoned bike behind a church. (The fat-wheeled, forty-year-old machine was one that only its fond owner could love; it's little wonder the thieves quickly discarded it.) After carefully checking it for damage, Marie treated the children to *tartines* in her apartment while they explained how they tracked down her beloved bike.

Marie enjoyed keeping children spellbound with accounts of her adventures. She knew how to speak to them on their level, her voice changing like magic, her face and gestures mimicking people they knew or fantastic characters.[50] It's possible that Marie did not so much want children of her own as to keep in touch with the lively child in herself. Her enthusiasm, her innocence, and her impulsiveness gave her the airs of a charming child and kept her from ever seeming old.

Godmother to a number of children, she had a particular tenderness for the daughter of her friend and cousin, "dear Marthe," young Jeanne Roumier (later Lhérault). She wrote Jeanne regularly, beginning in childhood: "My dear little Jeanne, Thank you for your letter, and congratulations on your progress in handwriting. I wish you a happy new year, and I send you a huge hug."[51]

Marie liked to bike from Nancy to Troyes to visit cousins (and she was still doing this at age eighty). She knew she'd find a family group with plenty of children, who awaited her with the anticipation they gave Saint Nicolas.[52]

In her book, *Les Années Folles de Chamonix,* Gaby Curral-Couttet told how Marie, who visited her parents for many years, took her and her sister under her wing, and scolded them when they did something rash. "Imagine," said Curral-Couttet, "She worried about us, when everything she'd ever done had been dangerous!"[53]

A journalist ended an article by saying he told Marie, "You have perhaps deprived France of several children, Mademoiselle, but you have provided thousands of young people with an important example."[54]

Beyond a love interest, most people need an emotional and relational life. In that respect, Marie's need for family, friends, and intimacy appeared to be much less insistent than most people's. Over and over, people said, "She knew everyone, and everyone knew her."[55] But she didn't really know anyone intimately, and, as her godson said, few people truly knew Marie.[56]

The untimely deaths of her mother and brother, as well as her grievous loss of Paul Écheman may have made her hesitate to relate deeply and meaningfully to anyone again. Or, by nature independent and action-oriented, Marie may have spent herself in physical and intellectual domains, leaving less psychic energy for the emotional.

Marie was much happier going somewhere, doing something, working toward a goal, than engaging in long friend-making talks with another person. She spoke intimately and regularly with only a few people: her father, until his death in 1916, who seemed to share or to have genetically bequeathed to her his emotional detachment; Paul Écheman, with whom she could share herself; Marthe Goupit Roumier, a year younger than Marie, who was a cousin of sorts (Marthe was the granddaughter of Marie's father's aunt); and Marthe's daughter, Jeanne Roumier Lhérault.

Marthe, who lived in Argenteuil, usually initiated contact. She was the one who sent Marie news of the family, invited her to gatherings, and kept stacks of newspaper clippings of her doings (usually sent to her by Marie). A representative letter from Marie dated April 17, 1937: "My dear cousin, Thank you for letting me know about the family celebration, which I was so happy to attend. I hope this card finds you in good health. With my best wishes to your parents, to Jeanne, and to you."

Marie's letters to Marthe and to Jeanne were usually accounts of what she had been doing, along with birthday greetings or comments about recent family happenings. There was rarely a sense of intimacy, give-and-take, exploring beneath the surface. After Marthe's death, Jeanne and her husband represented the family for Marie, and kept their ties alive.

Safi le 29-4-22
chez gal Bertrand
à Casablanca
Ma chère Marthe,
Suis venue au Maroc le 3 [illegible]
par voie aérienne Toulouse
Tanger (première, infecte sans
donnée) ai fait tout le voyage
avec Millerand. Conférencie
sur l'aviation avec succès
dans le sud. Vais partir
le 10 en colonne avec gal Poeymirau
pour suivre opérations
un point dissident [illegible]
l'Atlas. Ne serai pas
en France. Bien [illegible]
amitiés à tous et à Fleur
Marie

CARTE POSTALE
CORRESPONDANCE
ADRESSE

Madame Ronnier
5 av. du petit Marly
Argenteuil
Seine

Reproduction interdite. — Modèle déposé.

One of Marie's many postcards to her cousin Marthe.

Marie wrote hundreds of letters (she rarely forgot anyone's important life events) and she received a prodigious amount of mail. Charles Codron wrote that "her letters, in their idiosyncratic style, were true delights for her numerous friends."[57] Madame Reboul, wife of a telegraph operator in Nancy, said her husband was always taking handfuls of telegrams to Marie.[58]

Marie preferred postcards she made herself, with pictures of her latest adventure; the other side left only enough room for fairly conventional, if sincere, greetings. She took great pleasure in mailing off a stack of postcards, but there seemed to be more action than feeling involved.

Friends who often socialized with her told about dinner-table conversations with Marie: "Fascinating! Oh, how we laughed and talked!" They recalled the stories Marie told them, descriptions of places she'd been, humorous anecdotes from her adventures.[59]

But did she ask about their children? Did she keep up with their work? Did she notice what they had done around the house?

Blank looks. Why would they talk about such pedestrian things when Marie had been, well, everywhere?

Marie was constantly on the move, and relationships take time. In her youth, many of the people she saw most frequently died young in plane crashes or in the war. Logistical reasons, and her personal losses may all have played a role in the dearth of emotional attachments, but Marie was a woman of enormous willpower; had she been made for intimate relationships, she would have had them. Her nature was physical and intellectual. She had a big heart, but it embraced all humanity. She worked to serve others, but in the plural. She was almost constantly surrounded by people, but for her it was a case of audiences, banquets, events, and serial conversations. No one ever photographed Marie sitting in a café with a friend.

Marie and Paul Écheman seem to have been a perfect match, but fate never gave them a chance to see if their delight in each other could have survived domesticity or a longterm relationship. Their brief, intense relationship seemed, however, to have sustained her all her life in many ways, fulfilling a part of her that needed to say, yes, she had experienced that, too.

9

Making News, Breaking News

"How long did the interview last? Five minutes, half an hour, all day long? I don't know! Beginning with her first words, I no longer had any idea of time. All I could do was focus on listening, listening, and listening again, because the most marvelous story was being offered to me."—R.P., *Le Cri de Constantin* (1924)[1]

Marie's relationship with the press was an oddly harmonious and exuberant friendship that sustained her, in more ways than one, all her life. She was the news and the reporter of the news, the story and the story teller, and sometimes she was both, narrating her own adventures for the newspapers.

As the news, she was written about thousands of times, the unabashed darling of journalists. She did virtually nothing that failed to reach the pages of some newspaper. Sometimes her name appeared only to note she was back in town for the moment or had attended this event or that banquet.[2] Slow day in the newsroom? What's Marvingt doing?

Marie spent more time with reporters and photographers than with anyone else in her life. She knew them all, and the good ones could recite her accomplishments by heart. The press isn't always kind to those who swim upstream from convention, nor, as history often shows, to women who step out of line. But for the press, it's all about the news. And Marie was good copy. France, and the wider world, wanted to know what she did yesterday, what she was up to today, and what she had planned for tomorrow. Marie retooled conventions and created popular culture. That's news.

Sportswriters have occasionally admitted that the unwritten rule is to write about the best-looking women. Add the word daredevil, and the Fiancée of Danger was always good for a few column inches. Almost without exception, interviewers found her charming, animated, gay, and likeable. The short answer to her popularity with the press might be as simple as "charisma." Armand Rio, who interviewed her at length, wrote about her "good humor and high-spiritedness, crowned by a perfectly feminine gracefulness."[3]

For whatever reasons—her noteworthy activities, her personal magnetism, her vivacious good looks, her brushes with death—Marie was given the kind of publicity that people pay millions for today. Time after time, her name popped up, even when the event or article wasn't about her. She was at the head of every list, photographed next to the heroes of the moment, front and center in every group.[4] How did she consistently get that kind of notice?

A review of hundreds of articles indicates it was not a case of "ladies first," as might have been expected in that era. Nor was there a press agent or money involved. "La Marie" earned her star treatment the hard way, by confronting danger, by being better than almost everyone else at everything she did, and by conveying a genuineness that secured her immunity from spite and backlash.

As a journalist herself, she was a colleague of most of the reporters writing about her. This could have worked for her or against her. Competitive small-mindedness among her peers might have resulted in their neglecting to include her or to mention her in the last paragraph, which was often lopped off (too bad!) in the layout room. But that didn't happen. Simply, the press liked her. And it showed. For a public figure (idols today are not so gently treated), she was surrounded by the most laudatory, cheerful, and friendly publicity anyone could have desired.

Marie appeared to use newspapers as a combination personal assistant and public relations agent. In 1932, for example, when she was invited to be the guest speaker for the Geographical Society meeting on January 13 in Nancy, the local newspaper, *L'Est Républicain*, began on January 6 to describe the "sensational soirée" in store for the audience. On January 10, they added a few new details about the event. On January 12 they warned readers not to miss the "exceptional evening" to be held the next day. On January 13, the newspaper reminded everyone that tonight was the night! The following day, sounding fatigued, the newspaper informed readers that the conference the previous evening had been so successful and had continued so late into the night that they needed more time to write about it.[5] On January 15, a 700-word article detailed the sparkling evening, undoubtedly making all those who had not attended feel quite envious. The article listed the luminaries in attendance, and noted that the president of the Geographical Society scarcely needed to introduce Marie, who is "so well known and loved in Nancy, which is her own small country," but he did anyway, speaking of her enterprise, intrepidity, and many merits. They reported that she ended her talk by reciting several poems she'd written, inspired by the desert.[6] The audience applauded at length, and left the hall promising themselves to return for Mademoiselle Marvingt's next conference, to be held on January 25.

And so began the next round of articles, each one longer, more persuasive, and more vivid than the last: January 20, January 22, January 23, January 25 ("tonight!"), January 26 (the newspaper is tired again), and on January 27 a lengthy summary of what she said and how she said it entitled, "When Mademoiselle Marvingt Gives a Conference."[7] (The same pattern appeared in her relationships with North African newspapers.[8])

Similarly, that fall, for her particularly well-received talk, "Visions of the Orient," lengthy articles with photos, lists of slides, and details of her travels were published October 4, 6, 9, and 10.[9] (When Marie spoke, her event was often scheduled to begin "at 10:00 a.m. very precisely." The precision was Marie's.[10])

Before, during, and after these talks, and many others given to schoolchildren, the newspaper apparently allowed her to use it as her own personal message board: "Mademoiselle Marvingt wishes to thank everyone who signed up for one of the aviation leagues, and notes that you can still enroll, up until the end of the month, by sending your money to her at 8, Place de la Carrière."[11] Or: "Mademoiselle Marvingt is leaving Nancy this week, so she would like to thank all those who came to her talks, and to remind them that they can send their enrollments to headquarters in Paris."[12] Or: "Mademoiselle Marvingt would like to repeat her thanks to those who joined the league, to all the generous donors to it, to the French and American girls to whom she spoke, to the ushers who helped her Saturday evening. She would also like you to know that more of the signed prints are available at the Salle Poirel; she regrets running out of them Saturday night."[13]

Even when Marie wasn't in Nancy, the local paper kept track of her: "Many people have been wondering what Mademoiselle Marvingt is doing these days." They assured readers that she was happily engaged giving lectures in Tunisia, but was heading for southern Algeria and then to Morocco. From there she planned to fly to Dakar and visit parts of South Africa.[14]

During Marie's lifetime, her name appeared in newspapers, magazines, and books thousands of times, principally in France, Spain, Italy, Germany, Great Britain, Canada, Australia, the United States, and China. She was in all the big newspapers—*Le Figaro, Le Journal [Paris], Le Petit Parisien, Le Temps, The Times [London], London Standard, London Evening News, Sunday Times [Sydney], New York Times, The Washington Post, Boston Sunday Post, Los Angeles Times, Los Angeles Herald, Pacific Stars and Stripes*, and *European Stars and Stripes*—but she was equally good copy in small towns everywhere. The *Mercury* in Hobart, Tasmania, was talking about her in 1907.[15] In 1910, when she obtained her pilot's license, the news was reported everywhere from the *Weekly Sun* of Brandon, Manitoba, to the *Gleaner* of Kingston, Jamaica, to the Perth *West Australian*.[16] And when, soon afterwards, she set the first women's aviation records, people read about it in New York, London, and Pocahontas, Iowa.[17]

Many magazines and newspapers devoted lengthy, illustrated articles to her (she kept and annotated them, correcting errors and dates): *The Strand* (London); *L'Illustré* (Lausanne, Switzerland); *Forest and Stream* (U.S.A.; later absorbed by *Field and Stream*); *Verkusten* (Sweden); *Marie-France, Aviation Magazine*; *Point de Vue Images du Monde*; *La Croix*; *Les Anciens Combattants du Monde, Samedi Soir* (France), and more.[18]

Swiss journalist Georges Gygax explained his interest in her: "This past month, the French press published a curious bit of news. An 83-year-old woman from Nancy, Marie Marvingt, appeared to be taking pilot lessons on a jet-engined helicopter. A representative from *L'Illustré* was sent to verify the news. Marie Marvingt, her chest bespangled with decorations, received him despite a dizzying schedule: conferences, diction lessons, television, radio, press interviews, etc. The visit went on and on, with the visitor tossed into an existence so extraordinary, so exceptional, and so unexpected that he couldn't resist introducing her to the readers of *L'Illustré*." The resulting article ran so long and was illustrated so profusely that it appeared in two parts.[19]

In 1913, articles naming her the best sportswoman in the world circled the globe. Even in the years after her most noteworthy exploits, newspapers flagged any report with her name in it. In 1928, readers from Appleton, Wisconsin, to Fairbanks, Alaska, to Ogden, Utah, learned that a "French Woman Flier Predicts Air Travel Will Predominate."[20] Her arrival in Albuquerque in 1935 made the front page of the *Albuquerque Journal*.[21] She was particularly prominent in periodicals devoted to sports and aviation—she was featured on the covers of *L'Aéro*, *L'Aviation Illustrée*, and *La Vie au Grand Air*[22]—but she could be found smiling in the pages of popular dailies, weeklies, and monthlies covering everything from general news to mountain climbing to travel.

French newspapers have been integral to French politics, economics, and society since the early 1600s, but Marie benefited from an explosion of newspaper popularity that has not been seen before or since.[23] Circulation of the daily press in Paris went from one million in 1870, a few years before she was born, to five million in 1910, the year she won several balloon races, obtained her pilot's license, and set the first women's aviation records. *Le Petit Parisien*, which covered Marie extensively, had the largest newspaper circulation in the world at that time. In 1910, in the six months from June to December,

Marie was mentioned in twenty-two issues. The main dailies employed their own journalists who competed for news, which was then flashed throughout the world by the Agence Havas (now Agence France-Presse).

Marie wasn't always an easy interview. She became impatient when asked the same questions over and over, and she complained that she didn't have time for that sort of thing. Her efforts to make nice were sometimes almost visible; she remembered that publicity led directly to support for the causes she cherished.

She told journalist Armand Rio, "I don't like listing what everyone calls my 'exploits,' but how can I be quiet when I've agreed to an interview? And, so, because you condemn me to name my honors, I will do so, like a little girl reciting in the classroom."[24]

An Algerian reporter wrote, in 1924, that his editor assigned him to "first thing Monday morning, at 8:00, go interview Mademoiselle Marvingt at her hotel."[25]

"A woman up already and prepared to be interviewed at 8:00? Not possible," he thought. He went anyway.

"How long did the interview last? Five minutes, half an hour, all day long? I don't know! Beginning with her first words, I no longer had any idea of time. All I could do was focus on listening, listening, and listening again, because the most marvelous story was being offered to me." He claimed it would take several issues of his newspaper to tell everything he heard that day, and confessed, "I went to the interview with a very limited sympathy for female athletes, convinced that woman's place is in the home, not in the stadium, but I came away with an entirely different view."[26]

French sports journalist Frantz Reichel adored Marie, and wrote about her often. Reichel was an outstanding athlete, an Olympic gold-medalist, founder of the first international sports press association, and the first European journalist to fly a plane. In her autograph book, he wrote, "It is with great emotion, dear friend, that I outline your career, which is full of goodness, of useful and courageous gestures. You are the very soul of action, a woman of duty and devotion. By your words you have never stopped encouraging everything that is good, beautiful, generous, fearless. ... You are infinitely loved and admired."[27]

By 1910, when this photo was taken, she had been writing for the newspapers for five years.

When Marie died in 1963, hundreds of newspapers on six continents announced "Pioneer Woman Pilot Succumbs" or "French Sportswoman Dies" or "Famed Sportswoman and Aviatrix Dies." The *New York Times* published a long and laudatory obituary, covering her early career and her life as "the fiancée of danger." The piece closed with: "She was last at the controls of a plane in 1955, at the age of 80, when she piloted a jet-engined helicopter. During World War I Miss Marvingt, a college graduate and certified nurse, became a war correspondent. It made news when she reached the front line and fired at the enemy."[28]

The news of Marie's death rippled outward for a few days, a few weeks.

And then, the long silence.

Her death appeared to be more final than most people's. The curtain rang down, and there was a virtual absence of her name from every newspaper, record book, and compendium for the next twenty years. But that's another chapter.

As a reporter, Marie began submitting pieces at least as early as 1905, when she wrote for Nancy's *L'Est Républicain*. By 1907 she was hired by *Éclair de l'Est* and *Le Sport*, both of which referred to her thereafter as their colleague. (One of her most popular articles for *Le Sport* told readers how they too could become a pilot.[29])

She wrote for some forty publications, including general-interest newspapers and magazines (*Journal du Samedi Soir, Courrier de l'Est, Républicain Lorrain, Le Lorrain, Dimanche Éclair, Le Petit Journal, Daily Mail [London], The Times [London], L'Excelsior, La Croix, Lectures Pour Tous, La Vigie Marocaine, Le Petit Marocain, La Dépêche Tunisienne*) as well as aviation weeklies and monthlies (*Revue Aéronautique de France, L'Aérophile, Les Ailes, L'Echo des Ailes, Revue de la Société Scientifique et Historique de Documentation Aérienne*) and sports journals (*La Vie au Grand Air, Le Touring-Club de France, La Culture Physique*). In the United States, she said she'd written a piece for the *Los Angeles Times* on the Ethiopian situation, which was very much in the news at the time.[30] A number of anonymous articles from a "special envoy" were written about events that Marie attended, and appear to be her writing.

She bylined her newspaper articles "Myriel," which was also the name of one of the balloons she flew.[31] A curious choice, "Myriel" is seldom seen in France as a first name (and even less often as a last name). Most sources say the name's etymology and meaning are unknown.

The most likely Myriel for Marie would have been the fictional Bishop Myriel, the heroic personification of compassion and mercy in Victor Hugo's 1862 *Les Misérables*. Marie used the word "mercy" thousands of times in her talks, and referred most often to ambulance-airplanes as "the wings of mercy."[32] The choice of *nom de plume* may have had deep meaning for her. Then again, she may simply have liked the catchy one-word name.

Marie wrote under her own name for op-ed pieces, when relating her own adventures, and when reporting news to which she had a connection—for example, a 1914 on-the-ground report of the bombing of Nancy.[33] She wrote about Latham, Blériot, and other great personalities of the day because she knew more about them and had more personal stories than any other reporter.

Marie covered sports, aviation, celebrities, the military, and theater and literary events, attracting readers with her vivid, detail-rich stories studded with insider information—names, events, and places most readers wanted to know more about. Describing the first time she piloted a balloon, she brought to life the low-flying balloon, "the hungry express train approaching us with its huge luminous eyes," and the handful of sand she clutched in case she needed to escape the oncoming train.[34]

Marie visited airplane and balloon workshops to report on what was new. It was while researching an article on the builder Roger Sommer that she encountered flying bicycles[35] for the first time—and signed up immediately for lessons.[36]

She described the hayloft—"the nest of a new bird about to hatch"—where an aircraft designer stored the pieces of his new plane.[37] She talked him into carrying each of the parts (precariously, laboriously) down a ladder from the hayloft and reassembling them all in the street so she could take a photograph and explain the workings of a plane to her readers. She helped him carry and assemble, of course. But how did she persuade a struggling, part-time aviation designer to give up his day for her? The last paragraph of her article informs readers that if they would like to contribute to his work, donations could be left at the newspaper office.[38]

At the first international airshow in Reims in August 1909, she wrote: "This wonderful week finished with a superb finale. For the last day, we've had some incredible flights. Those big goldfish dirigibles, the *Zodiac* and the *Colonel-Renard*, captivated the crowds, with the latter winning the grand prize. As soon as the drag rope came out, I ran from the newspaper box to the hanger to photograph it."[39] There she interviewed its pilot (and her friend) the Count de la Vaulx, who was still limping a little from his last accident.

Professional, energetic, and connected, Marie knew everyone—inventors, owners, flyers, engineers, mechanics, workers. She brought the real-life, emerging aviation world to readers. Her enthusiastic, well-written, and useful articles helped win her a warm "welcome!" when she knocked on the next door for a story.

Her background and firsthand experience led to, for example, instructive, suspenseful articles about Alpine climbing and skiing. She explained unfamiliar terms and described scenes never experienced by her readers. Some of her articles inspired more than one new alpinist and popularized a number of Alpine winter resorts.[40]

In the summer months, the Alps were overrun by tourists. But as soon as autumn arrived, hotels closed their doors, stores shuttered, and guides and inhabitants slept in houses covered with snow. Choosing to visit the Alps in the winter months, much less hope to enjoy oneself, had previously been thought quite mad.[41] Marie, who spent the winters of 1908, 1909, and 1910 at the winter sports games in Chamonix, Ballon d'Alsace, and Gérardmer, was the ideal promoter for the novel concept of winter fun for the newly established resorts in the Alps and the brand-new sports themselves.

Her lengthy, two-part 1909 article for *The Times* of London, "The Alps in Winter," provided alluring, first-hand information on the new Swiss winter resorts, the sports practiced there (skating, hockey, curling, tobogganing, bobsledding, snowshoeing, and ski-running, as it was then called), the clothing and gear needed, and the climate ("some places, it is true, are sheltered, but at a price: the mountain that shuts off the wind also shuts off the sun."). She suggested readers send for pamphlets from the various resorts, adding, "They contain further indications for those who know how to read between the lines. Letters of recommendation, for instance, are likely to be signed by persons the hotel proprietor considers important: there won't be a letter from the bos'n about the admiral's inn."[42]

She further counseled, "In Switzerland, the hotel proprietors are aristocrats, and, like aristocrats nearer home, they are disinclined to sell for 10 francs what will fetch 15. Visitors may describe this characteristic as commercial, grasping, or ducal, according to their powers of invective, but they will do well to reckon with it when they make their selection. The price tends to rise with the altitude for a reason that may easily be appreciated; the higher the situation, the longer the winter." She warned visitors who went in November, February, or March to "rule out at once all the lower stations. There is no price so cheap as to make an Alpine village attractive in a thaw. "[43]

She advised not to spend too much on skating boots: "A champion would rip the stitches out of a pair that will serve the novice admirably, provided they fit closely at the heel and are loose at the toe—a tight toe is a cold toe." She offered a tip for a better fit: "Instead of throwing away your evening socks, when a toe comes through, cut off the toe portion; place them under your stockings and you may find that an old loose boot now fits tightly at the heel and is as easy as ever in front." She warned against the faddish "lion-tamers' boots": they "are not for recreant limbs that flinch at turns; they cost a lot of money, they take a long time to lace up, they excite unfulfilled expectations among the spectators, and their particular merits can only be turned to account by experts."

She thought of everything: "There is a beautiful strange superstition that you need to arrive in the Alps in a fur coat; but there are no grounds for it. If you are a motor-driver or a tenor and it is part of your professional equipment, bring it by all means, but it cannot be called a necessity unless you intend to sleigh over high passes or unless an aunt is paying for your kit. Besides, few people are strong-minded enough to leave a fur coat lying about; it has to be locked up safely three separate times a day." And fancy-dress clothes are unnecessary: A man "once sent for his tailor from London to make him a toreador's dress, and most people even then would have bet on the bull."[44]

Of an excursion sponsored by the French Touring Club, Marie wrote: "Everything was perfect. The lively receptions and cordial welcomes from the cities of Gérardmer, Saint-Dié, Bussang, and Remiremont will remain etched forever in my book of souvenirs. ... Beautiful weather—a nice surprise after certain grim forecasts—permitted sleigh excursions to Retrounemer, Donon, La Bresse, Cornimont, Orbey, Drumont, Ballon d'Alsace. ... Everywhere and always, there was joy, gaiety, and good humor. ... When we descended from the luxurious private train, fifty sleighs awaited us, to carry us off to one enchanting site after another, surrounded by friends and those who knew how to appreciate the glorious winter season. ... There was so much progress and so many more new 'players' than last year. Almost everyone, young and old, wanted to taste the intoxication of those flights down snowy slopes where luges and skis traced long sinuous trails. With joyous hearts and imaginations wonderfully stimulated, we arrived at nightfall in gaily decorated cities, received by local officials and enthusiastic crowds, music playing. We proceeded to our hotels where our luggage awaited us in our respective rooms. After evening banquets, where no one spent too much time speechifying, we returned to the hotel with torches illuminating the embracing mountains that surrounded us.... Despite the strenuous days, the balls, and the early-morning awakenings, we're not tired: the truly fit never are."[45]

It is possible that by "we're not tired," Marie really meant "I'm not tired," as her vitality was unparalleled. An Algerian paper once wrote, "The adjective 'energetic' seems very feeble when applied to Mademoiselle Marie Marvingt."[46]

She warned that the resorts had few satisfactory ballrooms. "The result is that the dancers melt. The alternative, that the chaperons should freeze, need not be considered: they pay the bills."[47]

Attending the official inauguration of the new winter resort area of Aix-les-Bains-Revard-Chambéry (which was competing with the established resort area of Saint-Moritz in Switzerland), she described "a day of enchantment, from the warm welcome to the friendly company, the temperature, and the snow."[48] She noted that the previous November eight skiers touched the snow there for the first time; now, in February, some five hundred were taking advantage of the new train chuffing them up to the "Eden of skiers." She admired the enormous work of laying nine kilometers of railroad tracks through snow, sometimes three feet deep, and noted that "the thin telephone lines that link the Revard hotel to the city of Aix have, thanks to the recent snow, metamorphosed into graceful ostrich feathers." She closed the piece by describing the happy skiers, waving to each other as they leave, saying cleverly, "Au Revard!"[49]

Marie did a great deal to popularize Alpine winter tourism. When she wrote, "I've stayed in all the French, Swiss, and German winter resorts, and there are none to rival Revard," other newspapers picked up her comments, and resorts used them to advertise.[50] Adding even more interest, she was often competing in the same winter sports she was covering for her papers.

Marie also did a great deal for tourism in North Africa, scouting the most attractive sites for her documentaries, and bringing stories of the area back with her to French audiences. While in North Africa, she filed stories with *La Vigie Marocaine*, *La Dépêche Tunisienne*, and *Petit Marocain*. For the latter, she wrote a lengthy piece honoring the memories of Lyautey and Poincaré, who had just died.[51]

Who else could reminisce about being invited by Poincaré to the presidential residence the day after his election? Or relay his thoughts on the Great War?[52] She reminded readers that great men often had great wives, and she spoke of the contributions of Mesdames Lyautey and Poincaré. Four days later, in *La Vigie Marocaine*, she devoted a long article to "Their Wives." Yes, they were still principally wives, but Marie posited that their contributions needed to be considered along with the king, president, general, or notable with whom their lives were linked, that they were worthy of admiration and visibility in their own right. Marie wrote about Queen Marie of Yugoslavia, Madame Barthou, Madame Poincaré, Madame Lyautey, and other outstanding women.[53]

Deeply patriotic, Marie familiarized herself with politics and current events as a matter of principle. Early in her journalism career, in 1905, she wrote about Wilhelm II, the last German Emperor and King of Prussia, telling readers about the time he visited the little town of Courcelles, since 1870 part of Germany but still mourned by the French who had lost the region.[54]

Some eighteen kilometers east of Metz, the Château d'Urville in Courcelles welcomed the sometimes bombastic, bellicose emperor, and Marie insisted on covering the visit. She had been warning readers and audiences for several years about what she saw as growing German aggression.

Her lengthy article described the schoolchildren carrying flowers and German flags who were required to yell "Hoch!" when the emperor passed by. Marie pointed out, however, that the children spoke in French while they waited for the emperor and their cries of "Hoch!" were barely audible. It didn't really matter, though, as the emperor and his entourage neither stopped nor acknowledged the villagers who had been obliged to wait for hours to see them.

Marie described three faces of Wilhelm II. While sitting near him in the church Sunday morning, thanks to the friendly Protestant pastor's assistance, she noted that he hadn't changed much since she saw him ten years earlier. He was a little grayer, but his mustache was as erect and proper as the emperor himself. His look around the church was cold, hard, severe.

Later, as Marie was walking on a country lane, the imperial carriage came toward her. The emperor, accompanied by an officer, stopped to chat. He was another man this time—gay, smiling, his triangular expression lengthening, chin and jaw softening. "Only his mustache remains the same," she wrote.

MEURTHE-ET-MOSELLE

Canton de Nancy

VILLE DE NANCY

SAUF-CONDUIT N° 193.132

Valable pour ... voyage(1)

du ... au ... 192.

Mode de locomotion autorisé (2)

Localités ou périmètres de circulation autorisés :

Mad.lle Marvingt Marie Français,

Profession de

né le 20 Fév 1875 à Aurillac

Domicilié à Nancy

est autorisé à faire usage du présent sauf-conduit, dans les conditions sus-indiquées.

Je certifie qu'à ma connaissance son attitude, au point de vue national n'a jamais donné lieu à remarque.

Fait à Nancy, le 24 Août 1920

Le Commissaire de Police.

SIGNALEMENT :

Age 46 ans

Taille

Cheveux

Sourcils

Barbe

Yeux

Nez

Menton

Front

Teint

Signes particuliers :

Signature du Titulaire :

M. Marvingt

Sauf dispositions spéciales, le présent sauf-conduit servira de permis de séjour dans les limites de date fixée. Si le sauf-conduit, une fois périmé, n'a pas été retiré au porteur à la gare de retour, il est à rendre par l'intéressé à l'autorité qui l'a délivré.

Copy of Marie's safe-conduct pass.

And then, the next morning, while leaving the village to return home on *Solidité*, her bicycle, she met yet a third emperor. On a superb piebald horse, he was dressed for hunting—saber on his left, long sword on his right, "perhaps a revolver hidden on him somewhere." He was neither the severe autocrat nor the charming man of the world. This time he was a lover of nature, entirely given over to the morning breezes, the beauty of the lilies of the valley, and other forest perfumes.

"When he passes me on his horse, he gives me a military wave and the same smile as yesterday. This morning, as a French cyclist, I respond openly, greeting this time the sportsman, the private man. The emperor's entourage, on nine horses, is in high good humor, laughing and talking; most of them call out 'Morgen, Fraulein,' to which I respond in French."

With her interest in horseback riding, she followed the group for a while on her bike, and appraised the emperor as a first-rate horseman, correct in every respect. After a woodcutter did a double take at seeing Wilhelm, his men, and the woman on the bicycle parade past, Marie turned around and headed back to French territory.

When 85-year-old Cardinal Luçon of Reims took his first airplane ride in 1928, Marie wrote about the event for several newspapers and magazines.[55] She relayed the Cardinal's grief that the Cathedral of Reims, the historic Gothic masterpiece where all the kings of France had been crowned, had been "unjustly murdered" by German bombs despite his repeated assurances ("on his honor") that the Cathedral served no military purpose. At the time of his aerial adventure, the Cardinal was in the midst of restoring the glorious stained-glass windows. Marie took up a collection among the airfield pilots, mechanics, and visitors, and handed over 3,000 francs to the Cardinal for his work.[56]

Before the Cardinal's flight, Marie leaned in the aircraft window and asked if he wasn't a bit nervous. He said, "I can certainly do once what you and your colleagues do every day. Twenty years ago I couldn't have done this because of what people would say. But today aviation has done wonderful things. And then, too, one has to keep up with one's times." After the flight, as she said goodbye, the Cardinal confided to her his most devout wish: that there be an end to war.[57]

She witnessed, and reported on, the 1932 Ierissos earthquake in Greece. Like many of her articles, this one made the front page.[58] Marie inhaled information with every breath and exhaled onto the page the telling particulars of what she'd seen; readers could almost relive her experiences. She seemed to understand precisely what readers wanted to know, and she gave it to them. The subjects of her pieces never complained about inaccuracies or negative spin. Journalism was a way for her to experience the world, to be part of events, and she took seriously her task of writing "a first draft of history."

One of her rather delightful pieces took Marie's readers behind the scenes of a magnificent royal banquet held at Versailles in 1938.[59]

At her first attempt to get into the kitchens, a guard told her, "No way, my little woman. Nobody gets in, especially not you annoying journalists."

She saw a dozen women, men, and children coming out of the kitchens.

"Wait! Nobody gets in? Who are those people?"

"That's different," said the guard. "Those are my friends."

Marie couldn't get past him. Wandering about, she spied a tall, distinguished-looking fellow dressed as a maître d'. Ah, Carlon's son-in-law. Carlon was the grand master of the banquet, overseeing 70 maître d's in wigs.

"I can't take you in myself," said the son-in-law, "but ask for Richard. Tell him Alex sent you, and he'll get you in."

"Richard" was apparently the magic word because, this time, the guard let her through.

In the kitchens, she interviewed the holy trinity (chef, sous-chef, sommelier). Knowing that few of her readers would ever get in the kitchens of Versailles, she gathered the details most likely to entertain them.

Each guest's place setting included twelve enormously costly Baccarat glasses for wines, champagnes, water, and cognac. The napkins and tablecloth (which was 148 feet long and 10 feet wide) were of double-thickness white satin. Intricate sugar sculptures, covered with fresh flowers and adorned with photos of the two young princesses, were placed before the queen and king. The china was Sèvres. One of the champagnes served was an 1895 Pommery, the year the queen was born. The cognac was Rémy Martin.

And the important question: "Does the queen eat bread?"

"No," said the grand chef, "I myself prepared little diamond-shaped toasts for her just like the ones she has for breakfast."

The description of each course was detailed and mouth-watering. After marveling at the years it took to create such an extraordinary chef, Marie asked him how the menu was chosen. He said the king and queen tasted everything being considered, but particularly liked the quail stuffed à la Talleyrand.

Marie asked if it would be indiscreet to request the recipe. Monsieur Carlon kindly gave it to her, and she passed it on to her readers.[60]

Years earlier, the great chef Curcio Morosini, a renowned restaurateur in Constantine, Algeria, gave Marie one of the volumes of recipes gathered and published by French epicure and gastronome Brillat Savarin—a rare book even then. She closed her article by saying that she left Versailles hoping to regale her friends with a few precious recipes tucked into the Brillat Savarin book.

Her war correspondent title was earned principally in World War I, when she covered the Italian front (often on bicycle), and in the North African Riff campaigns while accompanying Italian, French, and Spanish troops.[61] In World War I, she never went anywhere without her Ministry of War-issued little yellow press card with her photograph on it. It was, for her, a combination magic carpet and "Open, Sesame." Friends said that all her life she shamelessly used her press card to gain entry to some of the most interesting events of her day.[62]

The journalistic life was a perfect fit for Marie. She was an artful traveler and almost embarrassingly curious about everything. Moreover, she was often at the center of what most fascinated the French reading public at that time: sports, aviation, celebrities, travel, "firsts."

Her status as a journalist and her press contacts were priceless for publicizing her conferences and her causes. She believed in French tourism, in aviation, in her heroes, in her ambulance airplane—and she was able to convincingly advance their causes to the public through her articles and her newspaper friends.

Marie was still writing for newspapers at the age of seventy-four. A 1950 issue of *Le Lorrain* carried her piece about the "crusade of kindness" (one should smile constantly, even when alone; spare others pain; never show anger, unhappiness, sullenness),[63] while another, in *Dimanche Clair,* recounted the deaths of three Lorraines from the world of aviation.[64] A third introduced readers to a nun who had spent her life caring for orphans, and told how she managed to find food for her charges during two world wars.[65]

Early in her career, Marie got into the habit of dropping by the various Parisian newspaper bureaus—to chat, pick up breaking news, maybe suggest a story. An editor from the Paris office of the magazine *Télé-Paris* said that in her later years Marie—who was "kindness and generosity personified"—still stopped by from time to time, always with an outline for an article or an idea whose time she felt had come. He remembers that her projects usually described efforts to help others.[66] Aviation journalist Georges Houard of *Les Ailes* also mentioned Marie, "already in her eighties," biking across Paris to visit their offices.[67]

In the same way, in her later years, when Marie was more often at home in Nancy, there was scarcely a film, conference, or exhibition that she didn't attend, and review for the papers.

A small irony: on Marie's death certificate, she is listed as "without profession."[68]

10

Globetrotter

> "Mademoiselle Marvingt has just returned from fifty months in Africa, Southern Morocco, and Egypt. Altogether, she has covered more than 30,000 miles of territory, using the airplane, automobile, horse, mule, camel, as well as her own two feet."—*Berkeley [CA] Daily Gazette*, 1926[1]

A discreet want ad appeared on the last page of a 1901 travel monthly: "Mlle. M. Marvingt, 8, Place de la Carrière, Nancy, would like to exchange postcards, illustrated, from all countries, with the original stamps."[2]

The young woman who dreamed of the wider world would eventually travel thousands of miles across Europe, North Africa, Asia, and North America. In 1908, she traveled to the Arctic with a French expedition.[3] In 1911, she interrupted her work toward a military pilot's license to travel to Dalmatia, Herzogovina, and Montenegro for reasons she never explained.[4] In 1930, she flew over the Acropolis during a lecture tour in Greece.[5] In 1954, she attended the canonization of Pope Pius X in Rome.[6] In 1960, she was expecting a telegram to confirm a proposed visit to Argentina.[7]

Le Figaro journalist Maurice Londet claimed Marie "travels six months out of every twelve ... her proverbial energy and constant activity have earned her great public affection and admiration."[8] In 1926, under the headline "French Woman Has Record for Traveling and Sports," the *Berkeley Daily Gazette* wrote, "An old French saying which describes a Frenchman as a person who eats bread and never travels, must have been written long before the time of Mlle. Marie Marvingt, a Frenchwoman who holds a record that would make any American globetrotter feel envious."[9]

Another paper added provocatively that she traveled sometimes with official groups, other times with "a more or less reputable guide."[10] She was referred to in several articles as a "méhariste," which usually means a member of a desert-raiding camel corps, but in her case probably alluded to her familiarity with that means of transport.[11] When Marie wanted to go somewhere, she was opportunistic about her travel options. She once hitched a ride in North Africa to Tamanrasset with a convoy of fruit trucks owned by Tropical Transports.[12] And she was not above riding in a tank if it was going her way.[13]

During her life, Marie visited all of France, most of Europe (notably Belgium, Italy, Switzerland,[14] Spain, Germany, Great Britain, Wales, Turkey, Greece, Dalmatia, Herzegovna, Montenegro, and Monaco), and a bit of Asia (India).[15] She spent nearly two years in Greece and Turkey, giving talks, celebrating the centenary of Greek independence, and visiting both fabled and little-known places: the monasteries of Mont Athos, the theater at Delphi, Crete, Knossos, the isle of Delos, Istanbul, Smyrna, and many more.[16] As always, she found stories, lessons, and new acquaintances.

She toured the United States twice, in 1935–1936 and again in 1937, covering an estimated 75,000 miles in her peregrinations.[17] The first time, she was sponsored by the International Women's Aeronautics Association and was a much sought-after speaker by women's clubs and fans (more precisely, fanatics) of aviation. She essentially talked herself from one side of the country to the other.[18]

Her goals for her visits to the United States were several: to exchange information on aviation, outlining what was happening in France and discovering aviation's status in the United States; to promote, ever promote, her ambulance-airplane (she was concerned that the United States had not yet begun to develop air-rescue operations); to encourage Franco-American relations by bringing a sense of the French people to the U.S. and taking home with her a sense of what she had learned in the United States.[19] She repaid hosts, sponsors, and audiences with entertaining and informative speeches (she gave at least sixty-six[20]) about the debuts of aviation and her own adventurous life.

Marie on the steps of the New York Public Library, 1936.

As she had always done in France and North Africa, she visited numerous schools, which she viewed as the fountains of the future.[21] She made certain her audiences knew the name of the first American female pilot. Asking questions, as was her style in school visits, she found that only one of the thousands of students she saw was able to respond: "It was Miss Quimby, of Boston."[22] (In one of the air museums she toured, she pointed out that they had mistakenly named another woman as the first.)[23]

And she had other business. According to a 1935 issue of *Popular Aviation*, "Mme. Marie Marvingt, the first French woman to fly, is in the United States as a member of the French Legion and will attend the American Legion convention."[24] Described as "a gray-haired French aviatrix," she "was one of 315 Poilus greeted by Mayor LaGuardia yesterday as 'comrades.' The veterans, many wearing their country's highest decorations, arrived by train from Montreal and Niagara Falls for a two and a half day visit. They will participate in memorial exercises and feast with American veterans."[25] She then traveled to Washington, D.C., "to instigate interest in the convention of the American Legion to be held in Paris in 1937."[26]

The Washington Post wrote several pieces about her speech (delivered in a "rapid-fire flow of words") to the Alliance Française at the Hotel Carlton, introducing her as France's pioneer female flier who was also a sportswoman, inventor, teacher, poet, artist, journalist, and World War I veteran. "Mlle. Marvingt recently arrived in Washington, after a five-month air tour of this country. She came over to study American aviation, she said, and to find out 'how to stimulate interest in aviation in France.'"[27]

The article added, "With a number of inventions to her credit, she is proudest of all of the ambulance-airplane she designed. 'I have been told there are six ambulance planes in America now,' she said. 'But I have traveled all over the country without finding one of them.'"[28]

She apparently kept looking because when she met with General Westover in Washington, he responded to her query by explaining the United States had only one military air-ambulance, in San Antonio.[29]

She was naturally interested in American women who had become pilots; nearly 400 of them had pilot's licenses, but very few of them had their own plane. Many used planes owned by air clubs, and most no longer flew. This paralleled what was happening in France, so she wasn't surprised, but she had thought to find a better situation for female pilots in the United States.

Later, *Collier's* reproduced most of that Washington, D.C., speech, identifying Marie for their readers ("champion in all sports, inventor, poet, journalist, pilot") and adding wryly that "there are few women in the entire world with as many interests or who speak as rapidly."[30]

In Chicago, Mrs. Phoebe Bell, an excellent pilot herself and a member of the National Aeronautical Association, said that Marie's visit occasioned the first gathering of both the women's section and the men's section of the NAA.[31] Marie spoke to a number of groups on a variety of topics, reportedly in Chicago "on her tour of American aeronautical manufacturing plants and American commercial air lines. She is to fly to Los Angeles today and will return here next week to lecture before the Alliance Française."[32]

At that gathering the next week, "The valorous flyer was seated, and reading with machine gun-like rapidity the list of her flights over the North polar regions and the Sahara, while making a plea for the use of balloons and aeroplanes in carrying wounded off fields of battle. She is touring this country for that humanitarian purpose. ... She was heart and soul for her cause and full of energy and fire, and when I saw her handsome black eyes blazing with electricity, and heard her gay laugh, I realized that her brand of youth was that of courage and faith in unseen things, far removed from lipstick or the latest kink in the silhouettes. ... At the large luncheon in Piccadilly, following her speech, Consul General Weiller presided, with the French aviatrix and Amelia Earhart on either side."[33]

Chicago provided Marie with one of her fondest memories of the United States because of the time she spent with Amelia Earhart. The two had been invited to Chicago to present a dialogue about aviation for an audience of 6,000, preceded by a reception and followed by an elegant banquet. During the reception, Marie Marvingt and Amelia Earhart were photographed together, smiling, each holding a cup of tea. Earhart dedicated the photograph: "To Mademoiselle Marvingt, who has done so much for aviation."

The chance to hear two of the most famous women fliers in the world occasioned other invitations: "In Chicago, my young colleague Amelia Earhart and I spoke before more than 2,500 women about the need to be careful about all kinds of accidents. America," Marie noted, "is an absolute paradise for surgeons. I saw more accidents there than in my entire life, and I said so in a talk I gave on the radio. Happily, I returned to France in the same shape I left it, but filled with enthusiasm for all I had learned there."[34]

Marie arrives in Los Angeles, October 1, 1935.

In Los Angeles, headquarters of the International Women's Aeronautics Association, Marie gave more speeches, and stayed at the luxurious St. Regis Hotel.[35] She also visited the St. Francis Chapel at the Mission Inn in Riverside, where her signed and dated wings can be seen on the Fliers' Wall along with those of Orville Wright, Amelia Earhart, Eddie Rickenbacker, and other aviation greats.[36]

She apparently took advantage of her tour of the United States to sightsee: "Mme Marie Marvingt ... arrived in Winslow yesterday by Transcontinental Western Air, and left by another plane for the Grand Canyon."[37]

On her personal fact-finding mission, Marie visited airplane factories, hospitals, social work organizations, clubs, museums, airports, tourist sites, manufacturing plants, universities, and schools. She was likely to turn up anywhere, from an aeronautic exhibition in Providence, Rhode Island, to the Institute of Aeronautical Sciences in the Rockefeller building in New York.[38] She met its Secretary General, Lester D. Gardner, who later mentioned her at length in a piece he wrote for the *New York Times*: "Women Soon Took to Air: Their Firsts in Various Fields of Aviation Are Recorded."[39]

Her travels in the United States brought her tremendous satisfaction, partly because her reputation had preceded her, and the American press always treated her with the greatest admiration and enthusiasm.[40]

Upon her return to France, she published articles about her experiences, describing the types of aviation education available to young Americans, the air clubs she visited, the universities that encouraged aviation, the illustrated aviation magazines being published and widely read, the American Academy for Model Aeronautics, the glider demonstrations given in her honor by the renowned Abel, and the big aeronautical museums in Washington, New York, and Philadelphia. At the Franklin Institute in Philadelphia, she even found a replica of her white monoplane, the Antoinette.[41]

She was particularly impressed by the "Junior Birdmen of America," an organization sponsored by the Hearst newspapers to encourage interest in aviation in young people ages 18 to 25. Membership, then numbering 400,000, was free and included a badge. Each day the Hearst papers published aviation articles under the Junior Birdmen rubric.[42] Marie wished she'd thought of it ... and also owned a newspaper chain.

Gliders were big at the time, and Marie talked to French glider pilots about the Soaring Society of America and their publication, *The Gliding and Soaring Bulletin*.[43]

In short, as the competent journalist she was, she brought back to France all the news she knew would interest her audiences, along with a few things she felt they should know.

She also returned with souvenirs, notably hotel stationery, which she always thriftily tucked in her luggage. Responding to the 1936 letter nominating her to the Legion of Honor, she wrote on letterhead stationery (lightly crossed out) from The Willard Hotel in Washington, D.C., "Monsieur the Grand Chancellor: It was only on my return from America, where I have just spent more than a year, that I found your letter. … Enclosed are the two requested documents."[44]

Her follow-up letter to the Grand Chancellor, with information on her sponsor, was written on stationery from the Hotel Lotti in Paris.[45] Armand Lotti was a pilot and hero who had crossed the northern Atlantic from west to east, and the hotel, located on the historic and beautiful Place Vendôme, began as a residence (and Marie lived there for a while) after World War II, and remained Marie's hotel of choice in Paris; she was still staying there in her eighties.[46]

(The Lotti wasn't all happy memories. On June 29, 1937, her beloved bicycle was stolen from in front of the hotel. Marie handwrote the event and the date under a photograph of her standing in front of the Lotti that appeared in the newspapers.[47])

Marie spent more time in North Africa than anywhere except France. Her unwavering love for that part of the world began in 1922 when French President Alexandre Millerand invited her to accompany him and a handful of officials (she was the only woman) on a fact-finding tour of French military campaigns then underway.[48] At the time, women were not allowed on long airplane flights, but President Millerand insisted, and Marie was part of the group that flew from Toulouse to Casablanca on May 22, 1922.[49] That trip was foundational because it gave her an overview of North Africa's geography, leaders, and culture.

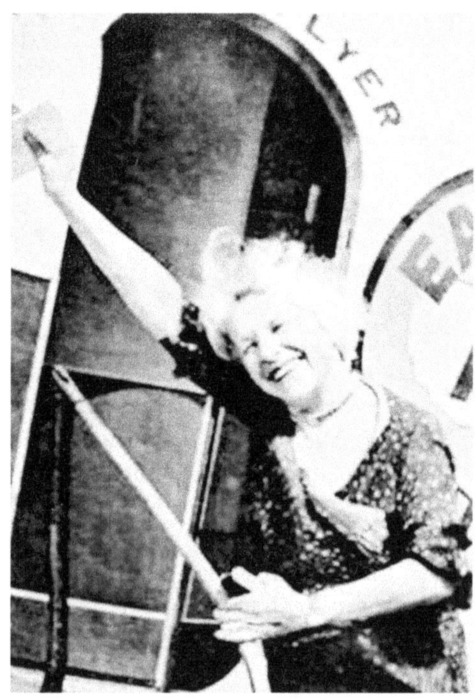

Marie arrives in yet another city on her 1937 tour of the United States.

She described one of many lighthearted moments:

"We arrived in Moulay-Idriss where an astonishing buffet was offered to us. There were no less than sixty-two different kinds of cakes—one of my friends, oh, the dreadful gourmand, amused himself by counting them—and the couscous was particularly delicious. After this huge spread was consumed, the president, very dignified, filled his melon hat with gold coins, which he then emptied into the coffers of Moulay-Idriss. Imagine the joy of our hosts! I don't know even today another president of the republic capable of repeating such a gift, not with gold at the price it is!"[50]

When the tour was finished, she did not return to France, but remained for nearly five years, flying home to France as needed. In North Africa, Marie found an outlet for her impulses toward action and service. Her travels fell into three major periods, from 1922 to 1927, 1932 to 1935, and from 1950 to 1953.[51] And she had her own agenda. Or agendas.

She had seen the need for air-rescue operations, particularly in the medically underserved outlying villages. Much of her time was thus devoted to establishing a functioning Moroccan air rescue organization, training personnel, and giving hundreds of educational and promotional conferences on air rescue. In fourteen months in 1925 and 1926, she traveled 40,000 kilometers.[52]

She also repeatedly told the history of aviation and emphasized its present-day importance. Flying still needed to be sold. Decades of persuasion and promotion were required for aviation to assume its place as a vital mode of transportation. Marie did a great deal of the early work in popularizing flight.

Another important goal for her in North Africa was to learn everything she could about its people and cultures, in order to introduce them to French audiences later on.[53]

Before embarking on the Millerand mission, Marie had never been to "the colonies." She approached her cousin, Dr. Pichon, and a school friend, Lieutenant Lalanne, who had both spent many years there. She was grateful for their overview of what she needed to know, and what she might expect, but she understood that she had a great deal to learn.[54] After waving farewell to the Millerand mission, she began visiting every major city and village accessible to her. Three times she made the long trip from Oudjda to Mazagan.[55] Wherever she went, she was received with respect and admiration. Marie confessed she was utterly enchanted by Si El Hadj Thami El Glaoul, the powerful chief of state in southern Morocco, as well as the pasha of Fez, Si El Paghdadi, and the sultan Mohammed V, who liked to bow to Marie rather than to allow her to bow to him.

She conceived an enormous admiration for North Africans, "these brave people who are always smiling and who are as witty as anyone you'd find in Montmartre."[56] Her summation: "They are never boring."[57]

She studied, observed, traveled, and learned. When she returned to France, she carried with her hundreds and hundreds of slides and photos in addition to impressive and useful information, which she reduced to manageable, entertaining anecdotes to enthrall French audiences.[58]

North Africans appreciated her: "Nothing stopped her, neither record-breaking temperatures in South Tunisia, nor auto accidents, nor smallpox and typhus epidemics when she nursed so many suffering indigenous peoples. With tireless perseverance, she documented our agricultural institutes, our factories, our industries, our hospitals, our mines, our hunting, our fishing. She devoted time to our young people in the schools, even those in the most faraway and isolated areas. She visited our ruins and stayed with indigenous people to study their mores, costumes, and dances."[59] In turn, she introduced her North African audiences to the country largely unknown to most of them; she was the first person from France some had ever met.[60]

Always the journalist, she wrote many articles for French and North African papers during her time there, describing politics, culture, and, her specialties, aviation and sports. And when a Congress on North Africa was held in Toulouse, France, July 23–25, 1927, Marie was present in her capacity as an expert on much of the area.[61]

Most of Marie's travels were impelled by her curiosity to know what was over the next dune, how people lived, ate, dressed, loved, and ruled. With the soul of an explorer, she found the people and scenery, so different from her native France, an endless banquet. And she never seemed to mind the intense heat.[62]

She entertained audiences, both public and private, for years afterward with the wonders she had seen. At a Moroccan *fantasia*, a riding and shooting festival held in honor of foreign dignitaries that showcased dozens of turbaned horseback riders deploying their expertise in swirling clouds of dust, Marie sat among blue-veiled women, barefoot children, and sturdy men in their long *jellebas*, studying the way the riders communicated with their horses.[63]

Marie met Bedouins in the desert, made friends with Berber tribes in the Atlas Mountains, and knew the chiefs of several *bleds*, managing to introduce herself into their family circles to satisfy her curiosity about their way of life. To the Arabs, Marie was known as "Lalla Tindouja," (Lady of Tindouja, a town in present-day Mauritania); "Lalla" was a title of respect of Moroccan Amazigh origin.[64] When she took off in a plane by herself, they marveled: "You're going up there alone?" In those early days of aviation, a woman at the controls was as magical as flight itself.[65]

A 1924 article reported that Marie was among those traveling by caravan from the south, that she had apparently eaten without any visible disgust the infamous gruel of the nomads, and that she was the only woman who accompanied the military columns across the North African Italian colonies.[66] Her travels were documented by numerous newspapers, and she was pictured at grand events, surrounded by generals, governors, officials, tribal chiefs, and children.[67]

From the golden beaches of Algeria to the streets of Casablanca and the magical cities of Algiers and Tripoli, she absorbed all the sensations North Africa had to offer: she heard the stringed instruments and the call to prayer five times a day, smelled the fragrant spices and incomparable fruits arranged in the markets, delighted in the embroidered caftans, the colorful slippers, the silver jewelry. She eventually felt completely at home in North Africa, and she seemed never to tire of its beauty, sometimes dramatic and appealing, sometimes cruel and demanding.

Innately driven to explore, Marie went wherever she found open doors. For a while her mail was sent care of the Soeurs Blanches in Géryville, Algeria.[68] In Casablanca, she lived at the home of General Bertrand[69]; in Boun, with Dr. Pichon[70]; in Hammamet, Tunisia, with Colonel Dangelzer.[71] She spent several months at the palatial home of the Bach-Agha Bengala,[72] and enjoyed many picturesque hotels: Tunisia Palace in Tunis, Hôtel Suau in Colomb-Béchar, Hôtel Gaulois in Rabat, Hôtel du Glacier in Fez, Hôtel Régina in Alger.[73]

In 1932, she lodged with the Little Brothers and Sisters, a community founded to carry on the work of her hero Charles de Foucauld. She stayed with them again in 1934 when recovering from a severe microbial illness, and again in 1951 while teaching Esperanto to members of religious orders from all over the world.[74]

Colonel Blanc, her longtime friend from Nancy, wrote, "She was known simply everywhere. If there was a military post anywhere she traveled, she went straight to the commandant, sure of a warm welcome. If there wasn't a military presence, she'd go directly to the local leader or chieftain of the area."[75] His explanation for Marie's universal welcome: "She was completely unselfish and uninterested in fame or money. She never profited from anything she did, which she very well could have, in my opinion."[76]

Even in places where she couldn't speak the language, she projected a childlike curiosity, innocence, and willingness to serve that needed no words. Accompanied by her outsized reputation, Marie, living unlike any other woman of that era, could travel alone with impunity, stay in a chieftain's château for a month if she wished, travel with troops and caravans or in the back of a truck without scandal or criticism.

She visited nearly all of Algeria, Tunisia, and Morocco, including the Atlas Mountains, which run through all three countries. She explored Libya, Egypt, Malta, and Sicily. Reports suggest she also ventured southward, visiting Dakar, Dahomey, Guinea, and Lake Tchad.[77]

One North African newspaper noted that Count de la Vaulx, balloonist and co-founder of the French Air Club, had given "39 conferences in our three colonies. Mlle. Marvingt has given ... 325."[78] This number eventually grew to nearly ten times that figure.[79]

Marie made herself popular with the Aeronautic League of France by signing up new members by the hundreds. She always ended her speeches on the importance of aviation by touting the organization responsible for its progress. And she lobbied constantly for young people to engage in sports, encouraging individual sports rather than team sports as she felt they were good for developing both willpower and a lifelong commitment to fitness.

Interviewed by the *Echo d'Alger* before one of her conferences, she explained that she felt she "knew and appreciated better than any woman in the world the marvels—particularly the unknown and hidden marvels—of North Africa."[80] At the end of the interview, she added, "I hope your newspaper can send me a good, and especially an intelligent audience." The reporter replied indignantly, "But all our readers are that sort of audience!"

She flew over the French and Italian fields of battle for five months during the Riff campaigns; she mentioned in particular seeing from the air the bombing of the camp of Abd el Krim. A journalist concurred: "She took photographs in which you can see the effects of the bombs on Abd el Krim and the area around the Bay of Alhucemas."[81]

Asked if she could speak Arabic, she said offhandedly, "Oh, I manage well enough to get along. It's always the same anyway. 'May I see your home? Are your wives doing well?' It's the Berlitz School sort of thing where you learn a foreign language in twenty easy lessons."[82] However, Marie gave classes on hygiene, medical basics, and first aid in small outposts and villages, where she also nursed as many as she could, so her language skills must have been better than she admitted.

Marie also enjoyed herself. She climbed several mountains, including the Moyen and the Grand Atlas; did some land yachting, which was a joy to her and popular in several places; swam and scuba-dived on the coast.[83] She wrote her cousin: "I'm hunting and I'm fishing."[84] She turned up in Timimoun when the international aviation rally, "Tour of Algeria and the Grand Western Erg," landed there.[85] And she did some gliding and watched several glider trials.[86]

Marie's good friend, Renée Enselme-Trichard, lived in Algiers from 1920 to 1946, so she attended the talks Marie gave and often invited her to dinner. The two used to dye fabrics for their clothing, and then take photographs of themselves in their new dresses in Enselme-Trichard's sunny garden.[87]

Marie would say that her most outstanding "first" in North Africa was the establishment of air-rescue operations. Less high-minded record-keepers point to her being the first woman to drive a car 1,750 miles across the Sahara and the first white woman to reach In Salah.

In 1922, Marie heard that André Citroën, the engineer and industrialist who introduced Henry Ford's methods to the European automobile industry, was sponsoring an automobile expedition to cross the Sahara from Touggourt to Timbucktu to open up that area to French colonization. The expedition aimed to test the possibility of going as far as In Salah, an oasis town in central Algeria. The Citroën convoy was headed by Georges-Marie Haardt and Louis Audouoin-Dubreuil, involved in previous military exploratory missions, and included cinematographers, chefs, and chauffeurs. In addition, Georgina Bingen Citroën, married to André Citroën, would be driving one of the cars, a late-model Citroën, which, like the other cars in their convoy, was fitted with caterpillar tracks around the wheels to get through the sandy terrain.[88]

Marie was certain she was a better driver than Georgina Citroën. She persuaded the newspaper *Le Petit Parisien* to sponsor her, and borrowed a car; identified in press reports as a Fiat 3549 with double tires: each of the two front tires was replaced by two thinner tires, making it a six-wheeled vehicle. In the publicity blitz that surrounded the expedition, notably the competition between the women, Madame Citroën was favored to win.

Marie later told a journalist: "Our poor little Fiat was set against the Citroën autos. We accidentally knocked over a few telegraph poles, but we had an electrician in our convoy, so we always quickly reestablished communication. That way, the papers could follow our progress. We arrived in In-Salah on a gorgeous morning. The Citroën expedition didn't get in until that night. Oh, was I ever happy!"[89]

Marie's triumph fortunately had a witness, Henri Thétard, correspondent for *Le Petit Parisien*. In a front-page story, he wrote: "I am now in In-Salah, as of 10:00 yesterday morning. Our arrival produced a certain sensation in the capital of Tidikeit because of the presence among us of Mademoiselle Marvingt, the first European woman ever seen here."[90]

He continued, "Last night, about 8:40, the second convoy arrived. … The three cars carried, in addition to their drivers, Monsieur and Madame Citroën, General Estienne, Monsieur Kegresse, and Monsieur France, who was operating the camera."[91] With a borrowed car and a skimpy budget, Marie had triumphed over the well-managed and financed Citroën group.

The first woman to drive across the Sahara arrives in In Salah February 18, 1923, with an electrician and a photographer.

Jean Renaud of the popular Parisian newspaper *Le Journal* gave the fairy-tale adventure a classic opening: "Once upon a time…. It is thus that I must begin the tale of one of the most grandiose spectacles I've ever seen. … But I wasn't dreaming. I am indeed in Ouargla, palm trees dancing, songs drifting on the air, fifes playing, tomtoms beating, shimmering gold and purple rugs hanging from the walls."[92]

He described the arrival of the Citroën convoy and the "unbelievable" crowd, waving palm fronds to greet the victors. In the midst of his purple prose, he looked across the square and there, among the crowds, stood Mademoiselle Marvingt, who had arrived ten hours earlier. Of course, a rival newspaper had sponsored her so she wasn't part of his story, but he gave her a mention for arriving first.

When it was mistakenly reported that Marie had crossed the desert with Madame Citroën, Marie's irritation showed. She wrote good friend Marcel Knecht, "A magazine has a huge error: it's not *with* Mme Citroën that I crossed the Sahara, but against her, with a 6-wheeled Fiat, and I beat her by quite a bit."[93] Those ten hours permitted her to wear the titles of "first woman to cross the Sahara in a car" and "first white woman to arrive in In-Salah," and she meant to have them.

Twelve years after her joyful arrival in In Salah, Marie marked another milestone, the first person to connect, by automobile, Tindouf with Zagora, an oasis near the Algerian-Moroccan border. And it was in Beni-Abbès that Marie's inventive mind led her to pursue her idea of metal skis.

When her ambulance-airplanes landed on the sand, their landing skis were sanded down, eventually to uselessness. Equipped with wooden "planks," she went to Beni-Abbès in 1923 and began testing different kinds of wood in the nearby desert under the watchful protection of an armored car. Her experiments coincided with sporadic desert uprisings (and she was in the dissident zone) but, armed with an automatic machine gun, she continued her work, stopping only when she needed to protect herself from attacks by marauding bands.[94]

She eventually realized that only metal skis could provide the necessary durability. She returned to Casablanca where she directed a foundry to fashion skis from an alloy, the forerunners of today's aluminum or titanium skis.[95] Her experiments satisfied her to

Marie experiments with metal skis in a Saharan dissident zone while armed bodyguards keep watch (courtesy Madame Coulet).

such a degree that she planned to equip all ambulance-airplanes with them, and to suggest them for use by Saharan planes and Arctic trappers; she is credited with being the first person to present a plan for equipping airplanes with skis for landing on snow or sand.[96] She also established a modified Alpine ski school in the desert for her Berber friends.[97] ("The Bedouins couldn't get over it!" Marie exclaimed years later.[98])

In 1923, she filmed a short documentary film in Mogador about skiing in the Sahara.[99] By 1935, the press was hailing as a great new tourist attraction Mademoiselle Marvingt's sport: skiing on the sand dunes along the Moroccan coast.[100]

Most of the souvenirs Marie brought back from North Africa were as priceless as they were intangible: memories. "Such beautiful places!" she said. "From Morocco to Erythrée, the Rummel gorges, the mountains of Djijelli, the ruins in Tizi-Ouzou in Alger, Bou-Saada, where I arrived in a six-wheeled Renault, the Canary Islands, Marrakech, Kenitra, the Fire Festivals, exploring the Middle Atlas with Captain Gillot, the wells at Inifel, the Water Festivals, the presidential voyage with Monsieur Millerand, so many souvenirs pile up in no particular order. But what joy ...! I saw almost as many marvelous spectacles as you'd find in the *Tales of the Thousand and One Nights*."[101]

Among her 2,000 photos were pictures of her at the foot of the Beni Selmane tower and others of her and a captain in front of the ruins of Goulimine.[102] In the conferences she gave later in France, she showed many of her slides, and spoke in the language of the passionate traveler of "the wild and exquisite island of Djerba," the casbahs, the sea baths in Monastir, the all-white holy cities of the Mzab region, the women of Falaint and the daily lives of Muslim women, the extraordinary dances of the Kerkemien fishers, the Troglodyte peoples, the so-called sorcerers and wizards of the desert, the rich culture of date trees, the miraculous fish of Azemmour, the hunting of wild beasts, the mysterious life of the city of M'Sat, the ruins of Tizi-Ouzou, the lifesaving oases, Morocco's modern cities built next to indigenous villages, the unprepossessing exteriors of Arab homes that revealed splendid interiors, rich phosphate mines, an unforgettable evening in Alssouas, a festive Friday in Hammann.[103]

Marie in the North Africa that she studied, admired and loved.

Audiences overwhelmed by the mere descriptions of her journeys must have been defeated trying to imagine actually visiting so many picturesque, inaccessible places. She brought back a few fetishes, autographs, pieces of jewelry for gifts, and several medals and honors.[104] Principally, however, she returned with stories. After hearing some of them, *L'Est Républicain* noted that "fifty-two months in the outback have evidently not taken the edge off Marie Marvingt's robust sense of humor."[105]

Out of respect for the culture, Marie often wore a veil or burqa.[106] This choice, along with nursing the sick, helping to eradicate locusts and malarial mosquitoes, educating villagers in first-aid basics, giving free talks to adults and children, and making the effort to learn languages, earned her a warm welcome wherever she went.

After her first lengthy stay in North Africa,[107] Marie couldn't seem to settle down in France. She was seen in Paris, she lectured in Pau, she told audiences in Nice about her travels, she piloted the prototype plane, *The Cardinal de Luçon*, from Reims to Nancy and back. But soon she was back in Tunisia.

Much of Marie's time in North Africa was spent with military operations in the ongoing "pacification" of North Africa. She was authorized to accompany French, Italian, and Spanish troops as war correspondent, nurse, surgical assistant, and pilot.[108] (For the French she also worked as an army medical officer.[109]) In Morocco, as the unpaid nursing assistant of a field hospital surgeon, she bettered her record: 36 hours without leaving the operating room.[110] Much in demand because of the rarity of qualified nurses speaking several languages, she also traveled as a volunteer surgical assistant with the campaign in Libya.[111]

Profoundly patriotic, Marie apparently never questioned the ethics of colonization. She appears to have felt that the French brought untold gifts to people who might benefit from them. She herself seemed to exhibit no trace of the arrogance often observed in representatives of colonizing countries. To her, the people she met were her equals, except they knew things she didn't know yet. She in effect traded her medical, sports, and aviation knowledge for information about their geographies, cultures, languages, and ingenious survival abilities.

She spent five months with the French columns of General Giraud and six months with the Spaniards in the Riff, where she often flew to survey the regions of Oudjda, Quezzan and Larache.[112] With the Italians, she traveled across Tripolitania, Cyrenaica, and Marmarica, visiting the outposts of Gerdi and Adgedabia. When Misrata fell, launching the Italian reconquest of Tripolitania, she worked in an aid station as a surgical nurse. From Tobruk, she flew along the Egyptian border.[113]

In 1935 she accompanied the Libyan campaign as far as Solum with General Graziani's troops.[114] She eventually worked under and with (she fit into no chain of command) Generals Huré, Trinquet, and Giraud in the "pacification" of Mauritania.[115]

When Marie wrote that "Count Volpi authorized me to circulate freely anywhere in Tripolitania,"[116] she assumed her readers knew that Count Giuseppe Volpi di Misurata, governor of Tripoliania from 1921 to 1925, developed utilities that brought electricity to Venice, other parts of Italy, and the Balkans; negotiated the end of the Italo-Turkish War; successfully dealt with Italy's World War I debt repayment to the United States and England; and pegged the value of the lira to the value of gold. He also founded the Venice Film Festival. No matter where she traveled, Marie consistently sought out and made friends with the most accomplished individuals in the area.

And when she wrote, "His Excellency the Commissioner Concellieri invited all the authorities to a banquet,"[117] her readers understood that she was of course one of those authorities.

"His Excellency General Bongiovanni permitted me to visit all parts of Cyrenaica. The Italians have shown me the utmost kindness and courtesy. Lord Plummer, Governor of Malta, has also accorded me every assistance, referring me to experts who can satisfy my curiosity about the history and culture of the country. Governor Basso of Sicily saw to it that his own friends attended my conferences."[118]

In July of 1925, Marie wrote to her cousin, Jeanne Roumier: "I write this from the back of beyond—I'm making a big tour through the Sahara as far as Timimoun, after which I'll return to Morocco where I'm scheduled to assist in a surgical clinic. It's past time for Marshal Pétain to arrive, and I'm just hoping he doesn't arrive too late and that Taza is not already taken by that time. I've masses of friends who have died. It's the beginning of the revenge of the Germans."[119]

When her old friend Colonel Trinquet summoned her to Beni-Abbès, in the middle of the dissident zone, she didn't hesitate, riding in an armored car through an area crawling with insurgents where several deadly ambushes had taken place. Armed like a soldier, Marie fired and was fired upon.[120]

She began her second North African tour in 1932. In July, she wrote from the Talifalet oasis where she was staying with the Soeurs Blanches: "I've just accompanied the soldiers to the borders of Mauritania. Very interesting, but difficult—five months without a bed! I established a civilian air-rescue operation in Morocco, and in just a few months injured individuals will have available to them air taxis equipped with a pilot and a certified nurse. I gave about fifty lectures and signed up 5,000 new members for my two leagues, and that's not counting thirty-two school groups. I'm bringing back some magnificent photos plus my 2,000 meters of filmed reels. I've recently been very ill during a month spent in Marrakech—a rare infectious disease that I picked up while marching with the soldiers. I've suffered rather horribly but I'm on my way to a complete recovery. I landed way in the south, over 125 degrees Fahrenheit in the shade. I'd never been to Tafilalet, and for good reason: it wasn't occupied during my last trip. Fascinating countries and customs. I'm going to go even farther south before returning to Algeria. I have no idea when I'll fly home. For the moment, I delight in the Saharan life. I've seen scarcely any Europeans. They've all fled from the heat."[121]

That same month, while Marie was roughing it without a bed in North Africa, her documentary, *The Wings of Mercy*, premiered in Paris at an elegant cinema on the Boulevard des Italiens, without the presence of its producer and star.[122]

During her third period of North African travels, 1950 to 1953, Marie covered tens of thousands of miles promoting her ambulance-airplanes and air-rescue organizations in, among other cities, Alger-la-Blanche, Tunis, and Bône. In 1951 she was in Morocco, endeavoring to ensure that each air club had one or more planes capable of being transformed rapidly into a rescue unit. In a newspaper article about her, Jean Boulet reported: "At the conference that she kindly gave yesterday at the Meknès airbase, there were in attendance, besides a large group of young pilots, base commander Colonel Hugo [Marie was a particular favorite of Henri Hugo]; Dr. Jules Colombani, president of the French Red Cross; Lt.-Col. Merle; commandants Vialle and Codet; M. Deliège, vice-president of the Air Club; Lt. Médelec, former head of the famous Tricolo patrol; as well as officers and the NCO of the hunting school.

"Dr. Colombani introduced the speaker to us in these terms: 'Mademoiselle Marvingt is a very old friend of mine. She has devoted her entire life to sports with, however, a certain predilection for aeronautics. She is, in addition, the first woman in the world to be licensed on a monoplane—Roger Sommer's Antoinette. I won't tell you her age, because one never gives a woman's age, but we were both born the same year. She has retained an extraordinary physical and mental vitality, which is an astonishment to me every time I see her. Throughout her life, her principle objective has always been to do good, and she's given everything she has to contribute to the wellbeing of humanity. In 1910 she invented the ambulance airplane. Listen to her now! You will see,' he adds with a teasing grin, 'she is not ordinary.'"[123]

Boulet's article reported Marie taking over the podium: "'I've traveled all over the world,' she says, "and I continue to do so. Sports have allowed me to preserve the vitality that my friend Dr. Colombani boasts of. And since he' (she glances with amusement at Dr. Colombani) 'wouldn't give you his age, well, I don't mind telling you mine. I'm seventy-six. A few weeks ago I biked seventy-five kilometers to visit a sick friend's bedside.'"[124]

In 1952, she wrote to cousin Louis Lhérault from the Tunisia Palace Hotel: "Your affectionate greetings, sent December 30, reached me only yesterday, saturated with the unique, marvelous fragrance of Tamanrasset, where they have been following me without catching up to me. … Last Wednesday, I was interviewed at some length on Radio Tunis about how I managed to fit so much into my life—and still do. They mentioned my fifty-one conferences in twenty-seven days in Algiers … and the fact that I'm nearly seventy-seven. My birthday, February 20, will be celebrated in Tunis where I'll be spending several weeks. Monsieur Herz asked me a final question about my current goals. I said I needed to persevere with the big tasks I've undertaken, like establishing air rescue worldwide, but I also want to fill the hearts and minds of my listeners with the desire to be of service, which is what it means, to me, to be a great Frenchwoman or Frenchman."[125]

Marie said farewell to North Africa for the last time at the age of seventy-eight. She wasn't, however, ready to stop traveling. People who wanted to visit her in Nancy had to write or call ahead because she was so rarely at home. She was certainly there in May of 1955 because she presided over a reunion of her former classmates from Sainte-Chrétienne in Metz.[126] January 19, 1956, she appears to have attended the winter sports festival in Cortina, Italy.[127] The next month she was lecturing again in Nice.[128] As Jean-Loup Nicolle wrote in his booklet about Marie for children, "a simple phone call was enough to send her flying to Lausanne, to London, or to New York to give a conference and to speak of the heroic era of the conquest of the skies."[129]

Journalist Achille Liégeois wrote: "Yesterday she arrived from Morocco. Tomorrow she leaves for the Indies. One cannot imagine her other than packing a suitcase or snapping the lock on a valise. She is the very epitome of a globetrotter."[130]

But even the term "globetrotter" didn't encompass Marie's aspirations: she was the woman who had signed up for a trip to the moon.[131]

11

Insatiable Curiosity

> "There is not an art, there is not a science to which this young woman, on whom Nature appears to have showered abundant gifts, is a stranger. In the midst of all her sports feats, she has studied medicine and art, theater and voice. She paints and sculpts. She writes, she speaks, she acts. Her example is admirable, extraordinary. But should we ever be astonished at what the soul of a Frenchwoman is capable of, and don't we always find in her the two greatest gifts: love of beauty and the passion to do good."—Armand Rio[1]

The offer was unheard of. The slender, auburn-haired Frenchwoman with the ready smile and adventure in her eyes had promised $10,000 to any man or woman in the world who could match her record of accomplishments.[2]

The year was 1922, and even journalists who had been filing stories on Marie Marvingt for years were astonished to realize all she had done. Each of them knew parts of the story, had covered this exploit or that, but none had appreciated the totality of her achievements.[3]

A French journalist said, "It leaves a person breathless! You want to holler 'Uncle!'"[4]

The idea of a personal challenge was original with Marie, perhaps inspired by her thirty-four decorations and seventeen championships or world records, or perhaps her impulse to say, "Look! Look what I've done." Marie viewed her abilities and successes with a childlike delight; she sometimes appeared as surprised and impressed as anyone to think that she had actually done things she'd only dreamed of doing.

Marie repeated her offer in New York in 1935, and again in 1948 in France (via the newest means of communication, television),[5] and again in 1949.[6] Despite widespread radio, television, and newspaper coverage of her unusual challenge, no one ever came forward.[7]

It's uncertain where Marie would have found the $10,000 in 1922 if someone had responded to her *défi* (French for "challenge," giving English *defy*, *defiance*, and *defiant*). However, she was scrupulously honest and would have kept her part of the bargain. She most likely accepted a friend's offer to back her, since Marie undoubtedly felt that the money would never be needed.

In addition to her well-known activities as a sportswoman, pilot, nurse, journalist, globetrotter, and humanitarian, she revealed intellectual accomplishments in her *défi* that were less well-known to the general public. Possessing the "'satiable curtiosity" of Rudyard Kipling's Elephant's Child, Marie was greedy to know and experience everything. Called "the eternally curious one" by newspapers,[8] she thirsted for knowledge, novelty, and expertise, and nothing fell outside her range of interests.

Although the 1880 laws establishing free public and private elementary and secondary schools included girls, their education was not mandatory, and it was aimed at a life spent in the home. Science, for example, was thought unproductive for this vocation, and was not taught to girls. Universities accepted women, although by 1900 only 500 of 15,000 students in France were female.[9] Marie was one of them, receiving her *licenciée ès lettres* (Bachelor of Arts) degree from the University of Nancy.[10]

She spent almost ten years at the University. In addition to the liberal arts curriculum, she enrolled in the law and medical faculties (in France women had been "permitted" to study medicine since 1868), acquiring a solid foundation in medicine and surgery. These served her well, even after her medical career was interrupted by her sports activities and, more definitively, by World War I.[11] She did, however, complete her nursing degree and later served with the Red Cross with the rank of major and later colonel.[12]

All her life, she retained a deep respect for the clinicians, medical professors, and physicians she trained and worked under: Theodore Weiss (an expert on wartime triage and surgery, eventually dean of the Strasbourg medical school); Hippolyte Bernheim (professor of medicine and neurology, famous in the history of hypnosis and psychotherapy—he met with young Sigmund Freud in Nancy in 1889[13]); Michel Gaston (professor of clinical surgery, only a year older than Marie, but she would attend his funeral in 1937); Jacques Parisot (idealistic, devoted to social action and preventive medicine, later Dean of the Medical School); Paul Rohmer (considered the father of modern French pediatrics); Jean-Louis Faure (noted surgeon and gynecologist); Theodore Guilloz (a year after Roentgen's discovery of the X-ray, he brought radiology to Nancy; losing some fingers to the effects of radiation, he was later a military hero); and Gauthier de Luxeuil (an eminent surgeon who chose Marie as his surgical assistant whenever possible).[14]

Women's participation in higher education was unusual at that time, and Marie's fruitful years of study and accomplishment at the University of Nancy were considered an unusual achievement. Her fascination with medicine and science led her to participate in several research projects.

Wealthy Joseph Vallot was an astronomer (he built an observatory on Mont Blanc), geographer, naturalist, mountain climber, and philanthropist. Marie's gift for meeting and establishing longterm friendships with the brightest minds of her day led her to this polymath, a man of her own kidney. Although the academic and scientific communities of his time never thought much of him, his work in botany, glaciology, construction, geology, photography, medicine, physiology, cartography, meteorology, and speleology (he married a speleologist) appealed enormously to Marie. Vallot's curiosity was the equal of hers, and she participated in several of his projects—in one case, because of her background in medicine and surgery.[15]

The "eternally curious one" in 1912.

After a friend died at Vallot's research center on Mont Blanc, an autopsy revealed what was defined, for the first time, as altitude sickness.

Thus, in 1913, Marie assisted Vallot and Dr. Raoul Bayeux with their experiments on the effects of altitude on guinea pigs and squirrels. In *Those Crazy Years in Chamonix*, Gaby Curral-Couttet tells how she often accompanied Marie to the Villa Vallot to check on the animals brought down from Mont Blanc.[16] Earlier, in 1905, Marie had assisted the Vallot Mission in the ongoing work of creating an exact topographical map of Mont Blanc; she accompanied them on several other expeditions because of her knowledge of the area and her experience on the mountain.

Another mind she found fascinating belonged to Serge Abrahamovitch Voronoff, a French surgeon of Russian extraction who gained fame for grafting monkey testicle tissue onto the testicles of men for therapeutic purposes. The details about the vogue he enjoyed, how he fell out of favor with the medical community, and then had his reputation restored a few decades ago is a story in itself. However, he was an inventive and original scientist, and Marie helped him with some of the first animal experiments on pneumothorax (collapsed lung).[17]

Lucien Cuénot, groundbreaking French biologist who proved that Mendelism applied to animals as well as to plants, was a friend and fan of Marie's, describing her as "an exceptional woman with an astounding resume, covered with honors. She was physically and morally a force of nature, blessed with a ferocious independence and unequaled generosity."[18] He even wrote a fairy tale for her, beginning, "When you were born, the fairies gathered around your cradle to decide your destiny." The first fairy, who was a bit myopic, couldn't decide if the infant was male or female, so she predicted that the child would become an outstanding athlete, of the kind known in ancient Greece. Another fairy, the sixth, bestowed on Marie a "good stomach," to be able to endure all the banquets and glasses of champagne raised in her honor. He concludes by saying there are still fairies today, but now we call them genes.[19]

In addition to the sciences, Marie embraced the arts. She studied acting for nearly ten years, some of it under French actors Sarah Bernhardt and Mounet-Sully.[20] She appeared in comedies of middle-class manners, romantic dramas, and some classics. She adored being on stage. In some sense, her entire life was played for an audience, and she always had the leading role.

Intensely patriotic, Marie often played the part of "Marianne," a nickname acquired by France in the early years of the Third Republic (1870–1940). Marianne is to the French what Uncle Sam is to Americans and John Bull is to the British. On every official building and for a long time on coins and stamps, France was represented by Marianne wearing the cap of liberty. Marie was proud of this new symbol, born nearly the same time she was. As Marianne, she wore a long white dress, with a wide banner in the red, white, and blue of the French flag crossing from her left shoulder to her waist. She appeared as Marianne for some time in Madame Morisse's review, *Visions of War*, performed for soldiers in service clubs across France.[21]

Marie attended the theater whenever she could. One evening when *Samedi Soir* theater critic Robert Kemp arrived a little late, he climbed over the back of his seat to take his place. The astonishment and raised eyebrows of those around him were doubled when an elegant, dignified-looking older woman in an evening gown did the same thing with remarkable agility.[22]

Marie knew many of the French performers of the day, sometimes slipping backstage at intermission to say hello. Her friend Colonel Blanc said, "In Paris, you might run into her at the opera, at a theater, anywhere."[23] When she was in Nancy, she saw as many films as she could. Blanc ran into her one night as he was leaving a local movie theater. He offered her a lift home but, no, she was on her way to see another film at a movie theater across town.[24] When she was eighty-one years old, she wrote to a friend that whenever she stayed in Paris, she attended the cinema every day.[25]

A natural performer, Marie was drawn to films all her life—a medium that had come of age with her. As with anything new, she was driven to explore it. She studied filmmaking under Max Linder, a renowned producer, director, and writer. Later she produced and starred in two documentaries about ambulance-airplanes: *The Wings of Mercy* and *Saved by the Dove*. She herself took much of the footage in both films, from the passenger seat of a second plane following the ambulance-airplane.[26]

Marie played cornet, and studied singing.[27] A former student at Sacred Heart school in Nancy told about the time a young priest directing the school chorus ambitiously decided to present "The Passion According to Matthew." The work included soprano solos, which no one in the group wanted to attempt. A professor at a local college who knew Marie asked if she'd help out. She did so most graciously and, as the alumna remembers, with undoubted talent. That was in 1909 when Marie was at the height of her popularity, and the students were thrilled to be working with her.[28] Her clear voice led to her later studying, and then teaching, elocution.

She spoke five languages (French, German, Italian, English, and Arabic[29]) and Esperanto, an international language developed in 1887 by L.L. Zamenhof as an easy-to-learn, politically neutral language to foster international understanding. Twelve years old when Zamenhof first published his language, and thirty when the first World Congress of Esperanto was organized in France, Marie described herself as "a fierce supporter." She added that she saw her faith in it vindicated the day Pope Pius XII received more than 10,000 Esperantists at the Vatican, and addressed them in Esperanto.[30]

Although today nearly two million people speak Esperanto to varying degrees, and some 120 countries belong to the World Esperanto Association, the language lacks some of the fervor and novelty it had in Marie's day. A 1910 Esperanto magazine listed over fifty Esperanto groups in and around Paris alone—and Marie was a woman of her time. Anything new on the horizon was worth investigating. She quickly joined the Association, learned Esperanto, and never missed an annual banquet in Paris. At the 1910 dinner, held at the Hôtel Moderne in Paris, the eighty-three guests were almost all well-known names including the Archdeacons, the Baron and Baroness de Menil, and the eminent mathematician Bourlet. While enjoying dessert they were addressed by several members, including Marie, in Esperanto.[31]

When she was in North Africa, Marie taught Esperanto to the Little Sisters working at the desert retreat of Père Foucauld. People came from all over the world to pray and meditate there, and Marie felt it would strengthen their spiritual bonds if they were able to speak with each other in a common language.[32]

Marie also painted, drew, etched, and sculpted.[33] And she wrote. In addition to the pieces that won literary awards, she apparently wrote a novelette, *Myriel*, and she told a reporter that some of her memoirs had been published in Canada in 1936.[34] In her later years, she was working on a book, *The Constitution and Character of Artists*, which was never published.[35]

She was not particularly gifted as a poet—and never claimed to be—but she often took to poetry to express her deepest emotions after experiencing dangers that allowed no time for thinking or feeling, only sheer action. Thus she wrote about seeing death—

tall, pale, unforgettably beautiful—in the midst of an avalanche.[36] When her balloon caught fire, she recalled Joan of Arc's words: "I give myself over to the skies." Her nearly fatal experience crossing the North Sea—bouncing from wave to wave, drenched, frozen, having lost everything but her life—is compressed into a few lines of poetry conveying both despair and hope.

But it was Marie's work as a journalist, a separate chapter of her life, that reveals most clearly her command of the language. Except for her published articles, most of her writings have been lost.

Marie embraced the concept of lifelong learning decades before it was popularized. She never stopped taking classes, studying esoteric subjects on her own, and apprenticing herself to distinguished experts in a variety of fields. *The Sfaxienne* observed that Marie's intellectual powers were as developed as her physical powers.[37]

She pursued taxidermy, astronomy (she took classes from Gaston Floquet, professor and dean of the College of Sciences at the University of Nancy), meteorology, phrenology, botany, writing analysis, physiognomy, topography (under Paul Helbronner, French alpinist and geodesist who pioneered cartography of the French Alps), astrology, aerology, and ballistics.[38] Charles Codron, her neighbor in Chamonix, listed her many interests and then wrote, with some astonishment, "And you could even discuss economics and politics with her."[39]

Some of her interests were passing fads, some were serious science. Years later, she said that learning different things (many of which were altogether new in her day) stimulated, relaxed, and entertained her all at the same time. She said that, looking back over her life, she had never been bored for an instant.[40]

With a mind that was constantly shuffling, filing, analyzing, reflecting, and investigating, Marie was a natural inventor. The air-ambulance was undoubtedly her greatest work. But her experience assisting surgeons led her to collaborate with Dr. Gauthier on a new surgical suture designed to minimize scar tissue.[41] And her invention of metal skis was an enduring one, a logical step in the development of new technology for helicopters and small planes.[42]

How many people were land-yachting in 1926?

Marie was probably one of the first women seriously interested in underground exploration. She studied under Norbert Casteret, a distinguished cave explorer, who was twenty-two years younger than she. She liked to say she was interested in caves before he was. She was evidently one of the first to explore the grottos at Lacade, and she often talked about the sixteen hours she spent caving in the Daya Chicker in Morocco.[43]

By nature an explorer, Marie examined everything that came her way. During her travels in the Sahara, she asked hundreds of questions about the fascinating and mysterious desert, and she studied its inhabitants, as well as its flora and fauna, with her usual lively curiosity. She spent time studying sea life too, and combined her nursing care of sponge divers in Tripoli with some practical learning experiences in harvesting sponges.[44]

Marie was an enthusiastic and tireless dancer—the foxtrot, tango, and modern dancing. The night she learned the samba, she wouldn't stop until the military orchestra put down their instruments. As an older woman, Marie confessed, "I love dancing, especially the waltz. Even today, my partners get out of breath!"[45]

As often as possible Marie combined her sports skills with her academic interests, as when she flew a glider to study meteorology and aerology, and used her mountain-climbing skills to help with the cartographic study of Mont Blanc. *Paris-Match* once said if Marie had a coat of arms, her motto would be "Sports, Arts, Sciences" and that she must be following Voltaire's advice: "Flee idleness, the rust that attaches itself to the most brilliant metals."[46]

Marie also sought answers to profound questions: Who are we? Why are we here? Where are we going? What's beyond the physical? She studied the history of philosophy and the attempts by humanity over the centuries to pierce the mysteries of the past, present, and future.

Her interest in "soft" sciences began when she studied hypnotism with Bernheim.[47] From her appreciation of hypnotism, it wasn't a big step for her to investigate mesmerism, spiritualism, magnetism, suggestion, palm-reading, phrenology, physiognomy, telepathy, graphology, astrology, psychometry, divination, precognition, and others. The so-called psychic sciences fascinated many, and it was a time of wide-ranging experimentation: no possibility seemed too far-fetched to be considered. During Marie's lifetime, apparent miracles of science and invention were emerging, almost more quickly than the public could absorb their wonders. Sorting out the enduring ideas from the others wasn't simple.

Being of a scientific disposition, Marie examined everything that was published in the field of the so-called psychic sciences, discarding some of its elements and embracing those she felt had merit. Toward the end of World War I, she wrote to her cousin about the talk she was attending that afternoon: "Mental Telepathy Among Soldiers During the War."

Eventually she took her own curiosities and experiences about the nonphysical world to her audiences, although she introduced these subjects only after several decades of her usual speeches.[48] Her listeners were as intrigued as she was, and turned out in huge numbers to hear her stories. In talking about psychic phenomena, Marie always welcomed other points of view or objections from her audience. According to multiple newspaper reports, only renewed applause followed.[49]

Along the way, Marie discovered in herself some modest psychic abilities, specifically, palm-reading and precognition. She explained that she could sometimes tell the future by a sort of intuition, not only for herself or for people she knew, but for others as well.[50]

She predicted, a year in advance, the September 9, 1954, earthquake in Algeria.[51] A 6.8 on the Richter scale, it is listed by the United States Geological Survey as one of the deadliest earthquakes in history. With thousands of dead and injured, ambulance-airplanes intervened for the first time in a major calamity. Marie had always emphasized the critical importance of air rescue, and she knew the area well. Her prediction may have been an inspired guess, or she may have genuinely "known," in a psychic sense, what was coming.

One evening in late 1910, a group of high-spirited, jovial pilots and Parisian journalists were gathered around a dinner table in Mourmelon, where the International Michelin Cup for long-distance flying was being held. At one point, the charming Hubert Latham, Marie's flight teacher and great friend, held out his left hand and said, "And me, Mademoiselle? What do you see in my life line?"

A quick look, and Marie attempted to change the subject. But he insisted.

She finally replied, "Accidental death."

"In a plane, naturally," said Latham, laughing.

Marie said, "Absolutely not in a plane. Far from your family, in the midst of wild beasts, in the next two years."[52]

Her prediction, published by most of the journalists present that evening (for example, Jacques Mortane, "The End Foretold," *L'Excelsior*), came true. Leading an expedition into an unexplored region of then-French Congo, Hubert Latham died during a buffalo hunt in June 1912. Although there is some doubt about the circumstances of his presence there (was he scouting for airfields?) and his death (was he mauled by a buffalo or shot by a porter?), there was no doubt that Marie had foretold his sad end.[53]

(Her psychic foretellings were different from her common-sense, knowledge-based predictions when, for example, she demanded the constructions of new airfields, which she said would be increasingly needed,[54] or when she claimed that "airplanes will seem as common as bicycles in another generation,"[55] or when she lectured architects about taking aviation into account when designing new buildings.[56] She was not always correct: In 1952, she predicted we would all be using helicopters to run errands in ten years.[57])

She sometimes combined hypnotism, suggestion, and auto-suggestion techniques with her nursing to help friends suffering from chronic or debilitating diseases. In January 1922 in Chamonix, where she had her ski school, she apparently cured, in front of numerous witnesses, five people seriously troubled with "organic" troubles, some by hypnotism, some by suggestion.[58]

In Tunis, at a popular society salon gathering, she convinced a number of skeptics by correctly identifying the birth months, characters, sicknesses, and accidents of people she was meeting for the first time.[59] Her cousins tell of her using hypnotism as something of a parlor trick, but maintain she did it only among family. Uncle Augustine ate a potato, convinced it was a peach. After a little talk with Marie in the kitchen, Grandfather Julien, who didn't even believe in those "diabolic matters," walked into the vestibule where Marie's bicycle was leaning against the wall and rang its bell repeatedly, having no idea why he felt compelled to do so.[60]

Marie told her newly married cousin Jeanne that she and Louis would have four children, one of whom would not be like the others. They had four children; one was a priest.[61] Once, Louis's sister wrote their mother about a little accident she'd had. The mother was worried that there was more to it than a simple injury, and asked Marie about it. "Show me the letter from Thérèse," said Marie. After reading it, she said, "Do not worry, it's not serious. She has an injury to her hip. Although she has quite a way to go to recovery, she's in no danger, and there will be no subsequent problems."[62] All true.

The combination of some solid science with her own intuition allowed her to make numerous correct predictions, more than twenty apparently verified in Tunisia alone.[63] Marie was captivated by autosuggestion, enunciated in Emile Coué's 1922 book, *Self Mastery Through Conscious Autosuggestion*, most likely because she had been mastering herself in a similar way for years. She believed profoundly in willpower as a force for action and self-discipline. Educating, training, and exercising the willpower was a high priority for her.

Almost everyone who knew Marie spoke at some point about her willpower.[64] A few called it obstinacy, saying she always went her own way and wouldn't listen to anyone. A friend who saw her often said, "When she wanted something, nothing stopped her. What a hard head she had. But that's how she was able to do everything she did, because she persisted. When people wouldn't do what she wanted, she pushed them, she pushed them: 'This is what I want.'"[65]

She seized upon the work of Alphonse Bertillon, French police officer and son and brother of statisticians, who created anthropometry (using physical characteristics to identify people), the first scientific system used by police to identify criminals. Bertillon's handprints were eventually supplanted by fingerprints, but Bertillon also created other forensics techniques, including forensic document examination, the use of galvanoplastic compounds to preserve footprints, ballistics, and the dynamometer, used to determine the degree of force used in breaking and entering. He also standardized the mug shot. Impressed by Bertillon's ideas, Marie began collecting handprints, which she used to document her predictions based on the hands, and then had them validated, signed, and dated by witnesses.[66]

The theories of psychical researchers on paranormal abilities convinced her that people who developed these abilities had a major advantage over others. However, she never failed to warn people against spiritualism and hypnotism in the wrong hands.[67]

From the barely credible to the highly promising, Marie's areas of fascination emphasize her awareness of the ideas of her day; as soon as something new appeared, she explored it, exhausting as many of its possibilities as she could. Adding luster to the Marvingt legend was that she lived in an era in which women rarely put themselves forward and even more rarely engaged in competitive, energetic activities. Marie's pronounced taste for sports, adventure, and intellectual investigations disoriented those who believed women's role to be "the Angel in the House," a concept glorified in an 1854 Coventry Patmore poem, and satirized in 1931 by Virginia Woolf.

Marie's curiosity was as much an imperative as her need to compete and excel. And, as Sarah Scott wrote in 1762, "Curiosity is one of those insatiable passions that grow by gratification."[68] The more Marie knew, the more she needed to know. Her curiosity, which occasionally led directly to risky activities, explained a great deal about Marie's life. "The curious are always in some danger. If you are curious you might never come home" (Jeanette Winterson).[69] Marie, it was generally agreed, spent very little time at home.

12

Psychological Portrait

> "We are free when our acts spring from our whole personality, when they express it, when they have that indefinable resemblance to it which one sometimes finds between the artist and his work."—Henri Bergson[1]

Marie's outer life appeared to faithfully reproduce her inner life, resulting in a pleasing coherence and integrity of character. Because her words and actions arose organically from who she was, those words and actions illuminate the woman behind them.

Marie benefited from a fanciful combination of nature and nurture. "Nature" provided the basics: innate physical and mental abilities that allowed her to excel in every arena she entered; an invincibly healthy body capable of enormous endurance and uncommon energy; innate drives toward sensation, service, and accomplishment; and a temperament perfectly suited to her iron will, her ambitious visions, and her refusal to accept limitations.

"Nurture" gave her opportunities and encouragement. Marie's cultural environment exploded with new frontiers in sports, aviation, invention, cultural freedom, and advances for women. Her personal environment contributed to her character in both negative and positive ways—the early death of her mother and the support of her sportsman father.

Examined with simple templates or with more complex temperament indicators, Marie's outsized life reveals remarkably well-defined traits and tendencies. Where individuals of more moderate actions and behavior might fit into a psychological category, Marie seems to be the very embodiment of the category.

One of the oldest classifications of temperament, used by Hippocrates to explain how individuals responded to life situations, was divided into four personality types: sanguine (optimistic and social), choleric (short-tempered and irritable), melancholic (analytical and quiet), and phlegmatic (relaxed and peaceful).[2] Individuals can be a mixture of several.

Although the four-temperaments theory is not embraced today with any enthusiasm, it's a charming lens through which to view Marie because she is an almost perfect model of the sanguine temperament. Since earliest times, this temperament has been associated with air. Sanguine individuals tend to be lively, sociable, carefree, talkative, and pleasure-seeking. They make new friends easily, are imaginative and artistic, and are filled with ideas. By contrast, Marie was rarely relaxed, quiet, or short-tempered, hallmarks of the other three temperaments.

The French term *sangfroid* comes from the same root as sanguine and was repeatedly used to describe Marie.[3] If her fiancé was danger, her sister was *sangfroid*. Over and over, she illustrated the cheerful, upbeat nature of the sanguine personality.

In *The Hero Within*, Carol Pearson describes the archetypes we live by: Innocent, Orphan, Martyr, Wanderer, Warrior, and Magician.[4] At one time or another, individuals experience all the archetypes, but over a lifetime, one or two tend to dominate. Marie's life exemplifies the Warrior and the Wanderer.

According to Pearson, Warriors have a horror of weakness, work intensely for success (which they expect to attain), and learn by way of competition, achievement, and motivation. Warriors are confident, courageous, and assertive, earning the respect of others. They tend to rely on fact rather than subjective information.

Wanderers are independent and autonomous, exploring new ideas in their own idiosyncratic way. Stoic, the Wanderer walks alone through life, becoming her own person and dealing with her emotions without sharing them. Wanderers will sacrifice money for freedom and independence. They live by the radical proposition that life does not consist principally of suffering but is, rather, an adventure.

Wanderers ignore traditional social roles; they prefer discovering who they are and what they want. Pearson says, "We are often aware of the Wanderers who externalize their journeys and either literally travel or experiment widely with new behaviors ... they definitely will define themselves in direct opposition to a conformist norm."[5]

In *Type A Behavior and Your Heart*, Meyer Friedman and R.H Rosenman proposed that the Type A individual struggles to achieve as much as possible in the least amount of time.[6] Type B is easygoing and lacks the sense of urgency of Type A. Originally constructed to explain coronary incidents, the theory was later amended to emphasize the presence or absence of hostility in the Type A. Some Type As are both driven and angry. Others (and Marie is a good example) retain the drive but lack the anger.

Type A people like Marie don't particularly need the love or affection of others, but they do need respect and admiration. They place value on the number rather than the quality of their achievements, and that number must keep increasing to appease their insatiable appetite for accomplishment.[7]

Marie's speeches, letters, and writings are peppered with numbers: she worked in the operating room for 27 hours straight; she gave 6,000 conferences; she spoke 5 languages; she gave 27 conferences in 19 days[8]; she wrote for 40 newspapers; she belonged to 91 organizations; she traveled 56,000 kilometers and lived 50 months in North Africa; she spent 40 years promoting ambulance-airplanes and air safety; she flew 900 times without a crash; she hiked 57 kilometers; she won 20 first prizes in winter sports; she received 34 decorations; she held 17 championship titles.

For the sake of precision, it is reasonable to use figures to describe one's life. However, Marie conveyed the impression that the numbers themselves meant a great deal to her. Marie also exhibited the Type A traits of competitiveness and an emphasis on drama in her speech and writing. She was not (to contrast her with a Type B) easygoing or tractable.

The following letter (written at 7:00 a.m.) might be typical of a Type A:

"My dear Marcel, Thank you for your thank-you, but you would be only too kind to send me the letter you promised me on October 22. I absolutely need it. Did Monsieur de Levis M. intervene with the Duke of L.F. and, if so, what *precise* response did he receive? I repeat, *this is urgent*. Yesterday morning I saw T. Berbey. He's dragging his feet. His son is in Holland following his schooling in Switzerland. Tonight is a big reception at our friends, F. Sanon. Last night was a splendid closing to the mission by Monsignor Lallier. More than 50,000 people attended. Quite lovely. Affectionately, M. Marvingt."[9]

An analysis of Marie's handwriting in that letter and many others confirms the main traits associated with her. Her firm, right-slanting handwriting with its sharply pointed letters shows a strong, extroverted, intelligent person with a great deal of self-confidence and drive.

Marie herself studied graphoanalysis and was enthralled with psychology, dating at least to her studies in hypnosis and suggestibility under Bernheim. She would most likely be intrigued by the explosion of research and new developments in psychology, psychiatry, neurology, and brain science since her time.

From a metaphysical viewpoint, human beings comprise four quadrants, each more or less developed in individuals according to temperament, interests, and environment: the mind, the heart, the body, and the soul. The ideal, rarely attained, is to develop each quadrant as fully as possible to achieve a harmonious balance.[10]

In Marie, the physical side was almost over-developed. Most of her energy seemed to be concentrated in exploiting her body's capabilities and endurance. But she also paid great attention to her mental gifts, achieving expertise in as many intellectual domains as were available to her. In the spiritual domain, she showed a curious mixture of belief-based faith and fact-based science, with a little superstition thrown in. Open to every possibility, she asked big questions, explored answers, and sought and found peace principally in meditation. "I am drawn to all that is vast and limitless. For me, there are four oceans: the seas, the mountains, the skies, and the deserts. These four immensities have provided me with a lifetime of thought and reflection."[11]

Does she look like a risk-taker? (courtesy International Marie Marvingt Committee).

Nobody has unlimited energy: what is spent in one direction cannot be spent in another. The quadrant that came up short in Marie appears to have been the heart. She was objective, detached, reasonable, logical, practical, almost emotionless at times. Although her writings show passion and she displayed great empathy toward those who were ill or in need, both these "heart" qualities seemed to originate primarily in her "head," in her principles and belief systems.

Gustave Flaubert and George Sand discussed the relative importance of reason to emotion. He felt she didn't know how to reason, that she knew only how to feel.[12] The opposite could be said of Marie. Her compassion for others, her actions on their behalf, and her devotion to being of service were never in doubt, but they grew out of an intellectual sympathy. Because she wasn't distracted by strong emotional involvements, she was able to act decisively, logically, and energetically to help others. Plural. For Marie, the one-to-one relationship was somewhat alien. She was most comfortable emotionally in crowds and groups of people. Novelist Patricia

Wentworth might have been describing Marie when she wrote, "You can't do such a lot and do it all so well and have much time left for the ordinary human feelings."[13]

A once-popular classification identified people by their body types: mesomorphs, ectomorphs, and endomorphs.[14] Most people are a mixture, but generally one type will predominate. Endomorphs are roundish, love to sleep and eat, and like their comfort. Ectomorphs are thin with long bones, are often nervous and sensitive, and have trouble sleeping. Mesomorphs have muscled, sturdy bodies, adore adventure, risks, and competition, and are aggressive, energetic, and active. It is obvious which type Marie was.

That Marie was known since 1903 as "the fiancée of danger," and that she lived a life so daring that *The Perils of Pauline* was modeled after her, points to a character trait found in a small percentage of individuals. Marie was a sensation-seeker.

Sensation-seekers crave novelty, complexity, and intense sensations. They love adventure for its own sake, and will sometimes risk their lives in pursuit of new experiences. Sam Gosling, psychologist at the University of Texas at Austin, writes in *Psychology Today*, "Easily bored without high levels of stimulation, they love bright lights and hustle and bustle and like to take risks and seek thrills."[15]

For most people, fear is an unpleasant feeling—anxiety raised from a yellow warning to a screaming red "Stop!" Fear functions to make us flee dangerous situations.

Some people are drawn repeatedly and inexorably to risky experiences. The expression "extreme sports" didn't enter the language until the early 1990s, but many of Marie's activities, in the context of her era, certainly qualify as extreme.

How does an attraction to dangerous pursuits override an analytic mind? Brain chemicals. With names like adrenaline, dopamine, endorphins, and serotonin, these happy-making neurotransmitters and neuropeptides stimulate positive, even euphoric feelings.[16] As always with our genetic makeup, each of us is constituted differently. A few of us are over-endowed with one or another of these particular brain chemicals.

An inventory to identify sensation-seekers asks, "Would you like to learn to fly a plane or would you prefer not to?" Most particularly in the early days of aviation, when airplanes were fragile, unpredictable, and deadly, that question would have nicely separated the sensation-seekers from everyone else.

Sensation-seekers prefer edgy people to comfortable people, originality to convention, novelty to familiarity, the unknown to the predictable, and intensity to comfort. They fear boredom more than any danger. Marie's life choices clearly define her as a sensation-seeker.

Journalist Georges Gygax wrote, "It is certain, the thing is evident: Marie Marvingt has lived in joy, in enthusiasm, in the exaltation of dearly acquired victories. Her life, despite her devotion and self-denial, especially in the war of 1914–1918, has kept her from being bored even one single time. She has lived intensely, and this intensity has never stopped delighting her."[17]

Marie agreed, saying she had never been bored in her life.[18] Nor, she said, had she ever felt fear.[19] Being so much more of a thinker than a feeler she very likely saw danger as a problem to be solved rather than something to react to emotionally.

Marvin Zuckerman, who pioneered the field of sensation-seeking behavior beginning in 1964, champions the biological argument: he has found that some babies as young as three days old seek external stimulation, while others avoid it. In *Psychology Today*, he says risk taking "is not just a behavior. It's a personality."[20] He says it's one of a handful of "core traits" that can describe human personality. According to him, for all the danger they put themselves in, and expose others to, sensation-seekers "personify—perhaps magnify is more precise—a human trait that is very much responsible for our survival as a species."[21] He suggests humans are a risk-taking species and that the drive to explore may be a vital part of our survival. However, sensation-seeking is good for the species only when it is in the middle range: too much risk-taking leads to early death, too little to stagnation.

Sensation-seekers are often attracted to speculative ideas, likely to participate in psychology experiments such as hypnosis, and more open to new and unusual experiences. They're less anxious about physical harm and have greater tolerance for discomfort and high-intensity stimuli. They tend to evaluate situations as less risky than others do. What would possess a person to dive from a dirigible into the Grand Canal in Venice?[22]

In her early years at least, she seemed to be aware at some level of trying to balance her love of action (and sensation) with her altruism. She had an unquestioned desire to be of service to humanity. And she did leave the world richer, with her invention of the ambulance-airplane, her tireless efforts to establish air-rescue operations around the world, her passionate patriotism and contribution to France's war efforts, and her years of nursing the sick and injured. She also indirectly but powerfully influenced the world of sports, young people, and women's issues. By any definition, she was an altruist (with money she would have been a philanthropist).

Marie seem to need to express, more than once, the importance to her of service. The French regularly use of a courtesy phrase offering to be "of service" to someone. Marie raised the notion to a career. On the other hand, she was from birth irresistibly attracted to risk, excitement, novelty, frontiers, obstacles. She repeatedly sought intense sensations by way of actions that required talent, strength, and nerve. She loved action, movement, jeopardy. (Amelia Earhart's autobiography was called *For the Fun of It*. She shared Marie's attitude: the risks they took were, for them, "fun.")

Time and the human condition eased the dilemma for Marie. As her declining athletic talents rendered her less competitive, she devoted more and more of her energies to service. She still craved adventure, danger, and the unfamiliar, but she responded to these urges less excessively—with travels, new experiences, outpacing her others her age, amassing records and awards, and (for example) flying helicopters in her eighties. She sought sensations of a different, less risky kind.

The most delightful personality inventory results based on Marie's life emerge from the Myers-Briggs Type Indicator.[23] This system classifies people according to four basic ways of perceiving the world and making decisions, resulting in sixteen personality "types." Marie never filled out the inventory, of course, but her type can be determined by responding to the questions with what she actually, and repeatedly, did throughout her life. Although the resulting psychological portrait is perhaps not particularly reliable, it does throw into relief some of Marie's most characteristic traits.

Marie seems to belong to the ESTP temperament type, in which E indicates that she was more extroverted than introverted, S that she learned primarily through her senses rather than through her intuition, T that she approached everything from an intellectual (thinking) perspective rather than an emotional (feeling) perspective, and P that she acted and thought more often in open, fluid, flexible ways, more interested in process than in closed, well-organized, rigid ways that are invested in results.

Primarily, ESTPs are women and men of action, oriented to movement, and to outdoing, surpassing, out-distancing, and triumphing over competition. They choose action occupations requiring precision, endurance, strength, boldness, and timing. According to David Keirsey and Marilyn Bates, highly skilled in analyzing Myers-Briggs results, one ESTP trait is "nerves of steel, engaging in what seems to others to be suicidal brinksmanship. ... A theme of seeking excitement through taking of risks runs through the lives of ESTPs."[24]

Keirsey and Bates would describe Marie as a "Promoter" type, which suits her perfectly: "Promoting is the art of putting forward an enterprise and then of winning others to your side, persuading them to have confidence in you and to go along with what you propose."[25] Her forty years of promoting air rescue, the ambulance-airplane, and training for air-rescue personnel—and her early struggles to overcome indifference and antagonism to her projects—exemplify the Promoter. (She also promoted swimming in the schools, the use of Esperanto, sports for girls, her metal skis for small planes, and more). Had she ever taken the Kuder Preference Record, she would have scored nearly 100 percent on one of the ten categories, "Persuasive."[26]

Family life comes second for most ESTPs. They live by the adage, "They go fastest who go alone," and rarely develop deep personal attachments. On the other hand, ESTPs are often extremely popular and know everyone. "They carry on amusing repartee, and laughter surrounds them as they recount from their endless supply of clever jokes and stories. Charm radiates from ESTPs." Friends report feeling comfortable with such people on first meeting but find that the friendship never grows or deepens, never becomes real intimacy, even over a lifetime.

Paradoxically, this type is also the most fiercely collegial: they understand "esprit de corps" and are loyal to their groups, battalions, organizations. But it is, once more, the appeal of the group as opposed to the appeal of the individuals in the group. This type is most at home in crowds, with the public, with strangers. Socially, they are refined, affable, and courteous. But, with their detachment, they are capable of surviving heartbreaks that would devastate other people. Setbacks defeat them only temporarily.

Because ESTPs are more easily able to leave social and emotional connections behind, they respond more quickly than any other type to the appeal of travel, and will abandon one activity or life to go off in search of new adventures.

Narrowing in on two parts of the ESTP (S and P) reveals more of Marie's marked traits: "Oddly, the SP seems to have endurance beyond that of other types."[27] These people put up with discomfort, deprivation, hunger, fatigue, and pain, and show courage in ways others do not. The SPs yearn to leave their mark: "To be without impact, to make no difference in human affairs, is like being deprived of oxygen." More than other types, SPs long "to perform some conspicuous function in their social context." A general who knew her well said that, for Marie, "her entire life was one long performance."[28]

History, even "prehistory," confirms that humans are notably attracted to symbols, gods, and tools. The SPs are not very interested in gods and they ignore symbols, but tools are their delight. Keirsey and Bates say SPs are driven to drive bulldozers, fly planes, fire guns, toot horns, wield scalpels, brushes, or chisels. Tools become an extension of the SP self, augmenting and amplifying the effects of their actions.

Marie's tools are quickly identified. She piloted balloons; flew airplanes, hydroplanes, gliders, and helicopters; drove locomotives and automobiles; sailed land yachts and sailboats; rode horses and bicycles. And then there were her canoes, skis, guns, épées, ice skates, roller skates, tennis rackets, billiard cues, and golf clubs. The horn she tooted was the cornet. She painted with brushes and sculpted with chisels. Her ultimate tool was, of course, her ambulance-airplane, and specifically her prototype, the *Captain Écheman*.

The SPs must have their freedom; they cannot be confined. When Marie closed her hospital for injured pilots, she wrote her cousins: "I've just closed my Refuge for Aviators, not without a certain joy—I can hardly wait to regain my freedom."[29] Once she said, "I love immensity. I have a horror of being imprisoned inside four walls. For me there are four oceans: sea, air, mountain, and desert. These four immensities have led me to meditation many times."[30]

They don't usually stockpile knowledge or accumulate power; they spend life as freely as possible. Action cannot be saved for tomorrow, and each day brings a need for new excitement, adventure, risk, testing one's luck. Resources are to be used, machinery operated, people enjoyed. They do not accumulate possessions, nor do they save for tomorrow. Besides requiring freedom, they need others to perceive that they are free.

They seem to attain perfection in action, and yet they never practice in the sense that others do. Somehow, caring about perfection and working for it are counterproductive for them; only the act itself can realize perfection. The SPs often excel without the practice others need. Writing in 1913, Henry Woodhouse pointed out that Marie "is very much a woman of impulse, liking to do things on the spur of the moment. In 1907, she successfully swam the Seine through Paris, without having previously attempted the course."[31] SPs don't like repeating themselves, they don't to rehearse or practice; they want to *do*. SPs follow their impulses and continue their actions as long as the urge compels. When it stops being compelling, they no longer "feel like" racing, climbing, flying. Marie sometimes moved from one activity to another without an apparent reason, captured by the next passion while the previous one died away.

They thrive on situations without rules or regulations, where the outcome is unknown, where there is freedom to test the limits. They love not knowing in advance the results of their efforts. William James wrote, "It is only by risking our persons from one hour to another that we live at all. And often enough our faith beforehand in an uncertified result *is the only thing that makes the result come true*."[32] SPs work best in crises, and the more severe the crisis, the more quickly and dramatically they respond. Marie's unprecedented and dangerous crossing of the North Sea illustrates this facet of her temperament.

Perhaps the simplest summary of Marie's psychology and healthy self-acceptance is seen in her lifelong expressions of inner joy and outer enthusiasms. When asked at the age of eighty what she wished for, she replied, "If, some day, a fairy appeared to me with her magic wand, asking me which I preferred—riches, beauty, or eternal youth—I would tell her that I am perfectly happy, young, and powerful. I wish for only one thing: to continue just as I am. I would choose the status quo."[33] The journalist concluded, "So saying, Marie Marvingt jumped on her Zéphirine and took the road to Versailles, where friends had invited her to dinner."[34]

13

All the Pretty Ribbons

"If you tried to name all the titles and medals that Marie Marvingt holds you'd risk forgetting half of them."—*L'Est Républicain*[1]

Asking *her* to list them wasn't much better.

"What good would that do? I couldn't possibly keep track. But perhaps you'd like to *see* some of them." And she held up one after another, naming them faster than reporters could write them down.[2]

Marie was the most decorated woman in the history of France, and still is today.[3] She received some thirty-four decorations,[4] most from France, some from other countries, and she held seventeen sports championships or world records.[5] (It appears that she is also the most decorated *person* of the twentieth century.[6])

The recognition most familiar to those outside France, and one to which Marie was very attached, was the Legion of Honor, the highest decoration in France (motto: "Honor and Country"). Established by Napoleon in 1802, this honor recognizes merit rather than nobility or high birth. Only after exhaustive investigation is it bestowed for outstanding civil or military conduct. Marie was named a Chevalier of the Legion of Honor in 1935, which required a minimum of twenty years of public service or twenty-five years of professional activity with "eminent merits." Marie was the only woman in that group of honorees.[7]

She was initially not considered for the Legion of Honor because it was felt she was too young to have earned it. When it was pointed out that, at the time, pilots hardly lived long enough to grow old, Marie received her honor.[8] The Ministry of Public Works, Transportation, and Tourism nominated her for her work as a pioneer pilot and as an incomparable proponent of aviation and the ambulance airplane.[9] In 1949 she was promoted to Officer of the Legion of Honor, which required at least eight years as a Chevalier and recent services in the interests of France.[10]

The file on her nomination for both grades was sealed until 1995 (it can be opened only 120 years after a nominee's birth date). The qualifying information on Marie included, among many others: obtained pilot's license in 1910; took part in two bombings of the German airfield in Frescaty; fired on the enemy from the front lines in World War I; war correspondent on the Italian front; five months in the French and Spanish Riff campaigns, several times flying over the fields of operation; won the 1911 Coupe Fémina; set the first women's aviation records in time and distance; participated in all the grand prix balloon races until 1914, winning first place in some; crossed 2,800 kilometers in the Sahara; worked in military hospitals as a nurse with exceptional knowledge of medicine and surgery; established the Captain Écheman Challenge for air-ambulances; honorary member of numerous French and foreign air clubs; holds licenses to fly balloons, planes, and hydroplanes; has devoted herself ceaselessly to the cause of aviation, giving hundreds of conferences on its behalf.[11]

Finally, the Minister of Air certified that Mademoiselle Marvingt's morality deemed her fit for admission into the Legion of Honor.[12]

When Marie was honored for her promotion to Officer, Minister of Air André Maroselli arranged to give it to her on the feast of Joan of Arc in a huge public ceremony because, he said, it was the only way to be sure she would be there for it. A newspaper piece concluded, "She is one of the most active women in the world, traveling everywhere abroad, giving conferences as she goes, and trying out every means of local transportation."[13]

One of her honors clearly corroborated her media-proclaimed status as the world's greatest sportswoman: the French Academy of Sports awarded her its gold medal in 1910 for excellence in *all* sports. Normally awarded for distinction in one sport, in Marie's case the gold medal was presented for outstanding achievement in all sports by unanimous decision of the Academy. To this day, Marie is the sole recipient in that category. *Le Figaro* announced the unusual award, saying she was "as brave as she is modest."[14]

Marie received the *Croix de Guerre 1914–1918 avec palmes*, a green silk ribbon with tiny red and yellow stripes and a bronze cross hanging from it, for her valor during World War I.[15] Although over two million of these were awarded (a passenger pigeon even received one), they hold high meaning for the French. Of particular value to Marie was the official recognition of her having flown a bomber, fought in the trenches, and served in an Alpine battalion to ski in provisions and ski out the wounded. She was entitled to wear two palms on her *Croix de Guerre*, one for her war work as a surgical nurse in field stations, the other for her two bombings of Frescaty in 1916.

Awarded in 1910, Marie's gold medal and certificate from the French Academy of Sports, honoring her expertise in all sports, wasn't given to her until many years later.

A recently discovered medal reveals the truth behind rumors that Marie worked in the Resistance during World War II. Small groups of armed women and men, the Resistance (referred to as the Maquis in rural areas) fought to overthrow the Nazi German occupation of France and to oust the collaborationist Vichy régime. In addition to guerrilla warfare activities, members of the Resistance published underground newspapers, provided first-hand intelligence, and maintained escape networks to free Allied soldiers and air troops trapped behind enemy lines. Resistance activities and names were closely guarded secrets for decades. Marie's precise roles remain for future historians to unearth, but because her medal was awarded *avec rosette* (with the little rose), her contributions must have been significant; only 10 percent of all Resistance workers received the medal *avec rosette.*

Parisian journalist Georges Gygax visited "La Marie" on the Place de la Carrière in 1958: "Among the trophies crowded into Marie Marvingt's little salon is a ravishing sculpture of the Victory of Samothrace. 'Ah, that one," she says. 'I like that one. It's recent, given to me three years ago by the National Aeronautical Federation.'"[16]

That imposing ceremony in January 1955, held in the big amphitheater at the Sorbonne, began with Weber's "Overture d'Obéron." Then Laurent-Eynac (Victor André Laurent Eynac went by his shortened name), president of the French Aeronautical Federation and Marie's good friend, distributed the certificates and prizes earned the previous year.[17]

Signaling to someone in the front row to join him onstage, Laurent-Eynac said their event would be incomplete without Mademoiselle Marvingt. As one newspaper put it, "It is then that one sees a woman spring up alertly, and quickly ascend the steps to the stage."[18] She was one of the few women in the auditorium. For the next hour, Laurent-Eynac reminded the audience of Marie's work in aviation medicine, in particular her invention of the ambulance-airplane and the fifty years she spent promoting air rescue around the world. He recounted triumphs and struggles, anecdotes and events. Then Laurent-Eynac presented to her the graceful bronze statue of the Victory of Samothrace that was to have pride of place in her apartment for the rest of her life.[19] Of course she liked that one.

She told Georges Gygax that she also particularly cherished the Grand Prix Deutsch de la Meurthe, which was awarded to her by unanimous vote March 29, 1954. Henry Deutsch de la Meurthe, wealthy industrialist, philanthropist, and avid supporter of early aviation, created the prize through the French Academy of Sports in 1910. Accompanied by 25,000 francs, the prize was given to just one person each year. In 1954 it was Marie's turn to be singled out for this high honor.[20]

France has specialized orders of merit, honors given for outstanding service in agriculture, social ministry, commerce, and other areas. Marie was named a commander in the field of sports. She was similarly honored in the fields of public health and education.[21] The Ministry of Air awarded her its aeronautics medal, given to civilians and military personnel who contributed many years to the progress or prestige of aviation. She was also awarded the *Palmes Académiques*, a purple silk ribbon with bronze palm leaves, for academic excellence.[22]

Receiving more medals on her 80th birthday. Good friend Marcel Knecht is at left.

When Marie wore all her decorations, as she sometimes did for important functions, she was impressive.

Two of her medals (her thirty-second and thirty-third) were presented to her on her eighty-third birthday, February 20, 1958.[23] In a ceremony at the Hôtel de Ville in Nancy, Marie sat in the chair usually reserved for the Senator-Mayor, the front of her suit covered with decorations. First the city municipal councillor gave her a medal naming her a Knight in the Service of Public Health. Then her longtime friend Colonel Blanc presented her with a medal of honor from the Air Force Health Service.[24]

In order to be sitting in that chair on that day, Marie had to turn down the honor of presiding at the banquet of the scientific press, which was being held in Paris that evening. She probably chose well. In Nancy, she was surrounded by a huge gathering of longtime friends, admirers, and officials. She was feted with letters, gifts, applause, and the overlapping flashes of the photographers. From her many congratulatory letters, she selected one to read, a letter of gratitude from a man wounded in the battle of Grand Couronné whom she'd cared for.[25]

The Nichan Iftikhar ("Order of Glory") was an honorary Tunisian order established in the 1830s, awarded to both Tunisians and foreigners for outstanding civil and military service; Marie was named an Officer. Awardees, who included Dwight D. Eisenhower, were given a handsome medal and a diploma that urged the recipient to "Wear it with joy and happiness!"[26]

In 1950, the mayor of Nancy presented Marie with the city's silver medal, to recognize her numerous services to Nancy and the surrounding areas of Lorraine.[27] The Aeronautics League awarded her its highest honor in a ceremony held on the Place Stanislas. Presenting the medal was another citizen of Nancy, General Vallin, Grand Cross of the Legion of Honor.[28] Other honors included a medal from the Ministry of War for being a first-class shot; the Peace Medal of Morocco; a gold medal for her work in physical education; a gold medal from the department of aviation medicine; an earlier bronze medal from the Aeronautics League; and a gold medal from the air force.[29]

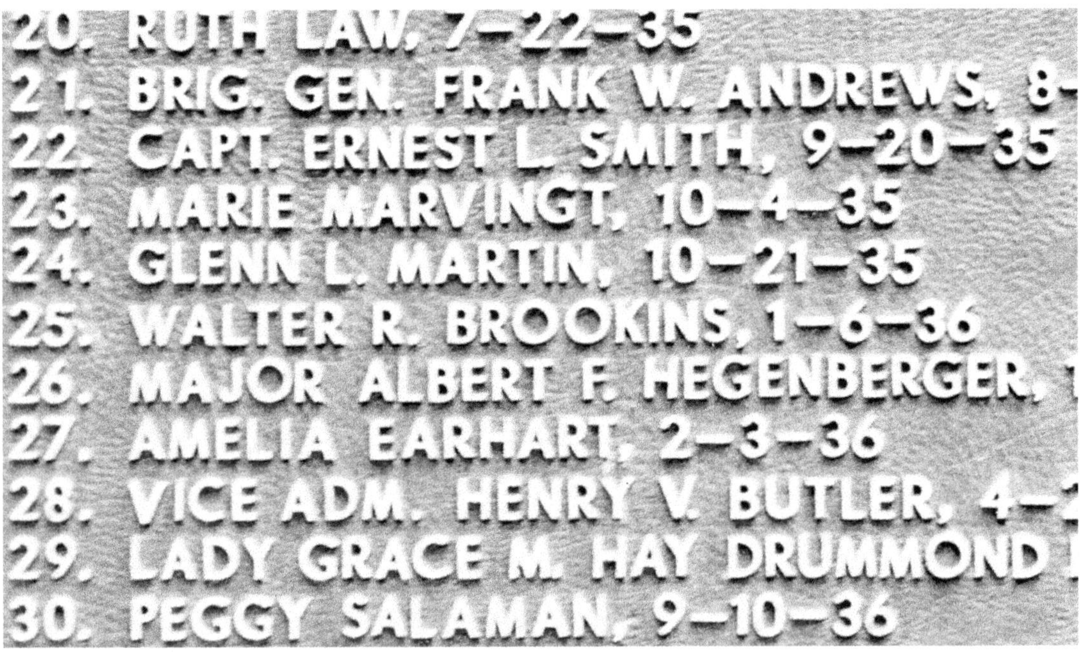

Some of the names on the fliers' wall at the Chapel of St. Francis, The Mission Inn, Riverside, California.

Marie's engraved wings.

Marie also prized a few offbeat medals. Having spent years in North Africa, often in Saharan military or rescue outposts with little to do outside work hours, Marie and her colleagues staved off boredom by creating imaginary orders of knighthood. A bronze bug, spider, or scarab would accompany the solemn presentation of diplomas awarded by, for example, "The Ancient Order of the Gold Bough of the Secular Trunk." She delighted in recalling the absurd honors: "The Tarantula of Tidikeit," "The Royal Kranfouss," "The Blatte" (a cockroach), and "The Laughing Wadi of the Sahara."[30]

Marie was honored in the United States as well. Her signed copper and enamel wings can be seen today on the famous fliers' wall in the St. Francis Chapel at the Mission Inn in Riverside, California.[31] (St. Francis is considered the patron saint of aviators because he befriended birds.) The wall is an international shrine, honoring those who made outstanding contributions to aviation. Marie later wrote, "During a lovely reception, the owners of the magnificent Mission Inn, Mr. and Mrs. Hutchings, invited me to add my name."[32] Her signature is engraved on a ten-inch pair of copper wings mounted on the west wall and dated October 4, 1935. She is in good company. The other 138 wings include those of Amelia Earhart, Eddie Rickenbacker, Orville Wright, and John Glenn.

A relatively small honor, but one she cherished, was winning first place two years in a row in the literature competition sponsored by the International Women's Aeronautics Federation of Los Angeles (with an honorable mention for her poetry). Founded May 23, 1929, when there were few women pilots, the federation established branches on six continents. In a 1949 letter, president Elizabeth Lippincott McQueen wrote that she "made many contacts with air-minded women and women pilots in these countries." She added, "The public press generally censured this outburst of 'feminism' in a man's world."[33] The federation's annual Aeronautical Literary Contest accepted articles, stories, and poems in all languages. To be part of such a forward-looking, inclusive group of female pilots was a joy for Marie. She won the prize in 1948 and 1949, once for "The Fiancée of Danger" and once for "My Crossing of the North Sea in a Balloon."[34] The day she found out about her second win, Marie celebrated by biking her favorite route around Paris ten times.[35] She was seventy-four.

14

My Dear Friend, the King

"I've been invited to visit my godson, the Infante of Spain, but I've also received an invitation to a river cruise in Russia for the same time."—Marie Marvingt[1]

"Marie Marvingt knew absolutely everyone!" Several friends said the same thing, summing up an entire way of life for Marie.[2] Another said she had a genius for knowing and being known.

Marie could sketch word pictures of one famous person after another, telling behind-the-scenes stories that fascinated her listeners. "You remember So-and-So?" she might say. "The chief of state, who was supposed to have killed himself? Well, no, I tell you it was murder, not suicide. I knew him well. Just look at this letter that I received from him a month before his sad end: 'My very dear friend.'"[3]

According to *L'Est Républicain*, it would be "futile to attempt to list all the great people she knew or met."[4] Marie was on a first-name basis with high-ranking military officers, politicians, the great sports figures of several eras, aviators, actors, scientists, writers, medical professors, religious leaders, and even some of the European royals. Every time she gave a talk, the press carried at least one paragraph naming the notables in attendance.[5] Throughout her life, she retained a childlike enthusiasm for people of action and accomplishment. An acquaintance from the early days of aviation described her as "a picturesque person who liked to surround herself with other picturesque people."[6]

"Myriel" wrote evocative, detailed articles about some of the famous people she knew. In a piece in which she brought to life for her readers Latham, Blériot, Farman, the American pilot Glenn Curtiss, and President and Madame Fallières, it was obvious she knew, and was known by, each of them.[7]

At the Grand Prix balloon race in 1912, *L'Aéro* reported: "Noticed among the select group gathered around the huge balloons were the Count de la Vaulx, Ernest Archdeacon, Legagneux, Bielovucic, Mademoiselle Marvingt, Georges Bans, Surcouf, and Rumpelmayer." Names straight from the headlines of the day.[8]

It didn't seem to occur to Marie to be awed by the celebrities she met. If she wondered how the little girl from Aurillac ended up being world famous, no one suspected it. Treated everywhere as a star, Marie assumed that other celebrities were kindred spirits. But it was her fathomless and gregarious curiosity that led her to enjoy with equal pleasure the company of foot soldiers and generals, schoolgirls and queens, mail carriers and movie actors. The woman who made her hats and the man who re-soled her shoes knew her better than most journalists did, and spoke of her like family.[9]

Marie's friendships with aviation heroes spanned some sixty years, giving rise to such odd juxtapositions in her autograph books as an inscription by Henri Farman, who flew the first rickety planes, next to one by Odette Rousseau, a French parachuting champion famous many years later. From early aviation pioneer Blériot to Elisabeth Boselli, world-renowned pilot of another generation, she knew them all.[10]

Sharing their passion for flying, Marie belonged to the same loyal and supportive club (with its tragically high dues) as the iconic greats from the early days: Santos-Dumont, Mermoz, Nungesser, Caudron, Blériot, Sommer, Farman, Latham, Voisin, Delagrange, Chavez, Nieuport, Védrines, Garros.[11] They either taught or were taught by each other, flew together once planes accommodated two, anxiously watched each other push limits and break records, spent hours together over dinner talking aviation, attended each other's funerals, and called each other in comradely fashion by their last names. Marie was "Marvingt" to a number of good friends.[12]

Aside from her adored Paul Écheman, the aviator she was closest to was Hubert Latham who taught her to fly, then became a wise mentor, and finally a dear friend. Even though she had predicted it, his odd death on June 25, 1912, dealt her a serious blow. Marie attended his funeral in Le Havre, amidst a grieving crowd at the Protestant church, his coffin all but disappearing under hundreds of floral offerings, wreaths, and crowns. Marie's modest contribution was duly noted in the newspapers because of their relationship.[13] A year later, one of the aviation articles mentioned in passing that Marie had just visited Latham's mother.[14]

Marie paid for her cherished friendships: she attended funerals all her life. In 1918, she attended the last rites for aviator Charles de Luynes, Duke of Chevreuse, her name listed in the newspaper between that of the Marquis of Pracomtal and the Marquise of Courcival.[15] When record-setting aviator Christian Moench went down in the Persian Gulf, his plane was found, but his body was not. Although thirty years younger and part of the second wave of aviation, Moench was a good friend, and Marie attended his memorial service in Nancy in 1938.

In 1915, Albert Féquant, a friend and former student, was killed in a battle with two German planes. He was the brother of a close friend of hers, Philippe Féquant. At the head of the funeral cortège were three military officers and Marie.[16] They were followed by fellow soldiers and pilots. She had been at the airfield the day Albert Féquant took off on his first military air run. She wrote about his death:

"One evening during the war when Nungesser and I were giving diving lessons in the Nancy pool, Latham swam toward me and said, 'Bousquet and Féquant are both dead.'

"Immediately I asked, 'Which one of the two?'

"'It was Albert.'

"Albert's pilot, the son of General Niox, had just landed in Malzéville that night, holding the steering wheel with one hand, and the body of Albert Féquant with the other. What a tragic homecoming. He'd been hit full face, and the cockpit of the bomber was full of his blood. By an extraordinary coincidence, Philippe Féquant was in that same sortie, and he had brought down the enemy pilot—he had no idea of this at the time—who had killed his brother a little earlier."[17]

At an October 1931 meeting of the Vieilles Tiges (pilots who were licensed before World War I), General Philippe Féquant wrote in one of her autograph books: "How pleased I was when the young and already-famous sportswoman agreed to be my flying godmother. Around her were gathered all the young comrades and friends, so full of ardor and gaiety who were just beginning their careers as 'cloud chasers.' So many of them have disappeared since, fallen in service to their country—and among them, my young brother. What faith and what enthusiasm I saw in that phalanx of young pilots!"[18]

Marie eventually wrote his obituary as well: "Philippe Féquant of the big heart, of the erudite brain is no more."[19] In the beginning days of aviation, two godparents were necessary for those attempting to obtain a pilot's license; Féquant, in 1911, was the first of Marie's many military pilot-godsons. "I was so proud of him," she said, "and I wrote to tell him so after every one of his promotions."[20]

She had trained a "multitude" (according to the papers) of pilots, including Giuseppe Néri, one of the first Italian aviators; Felice Nazzaro, the great Italian race driver; and Alfonso de Orleans y Borbon, Infante of Spain, cousin and heir to King Alfonso XIII.[21]

Alfonso the younger trained with Marie in 1910 to become one of the first and most distinguished aviators in the Spanish military; he was known as the father of Spanish aviation. They remained lifelong friends. She toured Spain by air with him several decades later,[22] and a photo shows Marie swimming in his pool while he and his family stand nearby squinting into the sun. She was eighty at the time. (On one of her visits, she received a dedication in her autograph book from Alfonso's cousin, the King of Spain.[23])

Captain, later Major, Julien Félix and Marie received their pilot's licenses at the same time. World record holder for altitude, he died at Chartres in 1914, trying out a plane with the new automatic stabilizers. Frantz Reichel (another record-holding aviator and later the journalist who wrote so admiringly about Marie) had flown the plane earlier and warned Félix against it. Marie attended Félix's funeral, held in Paris.[24]

Of the first seven pilots to die in air accidents, two were good friends. Both died in 1910: Léon Delagrange, record-setting pilot and accomplished sculptor, and young Jules Hauvette-Michelin. Delagrange, passionate about flying, was the victim of a plane with a motor too strong for its fragile body; a wing broke off, ensuring his death. Hauvette-Michelin was on the verge of abandoning aviation when he died. Marie was interviewed about his death under the headline, "This Latest Accident Does Not Discourage Mademoiselle Marvingt."[25]

"He was a charming comrade, an excellent companion," she said. "We spent time together at the Châlons camp." She emphasized, however, that his accident proved nothing about airplanes; he ran into a pylon. She herself was not deterred from her own plans to keep flying.[26]

Marie visited Alberto Santos-Dumont in his workshop at 5:00 one morning because she knew that was when the "persevering and audacious little Brazilian" (her description) would be there.[27] He welcomed her to his minuscule cloth hangar and introduced her to his Demoiselle. Weighing less than 200 pounds, the plane was so lightweight it could take off only in perfect calm. "As the curtains of his hangar are agitating strongly, the plane's rest will not be disturbed today," wrote Marie.[28]

Henri Deutsch de la Meurthe, an extremely wealthy petroleum magnate, was an early aviation fanatic. His daughter, Suzanne Deutsch de la Meurthe, was a talented, energetic nurse, pilot, and philanthropist who shared her father's passion for aviation. At most important aviation events, the guest list included at least two Deutsch de la Meurthes and one Marvingt.[29]

Ernest Archdeacon, a wealthy lawyer, was regarded as France's foremost promoter and sponsor of pioneer aviation: he co-founded the French Air Club, offered prizes (one of which Marie won), commissioned designs, and organized tests and events. He was also one of Marie's biggest fans. In 1910, he wrote a five-page, illustrated article for *La Revue Aérienne*, "Mademoiselle Marvingt: An Extraordinary Sportswoman," in which he claimed she was absolutely exceptional and admonished readers, "We mustn't forget that in similar sports victories, a woman deserves about ten times the credit as a man, not only because of their different physiques, but because to be there in the first place, and to succeed in the second place, the woman has to conquer a raft of prejudices, obstacles, and impedimenta of every sort that a man would never have to face."[30]

He went on to say that for a woman to have accomplished what Marie Marvingt had, she needed the "aex triplex" (the triple brass of indestructibility needed for battle armor), and that men needed to bow low when they encountered women like her.[31]

A charming photograph in *La Vie au Grand Air* showed Archdeacon, Latham, Marie, and Latham's other student-pilots sheltering under an overhang, "waiting for the calm" so they could fly.[32] As the only woman in the group, Marie stands out—and yet she doesn't. She seems as much a pilot eager for the winds and rains to die down as the others. By then, people were used to seeing her at the airfield, and they appeared to take her presence for granted.

That evening Archdeacon, Latham, and Marie are pictured dining at Voisin's table, with the addition of Raymonde de Laroche and half a dozen others.[33] Archdeacon admired Marie enough to appoint her one of the judges for an aviation prize offered by the Ligue Nationale Aérienne.[34] (Incidentally, Archdeacon and Marie also shared a passion for Esperanto, and Monsieur and Madame Archdeacon often sat with Marie at the Esperanto banquets in Paris.[35])

For Marie, the camaraderie among the early pilots was one of the most exhilarating facets of belonging to that elite but imperiled group. One postcard photo of Latham flying an Antoinette is signed by each of six flying buddies, including Marie, Roland Garros, and Louis Chatain. Where were they when they passed around the pen, and were they celebrating anything in particular?[36]

Wherever Marie went, she was liable to meet aviators she knew, or was known by. She hitched a plane ride with Eugène Gilbert in 1913, when she needed to get from Monaco to Ventimille, where she was giving a talk.[37] According to the newspaper *Gil Blas*, "6,000 people were present on the beach to cheer the landing of Gilbert and his charming passenger."[38] (Gilbert was licensed to fly a month before Marie was, and the next year, when flying over the Pyrenees, he was attacked by an angry mother eagle protecting her

Amelia Earhart and Marie Marvingt, Chicago, 1935.

young. He fired a warning shot; no eagles were harmed. He later became an ace pilot in World War I.)

L'Écho d'Alger reported in 1932 that the engaging master-pilot Cazeaux (once described as "an S-shaped man so tall that when you watch him climb out of the cockpit, you are never sure all of him will actually emerge"), was in the area and stopped by the small outpost of Ain-M'lila to see Marie. He enjoyed his time so much he stayed overnight, and the next day they gave plane rides to people who had never flown.[39]

One of Marie's favorite postcards shows her chatting with Amelia Earhart at a conference and reception in Chicago. Like Marie, Earhart was a nurse who was later drawn to flying. She was the first woman to cross the Atlantic when she flew as a passenger in Wilmer Stultz's plane in 1928. In 1932 she crossed the ocean alone, and she set more records in the next few years. In 1937 she and navigator Fred Noonan set out to fly around the world. With most of the trip completed, she left New Guinea for Howland Island. Somewhere in between, she radioed that she was out of sight of land, running out of gas, and lost. She was never heard from again. Earhart was part of the golden age of aviation, but she greatly admired those who had paved the way: "Behind modern women pilots stands another group who are the real pioneers. While comparatively few in number they must have had plenty of what my grandmother called 'spirit.' Their era was mostly from 1910 to 1919."[40] One of those pioneers was Marie Marvingt. At the time of Earhart's disappearance, Marie apparently submitted a seven-page, handwritten article to the papers, mourning the loss and conveying her admiration and friendship for the talented pilot, but it does not appear to have ever been published and its whereabouts are unknown.

Another of Marie's much-admired friends in the next generation of pilots was Maryse Bastié, twenty-one years younger and the winner of several international distance aviation awards. Her solo flight across the South Atlantic caused her friends hours of intense anguish, according to an article written by Marie: "I agree with Costes, who said in a recent meeting in Rome that crossing the Atlantic was pure folly. I've tried for several months to convince Maryse not to attempt it herself. She's told me she'll be careful. She and Jean Mermoz had flown together and he was an outstanding aviator. She said, 'I assure you, I'll only go if Mermoz okays it.'"[41]

Marie told her readers what they already knew: Mermoz wasn't there to give his okay. On December 7, 1936, flying across the ocean, he was lost at sea. Three weeks later, Bastié went anyway, making her solo, record-setting crossing of the Atlantic, from Natal to Dakar.

"And so," wrote Marie, "it wasn't until 1:00 in the morning, by a message from an Air France radio, that our anguish, which had lasted eighteen hours, was transformed into joy. There were many of us, friends of Maryse, who feared the worst. A journalist said, 'It's stupid of her. Maryse Bastié is committing suicide.' But fortunately, the competence, courage, and intelligent daring of our dear compatriot, like our wonderful American colleague Amelia Earhart, have been served by good fortune. I'm among those who are happiest for her well-deserved triumph. My young sister of the air is as simple, genuine, good, and modest as the amazing Lindbergh."[42] (Maryse Bastié died in 1952 while piloting her plane.)

Marie didn't really know that most famous pilot of them all, but she met Lindbergh on May 22, 1927, after he and *The Spirit of St. Louis* touched down at Le Bourget airport at 10:22 p.m. Paris time after completing the first solo nonstop transatlantic flight. The next day, when the President of France, Gaston Dumergue, pinned the Legion of Honor upon the lapel of Charles Lindbergh's borrowed suit, Marie Marvingt handed him the award.[43]

Marie also moved easily and familiarly in military circles. Of the eight Marshals of France (the highest military distinction) under the Third Republic, Marie had close relationships with six of them.

Ferdinand Foch was an academic who rose in the military (unevenly since his brother was a Jesuit and the government was extremely anticlerical at the time) to become Commander-in-Chief of the Allied Forces on the Western Front during World War I. A decorated twentieth-century hero, he is remembered for his keen military mind, and his military strategy: "My center is giving way, my right is retreating, situation excellent, I am attacking."[44]

Marshal Ferdinand Foch and Marie Marvingt on a Paris boulevard.

Before becoming immersed in the military, Foch's academic background and native curiosity led him to spend time in Paris with Gustave Doré, Sarah Bernhardt, Pierre Loti, Charles Gounod, and other intellectual and artistic wavemakers. This side of Foch naturally appealed to Marie. Her admiration only increased when, at the outbreak of war in August 1914, Foch's XX Corps successfully blocked the Germans short of Marie's beloved Nancy. They were similar in being individuals who believed in action. Foch used to say, "One must act, because it is only that which gives results."[45]

Marie and Foch ("who honored her with his paternal friendship"[46]) apparently met in the fall of 1913 when he took command of the XX Corps in Nancy. They maintained a warm, mutually respectful relationship for years. It was he who nicknamed her "Queen of the Air."[47] Less than a year before his death, he wrote in Marie's book: "To preach by example is the most eloquent form of preaching, and that is what Mademoiselle Marvingt does."[48] The photo of him and Marie on a Parisian boulevard is supposedly the only photo taken of him smiling.[49]

Although in the years before World War II, Marie had a number of official and social contacts with Marshal Philippe Pétain, later President of the French Republic, she did not claim acquaintanceship with him, undoubtedly because of her lack of sympathy for his position during the Vichy years. He knew her, however, and generously wrote in one of her autograph books: "Men have made of the airplane a machine of sport, commerce, and war. Mademoiselle Marvingt has enrolled aviation in the service of humanity."[50]

Marshal Franchet d'Espèrey wrote something similar: "How much gratitude the Army owes to air rescue and to its godmother, Mademoiselle Marvingt."[51] Their paths crossed in unlikely places, as Marie related: "I had run into him crossing the Sahara, when he was taking a group of politicians on a tour. Almost all of them were ruined, exhausted, wrung out, sick; only the intrepid Franchet d'Espèrey was fresh and ready to keep going."[52] The next time they met, the two of them were presiding over a banquet in Tunis for war veterans. Marie marveled that even after serious injuries from an automobile accident, Franchet d'Espèrey was the youngest one in the room with his verve, good humor, and spirit.

"The last time I visited him at his office in the *Invalides* to ask after his health and his impressions during his recent visit to North Africa, he said to me: 'Last week, thanks to the airplane, I lunched in Oran, took tea in Tabelballa, and dined that same evening with Colonel Trinquet in Colomb-Béchar! We are waiting for your visit, Mademoiselle, in Tabelballa.'"[53]

Franchet d'Espérey had been one of the few leaders who believed in the airplane from the beginning, thus his subtle emphasis on how well their common dream had turned out.

Of all the Marshals, it was Hubert Lyautey of whom Marie was most fond, and with whom she spent the most time. They met in 1897 as part of a literary association that she and newspaper publisher Marcel Knecht had founded.[54] Lyautey was their honorary president while Maurice Barrès served as acting president, and the meetings were sometimes held at Crévic, Lyautey's home. Or, said Marie, they took a small motor boat to Charmes to meet at Barrès's home. In any case, she immediately fell under the spell of his "intellectual charm."[55] (His brother, Colonel Lyautey, was Marie's neighbor on the Place de la Carrière.[56])

For nearly four decades, Louis-Hubert Lyautey and Marie Marvingt worked together and kept in touch. His motto, borrowed from Shelley, expressed their similar outlooks: "The soul finds joy in action."[57] As Governor-General of Morocco, he possessed a deep love for Morocco and an abiding respect for its inhabitants, working tirelessly toward North African independence. He even requested to be buried there rather than in his native land, and he was.[58] It was Lyautey, along with then-president Alexandre Millerand, who invited Marie in 1922 to join a group of Parisian officials on a fact-finding inspection of French military campaigns then underway in Morocco. She was the only woman in the group. Later, Lyautey asked Marie to come to Morocco to establish a system of air rescue there.[59]

Marie went to hear him speak whenever she could. She was in the Chamber of the Deputies when he gave his first speech as Minister of War, and later at his induction into the French Academy. Marie used to bike from Nancy to his castle in Thorey to visit.[60]

She described a five-hundred-guest banquet she attended with him in North Africa in the 1930s: "All the great chiefs of Morocco were there. It was one of the most brilliant soirées we'd ever had. Lyautey was superb, evoking in a vibrant fashion the whole history of his beloved Morocco. Among the vast panoply of white burnooses, interspersed with the red ribbons on the numerous tables, many of our great Muslim friends didn't understand French, but all were engaged by the flaming love for Morocco that Lyautey was so clearly conveying. He was unanimously applauded by all the representatives of his dear Morocco. I have never seen the Marshal as completely happy as he was that night."[61]

The last time she saw Lyautey was a little more than a year before his death, just before she left for Morocco. Something told her she might not see him again, so she asked him to jot a note in her album of souvenirs.[62]

Created a Marshal of France posthumously, Michel Joseph Maunoury was one of the few generals invited to sign the Treaty of Versailles. Marie kept in touch with Maunoury and his family. According to her friends, Maunoury's grandson was a director of the mines in Khouribga. He invited Marie to give a conference there and added that she would, of course, be welcome to stay with him and his family. Her talk duly took place one afternoon. But Marie enjoyed their hospitality so much that she stayed for fifteen days. Her hosts became quite frantic after the first week, with no end in sight, but Marie, always smiling and gracious, made it impossible for them to hint that she had overstayed her welcome.[63]

As for Marie, her memory of that visit with Maunoury's grandson was a little different: "I recently had the chance, after a conference, to read his hand and saw that this young engineer is a worthy descendant of his glorious grandfather."[64]

Marshals, generals, lieutenants, captains, and foot soldiers were all part of Marie's world.

General Maxime Weygand, second in command to Foch during and after World War I, and in charge of the Polish resistance in 1920, was a great friend of Marie's. It was he who circulated the story about Marshal Foch referring to Marie as "Queen of the Air."[65] Marie knew Weygand from the World War I years, and their friendship endured. In 1947, he gave her a copy of his newly published book, *Foch*, and dedicated it inside: "To Mademoiselle Marie Marvingt whose example and fervor are not forgotten by the world of French aviation. With most respectful homages."[66] Weygand said, "Eighty years aren't really eighty years when it's Marie Marvingt who has them!"[67] And several years after that, he was looking through one of Marie's autograph books and saw Foch's dedication to her. He wrote underneath: "Add to what Foch said? Only words from Foch himself would be worthy of appearing here." He then related in writing the entire "Queen of the Air" story.[68]

In 1934, Weygand was being considered for nomination to Marshal, but his age was against him. Arguing ardently in print for the promotion, Marie described him as "one of the brightest, most balanced, most educated minds of our day."[69] After shaking a verbal finger at the deal makers ("you perhaps don't know, gentlemen, that Weygand ..."), she listed his considerable achievements and all he had done for France. "Messieurs, do the right thing. You who have bestowed eight marshal batons, hand out the ninth. You allow other military officers to maintain their rank despite their years. The Weygands of this world are, alas, rare and now, more than ever, in the face of the dark horizons ahead, France has need of their vital light."[70] Loyalty was a two-way street, and Marie was fortunate in having good friends, and being one.

When General Louis Edmond Grange, accompanying the XX Army Division, made his triumphal entry into Nancy after World War I, crowds massed on both sides of the street to see the popular hero. Greenery and flowers were threaded through a hastily constructed archway, and the tricolored French flag hung from nearly every window. Police barricades attempted to keep the crowds of jubilant citizens away from the columns of soldiers advancing on horseback, bicycle, and foot.

Unable to get through the crowds, Marie told a police officer, "Look here, I'm Marie Marvingt, and I know the General would like to see me." The harried policeman glanced at the pretty young woman standing beside him and irreverently snapped, "If you're Marie Marvingt, I'm the Pope!"

Marie wasted no time convincing him otherwise but, demonstrating her aptitude for gatecrashing, she snaked through the crowds of feathered hats and straw panamas to reach General Grange. He was delighted to see her and held her arm firmly while he introduced her to nearby officials, explaining her many contributions to the war effort. Marie wore a cape that day, information furnished by the niece of her dressmaker. She had an autograph book either under her cape or in her purse, because General Grange wrote in it that day.[71]

Jean Joseph Anne-Marie du Plessis de Grenédan was a French navy officer, commander of the dirigible *Dixmude*, one of two French zeppelins on which he set world records. The *Dixmude* disappeared in a storm over the Mediterranean near Sicily in late December 1923. Seven passengers and forty-three crew members were lost. Locals fishing off Sciacca found du Plessis's body in their nets on December 26. In his pockets was a rosary, several medals, a billfold, a relic of a saint, a Saint Christopher medal, and a watch that had stopped at 2:27 in the morning. This tragedy marked the end of the military use of dirigibles.[72]

Marie had flown to Palermo during the storm that brought down the *Dixmude* (a newspaper reported that she "miraculously escaped the tempest"[73]) and was therefore in Palermo when the body of the heroic young commander was brought by train from Sciacca. With French and Italian officials and civil authorities, the only woman, she followed the massive funeral cortège that paraded for two hours in a driving rain from the opposite side of Palermo down to the port, where his coffin was put on an Italian ship to be taken home.[74]

After the ceremony, Commander Joubert asked Marie to tell her audience for her upcoming talk how grateful France was to Italy for its kindness to one of their own. When Marie arrived at the conference hall to give her talk, she was surprised and pleased to see His Excellency General Basso and most of the officers from the funeral seated before her.[75]

Another Italian, General Luigi Bongiovanni, governor of Cyrenaica (the eastern part of Libya at that time) from 1922 to 1924, showed Marie every hospitality and authorized her to travel anywhere she liked in Cyrenaica. She reported, "The Italians were exquisitely polite and devotedly courteous to me."[76]

General Marty wrote an homage to Marie that appeared in the *Ouest Éclair* in June of 1928: "During the operations in Morocco in 1925, everywhere I went, your path and mine were close but never crossed. And everywhere I went, I heard about the superb daring and the complete pureness of heart of Mademoiselle Marvingt. I wanted badly to meet you. Only this year, in Toulouse, was my wish granted. When I heard your conference, I realized your talents as a speaker were not inferior to your heroism as an explorer."[77] He ended by saying he had never forgotten how Si-El Hadi Omer Tazi made available for Marie's own use his prize possession, the brand-new Caudron Super-Phalène plane which he and the Pasha of Marrakech had just bought together.[78]

Among the many other military officers with whom Marie had cordial and mutually admiring relations were General Auguste Edouard Hirschauer; General Jean-Etienne Cheutin (they were co-vice presidents of the air rescue organization); General Henri Giraud; General Antoine Huré; General Mazaud; and Captain Rezzot (he invited her to be the first to use the new guest room at the Saharan post he had just established).[79]

Marie also knew many political leaders, including Georges Clemenceau (who led France during World War I and was Prime Minister 1906–1909 and 1917–1920), and six

French presidents: Armand Fallières (1906–1912), Raymond Poincaré (1912-1920), Alexandre Millerand (1920–1924), Albert LeBrun (1932–1940), Philippe Pétain (1940-1944), and René Coty (1954-1959).

Marie was introduced to Raymond Poincaré at a reception in 1909, but first spoke with him in 1910 at a Bar-le-Duc banquet for the new airfield, when their conversation consisted principally of him questioning her about whether she liked planes more than balloons, how she came to fly, and what it was like.[80]

In 1913, when a delegation from Lorraine presented a bronze plaque to President Poincaré, who was originally from Lorraine, Marie was chosen to do the honors. Amidst numerous local and national officials, Marie handed Poincaré an artistic plaque engraved with the words to the national anthem, the "Marseillaise," and the inscription: "Respectful homages from the children of Lorraine to their illustrious compatriot." She then read a poem she'd written addressed to him ("A joyful Lorraine salutes one of her sons.").[81]

The first time she was invited to the presidential palace in Paris, it was at Poincaré's request, the day after his election.[82] At that time she met Madame Poincaré. Later, she and Poincaré's brother were both active in the air-rescue league, and the two brothers hosted Marie at a conference in Menton when the president was there recovering from surgery.[83]

Marie once asked Poincaré: "What is your most vivid memory of the years 1914–1918?"

He responded, "It was the day after the Armistice, when the great Marshal Foch and I made our official entry into Metz, which was once again French land, and we were surrounded by the enthusiasm—overflowing, sincere, unanimous, grateful—of the people of Metz liberated from those who had oppressed them since 1870. Those cheers, Mademoiselle, I will never forget them."[84]

Marie's own most touching memory of Poincaré was of him in the trenches. "At that time, 1916, I was a soldier in an infantry regiment, 60 meters from the German lines, when the President of the Republic came, as he often did, to visit the front lines. Wearing a flat helmet, a bloodied vest, and black leather leggings, he stopped to speak with a captain who stood right next to me. My heart beat faster than it had when we were being shelled: would the president recognize me? But nobody had so far, not even my cousin a few days earlier. Poincaré never knew that under the muddy uniform—and how very muddy it was!—stood that long-ago friend from Bar-le-Duc."[85]

When Poincaré died in 1934, following Lyautey who had died a few months earlier, Marie wrote a lengthy tribute to "two great Lorrains reunited in death."

A decade earlier, at Lyautey's suggestion, then-President Alexandre Millerand invited Marie to accompany a small group of officials on a fact-finding mission to North Africa. It is no coincidence that Millerand, while serving as Minister of War under Raymond Poincaré, gave military aviation a place of importance for the first time, and that he was an early advocate of the rights of women.[86] The reliably acerbic Georges Clemenceau described Millerand as looking "like an idiot, an evil one ... a square head, closed to everything, the eyes of the near-sighted, and yet, and yet, he shows glimmers of common sense." Marie admired Millerand's support of women and aviation, and was deeply grateful to him for her invitation to North Africa—she learned much and she made useful contacts in the military and newspaper worlds.

In 1939, newspapers that regularly tracked the Elysée (the presidential palace) noted that the afternoon before, the President (Albert LeBrun) had visited with a high-ranking transportation executive, two political leaders, and Mademoiselle Marvingt.[87] Other politicians in her orbit included Lord Plummer, the Governor of Malta, who introduced her to the Malta intelligentsia, military, and others with whom she had common interests.[88]

When she required a sponsor for her nomination to the Legion of Honor, she chose then-Minister of Air André Maroselli, a longtime friend and admirer, who duly signed off on the paperwork for the award.[89]

Marie sought out North African leaders during her years there. She was fêted and received by Glaoul, the powerful ruler in the South, and by Sultan Mohamed V who used to kneel to her, in good-humored admiration.[90]

General Franco is reported to have signed one of her autograph books, but their whereabouts are unknown, and it's unclear how she would have met him, given her longtime relationship with the Spanish royals. In addition to her friendship with Alfonso, Infante of Spain, she was in touch with his cousin, Alfonso XIII, and whenever she had a letter from him, it was always "My dear friend, the King…" Of another royal friend, she used to tell her neighbor, "I have just received a letter from my little Rainier," referring to Prince Rainier of Monaco.[91]

Mario Brun, a journalist from Nice, recounted in a 1956 issue of *Nice Matin* how he stood on a sunny Monday morning, waiting at the formal entrance outside the Palace of Monaco for the Prince and Princess of Monaco to emerge. In the front row of the huge crowd, he noticed a little figure in a felt hat and a Legion of Honor rosette who seemed to be having trouble keeping her place. As she pushed her hands into the pockets of her man's overcoat, which reached to her boots, he was struck by her alert and curious look, wondering who that uncommon little person could be. What was she doing there, elbow to elbow with a group of excellencies, ministers, and even the bishop of Monaco?[92]

Brun was still watching when the man standing next to the little woman apparently saw her for the first time. He shouted and took her in his arms, hugging her so enthusiastically that her toque almost fell off. It was Maurice Lozé, the Plenipotentiary Minister of Monaco to Germany, who covered her with effusive remarks, laughter, and obvious affection. The two of them spoke animatedly until the trumpets announced the arrival of the royals.

Curious to know who she was, Brun sought her out after the parade, just as she was opening her arms to a man in an opera hat. The journalist recognized Monsieur Powilewicz, Finnish consul in Nice. As he put it, "As they embrace, I see a rich past mounting to their lips."[93] He eventually reached Marie and discovered that Prince Rainier was currently honorary president of the Air Rescue Association, and Marie was acting president.

Rumor still has it that Marie was invited to Grace Kelly's wedding to the Prince. Further, they say, she arrived at the wedding on her bike. However, it appears that the Rainier wedding was confused with another famous one.

On May 10, 1951, the Archduke Otto of Hapsburg, heir to the Duke of Austria and descendant of Francis III, last duke in the Lorraine line, was married to Princess Regina of Saxe-Meiningen. The glittering, outlandishly expensive wedding was a fairy-tale love match. Although Otto was living in Quebec at the time, he chose to be married in Nancy because of his Lorraine connections.

The city of Nancy prepared for the wedding for many weeks, and the local newspaper was full of the most minute details of the arrangements. The municipal sniff was almost audible when it was reported that flowers for the royal wedding were being brought in from Paris even though local florists would have been more than capable of rising to the occasion. Hurt feelings were somewhat assuaged when the privilege of making the royal wedding cake was granted to a local bakery. All Nancy read that this many-layered cake was to embody the twin themes of Love and Lorraine and would be decorated with allegorical references, symbols of love, horns of plenty, and historical motifs.[94]

The entire Grand Hôtel on the Place Stanislas was reserved for guests, including most of Europe's royalty, many heads of state, and journalists from nearly every country in the world. Some 1,200 Austrians played accordions, and 120,000 local people crowded the streets to catch a glimpse of the bridal couple and their famous guests.

Marie had an invitation for one of the reserved seats on the Place Stanislas. But she wanted to attend the church wedding and sit with all the highest-ranking attendees, including U.S. Ambassador David K. Bruce. Despite rigid security, Marie managed to get inside the church. She took mental notes of details that did not appear in the press, and regaled friends with them for years.[95]

In 1953 she wrote a friend, "I attended the marriage of Charlotte of Luxembourg. It was magnificent!"[96] Other royalty Marie was either acquainted with or possessed autographs of included Queen Elizabeth II of England,[97] King Albert I of Belgium, Queen Astrid of Belgium, Queen Marie of Yugoslavia, and King George V of England.[98] After being the first woman to climb the Dent du Géant in 1903, she received a medal from the hands of the Queen Mother Marguerite of Italy, a moment she savored for years.[99]

The person Marie most admired, however, was not, in a sense, entirely of this world: she spoke warmly and often of Pope Pius X. Around 1910 she had three audiences with him, saying, "He gave me permission to land my machine in the Vatican grounds."[100]

She once told a reporter, "I am protected by His Holiness the Pope."

"What?"

"Don't interrupt me," said Marie, "or I won't tell you any more. I am under the protection of His Holiness Pope Pius X. One day when I was lunching at the Vatican, he offered me a small medal and, as he held out his hand to me, I immediately saw his destiny in the lines of his palm. I predicted to him that he would be elevated to sainthood."[101] She went on to say that many years ago a fortune-teller told Marie that she appeared to be under the protection of a saint.

"Don't laugh!" Marie said to the reporter. "What I tell you is very serious. His Holiness *has* been named a saint."[102] Her attachment to this holy man led her to travel to Rome for his canonization on May 29, 1954.[103]

As a would-be artist of deep interest, moderate talent, and little time, Marie appreciated the creative people she encountered. With artist, portrait painter, and sculptor Emile Friant, Marie shared a passion for sports, aviation, art, and the city of Nancy. He was the founder and first president of the Eastern Air Club and she was a member and later vice president.[104] They met at air meetings, exhibitions, and local civic events. It was he who sketched the widely circulated drawing of Marie tending an injured soldier next to her ambulance-airplane, the *Captain Écheman.*[105]

Another friend from Nancy was Louis Guingot, painter of the School of Nancy who invented military camouflage in 1914.[106] Georges A.L. Boisselier, well-known academic portrait painter, was her friend and neighbor on the Place de la Carrière. He apparently painted the full-length portrait of her in an evening gown, shown on the wall of her apartment in some press photos; its whereabouts are presently unknown.[107]

When Maurice Daum of the internationally famous art nouveau glassworks family died of a climbing accident in the Alps, she wrote about him and his family; Nancy is very proud of its art glass.[108]

Samivel, French writer, artist, film-maker, and ardent conservationist, recalls one of his encounters with Marie. Among his many achievements, he created a new film style of evoking the past; his most famous film of this type was *Trésors de l'Egypte*.

"At one of the first viewings, I was about to go onstage to introduce my film at the Salle Pleyel in Paris when suddenly a thin woman in trousers burst into the foyer.

"'Monsieur,' she said to me, 'I've just biked across Paris to hear you. Can you find me a seat? My name is Marie Marvingt.' I started a little and said, 'Marie Marvingt! Is it you, Madame, the famous sportswoman, aviator, skier?' I saw her expression relax: 'Ah, you know me.' It was a near thing that she didn't add, 'Of course you know me.'"[109]

Marie's autograph books included dedications by Juliette Gréco and Gaby Morlay. Morlay, popular stage and film actor, was also licensed to pilot dirigibles, making her doubly admirable to Marie. Backstage after a performance of *Les Joies de la Famille*, Morlay signed one of Marie's autograph books in typical dramatic fashion: "Darling, I think you are marvelous, and I admire you so much!"[110]

All her life, Marie retained a reputation for being seen everywhere: "She never misses a ceremony, an inauguration, a spectacle, anything that's worth seeing."[111] The Nancy newspaper reported that she traveled the circuit of events.[112] In Paris, she was seen at the Opera and the theater, at political and military functions, anywhere, in fact, where she could be part of lights and people and ideas and laughter. Adds a friend: "And for Marie, everything was free."[113]

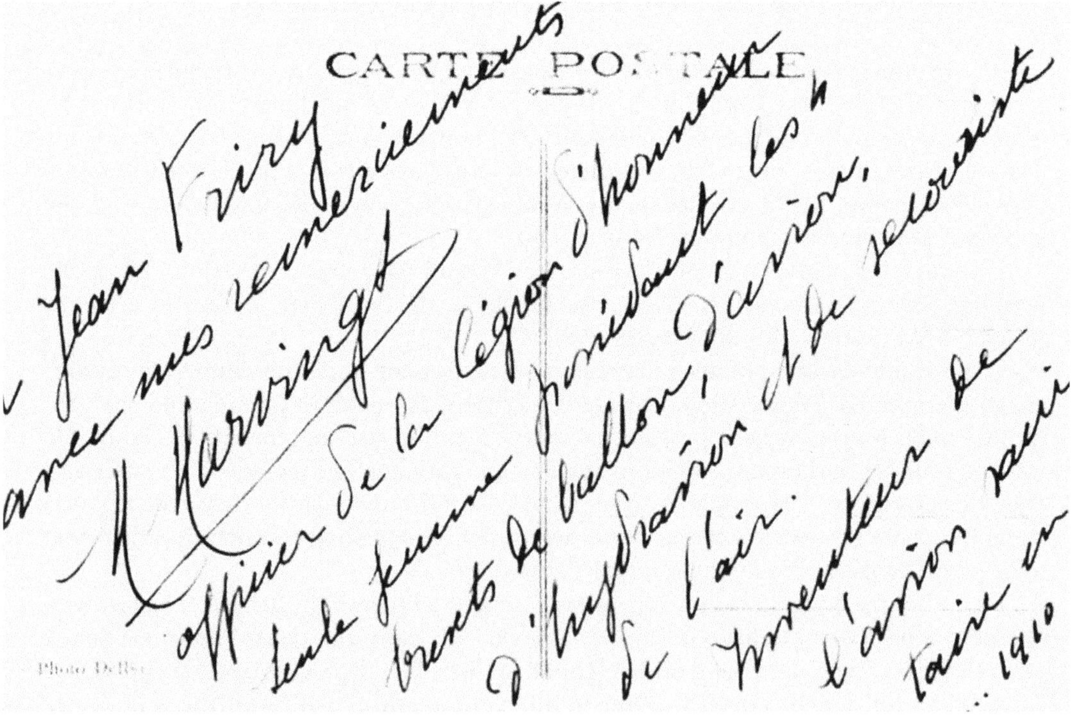
CARTE POSTALE

à Jean Friry
avec mes remerciements
M. Marvingt
officier de la Légion d'honneur
seule femme possédant les
brevets de ballon, d'avion,
d'infirmière et de secouriste
de l'air.
Inventeur de
l'avion sanitaire
1910

Postcard to Jean Friry, good friend and shoemaker to Marie.

Her social life was as extravagant as everything else she did. Balls and banquets, lunches and dinners, resorts and hotels, clubs and organizations. Marie was outrageously gregarious, and nothing invigorated her more than to spend time with people. Not for her was the one-on-one chat over an apéritif. She wanted to be surrounded by people, many people.

Le Figaro announced in the summer of 1914 that "the season" was underway in Vittel, the latest chichi hot spot. High society from France and abroad was settling into the new Grand Hotel that very week. Listed among names taken from the society pages was that of Marie Marvingt.[114]

An extrovert of the keenest sort, Marie enjoyed one of the perks of belonging to the small world of early aviation: socializing. She was part of all the monthly dinners and banquets and meetings and committees and boards and congresses and fundraisers.[115] In that world, it wasn't money or social standing that counted—Marie had neither—but reputation, daring, and accomplishment.

Eventually there were air clubs throughout France, united under the umbrella of the Paris-based French Air Club, and she was welcome at all of them. The only way to keep abreast of one's field in those days was to show up at the airfields during the day and to socialize at night to hear about the newest designs, records, inventions, and techniques.

In the 1920s, 1930s, and 1940s, when Marie wasn't spending months at a time in the United States, North Africa, or Asia, she kept a demanding schedule.[116]

She attended every ball held in Nancy,[117] including the Gala of the Lions sponsored by the artistic group, the Léophiles[118]; the annual balls of the Union des Femmes de France[119] ; the Press Club balls (which were sometimes costumed)[120]; and the balls sponsored by the Faculty of Letters of the University of Nancy.[121] The word "elegant" was invariably attached to the word "ball" in the newspaper pieces of the era, so Marie must have put her one decent ball gown to good use.[122]

She was also seen, and reported to be seen, at dance, theater, art, and jazz events ("Mademoiselle Marvingt is just back from Cairo"[123]), most often in the front row or in the loge reserved for civic leaders.[124] She heard almost every speaker who came through Nancy.[125] She attended numerous funerals during these years[126] (including that of her old professor, Dr. Gaston Michel,[127] and of her friend and tireless co-worker, Robert Charlet[128]).

All the relationships that arose from her astonishing social life took concrete form in stacks of photographs, her autograph books,[129] and boxes and boxes of letters and telegrams. And invitations. From North Africa she wrote a friend: "I'm offered hospitality on all sides: Côte d'Azur, the chateaux of the Loire, Brittany, etc. I'm not sure yet what I'll do."[130]

Her correspondence must have been enormous. Only a small fraction of it survives today; her voluminous collection of letters was lost or destroyed. However, recipients sometimes retained communications from her, consisting of either postcards, which allowed her to be brief, or one-to-two-page letters written on the stationery of various hotels.

In 1903, the French Minister of Commerce authorized a new type of postcard that featured a photo or illustration on one side, the other side divided between the address and a bit of correspondence. From that time forward, Marie had postcards made with pictures of her exploits. She then had only to dash off a "best wishes!" note and her signature on the back, allowing the recipient to see what she had been doing lately.

Postcards track Marie's activities for decades, showing her in planes and balloons, on skis, with a camel, in North African dress, dressed as a nurse, soldier, sailor, in a helmet, in pearls, in a plumed hat.

Marie rarely lost contact with her legions of friends and acquaintances. At age eighty-seven, during a conversation taped at a friend's, she spoke of a letter she'd just received from Roland Dorgelès, novelist and member of the Goncourt Academy; they'd been talking about the anniversary of the death of his first wife.[131]

She cherished above all her numerous photographs. A journalist wanted to borrow several for an article.

"What? My photographs? You want to borrow some of the pictures in my collection? Impossible! I am too attached to them. I have thousands of them, which I haven't yet had the time to organize. To lose even one of them would plunge me into despair. Every one of them reminds me of an event, an important date, a triumph. And, besides, newspapers never return any photos you lend them."[132]

She liked to pull out a diploma, document, newspaper article, or photo to back up her stories. She had a phenomenal memory for names, places, and dates, but she knew it was more effective to furnish a bit of proof when describing the extraordinary things she'd done and people she'd known.

Her autograph books were almost as important to her as her photos. She was not the type to request an autograph from a celebrity she spotted across the room. She cherished her seven (some reports indicate nine) gold-embossed, leather autograph books because they held warm personal notes from people who played an important role in her life.[133] Although dedications in autograph books are notoriously exaggerated and uniformly complimentary, they provided Marie with a way of documenting her life that didn't become popular until the era of selfies, Instagram, and daily postings of one's status. Irreplaceable, regrettably lost, her autograph books would delight a historian, representing a who's who from 1895 to 1963.

Marie belonged to over ninety organizations, and she was a delegate, officer, board member, or committee member of many of them.[134]

And the banquets. Oh, the banquets. Because that's what they did then. Many of the organizations to which she belonged reveled in their annual or semiannual banquets.[135] Among the scores of glittering affairs was the five-hundred-guest gathering at the Hotel Claridge to honor Laurent-Eynac, Prime Minister of Air[136]; the annual banquet of the Vieilles Tiges at the lovely Hotel Lutetia, which she rarely missed[137]; the banquets of a group of Algerian gourmets-gourmands ("keen forks and forthright drinkers")[138]; the majestic banquet and numerous receptions in London offered by the Franco-British Alliance[139]; many banquets celebrating the finale of international aviation, balloon, or air rescue events, where Marie usually spoke or was recognized.[140]

Underscoring the impact of an unescorted woman attending banquets at that time is an excerpt from *Les Ailes* describing a "brilliant" event for five hundred diners: "To list everyone who attended would be impossible. Still, we wander from group to group, from table to table. What a gathering! What a gathering! And then, heavens! What's this? It's not possible. A woman! Wait, two women, each in a gorgeous evening gown. It's the revolution! But it's true: the French Air Club, which has always refused to admit women to their affairs, has, for the first time, opened their doors. And that has given us the pleasure of seeing several ravishing dresses enlivening a sea of black suits."[141]

One of the women was Marie and there were, in truth, at least half a dozen women in attendance that evening. The reporter commented that Mademoiselle Deutsch de la Meurthe was once allowed to speak at a dinner honoring her father, but that was the exception. "From now on," finished the piece, "it will be the rule."[142]

If asked to name her favorite banquet, Marie might well have chosen one held December 14, 1911, at the luxurious headquarters of the French Automobile Club on the Place de la Concorde, where she was awarded a gold medal by the Academy of Sports for her excellence in all sports, the only such medal ever given.[143]

She regularly attended a gathering of intellectuals that met at the Club du Faubourg in Paris. They held big banquets that honored writers, diplomats, or others in the news, and they debated issues of the day. The topic on December 16, 1936, was the place of women in Islam. Marie attended that evening along with the actor Régina Carmier, Commander Bernard Frank, and the Princess of Uzès.[144] On January 9, 1937, they compared the status of contemporary Frenchwomen with Frenchwomen of the past.[145] On October 27, 1937, the novelist, playwright, and lyricist Francis de Croisset presided over the banquet (he died eleven days later). Marie, along with Chinese pilot Metie Tehern and, according to *Le Journal*, "a number of other well-known personalities," discussed marriages between "whites" and "yellows," the poetry of Indochina, and travel writings.[146] She met journalist Jean Nocher at one banquet, although it wasn't she, but he who remembered, and mentioned it in a stirring radio piece about Marie after her death.[147]

The social life Marie liked best involved like-minded people, preferably, pilots and adventurers.

One article started soberly enough: "A new organization has been established recently, although up until now, it has not been mentioned in the press."[148] In 1932, pilot Maryse Bastié started a club for female flyers, including Marie, whose headquarters was the Pilot's Bar at Orly Airport. The bar owner posted the club's rules in a prominent place: Each member put 50 *centimes* in the kitty, with a maximum of 3 francs, for each aviation-related word said. When there was enough money in the kitty, they treated themselves to dinner. According to *Les Ailes*, "The group usually doesn't even get through the hors d'oeuvres before everyone's maximum has been reached. If they didn't have an upper limit, all the regulars at the Pilot's Bar would be broke today."[149]

And so this picturesque woman moved among other picturesque people, feasting at a perpetual banquet of talented, vibrant individuals of every rank, stripe, occupation, and personality. And she relished every moment.

15

Who Was That Woman?

"It's important to grasp the zeitgeist of those early years of the twentieth century to understand the significant challenge to the status quo that was Marie's entire life. At that time, it was considered proper for women to faint at the least provocation, women didn't vote, didn't drive, didn't smoke, and defined very nicely the phrase, 'the weaker sex.'"—Michel Daurat, *Lorraine Magazine*, 1955[1]

Marie moved so quickly through her storied life that it was difficult to distinguish the woman at the center of the tornado. By the time one person said, "Here she comes," another was saying, "There she goes."

Marie Marvingt was 5'2½" tall ("a little bit of a woman"[2]), with dark brown eyes ("magnificent eyes, which drew people to her," according to a contemporary[3]), and dark hair, variously reported as reddish, chestnut, or auburn. Her weight was mentioned only once, in 1911, when Roger Sommer took six passengers up in his plane, an outrageous risk at the time. Each passenger was weighed, in their flight gear, to estimate the safety of such a loaded airplane. Marie weighed 147 pounds.[4] She was several times described as curvy. Ultimately, she was the perfect size for what her life required of her.

When Charles Codron, Marie's neighbor during the seasons in Chamonix, was asked to describe her, he said, "She is 24 to 28 years old, with a Greek profile, unfathomable eyes, and tanned skin, usually wearing a straw hat and dressed for the mountains."[5] (Marie was actually thirty-three at the time, but her age was consistently underestimated for most of her life.) "What a woman! She was raised on sports, which made her slender, supple, and powerful without taking away any of her gracefulness."[6]

Some people found her captivating, alluring[7]; one who knew her well said "her liveliness, gaiety, and kindness charm everyone who meets her."[8] A friend said "she flirted, but in the good sense of the word."[9] A Canadian newspaper referred to Marie and Hélène Dutrieu as "handsome young women, lively on the ground, and as cool as cucumbers in the air."[10] In her 2009 book *Gloires de Sport*, Monique Berlioux said that "for a beautiful woman in the nineteenth century, Marie Marvingt was an exception to all the rules."[11]

Dr. Colombani, who worked with her for years on the establishment of air rescue organizations, described Marie so well that his description was repeated in a number of articles: "Her face reflects an intense vitality underlined by a searching regard imprinted both with an infinite sweetness and an unalterable willpower. Everything in her breathes decision and energy, and one understands better why she has been baptized the Fiancée of Danger."[12]

Marie smiled and laughed more often than she did anything except talk. Even at the age of eighty-eight, on a tape-recorded visit with Dr. Michel Royer, her laugh punctuates nearly every sentence.[13] Marie was said to have a calm, gracious, poised demeanor, except when she was talking about one of her many passions—and then her speech was as swift as it was theatrical. However, due to her diction lessons, no matter how rapidly she spoke, she was always intelligible because of her careful enunciation. People mentioned her warm voice, the eloquence of her speech, and her rich vocabulary,[14] and at the thousands of conferences she gave, she was met with near-adoration. Her speech, "clear, precise, and elegant," and "her sympathetic, appealing personality left no one unmoved."[15]

Unable to experience that personality firsthand requires deducing it from others' impressions or from her own words and actions. According to Gaby Curral Couttet, Marie was "a rare individual with a strong personality, an astonishing woman, interested in everything, succeeding at everything, and admired by everyone."[16]

A distant cousin, Madame Brun, agreed: "She had a great deal of personality."[17] But was such a comment ambiguous? Was it, perhaps, too much "personality"? Everything about Marie was larger than life: she laughed loudly and often, she rarely reigned in her dramatic side, and she thought little of social norms. For the most part, however, she was described as endearing, winning, attractive, curious, dynamic, gracious, and lively.

Despite her competitive and unconventional activities, generally left to men in that era, she was said to be "a real lady." Although "ladies" did not generally engage in sports, fight in wars, or travel alone in the desert, she evidently conveyed something that her contemporaries considered "womanly."[18]

A journalist known only as R.P., wrote in 1924: "As she spoke, I saw a flame of energy in the frank, clear eyes, and the willpower—sheer stubbornness you could say—inscribing a vertical line on the forehead of this face, still however so full of kindness and delicacy."[19]

Marie-José Lionel-Pellerin, physician, senator, mayor's wife, knew Marie as well as anyone and probably spent more time with her than most of Marie's friends. She was one of Marie's most loyal and knowledgeable supporters, and she wanted to be sure the record showed Marie's thoughtful, sensitive, and spiritual side. The public, by way of the newspapers, saw only the exploits, triumphs, travels, and action photos. A few people, like Pellerin, saw the gentle woman in love with all the world and determined to do as much good as she was capable of.[20]

Her peers would not have used the terms, but Marie was both avant-garde and a free spirit. She was in the forefront of every new movement, and she lived to please herself. As Achille Liégeois put it, "She spends very little time worrying about protocol."[21] She conformed to others' wishes only to obtain what she wanted.

Her patriotism was legendary,[22] and she was capable of great self-discipline.[23] Although she cared little for others' opinions, she greatly disliked being thought self-indulgent or arrogant.

Marie possessed enormous self-confidence, and as an intelligent woman, she appreciated the value of her achievements, the singularity of her exploits, and the merits of her contributions to humanity. Always direct, rarely coy, Marie had a fundamental honesty that demanded she avoid both glory-seeking and false modesty. Pellerin described Marie as wholly unpretentious, but others felt she thought too much of herself: "She did fabulous things, but she felt she wasn't getting enough attention, wasn't sufficiently respected. Everything had to be about her. When she was there, there was no air for anyone else. That's the effect she had on people. 'It is I, Marvingt,' *n'est-ce pas*?"[24] And, "Women just didn't get themselves talked about in sports in those days."[25]

Others spoke of her egocentrism, her way of imposing herself onto events, her abruptness, her sermonizing, even her "habitual truculence."[26] Another said, "When she went to the theater, she expected a box seat. She would say, 'It's the least of things considering all I've done.'"[27]

Jacques Coulet, a good friend, said, "When faced with the enormity of the things Marie described, people had a tendency to disbelieve her. And yet it was all true. There is proof! If she bragged, she had a right to."[28]

When asked about her activities, Marie might mention what she had been doing lately, but she usually left out many of her exploits, partly because her mind was already running on to the next item on her agenda. In fact, she once complained, "I've heard too much about what I've done. What's important is what I'm actually doing."[29] Even when she understated her achievements, they were so numerous, so varied, and so extraordinary that she appeared to be self-important.

In fact, Marie was one of the most interesting people she knew. In her childlike candor and enthusiasm, she liked sharing who she was and some of the fascinating and fun things she had done. It was not a question of "Aren't I wonderful?" but rather, "Isn't it all amazing?"

One aspect of Marie's personality, if one can situate it there, was her constant activity.

Armand Rio wrote in 1913, "Unless one has the gift of ubiquity and can be in Gérardmer, Chamonix, and the airfield all at the same time, there's nothing worse than trying to interview Mademoiselle Marvingt."[30]

Marie "seemed to have the gift of being everywhere at once,"[31] and she missed nothing, even if she had to double-book herself. On Saturday, March 5, 1938, she was seen (and reported by newspaper articles to be in attendance) at an 11:00 a.m. wedding of the Viscount Jean de Suyrot, a young aviator friend of hers; at the General Assembly of the Eastern Air Club at 2.30 (she was an officer); and at the fiftieth anniversary celebration of the Union des Femmes de France at 4:00.[32]

In 1911 in Chamonix, she won or placed in six winter sports events in one day.[33] In twenty-four hours in 1912, she spent three hours in a balloon over Nancy, flew her own airplane for an hour at Compiègne, drove to Paris to attend a banquet given by the Academy of Lorraine, and returned the next day to give a conference on mountain-climbing in Bar-le-Duc.[34]

Later that same year, she flew her Deperdussin one Saturday morning in Reims, then gave a fundraising talk in Gérardmer that afternoon. The next morning she drove to Bussang where she gave another talk on Monday, after which she biked back to Gérardmer, then left by auto to reach Aurillac to give another conference, and returned by automobile to Reims.[35]

An article in *Le Petit Parisien* described Marie's busy year: "From May to December 1912, she took to the air in her Deperdussin 717 times with no accidents; she made 15 balloon ascensions and took 32 neophytes for their first flight; she gave 89 conferences."[36]

Achille Liégeois, a reporter who wanted to interview Marie, showed up late one evening at her door. After introducing himself, he said, "Mademoiselle, please excuse my temerity in dropping by like this, but I heard you were back in Nancy and, well, I know that you don't put down roots anywhere, so I thought I'd better come right away."[37]

J.J. Leblond wrote in the *Dauphiné*: "Our famous citizen Marie Marvingt has returned from a long stay in North Africa, and she doesn't miss a single event here—not a ceremony, inauguration, or spectacle that's worth seeing. She is everywhere, attending and participating in local events with her usual fruitful curiosity."[38]

In his book on female aviators, Bernard Marck explained under the photos of three women that the first woman piloted balloons, gliders, airplanes, hydroplanes, and helicopters. The second woman was "a medalist who made a clean sweep in the entire range of sports: swimming, gymnastic, skiing, walking, fencing (saber, foil, and épée), polo, bobsled, boxing, ju-jitsu, shooting, equitation, ice-skating, rock climbing, and more." The third woman was an inventor of several useful items. He concluded, "The whole is just one woman: Marie Marvingt."[39]

Marie's multiple successes in multiple domains inspired newspapers to refer to her as "the Admirable Crichton of her sex" or "the female Admirable Crichton."[40] Known as "the Admirable Crichton," Scottish polymath James Crichton was a sixteenth-century genius and athlete, known for his extraordinary accomplishments in numerous fields. A 1902 stage play by J.M. Barrie, The *Admirable Crichton*, had resurrected the name in popular culture shortly before Marie began appearing in the headlines, and many found it a fitting sobriquet for her.

Reporters invariably asked Marie how she was able to accomplish so much.

On a practical level, during some periods of her life, Marie slept only one or two hours a night. At other times she slept up to four hours, but apparently never more than that. Neighbors who woke in the night saw Marie's light shining. When she traveled, she carried piles of books, sometimes leaving behind an extra pair of shoes to make room for them. People who said good-night to her one evening could discover over breakfast that she was now newly versed in some aspect of her current interest. The majority of her blithe postcards dashed off to friends and relatives were penned in the wee hours of the night and usually signed "In haste."[41] The night before her eighty-sixth birthday, she sent a letter datelined 2:00 in the morning.[42]

Over her lifetime of eighty-eight years, Marie had at least an extra 8,000 hours that others didn't have. Most people assumed Marie was stretching the truth about how little she slept, despite her accuracy in other matters. It wasn't until long after her death that research showed that about 1 percent of the population, known as "short sleepers," need a maximum of only four hours of sleep per night. These people also tend to be physically active and optimistic, and they accomplish more than most people.

Marie had virtually no family—no husband, no children, no living siblings after Eugène's death in 1897. When she died, her nearest relatives were three fifth cousins on her mother's side. On her father's side, a search was made up to and including sixth cousins without finding a living relative.[43] (Those whom Marie called "cousins" were genealogically distant.) She thus had time and energy that others spend with a partner or raising children, caring for aging parents, entertaining relatives. Family ties never prevented Marie from taking off for any part of the world or doing anything she pleased.

Marie also began her career very young; by the age of fifteen, she was already winning prizes and learning dozens of skills. And she seems to have been efficient: there are no stories of missed deadlines, forgotten appointments, or doubled-up speaking engagements.

The best explanation, however, of how Marie did everything she did is most likely her extraordinary, almost terrifying energy, fortunately coupled with a phenomenal mind. Her activity was chosen, focused, productive.

What she was doing was always more important to Marie than how she looked. When it came to her wardrobe, she simply wasn't interested. Ginette Mechelle's mother was Marie's dressmaker while Ginette's aunt, Madame Rollin, was Marie's hatmaker. Mechelle said, "My mother admired her a great deal, and would have made her clothes for nothing, just for the pleasure of hearing stories of her adventures."[44] Marie was a faithful client, relying on them whenever she bought clothes or hats. In return, they worshiped her. After her crossing of the North Sea, Marie sent a postcard to Madame Rollin, which was respectfully ensconced on the mantelpiece for decades afterward.

"She had hats for every season," said Mechelle. "At that time, it was necessary to be *chapeauté*. She often wore a fur hat, a toque." A pause. "She was not very elegant, you know." Caught between near-idolatry and despair that anyone could so badly neglect their wardrobe, Mechelle burst out, "Do you know how she cleaned her hats? She dipped them in airplane gasoline!"

Clothes were little more than tools for Marie. Georges Gygax said, "She wears the overalls of an aviator, the uniform of a soldier, the costume of an actor, and the fancy dress of a gala guest, all with the same joy."[45] And with a certain indifference. Mechelle remembered "one time when she went to a wedding she arrived on her bike, which she propped up against a pillar (it was at Saint Léon) and she was wearing a cap (she most often wore caps or berets). Oh yes, she always dressed like that. She might be in an evening gown, but she had her cap and her bike with her."

A 1913 photo of Marie in her little black suit at Chalons-sur-Saone balloon festival.

Her basic wardrobe consisted largely of neutral, well-made, durable outfits (dark skirts, off-white wool sweaters). "Practical" seemed to be her principal style criterion. On the other hand, one of her outfits was both smart and appropriate, and appears curiously contemporary today: a dark velvet two-piece suit with matching jacket and floor-length skirt. A white ruffled blouse was worn under the jacket with a small black tie at the collar. It was attractive, photogenic, and allowed her to move freely in the gondola of her balloon.

In a 1912 article by Annie Wanss in *La Revue Aérienne*, Marie responded to a request for advice: "You ask my opinion on the most gracious and practical clothes and headgear for flying. First of all, let's put that second adjective first. Everything you wear must be practical. In flying, wearing the wrong clothes can be dangerous."[46]

She explained:

> A hat that isn't secured on your head can cause an accident, as a certain passenger will remember because hers managed to get caught in the propeller and broke it. It all depends on the season, how long the flight is, and what kind of a plane you're in. I usually wear a tailored suit, with a culotte skirt, which is practical, allowing you to get in and out of your machine without difficulty. When it's cold or the flight is longer, I add a leather or fur coat and sheepskin-lined foot muffs, with a cord around the top. These are warmer than leggings yet don't interfere with the movement of your feet on the direction bar. For my hands, I wear long mittens and fur-lined gloves.
>
> Opinions on goggles are divided. Myself, I prefer wearing them. However, you have to know how to get along without them in case of, for example, rain or an oil leak. Oil will blow back and smear the goggles.
>
> Balaclavas and fur hats are warm and practical, and never fly off your head. In summer, I wear something similar, but much lighter weight. I rarely wear a metal helmet. Latham, my first teacher, didn't think it suitable for the Antoinette monoplane, and I agree. On the other hand, metal helmets have saved the lives of several pilots of biplanes.
>
> In planes where you're less sheltered than you are in the Antoinette, particularly student planes, women might want to wear the cloth overalls used by men since you often get covered with oil.[47]

In the same article she described clothing appropriate for ballooning:

> Ballooning being a more peaceful sport, it doesn't much matter what a woman wears if she's just a passenger. But when a woman is piloting a balloon, especially in a distance competition where she'll spend the night up in the air, I recommend warm, practical, sporting clothes. I can't see such a woman trying to fold up her balloon wearing a lace or silk robe all stuck together by the night's humidity, and fighting off a huge, outsized hat.
>
> In a balloon you have to be able to move freely and lean through the cords. You don't want things getting caught on something. For your head, a balaclava helmet or a supple felt cap, depending on the season. When I crossed the North Sea, I was wearing a tailored suit with the divided skirt or culotte (nothing like the ones that came out last year), which I've been wearing for a dozen years for my bicycle trips and for horseback riding. I also had a heavy fur coat and a balaclava. On long trips, wear heavy, well-made shoes in case you have to walk a good distance from your landing spot or in case you land in a plowed field or in the snow. Speaking of which, I remember how useful my umbrella was one night, as we flew for several hours in a pouring rain.[48]

The Belfast London collection 2019-2020 featured a black leather jacket named the Marvingt Coat.

Marie had evening wear for formal occasions (described several times as "an elegant long gown"[49]), and one string of good pearls that she wore with small pearl earrings. Her other choice was a jeweled-cross necklace and a pair of dangling earrings. In later years, she was covered with decorations, which meant that was all most people noticed about her appearance.

She had one good, heavy fur coat, probably bought near the beginning of her career for driving automobiles. Cars were open and windy, and fur coats were standard. Other items in her wardrobe included hiking boots, an Alpine beret, scarves for all seasons, perhaps one other long gown, an often-used cape, an aviator's fur cap with ear muffs, thick gloves, dress gloves, helmet, and goggles. And of course, a divided skirt or two.

Many credit Marie with the introduction of the culotte.[50] She was pictured at the top of the Trélaporte wearing one in 1903,[51] preceding by years the date given in the *Figaro*: "February 19, 1911, the culotte makes its appearance."[52]

Madame Paul-Franz Namur-Vallot (her first name seems never to have appeared in print), daughter of the director of the Mont Blanc Observatory, wrote that she had been skiing with Marie at Chamonix since 1905. "She was one of the first women to wear pants. I had quickly seen that the skirt was incompatible with a sport where you often had to do acrobatics to untangle yourself from the snow. That's why, like Marie Marvingt, the famous pilot and mountain climber, I courageously wear the culotte."[53]

When Marie defied conventions, as she often did, it was to live as she needed to live, never for the sake of shocking people, or even to change society. When she wore culottes, for example, she rarely gave a thought to the effect this would have on others; she knew only that she was a lot more comfortable.

A fact that didn't emerge for many years was that Marie simply could not afford a large wardrobe. She wore one maroon outfit for many years; people must have assumed she really liked it, or credited it to a personal quirk.[54]

She wasn't completely devoid of interest in her appearance. Old photographs show a smiling woman—vibrant, attractive, and appropriately dressed. In 1903 she had her portrait taken at the Steiner Studio on the Place Stanislas. The striking photograph, bordered with flower designs, shows a lovely woman with long hair pulled back, revealing her sculpted features. She sparkles in other photographs—on skis, hanging by her knees from a bar, on a bicycle wearing knickers, in the desert with a camel, strolling on a Parisian boulevard with Marshal Foch, at the wheel of her plane. A large, handsome oil painting in an ornate gold frame used to hang in her apartment; she is around twenty, strikingly beautiful, dressed in an evening gown and holding a fan.

Marie's shoemaker, a longtime admirer and good friend, always re-soled Marie's shoes. He says she complained bitterly about how the repairs were revealed when she knelt at the church altar to take communion. It wasn't vanity on her part, he reports, but pride; she didn't like people knowing she couldn't afford new shoes.[55]

Marie was more interested in good health than in good clothes. And her health was magnificent, perhaps partly due to the genetic heritage from her father, the sports champion who lived to eighty-nine in good health. Her physical vitality was so obvious she was often asked about it. "The secret of my good health? I live a regulated life. Every six months I have a checkup with a competent physician. Everything: nose, lungs, stomach, everything! How many times during an operation have I heard a surgeon say to his assistants, 'Take him away, it's too late.'"[56]

Following a reception to celebrate Marie's eighty-third birthday (as well as to award her two new medals), a journalist wrote: "Brimming with spirit, vitality, and eternal youth, Mademoiselle Marvingt furnishes us, in a firm voice, with the secret of her exuberant good health. She never smokes, eats small meals, sleeps only one or two hours a night, and rides her bicycle every day."[57]

At the age of eighty-six, she informed an American writer who had admired her fitness that she hadn't seen a dentist in years, that she had perfect hearing, and that she still read without glasses. She added that she never slept more than four hours and refused big meals ("bad for the liver"), preferring six or seven snacks a day. She also biked over 1,000 kilometers a year "just to keep in shape."[58]

On her eightieth birthday, she gave a curiously modern recipe for good health: "No sauces, no boiled beef. Eat raw carrots grated with lemon juice, grate and squeeze celery for its juice, don't cook with fat but rather with olive oil. Eat cheese, especially Roquefort. I never drink while eating, and I never smoke. Get a medical check-up every six months and, upon waking, do a quarter of an hour of exercise."[59] Another time, she answered the question by attributing her perfect health to "very simply, willpower. And exercise." Genetics or common-sense living, Marie was still dealt a very good hand when it came to her health.

When Marie was eighty-three, a reporter exclaimed, "What youth! What vigor! What vitality!"[60] Marie responded, "I'm going to astonish you still further, Monsieur." She stood on one leg with her other leg and her body stretched out horizontally, while she balanced a bottle of mineral water on her head. She slowly counted to twenty. She was as rigid as a statue, a tribute to muscles that had been rigorously trained and conserved by fierce willpower and the discipline of the true athlete.

"You try it," she said when she had finished. "You'll see if it's easy or not!"

No one ever saw Marie take medication of any kind but it is highly unlikely that she would have allowed anyone to do so, out of pride.[61]

What was she really like?

A 1911 story in *Le Sport* entitled, "A Good Example of Camaraderie in Sports" took readers to an airfield in Turin, Italy.[62] Aviators flying in the Turin airshow signed contracts saying that they would receive their stipend only if they flew a minimum of two hours. The sponsors needed to guarantee enough aerial entertainment for the crowds. One of the aviators, Paul Van Gaver, had encountered a series of misfortunes—among others, a defective motor and a landing that damaged his plane—and the deadline for getting in his two hours in the air was approaching. As the magazine put it dramatically:

"'I will be able to do nothing today,' cried Van Gaver in a strangled voice.

"His persistent ill-chance seemed a bad omen. He had little money and four small children to support; he began to think the Fates were telling him to stay on the ground. He needed the money, but risking his life seemed criminal. He resigned himself to giving up and writing off the trip to Turin. Just then Mademoiselle Marvingt (the journalist Myriel) said there might be a way to avert disaster. She spoke with the exhibition officials, who said, 'We see no problem with that.'

"And so, as soon as his plane was repaired, Mademoiselle Marvingt took off in it and flew for two hours, to the enthusiastic plaudits of the crowd. Van Gaver was given his payment."

Another trait that endeared people to her, and defanged her few critics, was her effervescent generous nature. Not once in millions of written and spoken words did she belittle anyone or show any small-mindedness. In fact, she was routinely complimentary, encouraging, and enthusiastic. She seemed to enjoy other people's successes, and she tended to find the best in people.

While traveling in the United States, newspapers quoted her commending American pilots: "I am impressed," she said, "by the precautions taken by your pilots on passenger planes. They are excellent aviators and are extremely considerate of their passengers."[63]

After giving a conference in the French city of Pau, she was interviewed saying, "I've traveled a great deal, have seen a great deal, but I have rarely enjoyed a more beautiful or peaceful view than the one from my windows at the Hotel Gassion in Pau."[64]

In her hundreds of talks, she frequently mentioned Raymonde de Laroche's talent and the fact that de Laroche was the first woman to obtain a pilot's license. In the United States, she made sure the schoolchildren knew who obtained the first American female license.[65]

In 1919, Louis Bréguet founded the airline company that evolved into Air France. Marie flew one of his planes to Morocco and she wrote immediately afterward to compliment him.[66] New companies and their entrepreneurs appreciate such gestures. And the people of Pau invited her back.

When interviewed, she often insisted first on recalling the late pioneers to whom aviation owed so much: Wright, Delagrange, Chavez, Ferber, Nieuport, Kimmerling, Echeman. "It's up to us," she said, "the survivors of those early days, to make sure we don't forget them."[67] These examples are standard Marvingt. She never failed to give credit to those associated with any of her endeavors. She included the names of the porters when she climbed mountains, she used her journalistic keys to open doors for women and for those who might otherwise go unnoticed. She generously praised those considered her competitors. According to neighbor Charles Codron, her friends were "beautiful," the celebrities she met were "marvelous," their deeds "incredible."[68] Simply put, Marie had a happy nature, some of it innate, and some of it based on a belief in love and service to others.

As a Roman Catholic, Marie attended Mass regularly, often daily, at her parish church, the beautiful neo-Gothic St. Epvre. Many church-goers remembered seeing Marie approach the communion rail, rosary in hand. She was well known to the priests there too: she thought nothing of stopping in to give them counsel. When she had an argument with another woman over a matter of church doctrine, she persuaded the priest to give a sermon on the subject the following Sunday.[69]

Members of the Red Cross in Nancy still tell the story about Marie's participation in retreats organized at St-Epvre Church for Red Cross nurses. During times set aside for silence and meditation, the nurses scattered throughout the church pews or on the broad steps outside. But organized, punctual Marie always brought with her, in her large black handbag, an alarm clock that announced the end of reflection time. It apparently sounded like the horses were coming out of the gate. The nurses said they never got used to it.[70]

Marie repeatedly said that her most memorable moments were those she spent at Tamanrasset, a place of prayer deep in the Sahara founded by the apostle of the desert and one of her heroes, Père Foucauld.[71] She visited Tamanrasset several times. From North Africa she wrote a cousin that she was

> longing to see Tamanrasset again. One receives communion kneeling on the sand. Often in the morning I was awakened by the joyful songs of a minuscule bird. At that early hour, he and I were alone in the immense desert.

> In addition to falling under the spell of the Sahara, there is an extraordinary atmosphere there in Tamanrasset that swallows one up whole, body and soul. Near the place where Père Foucauld lived for six years, some Sisters of Montpellier have moved into a house offered them by an old woman of the Sahara. The latter is the only desert dweller who attends daily Mass. She, as well as Brother Jean-Marie Certade, visit the poor and sick Tuaregs of the region. She says she finds it very expensive renting her camel (900 francs a day!).
>
> The door in front of which Père Foucauld was assassinated is still there, just as it was then.
>
> I have spent some marvelous hours meditating both day and night there, in front of some extraordinary landscapes, when the moon silvered the black shadow of the fantastically shaped hill I could see against the horizon. I have never understood better the voice of silence, the call to sincere, profound adoration as I have experienced it there.[72]

She sometimes replied to queries about her fearlessness with, "When one has a deep faith, one doesn't fear death, and I am convinced that beyond this life there is another. So?"[73]

When a newspaper article said that "Destiny" protected her, she crossed it out and wrote "God."

Gaby Curral Couttet said, "She was very devout." But she also remembered Marie visiting her parents, dressed in the recurring maroon outfit and wearing on a chain both a medal that Pope Pius X gave her and a tiny skull.[74]

Robert Marchand noted that, in addition to Marie's bicycle and her large black handbag, the two things she would never be separated from were the skull and the medal. Marie explained to him, "The skull was my only good luck charm. It had protected me from innumerable dangers. I went to the Pope [Pius X] to ask him to bless it. He did, and then offered me one of his medals. These are not just two good luck charms, but two pious souvenirs because I really believe in Providence and in protection from above."[75]

She embraced both Church doctrine and tiny skulls in the same way, perhaps, that she appreciated both astronomy and astrology.

Was Marie Marvingt a feminist? Defined as someone who believed in equal rights for all, absolutely. Defined as someone who actively campaigned for women's rights, no.

An article that described Marie ironically as "the woman included in the ranks of great men" said, "She has not campaigned for women's rights or worked for the nomination of a Minister of the Condition of Women, but she's given to France and to the world an astonishing image—explosive for our times—of what a woman without limits can achieve."[76] In his book on "scandalous" women, Philippe Valode situates her as "the most liberated woman in France."[77]

In 1922, a pilot-journalist devoted many column inches to complimenting Marie and listing her achievements. "I could fill even more columns, but I'm a little humiliated to see in a woman so many qualities that I myself lack. I think it's like that that Mlle. Marvingt understands feminism. Her way of being a feminist is surely better than that of Monsieur Lloyd George's suffragettes."[78]

Never mind that both methods—fighting for the vote and being a woman who accepts no inequalities—are both valid. (The writings of the period understood the word "feminism" as it is understood today, and Marie was often referred to as a feminist, but because of her actions rather than her words.) Several writers[79] feel that Marie's goal in life was to prove that women were as capable as men: "At a time when women's emancipation was one of the popular themes in books and the press, she undertook to demonstrate that the beautiful sex had nothing to envy in the strong sex."[80] But that's not accurate. She never thought of "demonstrating" anything. One writer said she wore culottes to "act like a man."[81] No. She wore culottes so she wouldn't die when her long skirt caught on a ski or under the plane rudder or in a climbing rope.[82]

As one admirer put it, "She never spoke of her rights. Serenely and without hesitation, she simply claimed them."[83]

Without being an activist, Marie changed the established order forever, opening the route to independence for many other women. One journalist wrote, "By her actions, she has done more for women than a hundred passionate discourses."[84] Françoise Baron Boilley concludes that Marie's activities redesigned the contours of feminine gender as they were defined at the end of the nineteenth and beginning of the twentieth centuries; she transgressed the rules and codes of behavior for women.[85] In today's popular culture parlance, she was "a world-class badass, by almost anyone's standards."[86]

General Barthélémy, who knew Marie both professionally and socially, said, "She wanted to be the first to do things that no woman had ever done, but she wasn't a feminist in the strict sense of the word. She was, however, very proud of her own successes and the triumphs of other women, particularly in aviation. She took great pleasure in reminding people that between 1910 and 1930, more than 600 women around the world had gotten their pilots' licenses, and that a great number of them had bettered records, held important positions, or broken barriers in aeronautics."[87]

Marie lived at a time when women were subordinated to male domination at all levels—political, economic, cultural, ideological, and intellectual, a subordination underwritten by medical, religious, and philosophical arguments. Articles that appeared in the popular press between 1890 and 1909, a period when Marie was much in the news, included titles such as: "On the Absence of Creativity in Women," "The Defective Sense of Humor in Women," "The True Nature of the Inferiority of Women to Men," "A Monstrous Regiment of Women Is Looking for Work," "Why Women Don't Need to Work," "Women Are Hare-Brained," and the big question: "Are Women Really Human Beings?"

Marie's pronounced taste for energetic physical activities, adventures, and intellectual explorations was bewildering to those who felt women belonged in the home, occupying themselves with their children and their domestic responsibilities. In most people's view, woman was made to embellish the existence of others. To live for herself went beyond comprehension or acceptance.

One version of her culottes, 1912 (courtesy Mireille Bertrand).

Great emphasis was placed on what was "natural" for a woman. The French attitude toward women had been institutionalized in 1804 with the Napoleonic Code, which guaranteed that women would remain powerless under the law. Seventy years later, the subject of women's rights was declared subversive and immoral. Women did not get the vote in France until 1945, and it wasn't until two years after Marie's death that women could open a bank account without their husband's permission.

Marie met sex-based barriers and inequality constantly, from small obstacles (the first international requirements for commercial pilots required female pilots to be medically examined every three months, male pilots every six months; pilots hired for special events had to be twenty-one years if male, thirty if female[88]) to big ones (pilot and filmmaker Michèle Larue said, "As long as aviation was just a sport—as was the case in 1910—no one minded women's participation, but as soon as aviation became a business, and women were seen as competitors for jobs, they were no longer welcome"[89]).

The era's sexism reveals itself in a much-reprinted 1910 article called "The Aviatress":

> This is a new word, but there is a meaning behind it.
>
> At Paris Mlle. Marvingt, flying for the 'Coupe Femina,' has kept in the air for thirty-three minutes. If no woman beats that before the end of the year, the 'Coupe' will be hers.
>
> How is Mademoiselle, aloft in an Antoinette, regarded by her male rivals?
>
> One has but to recall the case of the 10-year-old boy who climbed up to the ridgepole of his home to perform antics for the entertainment and surprise of two little girls on the ground. Hearing a noise behind him, he turned to find that one of his admirers had come up to imitate and emulate him. Ignoring her, he descended to reap the plaudits of the second little girl below.
>
> So true is it that what man asks of woman is not emulation but appreciation.
>
> The satisfactory woman is she who remains in the grand stand with her gaze on high. Then when the airman comes down she is there to comfort and applaud. The girl on the ridgepole was "unwomanly," and the boy felt it. …
>
> Yet who will find a way to confine the ladies to the grand stand?[90]

When women were heralded in 1911 for being pioneers in flight, one popular headline read, "Flights of Women: Steering Aeroplanes As Though They Were Perambulators."[91] They might have been steering airplanes as though they were, well, airplanes. But reported at the same time were some early results of women's daring: at a course of public lectures on aviation being given at that time in Paris by the Polymathic Society, most of the students were women; municipal statistics showed that already one hundred and thirteen "lady students" attended lectures on trigonometry, political economy, and aviation, while classes for painting, music, and singing were attended by barely a score of "girls."[92]

Marie did much to counteract the negative image of women who dared to act in the public sphere or live lives of their own choosing. It was hard to be critical of a much-decorated patriot who had defended her country and brought glory to France through her international sports championships and her invention of the ambulance-airplane. A number of supporters at the time hastened to assure readers and audiences that, although Marie could do anything, she was still possessed of those feminine graces that made her a real woman.[93] Marie was somehow palatable to the general public, modifying the frightening image many people had of the New Woman. Today, she can be described as "the allegory of feminism and the emancipation of the woman at the turn of the twentieth century."[94]

One of the mysteries of Marie's life was her immunity from public ridicule, criticism, and exclusion. Women who stepped out of line, who drew attention to themselves, who assumed everyone understood that of course they were the equal of anyone, usually paid dearly.[95] Marie's patriotism made her a tricky target: "as a famous sportswoman, [she] was heralded as a national hero for her role in the trenches and in the air, rather than condemned for acting against her sex.[96] Her lack of didacticism, arrogance, and ego helped deflect negative views of her. As *Le Figaro* put it in a 1909 page-one story about her, she was "a charming example of energy and good humor."[97] She was also one of a comradely group of journalists, a colleague of those who wrote about her.

A Marine once commented on Pearl White, the star of *The Perils of Pauline*: "We like Pearl White ... because she can do stunts that would stump the best of us, and she comes up smiling after them."[98] Something of this good-natured approval clung to Marie's legend too.

Amélie Gayraud, a teacher, was curious to know how this New Woman of 1914 was regarded among young women. She conducted a poll among her female students and other young women ages 18 to 25 about love, marriage, women in society, religion, and sports.[99] Most of her respondents were excited about sports. The girls expressed passionate admiration for the Fiancée of Danger. Many of them said they had begged their parents for permission to ride in an airplane. ("Their parents sensibly said no.") The young women added that sports provided the best way to meet young men in the spirit of camaraderie that Mademoiselle Marvingt had popularized.

In his book on gender and education in China, Paul J. Baily noted that in pre–World War I China, women were reading magazines such as *The Ladies Magazine* and *Women's World*, which referred "enthusiastically to a 'new world trend' of active women driving trains, flying planes, and engaging in competitive sports. The exploits of the Frenchwoman, Marie Marvingt (1875–1963), described as the world's first great female athlete, who performed acts of derring-do such as climbing mountains and flying in air balloons, were especially singled out."[100] Young women read, and their worldview shifted, if only slightly.

People absorb models of what it means to be female or male from many sources. In Marie's day, the growth of newspapers available to everyone who could read was most likely a major factor in transmitting her way of being a woman to the general public. Habituated to getting the news every day, and to believing what they read, readers didn't find it much of a stretch for the new reality Marie was so publicly exhibiting to become something of a norm, or at least acceptable.

Feminine magazines of that era, like *Femina* and *La Vie Heureuse*, did their part by melding the activities of the New Woman with the activities of more traditional women. In this way, they mitigated fear of the masculinization of women or the unspoken horror that women might someday feel they were the equal of men or, unimaginably worse, treat men as men had been treating women for centuries.

Today, Marie is studied for her influence on her times. In their chapter of *International Review on Sport and Violence* (2014), Evelyne Combeau-Mari and Valérie Boulain write, "Aware of the risks and the experience of an imminent death, Marie Marvingt expressed in a form of ultimate violence towards herself and society her refusal to comply with the interdictions imposed on the women of her generation and condition. Perhaps without having really wished it, her going beyond the limits in sports and aviation then became a political stand."[101]

Carolyn Heilbrun, writing about independent women, says that some women seek risk and personal accomplishment in addition to, or instead of, the traditional conjugal life. A study of women who have lived nontraditional lives indicates they recognized early on they had gifts they could neither name nor define.[102]

Marie may have been an unwitting role model, but she was very much a supporter of women and their activities. She gave hundreds of talks to schoolgirls and young women, urging them to adopt sports, follow their dreams, and cultivate independence. She made a point of seeking out girls' schools, and was particularly effective there because of her celebrity. She was living proof that women could do anything, and she wanted to show them a world with more possibilities than they were being raised to expect.

She created a minor sensation when she spoke in Salonika, Greece, because she was the first woman to speak there in public.[103] She kept in contact with the groups she met, encouraging them, making helpful connections for them, and publicizing their efforts. She donated money for an award given to the two best female athletes at the Women's Athletic Fete on September 16, 1941 (Marshal Pétain signed the letter acknowledging her donation of 1,040 francs).[104]

Marie received public appreciation when she agreed to be part of the *Comité des Dames*. An auxiliary to aviation groups, it wasn't the sort of organization she preferred (it held many fundraising sales), but they needed her, they said, "to appeal to the female public, too often inclined to judge something's worth by its value to the other sex."[105]

And when women did something important, she outdid any male journalist in her detailed, accurate articles about them. As she pointed out in one long 1937 article about Maryse Bastié, "The same night that Dieudonné Costes and Maurice Bollonte crossed the Atlantic, Maryse Bastié, alone in her plane, flew three hours longer than they did, thus basically doing the same thing."[106] She adds that she's told that story hundreds of times before thousands of listeners because, although they're familiar with the first story, they have never heard the second.

She wrote several articles about the aviation successes of women from around the world, including names of women who owned their own planes, those who had died in flying mishaps, those who were training pilots, those who ran an airport staffed entirely by women, those serving as aeronautical engineers, pilots on commercial airlines, and glider pilots.[107]

As often as she could, Marie attempted to redress the invisibility of women. In her 1934 piece, "Their Wives," she emphasized the importance of women generally treated as appendages of the great men to whom they were married.[108] Few people had any idea of what these women were contributing.

Marie's feminism showed as she wrote about Queen Marie of Yugoslavia, granddaughter of Queen Victoria and daughter of Marie of Romania, also a well-educated woman who established numerous humanitarian projects in and around Belgrade. Never one to omit mention of her impressive acquaintances, she noted, "She and her mother both love Paris, which is where I met them some time ago at a number of official receptions."

Marie's success as a journalist was often due to her knowing the people who made the headlines and being wherever anything was happening. Some of it was luck; much of it was planned. She admired Henriette Poincaré, who was married to Raymond Poincaré, president during the Great War, and worked closely with him. Marie says one never appealed in vain to the distinguished and highly intelligent Henriette on behalf of any worthy charity.[109]

The Poincarés left the Elysées Palace the first time in 1920. Marie said she often heard Poincaré say that he would never take up office again unless France was in danger.[110]

A few years later, on a Thursday in 1926, the day Henriette Poincaré always received her friends at home, a woman came in and whispered in Henriette's ear. She stood, with her back to the fireplace, and said, "I ask you for a moment of silence. I've just found out that the present minister has been overturned."

Marie wrote, "We all threw ourselves around her to congratulate her because we all knew that for some time, France had been in trouble. The franc was about to go under. But with the return of Poincaré would come confidence, and rescue for the franc."[111]

Marie admired and wrote about her longtime friend Inès de Bourgoing Lyautey for many reasons, but Lyautey was, most importantly in Marie's view, a registered nurse. In Morocco, she founded numerous maternity hospitals, daycare centers, a 300-bed hospital, orphanages, preschools, clinics, anti-tuberculosis dispensaries, and nursing schools. In Nancy, she spearheaded many health centers, was a major in the Red Cross, and supported Marie's work with air rescue. She was particularly involved in preventing and treating poison-gas injuries during the Great War. It's little wonder that Marie was outraged at the lack of credit her friend received.[112]

She once offered tongue-in-cheek advice: "Above all, women should avoid advertising their strong will!"[113] Of course, she herself made no secret of hers.

The day after she set the first international women's aviation records in 1910, the question came up in interviews: "Will women make good pilots?" In many newspapers around the world, Marie's answer was headlined simplistically "Women Not Good Flyers" (*The Washington Herald*), "Flying Not the Sport for Women" (*El Paso Herald*), "Says Women Will Not Excel in Air" (*Trenton Evening Times*), "Not Suitable for Women" (*New Castle [Pennsylvania] Herald*). What Marie actually said was, "Flying requires quantities of *sangfroid*, initiative, and prompt decision, which not one-quarter of women possesses. None but those who have been active in sports all their lives can ever hope to fly an aeroplane alone."[114] The same could be said of one-quarter of men, but her qualified answer would not have made a catchy headline.

Marie appeared at the best possible time to make an impact on women's and girls' lives. Had she lived earlier, she might not have had an education; the laws allowing girls to attend school were passed five years after she was born. She would have missed many of the new sports, inventions, and, most grievously, the airplane. She was fifteen years old when Clément Ader invented the word "plane" and succeeded in flying the first heavier-than-air. For Marie, the timing was perfect.

The doors were creaking open a little for women—not much, but enough for Marie to push through. Earlier, most doors were shut, some were locked. In addition to the opportunities that Marie forged for herself, the climate of the Belle Époque (Raymond Rudorff describes it as "the last of the 'good old days' before world wars, revolutions, and galloping technological progress transformed the world out of recognition"[115]) encouraged those who could entertain others, whether by artistic creations, daring exploits in the air, or proficiency in sports. There was a live audience for public events, and the public craved heroes and idols. It was the best time to be one.

Sports today is dominated by people who train rigorously from earliest childhood for one particular sport. Marie won a swimming championship one year, a ski championship the next, while setting records in aviation, aeronautics, bicycling, riflery, and other sports. Meanwhile, she filed stories as a journalist, gathered supporters for her ambulance-airplane, and learned Esperanto. Gordon Ackerman wrote, "Never able to decide on a single area of endeavor, Marie decided on all of them, and has been described by a friend as 'a Jack-of-all-trades, and master of most.'"[116]

Today there is a deepening and narrowing, rather than a broadening, of interests.

Achieving top billing in one's chosen field is a rare and short-lived achievement. One newspaper said that when writing of Marie Marvingt, a sentence invariably appeared: "She was the first ... [fill in the blank]."[117] Being a "first" in any domain is no longer likely. And the athlete who wants into the record books knows that switching from baseball to golf to track to swimming isn't the way to get there. The early years of the twentieth century was one of the few times in history when one could excel at everything. Aviators were also sculptors, skiers were inventors, and talented people of both sexes could expect several successful careers, one after the other.

Marie crossed lines, broke barriers, erased categories. And her era tolerated her, perhaps because she was a familiar embodiment of it. Marie lived her life out to the edges. She was the best Marie Marvingt she could be.

Was she always success incarnate and an unmitigated joy? Not particularly.

Marie rarely took herself seriously, but nothing annoyed her more than the mispronunciation of her last name as "Marveen."[118] And "annoyed" would be an understatement when she was mischaracterized in the press. After she crossed the North Sea, a letter to the editor appeared in *The Times* of London, written by a Professor Spiers, of King's College, London. Although his connection with her is unknown, he said he was writing on behalf of Mademoiselle Marvingt "who, as described in these columns on October 28, crossed the North Sea in a balloon and landed on the Suffolk coast, to correct the statement that she is a professional swimmer." Mademoiselle Marvingt, he wrote, "is a lady of independent means and a journalist. The statement that she jumped from the car, which appeared in several newspapers, but not in *The Times*, was also incorrect."[119] She asked for corrections in other newspapers, too.[120]

Two years later, several English-language papers reported that "A woman aviator named Marvingt, while monoplaning at St. Etienne yesterday, became scared by the spectacle of a crowd in the grounds of a restaurant. She lost her nerve, and collided with a tree, falling among the tables, and scattering the diners and waiters."[121] Whether due to something lost in translation or to sexist assumptions, the report was inaccurate. Marie "scared"? "Lost her nerve"? Having a passion for accuracy herself, it was probably the errors that outraged her more than the insult to her oft-confirmed status as fearless and nerveless.

Frustrations with an inaccurate press would fade to nothing against the frustrations of her later years. In the way that "old age makes caricatures of us all" (P.D.James),[122] only in the last fifteen to twenty years of her life were her faults remarked upon. Perhaps she'd always been a bit impatient, a bit abrupt, a bit imperious, but with her record of accomplishments, who noticed? Marie's foibles showed up more clearly as she struggled against the disconnect described by Doris Lessing: "The great secret that all old people share is that you really haven't changed in seventy or eighty years. Your body changes, but you don't change at all. And that, of course, causes great confusion."[123]

16

Winter of the Grasshopper

"At the age of 80, Mademoiselle Marvingt is the startling proof that time withers only those who permit it, and that youth is a phenomenon of the heart rather than a number."—Michel Daurat, *Lorraine Magazine* (1955)[1]

"Last January, during one of the bitterest cold spells of the winter in Europe," wrote Gordon Ackerman in the June 26, 1961, issue of *Sports Illustrated*, "an 86-year-old woman in Nancy, France, bundled herself in a heavy overcoat and three wool sweaters, pulled a pair of goggles over her eyes, mounted a bicycle, waved to her neighbors, and pedaled off toward Paris, 280 kilometers away. She rode ten hours daily, across the frigid plains of eastern France, over mountains, through cities and villages where she attracted the whimsical attention of townsfolk. Every two hours she stopped to rest in the countryside, leaning her bike against a tree and spreading a blanket on the frozen earth and nibbling cold cuts and vegetables from the 30-pound knapsack strapped to her aged shoulders.

"At dinner time she stopped at the best restaurant in the area, where she stood inside the entrance—as erect and alert as an Air Force recruit—until she was recognized by the maître d'hôtel. The maître came over to her and said in French, "You are … that is, do we have the honor of … it is Mademoiselle Marvingt, is it not?" Then she seated herself at a choice table and consumed a light dinner, interrupted by a dozen customers seeking her autograph. There was, of course, no bill to pay. She was the guest of the house, just as she is the guest of most good res-taurants in France.

"For six days she pedaled, until one morning she reached the city limits of Paris, coasted slowly around the Concorde and pulled up in front of the Ritz Hotel—the city's best—where a suite overlooking the historic Place Vendôme awaited her. That afternoon she was at an air base outside Paris—making a solo flight at the controls of a helicopter.

"Unusual? Not for Marie Marvingt."[2]

Ackerman then detailed Marie's accomplishments and honors, saying, "She is France's greatest living adventurer. Seldom have the qualities of bravery, endurance, spirit and a love of danger been so clearly evident in a single person." He ended the lengthy piece by quoting Marie: "When I die, the city is going to build a museum to hold my trophies. When I was ill many years ago, they decided my time had come and that it would be nice to inform me of their plans to build a museum in my honor after I was gone. So every few days since then somebody has come to look in and see if Marvingt is still around, and if they can start work on the museum. This has been going on for a long time. They are starting to lose interest."[3]

Marie evidently outlived the city's interest in her, for to this day there is no space devoted to her in local museums.

Marie's later years tell the real story of who she was.

Gabriel Bichet, a journalist from Nancy and author of a number of articles about Marie, described her as "an unusual, fascinating figure, toward whom Time, taken aback, has made an exception." Then he wrote movingly of her willpower, "willpower that allowed her to lead to the very end the kind of life she had always lived. This old woman refused to grow old. In her old age, she refused to conform, to give up her pastimes, to wear dreary clothing, to sit around drinking tea and ruminating on useless regrets. She reminds one of those amazing older people, Made in U.S.A., supple, jovial, colorful, who die suddenly without ever having known—between the time they become an adult and the time they die—the demeaning affliction of old age."[4] (Marie asked Bichet to be her biographer, but he was unable to accept.[5])

Journalist Georges Gygax interviewed Marie when she was eighty-three: "While she's talking, underlining her words with eloquent gestures, jumping up from her seat, pivoting around, sketching a little waltz step, we are fascinated by the warmth of her voice, the elegance of her speech, the richness of her vocabulary, and by an eye that sparkles with both wit and kindness. Eighty-three years old! She looks no more than sixty-five. Her hands are the hands of a woman of forty. Proudly—and she has good reason to be proud—she places her hands next to a plaster cast of them made more than half a century earlier. Where is old age?"

He concludes, "Mademoiselle Marvingt has lived in joy, in enthusiasm, in the exaltation of victories sometimes painfully acquired. Her life—in spite of having little in it materially—has been such that she hasn't been bored for a minute. She lives intensely, and this intensity never ceases to entertain her. With a last smile accompanying the handshake of a weight-lifter, Marie the Unsinkable takes off on her bike in the direction of the Place Stanislas, where the round-eyed pigeons watch her passing with a certain degree of awe."[6]

Another journalist fan of Marie's, Jacques Boetsch, sketched an animated portrait after he encountered her at the hotel on the Place Vendôme where she usually stayed in Paris:

> Despite the comfortable armchairs, the ornate pedestal tables, the flowers and green plants, it was not possible, even in that first moment, to mistake her. The white-haired woman sitting across from me, Legion of Honor medal pinned to her coat, was still the intrepid little person balancing on one foot on a mountain top, thumbing her nose at onlookers as well as at danger—the woman whom the Americans called "the most incredible woman since Joan of Arc."
>
> The public has applauded her exploits for half a century now, and although celebrities today are pitilessly forgotten, Marie Marvingt remains in the spotlight, ceaselessly adding to her accomplishments.
>
> As I listen to her, pinned to my chair by her dizzying monologue, searching desperately for the main thread of her tortuous detours through the labyrinth of recounted perils, that mischievous little silhouette perched on one foot on the point of a mountain sits across from me, describing thrills and menaces, and pulling me along into the country of the impossible.
>
> It's not enough to say that at the age of eighty, Marie Marvingt is one of the youngest women you'll find in Paris. She's also one of the most multiply talented and most unusual. Rising early each morning (she never sleeps more than four hours a night), she bikes out to the military aviation field to chat with former students or old friends, then gives a conference somewhere in Paris, returns to her hotel to freshen up, and spends the evening at the theater, visiting with the cast after the production. And she does all this on her bicycle.

> This small woman with the quick eye more than once hammers home her opinions with a gesture and a level look at me—not that I would dare contradict her—before realizing that she's quite forgotten the time and needs to end the interview.
> "Now that I'm an old woman," she says, "I have to hurry. There's still so much to do."
> And, impatient with herself for accomplishing virtually nothing this morning, Marie Marvingt mounts her bicycle and, with great strong strokes of the pedals, flies off toward her daily adventures.[7]

The stories were endless: "Hardened as they are to acts of daring on the traffic-jammed streets and avenues of the capital, Parisians still turn to take another look at a dignified, elderly woman on a cycle, powered with a small auxiliary engine, zigzagging in and out of the traffic with the agility and aplomb of a newspaper delivery boy. Wearing an expensive Persian lamb coat, and a stylish, close-fitting toque, she sits her cycle as if she was riding to the hounds, and looks a young sixty. Astonishment would be even greater if Parisians recognized in this audacious cyclist the woman who won for herself the name of Danger's Sweetheart at the beginning of the century, or knew that last week she celebrated her eightieth birthday."[8]

It was the same in Nancy, said Michel Daurat: "Everyone here knows this extraordinary figure who pedals swiftly around town. Police salute her, passersby smile at her. They all know it's Marie Marvingt but they have no idea of her age. If they did, they'd think she was old but the truth is, she's younger than they are."[9]

If Marie ever walked anywhere, there were no witnesses to it. She pedaled around Nancy on her homely errands—getting a pair of shoes soled, visiting a sick friend, buying produce at the open-air market in front of St. Sebastian's. Just as tirelessly she pedaled home again, the morning paper and a baguette from the bakery in St. Catherine's Street sticking out of her basket, a string bag holding a few potatoes, onions, and beans hanging from one handlebar, and her umbrella and bulky black handbag swinging from the other. Her bike (instantly recognizable to most of Nancy's inhabitants) could be seen leaning against the side of St. Epvre Church while she attended early morning Mass, or parked carelessly in front of her usual stops: the town hall, army headquarters, the newspaper office.

Marie bicycling in Paris in 1949.

Adds Achille Liégeois: "As she bikes through the streets of Nancy, she hastily waves at neighbors and friends without slowing down, as if she were in perpetual fear of missing her train."[10]

When she was eighty-three a journalist wrote, "Marie Marvingt is certainly the most astonishing and dynamic woman I know. I see her everywhere—at the biggest sporting events, theater performances, festive Masses, public gatherings, in short, anywhere there's something going on."[11] That same year, Georges Gygax wrote, "The activity of this woman is difficult to summarize in a few lines. She is interested in everything and even today, she remains indifferent to nothing. She has been everywhere, and everywhere, she has distinguished herself."[12] Sometimes she looked it. Colonel Blanc said, "I can still see her in evening dress with her decorations at the Governor's Ball. She was dignified, compelling."[13]

In many ways, Marie was regarded as a civic treasure in Nancy. But other stars grew brighter, and, as she aged, hers faded and became less visible. The movers and shakers somehow forgot to invite her to important events. Friends who guessed something of the truth about Marie's circumstances and who knew how proud she was made sure that her name was on as many invitations as possible. However, when a request for her presence wasn't forthcoming, Marie usually decided they would want her there, so she mentally forgave the oversight and attended anyway.[14]

In her later years, she became a skilled and inveterate gatecrasher: she couldn't afford to pass up the edibles she might find at a reception or official gathering.[15] Perhaps more important, she couldn't bear not being part of movement, gaiety, crowds, occasions, culture, of life itself. As one person said bluntly, "That was the way she was. She barged in anywhere, and usually she was welcome."[16]

Marie elevated the art of gatecrashing to a science, practiced at civic receptions, weddings, military gatherings, and reunions. Most people felt that her contributions to France entitled her to do whatever she liked. A few groused that they didn't mind her gatecrashing, but why must she always leave her great, hulking bicycle in the entryway where everyone tripped over it?[17]

Marianne Jansen wrote a beautifully telling interview for *Écho de la Mode* when Marie was eighty-six:

> She opens the door herself to her dark and crowded apartment. Tiny, alert, with a quick eye, and an imperious tone, she says before I can get a word in:
>
> "Ah! There you are. I was waiting for you. But, you know, I have not got very much time to give you. I hate losing my time with journalists. They write the most foolish things. They're the ones who nicknamed me the flying grandmother. I ask you! Grandmother! Me? I, who have never been married? You think that makes me happy?"
>
> I try to apologize in the name of my fellow journalists, but she isn't listening. She continues combatively:
>
> "I could teach you to write. Twice I won the international aeronautic literature competition."
>
> The evocation of this literary laurel seems to calm her aggressive mood. She notices me, although I am trying to make myself very small in the face of this storm.
>
> "Sit down, if you like," she tells me in a grumpy voice, pointing to a chair on which sits an impressive but unsteady pile of papers.
>
> I clear the chair carefully and sit down while Marie Marvingt, relaunched, says, "That colonial helmet that you see on that bust (which is my work because I'm also a sculptor), I brought that back from my historic crossing of the Sahara! Were we ever hot! I went back later and invented metal skis which were as useful on the sand as on the snow. And I founded a ski school in Morocco. The Bedouins couldn't get over it."
>
> I must admit I feel a little like the Bedouins.
>
> "I'm thinking of getting a diving suit. I'd really like to be a frog-woman."
>
> I look at her closely. Is she making fun of me? No, she is perfectly serious.
>
> She suddenly aimed a furious look at me.
>
> "They wanted to refuse me the renewal of my pilot's license. They said I was too old. And you? Do you think I'm too old?"
>
> I quickly denied it, assuring her that her energy and vitality left me breathless, which was the truth.
>
> "Old age," she said, "What does that mean? I know people of twenty who can't ride a bike. I, at my age, I go everywhere on my bicycle. Especially in Paris, it's sensational. I just snake through all the worst traffic jams. People ask me my secret. Well, I'm going to give it to you: Willpower, cheerfulness, and action—and never stop!

"Write that down just as I said it. Don't make me say something I haven't said," she commands, while eyeing me suspiciously as I take notes.

Suddenly, she gets up. "And now," she says, "you'll have to go. I have lost enough time here. I must prepare for my next series of conferences. I've got work to do. And life is short."

To say that after eighty-six years filled to the brim … what optimism!

I found myself outside, a little dumbfounded.

Leaving the building, I bumped into a bicycle. The soldier on duty at the door of the barracks said, "That's Mademoiselle Marvingt's bike. Have you just been to see her? She's not easy, is she?"

He smiled and added, "But she is incredible. In Nancy, she is as famous as the Place Stanislas."

I can well believe it.[18]

Marie still lived only a few steps from the beautiful Place Stanislas in the historic and noble building that formerly housed King Stanislas's pages. But after her father's death in 1916, she moved to one of the smallest apartments in the building, three rooms on the top floor.[19] Although it had a tiny iron balcony, it was situated at the end of a long gloomy hallway. Marie's visitors had to cross the sunny courtyard, pass under a large arched passageway, turn left, and then climb to the fifth floor.

Even so, her little apartment was described as "an Ali Baba's cave of treasures."[20] A large panel was covered with medals and ribbons. Piled on every surface, spilling from boxes, dangling from every wall were her trophies, loving cups, the ornate plaques so popular then, statuettes, awards, more than 5,000 photographs, letters, testimonials, keepsakes, clippings, and autograph books.

But the dazzling gold and silver symbols of Marie's successes were not the sorts of riches you could spend, eat, or wear. The elegant high ceilings and thick walls of her apartment were scabby with flaking paint. The ornate, mantled hearth was lifeless and cold; firewood was expensive. Some days the kitchen cupboards yielded little more than a canister of tea.[21]

Marie in her apartment on the Place Carrière.

The woman who had shown her daring in a hundred ways now summoned the courage to grow old, poor, and forgotten. Friends said, "Can you imagine her going home at night to that tiny, bare, cold apartment after having been the center of attention at glittering events?"[22]

French fabulist Jean de la Fontaine told the story of a grasshopper and an ant. The ant worked busily all summer saving up for winter, while the grass-hopper sang songs to give people pleasure, with no thought of tomorrow or of its own needs. When winter came, the ant—

a hard-hearted, greedy creature—refused to share anything with the grasshopper and said cruelly, "You've sung all summer, now you can dance all winter."

Marie Marvingt had sung for others all her life. Unlike the grasshopper, however, she never asked for help when she found winter had come and she had nothing to live on. In the last years of her life, Marie often didn't have enough to eat. She was nearly evicted from her apartment several times because she paid her rent only when she could.[23]

At the time, no one saw Marie as deprived and needy. She was enthusiastic and optimistic, and she refused to play the role of "poor old Marie." With her customary willpower and love of theater, she instead played a graciously aging but still vigorous celebrity who used all the hours God sent to keep up with her voluminous correspondence, give conferences, submit to being interviewed (yet again), attend lectures, receptions, meetings, and events. An object of pity? Certainly not. Even during her most miserable days, few people suspected Marie's situation.[24]

How could a world-renowned celebrity be destitute?

While her father was alive, he and Marie lived on his pension from the postal service. He left her a small nest egg when he died, but she lost that—as did so many French people and their heirs at the time—when annuities were devalued.[25] The principal reason Marie had no money is that she never made much. Her only income-producing work was journalism. Even today, free-lance journalists do not become well-to-do on their fees. Oftentimes Marie received nothing or only a small sum for her articles.

The fabulous records she established or broke, the competitions and prizes she won, paid little or nothing (among her papers are receipts for the 100 francs she received when named a Knight of the Legion of Honor, and the 600 francs when she was promoted to Officer).[26] Amateur sports have never been lucrative, especially not in their early years, and especially not for women.

The hours Marie spent in the air brought her success and joy. But there was no money in it. Maryse Bastié once told aviation journalist Robert Marchand: "What a sad profession ours is. If you like champagne, if you like flowers, it's wonderful—they spoil you. But rent, bread, gas? It means nothing. It never occurs to these admirers of the marvels of aviation that a woman might like to earn a living at it."[27]

In a 1910 piece called "The Price of Heroism," *Le Figaro* wrote, "The public looks at flyers and thinks, 'Yes, they risk their lives, but they can earn a fortune!' A fortune! The truth is a great deal less golden that you'd think!" By that time, Louis Bréguet, inventor, builder, and pilot, had earned a total of 143 francs.[28] In today's money, he would be lucky to find that much under a couch cushion.

Among the female pilots named by *Le Figaro* (Laroche, Frank, Niel, Dutrieu, and Marvingt), the biggest earnings went to Laroche, who won 100 francs in Rouen, but the paper pointed out that one needed to factor in the dreadful crash she was still recovering from.[29]

The interest in aviators' earnings showed up in Montana, where the *Anaconda Standard* revealed figures up to September 10, 1910: "In 13 months from the first aviation meeting at Reims, in August 1909, aviators have won in prizes $712,019.53, and in the same space of time 24 aviators have paid the toll of their lives to the new form of locomotion. … For many fliers these sums have had to cover expenses, and if their career is glorious and at times profitable, it is always costly. The purchase of a machine, lessons, the hiring of a shed, wages of at least two mechanics, pieces for repairs, parts and wood—which in aviation are expensively dear—run away with large sums."[30]

Only once did a great honor for Marie come with a fair amount of money. In 1955 she was the winner of the Henri de la Meurthe Prix, which carried with it a prize of 25,000 francs.[31] She was eighty years old at the time, and probably decided not to spend it all at once. It was worth approximately $70.

How did Marie manage to pay for her airplanes, balloons, bicycles, skis, ice skates, cameras, and globetrotting, let alone rent, food, and clothing? She evidently knew something about making bricks from straw because after the loss of her father in 1916, along with his pension, she never again had a regular income.

A careful look between the lines reveals a thrifty, quick-witted woman with a great many friends and admirers who were eager to invite her on expeditions, lend her a car, fly her to her next conference, or subsidize her adventures. Based on the reality of her enormous fame and the assumption that because of it she must be well off, she created the illusion of someone who needed nothing, who thought her benefactors were quite wonderful for thinking up such delightful pleasures for her.

What little she made from nursing and journalism went for necessities. The rest was a glittering legerdemain of cobbled-together resources that she contrived without ever appearing to contrive any of it. She often flew to and around North Africa in postal planes, seated as best she could among the sacks of mail.[32]

As she grew older and made virtually nothing from her writings and nursing, her financial situation went from thin ice to cold, dark waters. She had kind and interested relatives (the closest were fifth- and sixth-degree cousins), but they lived in other parts of France and were not involved enough in her life to suspect that she had trouble meeting her basic needs. As postcards from Marie were uniformly cheery and full of news about what she was doing, these distant cousins assumed, as did so many people, that anyone as famous as Marie Marvingt must be comfortable.

She sold what she could—and bought what she needed—at thrift stores (her "good" black overcoat with the fur collar had belonged to a former mayor, then deceased). She practiced petty economies such as mailing her letters in used envelopes,[33] selling back-issues of her magazines in the want ads,[34] accepting medical supplies from the kind pharmacist, Monsieur Coulet.[35] She attended the second, less expensive showing of the movies she saw.[36]

Knowing she could turn it into a tasty Croque-Monsieur, she asked her butcher for "that bit of ham you won't be able to sell."[37] She re-gifted remembrances and presents from others. Just before her final hospitalization, she made herself a vest and skirt from a length of sailcloth she had taken off someone's hands.[38]

Gény Didier, a translator who often shared a simple lunch with Marie at the Red Cross headquarters in Nancy, said she once brought Marie some laundry bluing to whiten her hair, which they did in the sink there at the Red Cross.[39]

In the 1950s and early 1960s, Marie gave medical injections at home to earn a little money, and she taught diction,[40] crediting her early training under Sarah Bernhardt and Mounet-Sully for her expertise. She sold perfume to the officers' wives on the military base in Metz in the spirit of one friend doing a favor for another.[41] Both Marie and the officers' wives pretended to be unaware of the cruel necessities for such a transaction. A friend related, "Sometimes I drove her to the American military base where events were taking place, but when I brought her home, I noticed she had no fire. She couldn't afford it."[42]

Friends say Marie could have been rich. She could have charged for her lectures, or kept for herself some small portion of the money she collected for her ambulance-airplane or other fundraising causes. She could have accepted the offers of wealthy friends to sponsor her.[43] But Marie was nothing if not self-sufficient—and proud. As long as she had enough for today, she lived out her own inner imperatives while glorying in the abundance of her talents, dreams, and bottomless energy. A friend said, "She was very poor. But money, to her, was nothing. She paid no attention to it. And she never spoke of money, or of her difficulties. It was only when she became ill that we knew."[44]

Indeed, since the dangerous early days of aviation, "tomorrow" had not been part of Marie's thinking. Having arranged her own funeral many times, "tomorrow" was ephemeral and possibly dreary, whereas "today" always brought adventure, novelty, and joy. "Today" was in her own hands. "Tomorrow"? Not so much.

In the same way that the glitter of her apartment diverted visitors' attention from the shabbiness that lay underneath, Marie's quick gestures, flashing smile, and animated conversation distracted people from her threadbare clothes, her unheated rooms, and the embarrassingly few coins in her purse.

Madame Bouchet, married to an aviation general, was appalled when Marie opened the door to her dressed in her pilot's overalls and her old leather flying cap because it was so cold in her apartment. Madame Bouchet notified the head of the Eastern Air Club who sent two young mechanics to Marie's apartment with a load of coal. Marie told them that coal was filthy stuff and she wouldn't have it. The young men obediently took away the coal and brought her wood.[45] That exchange was reminiscent of one in the same doorway at the end of World War II, when a parachutist came knocking on Marie's door at 8, Place de la Carrière. Attached to the liberation forces in Nancy, he was carrying a requisition order, which allowed him to commandeer a resident's home or apartment to billet soldiers. Seeing a small, smiling woman, he was probably unprepared for the reception he received. Marie took one scornful look at his stamped, signed order and slammed the door in his face, yelling after him, "The Germans couldn't kick me out. Neither will you!" And they didn't.[46]

When part of her historic building was later transformed into a barracks, Marie was politely asked to leave to make room for the military. She confronted the colonel in charge: "When I came to live here in 1889, you weren't even born. So do not think you're going to throw me out. Marshal Foch would never have countenanced such a thing. I knew him well. Look here." Digging into her handbag, she pulled out a picture of Foch dedicated to her. Marie stayed. Later, when the military tried to evict her for nonpayment of rent, she resisted. The military was eventually obliged to give up.[47]

Jeanne Willotte, who lived at 10, Place de la Carrière, said that even in her last years, Marie always had her bike with her. "She used it as a cane. It was more sporting like that," said Madame Willotte. "She would hang her shopping bag and her umbrella from the handlebar."[48] Her voluminous black handbag hung on the other handlebar. Inside were photographs, letters, postcards, newspaper and magazine clippings.

Mademoiselle de Lalanne, a social worker who lived not far from Marie, invited her to speak at a conference. During her talk, Marie passed around some of her photos and letters. The next day, Mademoiselle de Lalanne was confronted by an indignant Marie: "You lost one of my papers!"[49] Each item in the bulky black bag stood for an important moment in Marie's life. Together, they were proof of who she was and what she had done.

Several people told the same story: "She used to drive people crazy. She always carried a large purse with her and she'd tell story after story about the pictures and letters and clippings she'd pull out of it like a magician with endless rabbits. It was all very interesting, but nobody had the time, you see."[50]

By carrying her purse with her everywhere, Marie managed to hang onto a strong sense of self. She not only kept body and soul together in her own fashion, but she maintained her outsized pride intact. She never lost her military bearing either,[51] and until her death she remained a hero to herself.

How, then, on a practical level, did she survive?

Colonel Blanc and his wife often welcomed Marie to their elegant mid-century apartment. Smiling in memory, they enthusiastically describe Marie as a loyal, generous friend and a dazzling conversationalist.

"Oh, yes, we saw her quite often," said the colonel. "She was always welcome in our home. Her stories, you know ... she entertained us for hours." He remembered that Marie used to drop in around dinner time. "Naturally, we invited her to stay and dine with us," he recalled. "She came most often, it seems to me, in the winter."

"Yes, yes," agreed Madame Blanc. "We had such lively dinners together."[52]

Madame Coulet and her pharmacist son Jacques missed their old friend and were only too glad to talk about her.[53] "She was always in the newspapers, you know," said Madame Coulet. "At least in the earlier years. Whenever she was in town, she came by to say hello."

"She very often stopped by around noontime," added Jacques Coulet. "I think she hoped to be invited for lunch. She had nothing to eat, you know." His mother nodded. "We suspected how poor she was, so we always encouraged her to stay and have lunch with us. How we laughed and talked!"

The Coulets, Blancs and others who knew that what Marie ate at their table was probably the only meal she might have that day said Marie was always gracious and well-bred, eating only enough to satisfy her hunger, and never giving the slightest hint that she would like to take home that bit of leftover.[54] Because she accepted small favors with so much charm and appreciation, her benefactors felt it was they who were in her debt.

A young journalist from Nancy described his parents' bakery and tea shop in St. Catherine Street, not far from Marie's apartment. Next to the counter showing off the flaky Napoleons, Baba au Rhums, strawberry tarts, and cream-filled cakes was a stack of small plates and dessert forks. Customers could take their purchases home or enjoy them with coffee at one of the tiny round tables. Marie came in most mornings and either took a small string-tied white box of croissants or petits pains home with her, or selected a pastry to eat in the shop while she chatted with other customers, glanced over someone's shoulder at the morning paper, or indulged her insatiable curiosity about life and people.

The journalist was a schoolboy at the time but, like everyone else in Nancy, he knew who Marie was. He remembers the way she approached the cash register before leaving the bakery. Inclining her head, she said in a firm voice, "I am Marie Marvingt, you know." His parents always nodded and smiled and never collected a cent from her.[55]

"She used to get herself invited for meals, arriving just before it was time to eat," confirmed a neighbor. "But she kept her hosts entertained for hours on end with her marvelous stories."[56] People enjoyed having her at their table. She used to dine about once a week with her friend Madame Bertrand, and she was often invited by the prominent families of Nancy such as the senator-mayor and his physician wife or the families of various city officials.

Military friends and acquaintances were particularly faithful, inviting her to receptions and social functions where she not only gloried in reminiscing and talking shop, but also found food and warmth.[57] Whenever she was in Nancy during the last twenty years of her life, she had a standing invitation to eat at the officers' mess at the military base in Metz.[58] For Marie, this was nourishment not only for the body, but for a spirit that craved exchanging news and anecdotes with those she considered her peers.

As Marie used to say, "All the same, you know, I have some lovely friends."[59] And it was true. A friend who knew she loved dates left them on her doorstep as often as she could.[60] In her later years when she stayed in Paris at the Ritz, Marie was sometimes the guest of the hotel, sometimes of a friend. All her hosts, voluntary or involuntary, ended up feeling the same: "It was a very great pleasure for us to have her. Between friends, you know, it is never a question of charity."[61]

Gilbert Grandidier, who worked then at the Nancy train station in the baggage department, told how he met Marie. "In 1960, I was new there, working nights, and I found a bike on the local train. No name, no address, so I put it in the consignment area. Toward midnight, an older woman demanded her bike, which I had had the nerve to put in the consignment section. Another worker knew her and paid the fee. The next night when I got to work, the chief of the station and someone from the police prefecture met me and explained that Mademoiselle Marie Marvingt was a great woman, had been an amazing sportswoman, and had rendered great services to the nation. 'She doesn't have much money,' they told me, 'so we don't charge her for anything here.'"[62]

Voilà! Breakfast, lunch, dinner, a few amenities, Marie Marvingt style. And she did have style. She returned kindnesses with charm and loyalty. She was generous with her friends whenever she had anything—time, favors, free tickets—to be generous with. She was quick to give away just about anything except one of her precious photographs or souvenirs. Even toward the end of her life, when she had very little, she never hesitated to invite someone to share her bit of food, or to press a coin in a visiting child's hand. Once when admirers presented her with a bouquet of roses, she took them to a sick friend. When someone protested, she said, "One has given nothing until one has given everything."[63]

With several rare exceptions, Marie refused to accept any obvious gifts from others. One night when she'd lost her keys, she asked her neighbors if she could sleep on their couch. They tried to get her to use their guest room, but she refused. She didn't like imposing on others, she said.[64] Her pride kept her from enrolling in the welfare program, but she accepted a little "public assistance" in her later years. "She didn't have a cent," said one social worker.[65]

Gabriel Bichet described an incident from the last days of her life: "I saw Marie downtown, sitting on the steps of a monument. She didn't look at all good, and was apparently feeling ill. A friend of hers who owns a pharmacy ran from his nearby store to bring her a tonic of some sort. Two passing police officers stopped to express their concern and offered to drive her home.

"'No. No, I'm fine,' she told them. 'I'll just go home.' It was useless to insist. After a few minutes she got back on her feet. She climbed on her Zéphyrine, wobbling just a little, and headed off toward the Place de la Carrière."[66]

Marie's cousin said that "Marie wasn't in good health toward the end but, as always, she remained stoic." Just before her eighty-eighth birthday, while visiting with Dr. Michel Royer, a longtime friend, she commented, "Something worries me a little and intrigues me. My arms seem to be getting thinner."

The doctor said, "All older people have that."

"Yes," said Marie, "but it seemed to have happened in four or five days, just like that."

"That's nothing to worry about," said Dr. Royer. "I wouldn't even think about it. Everyone over eighty gets that. It's nothing."

Marie said sharply, "That's easy for you to say. You make light of it because it's surely never happened to you!"[67]

Jehanne Plauche-Gillon, alerted by the Willottes that they hadn't seen Marie for a few days, went to visit her, and found her with a towel wrapped around her head for warmth. Horrified, she wrote to Marie's nearest relatives in Argenteuil: "She can no longer live alone. She has become extremely weak, doesn't eat, and is living without fire, without water, and without light."[68]

Jeanne Willotte described the beginning of the end: "We were neighbors on the Place de la Carrière, and Marie was a friend of my parents. One day she felt so ill that she called out the window to a neighbor to go find my husband because she wanted him to write her will. My husband went to see her, but found her in bed, in such a sorry state that he came home and said she needed me more than she needed him.

"I went over and promised Marie I'd do whatever she wanted. I called in a nurse and the two of us took care of her, and made up her bed with clean sheets because she said a priest was coming. When he arrived, we left them alone.

"My brother was a doctor and when I told him the situation, he sent his son, Michel, who was also a physician. My nephew said Marie needed to be hospitalized for further tests."[69]

Marie agreed to be hospitalized only on the condition that she be transported by a Red Cross ambulance. And so she left her home of seventy-five years in the way she chose to leave and was admitted to the geriatric hospital, Maison St. Charles, on Quatre Églises street.

Madame Willotte remembered that "when she was first hospitalized, Marie and I gathered her billfold and her jewelry, which I found in the kitchen, to give to my niece who had a safe. After the funeral, that was turned over to the relative in Paris." Willotte never forgot her shock when she saw that the heroine of the skies, the darling of millions, the great Fiancée of Danger had so pitifully little.[70]

Renée Enselme-Trichard visited Marie every day, taking her some small treat like *bergamotes*, the Earl-Grey-flavored candy made only in Nancy that Marie liked so well. "She seemed to be better once she was cared for. I brought her mail and read it to her and answered a few letters for her. When I had to go to Brittany to visit my children, Madame Coulet, the pharmacist's wife, and Madame Lionel Pellerin, the wife of the former mayor of Nancy, took over."[71]

Madame Plauche-Gillon said Marie lost her hearing, and her sight was too poor for her to read or write. "She was still overjoyed when she had visitors, however." According to Colonel Blanc, friends visited her often, but she became confused and the Maison St. Charles transferred her to the psychotherapeutic clinic in Laxou. Admitted on November 18, 1963, as a voluntary patient, with no fees to be assessed, Marie was in a dreamlike state and had cardiac arrhythmia. Her condition was "mediocre," according to Dr. Michel Royer. The prognosis: "reserved."[72]

Experiencing visual and auditory hallucinations, she mistook an intern for the pastor at St. Epvre and a nurse for the mayor of Nancy. She also, according to the medical records, "saluted militarily at any instant."[73]

"We went there to visit her," said Colonel Blanc. "The director was wonderful. He even telephoned me to reassure me that he was taking good care of Marie. She was much better off there."[74]

The pastor of Saint-Epvre said that Jehanne and Pascal Plauche-Gillon were the most faithful of Marie's visitors. A cousin, Madame Thévenin, knew how much Marie loved the active life, and asked if she weren't bored at the clinic. Indicating the rosary in her hands, Marie responded, "No, I think and I pray."[75]

On November 29, Marie suffered a stroke and was in a coma thereafter. On December 3 Dr. Royer noted that her death could be expected in the days to come.

Willotte said, "Several days before her death, I saw her. She was in a semi-coma. She talked as if she were still piloting an airplane. It was very touching." It was Willotte who caught Marie's last smile: "I had been called to her bedside by the doctor there. Eyes closed, she recognized my voice and the way I caressed her hand. She opened her eyes, which no longer saw anything and said, 'My dear.' Pink colored her cheeks and a smile relaxed her face. After that, she didn't move again, and went to a world more just than ours."[76]

It was 7:20 in the morning, December 14, 1963.

Willotte said, "I think it was the city of Nancy that took over the costs of her hospitalization and burial. My husband and I were there when she was placed in her coffin, and we accompanied the hearse that took her to St. Epvre for the funeral service." Willotte reflected, "She only knew how to grow old instead of dying young, despite all the risks she took." According to local rumor, the director of the Eastern Air Club had Marie's body taken very quietly to the church so that no one would know she died at the clinic in Laxou, the "insane asylum," as it was known then.[77]

When Marie died, every newspaper in France and most newspapers around the world carried the news:

"Death, with which Marie Marvingt flirted for 88 years, finally arrived to end the career of the one of the most extraordinary women who ever lived."[78]

"Marie Marvingt is no longer! The news exploded yesterday like a huge cannon, and the sound still echoes."[79]

"Marie Marvingt was one of the most extraordinary women of the century."[80]

"One of the great figures in aviation has left the scene."[81]

"With the death of Marie Marvingt, at age 88, disappears one of the most original women of our era."[82]

"Champion at everything, Marie Marvingt was one of the most extraordinary personages of the early twentieth century."[83]

The Times of India called her "one of the most outstanding personalities of her day."[84]

And according to scores of articles, a newspaper in the United States announced the death of "the most extraordinary woman since Joan of Arc."[85]

The most fitting epitaph, however, might have been the article that Gabriel Bichet wrote for *L'Est Républicain*. It read in part:

"We won't be seeing Marie Marvingt any more. Here in Nancy, where Marie was surrounded with an affectionate and almost familial deference, we speak of her in everyday terms, as we would for someone very close to us. She was very much a part of us all. For a long time, survivor of an era that she had lived intensely, she walked in the company of ghosts. All those whom she knew, who had been her companions, had—one after another—disappeared from the scene where many of them had played major roles. Even in the face of her solitariness, Marie kept smiling.

"But this smile was, above all, the supreme expression of her extraordinary willpower that had made of her what she was, a willpower that, at an age and in an era when young women learned how to embroider, made her choose to be a daredevil. Today she would be the female counterpart of a stuntman.

"Willpower supported her in the inch-by-inch struggle against her deplorable end ... having lived at a time when danger was not yet commercialized and no one thought of the financial possibilities of publicity, she had virtually nothing to live on.

"She subsisted on packages brought to her by her friends. Thanks are due to the military circles at whose tables she was always welcomed. Thanks are also due—and one can say it because it is the scandal of a dreadful social system—to the buffets of the receptions where, sometimes, she used her notoriety instead of an invitation to get in.

"She had willpower. Willpower—hasn't that struck you?—the willpower never to complain. Willpower that kept her fit physically and sustained her military bearing. She was a sort of female soldier, a feminist before the fact ...

"Unfortunately, willpower has no effect on the sway of mortality in this world. And so Marie Marvingt is dead. She was, according to the papers, the Fiancée of Danger. Poor Fiancée of Danger: How many times she fooled him. But he has had his revenge. It is always like that."[86]

The following Tuesday, December 17, a funeral Mass for Marie drew crowds to the neo-Gothic St. Epvre, with its three huge portals and high pointed steeple overlooking the fountain in the square. Magnificent wreaths flowered on the steps of the altar and lined the aisles—offerings and tributes from many of those present: members of aviation groups and sports organizations, nurses, Vieilles Tiges, military officers, cousins, friends, neighbors, acquaintances, townspeople, old flying comrades, government officials, police officers, and chamber of commerce dignitaries.

Marie's Legion of Honor cross was pinned to the flag covering her coffin. Aviators and members of the local Red Cross formed the honor guard while a military chaplain provided music. Abbé Berger, the pastor of St. Epvre, told the gathered mourners that while they had often seen Marie at official functions and civic events, he had seen her just as often in church. "She was no stranger to parish life," he said, describing her deep and enduring faith.[87]

After the funeral Mass, Monsieur Thirion, secretary-general of the Eastern Airclub, addressed the crowd from the esplanade in front of the church.

"At the age of eighty-eight, Marie Marvingt has left this world having been privileged to make history during her lifetime," he said. "One of the greatest figures in aviation has disappeared. Our illustrious compatriot leaves behind for present and future generations the memory of an extraordinary woman whose resplendent heart and mind were allied with a fierce will and undaunted perseverance ... engraved deeply in our hearts is an outstanding example of unquenchable courage in the service of aviation."[88]

That day's edition of *L'Est Républicain* reported, "Our distinguished citizen certainly had the send-off she would have wished. Gathered together were all her old comrades, her friends, the people she met on her many travels. Lined up in front of the Air Force flag and the Red Cross flag, they were all there: generals, colonels, majors, captains, the Mayor, even a baron—rows upon rows of Nancy's finest."[89] Marie's extravagant funeral was an unequivocal tribute to her remarkable life.

She was buried in the cemetery of Préville where, years earlier, she had bought an imposing, elegant granite tombstone with a large free-standing cross for the Marvingt plot. Her father and brother were buried there, and Marie had expected to lie there herself some day. But she could afford only to lease the plot. Since, at the time of her death, she was indigent, the lease was continued for twenty years at the city's expense.

Engraved on the granite tombstone along with her father's and brother's names was:

MARIE MARVINGT
1875–1963
Officier de la Légion d'Honneur
Croix de Guerre 14–18
Pionnière de l'Aviation Sanitaire

Twenty years after Marie's death, the lease expired. Marie once more faced eviction—of a particularly undignified sort. Not having received from anyone the 30,000 francs for a renewal of the lease, cemetery officials decided that Marie, her father, and her brother would be moved from their burial plot to a common grave, thus allowing the cemetery to sell or re-lease that ground. Outraged residents successfully petitioned the mayor to allow Marie to lie undisturbed.[90]

Marguerite Erbstein wrote, "Marie, empress of the skies, quitted this earth poor, forgotten, and the object of indifference. Thus goes the glory of the world."[91]

After Marie's death, her extensive collection of memorabilia disappeared. So too did her unfinished memoirs, scripts and notes for her two documentaries, and drafts for her journal articles. Dozens of journalists had described her salon prodigally populated with piles of photos teetering on chairs, clippings pinned to the back of the sofa, collages of medals. A photo of Marie sitting on the floor sorting papers gives a glimpse of what she considered her riches. And it all disappeared.

Some people said that an auction house came in and took whatever they wanted (the assumption being that proceeds would help pay for Marie's burial). Others remembered seeing boxes of Marie's affairs on the street one day, available for picking over, with a heavy rain destroying whatever was left the next day. There was also talk of a fire accidentally set by workers cleaning out the apartment. Whatever was left was tossed in the garbage.

In the end, a combination of auctioneers, rain, fire, and indifference seems to have been responsible; the bureaucrats or building owners who authorized this distressing loss have never been identified. According to a notarized document drawn up after Marie's death to deal with her "estate," no inventory was taken after she died, making it impossible to even guess all that had disappeared.[92]

In addition, the notary's office found that Marie had no descendants and thus no responsible person to take charge of her effects. Three distant cousins were named Marie's heirs, although two renounced their claims, leaving Madame Gaymard from Paris as sole heir, although it's unknown what in fact she inherited from Marie other than the bits of inexpensive jewelry.[93]

A few of Marie's hand-signed postcards found their way to antique stores and thence to collectors or admirers. The Lhérault family has conserved a wonderful collection of newspaper clippings. Someone has one of her trophies, but this information is kept secret—apparently not even a photo can be taken of it.

In 1990, when the Zonta Club, under president Annie Becquer, was preparing a Marie Marvingt exhibition, they turned Nancy upside-down looking for Marie's cherished bicycle. They discovered it had been preserved in the basement of the sporting goods store Noirtin for some twenty-seven years. Madame Noirtin donated it to the International

Marie Marvingt committee. The Deperdussin she flew is in the Musée de l'Air at Bourget, and the metal skis she designed can been seen in the Musée Dauphinois.

The International Marie Marvingt Committee, as well as a handful of other devoted admirers, are slowly gathering books, materials, artwork, and memorabilia in the hopes of establishing a Marvingt room in some museum in Nancy.

Marie would like that.

17

The Final Danger

"Heroing is one of the shortest-lived professions there is."—Will Rogers[1]

Near the end of 1963, Jean Nocher said to listeners of his popular Parisian radio show, "I've got a riddle for you, for all you champions of radio and television games, for all you history buffs, for all you teachers who want to hold up to our youth the best examples of contemporary heroism."[2]

He asked his listeners if they could identify by name "the third woman in the world to obtain her pilot's license and incontestably the greatest female aviation pioneer who has just died in total obscurity and poverty after having enjoyed one of the most brilliant, the most exceptional careers of any woman in history."

Nocher (real name, Gaston Charon), whose radio show ran for nine years, went on to give clues, telling his listeners that at the height of her popularity she was more famous than the astronauts of the early 1960s. He lists many of her accomplishments and awards and adds that she was the most eloquent speaker he'd ever had the good fortune to meet. "Who among you remembers her?" he asked. "Do historians remember her? Sociologists? Philosophers? Who can answer my question, except for a few people from Lorraine or the nurses in the hospital who received, in the greatest silence, her last sigh?"

He challenged his audience: "Admit that it's intriguing, engrossing, and even fascinating. Who was she? Is there one young French person out of ten thousand who could tell me her name? Try it for yourself. Ask people you know if they remember her. It's a cruel game, but it will tell you a great deal about today's society."

He noted that she had been on his program the year before, when she was eighty-seven, still with a marvelous voice ("courage preserves a person," he said). She encouraged young people to be committed to doing good and staying physically active.

After he revealed her name, Nocher said, "Don't you find it inconceivable that we have responded with silence and ingratitude to this heroine who brought so much honor to her country and to her sex?" He suggested that a few streets here and there in France be named after her. "That's not asking much, is it? I'm not proposing that we rename one of our great monuments after her. It appears that would be very expensive. But doesn't it cost us more, in the long run, to forget such a marvelous role model as Marie Marvingt?"

Twenty years after that broadcast, Marie was still so little remembered that there was no civic or national recognition of her life—not a plaque, not a street, nothing.[3] At that time, *L'Est Républicain* wrote, "Except for several medals, some recognition, and a lovely funeral, Marie Marvingt has not been greatly honored by her city." Twenty years after Marie's death, the concession on her burial plot expired. Marie was to be moved to a common burial ground.[4]

How could she have been so completely forgotten, her flame so carelessly extinguished?

One possible reason is profoundly ironic: Marie was too good in too many different domains. She should have specialized.

Florence Nightingale was an iconic nurse, and nurses the world over have made her their patron saint. Amelia Earhart was an unforgettable figure, and women pilots still look to her as their foremother. Sojourner Truth will always be associated with abolition, Marie Curie with science, Babe Didrickson Zaharias with sports, Gloria Steinem with feminism, Frida Kahlo with art. Those women whose names we remember carved out distinctive niches in specific fields, even though they may have had other interests.

To whom does Marie belong? Which group holds *her* up as their hero?

She too was a nurse, a pilot, a sportswoman, a feminist-in-action, but she did all those things and more besides. Had Marie only invented the ambulance-airplane or had she only crossed the North Sea, she might have found a place in the history books. Everyone's woman became no one's woman. One of Marie's lines in a 1909 article she wrote for *The Times* of London seems germane: "What is everybody's business is nobody's business."[5]

Another reasonable possibility for her obscurity: she lived too long. Cowboy-philosopher Will Rogers, who died in a plane crash himself, once said, "This thing of being a hero, about the main thing to do is to know when to die."[6] He wasn't the first to think so. Marie de Rabutin-Chantal, Marquise de Sévigné, wrote in 1675, "Long life will sometimes obscure the star of fame."[7]

Had Marie died, like the aviator Hélène Boucher, at the devastatingly young age of twenty-six, she too might have been buried in a burst of national pride and affection in the *Invalides* in Paris, site of Napoleon's tomb—a glorious and unprecedented tribute.[8]

Had she died at the age of thirty-nine in mysterious circumstances, as Amelia Earhart did, she might have earned an enduring place in the public memory. Novelist Ellen Glasgow wrote, "Life has taught me that the greatest tragedy is not to die too soon but to live too long."[9]

Most female aviation pioneers whose names are still recognized died young. Raymonde de Laroche, first woman in the world to be licensed to fly, died at thirty-seven while co-piloting an experimental plane. Harriet Quimby, who earned the first U.S. women's pilot license, was ejected from her plane in front of a crowd in Boston, also at the age of thirty-seven. As Alice James pointed out in 1891, "The success or failure of a life, as far as posterity goes, seems to lie in the more or less luck of seizing the right moment of escape."[10]

Marie's friends agreed that while no one intends to die young, it is one way of assuring oneself of immortality. Madame Willotte remarked, "If Marie had died while flying, they would have erected a monument to her, but she simply became, after all she had done, an old woman, always smiling and living her solitary life discreetly."[11]

Marie's little anecdote about the museum that kept checking to see if she was still around is revealing. "They're losing interest," she said. Had she died at the height of her successes, it seems probable that a local museum would have moved quickly to house her collection of medals and papers. But Marie died quietly in bed at the age of eighty-eight. For the first time in her life, something she did was not good copy. The museum people were not the only ones who lost interest.

Marcel Cordier says, "One rarely forgives a champion for dying in bed."[12] And Ralph Waldo Emerson believed that "every hero becomes a bore at last."[13] By living too long and dying in bed, could the fiancée of danger possibly have become a bore?

Keeping alive the legend of a dashing, beautiful daredevil of a woman is decidedly difficult when the real thing wears a raggedy bathrobe, rides a clumsy old bike, and backs you into a corner to show you her latest autograph of King So-and-So. In her later years, Marie was as familiar a sight in Nancy as the market on the Place de Saint-Epvre or the Craffe Arch leading to Old Town.[14] She came to be viewed as part of the scenery. Stripped of mystery, headlines, and youth, she inspired the same unthinking and affectionate reaction as any local landmark. (She once said, "Here at home, nobody bothers me. But in Calcutta, in America, and everywhere else it's always the kiss-kiss."[15] She liked being ignored. She also didn't like being ignored.)

Georgette Pessel was sixteen years old, and had a crush on the outrageously daring and famous Marie Marvingt. "One day, getting up my courage, I went and knocked on her door. She opened the door to me herself. The landing was dark and the only light came from behind her. She was dressed in an old dun-colored robe with holes held together with safety pins. I could see that her apartment was kind of a mess and maybe for that reason she didn't invite me in.

"I congratulated her on her exploits and expressed to her my deep admiration, all while apologizing for bothering her. She told me rather brusquely that it was nothing, that in this world everyone does what they are destined to do. What perhaps seemed to me like exploits were for her no more than whipping up a mayonnaise.

"At my age, I would have liked to have met a beautiful and flamboyant hero. That didn't happen but she unknowingly did me a great service. Since then, I've never idolized anyone."[16]

In the last years of her life, when people came looking for the intrepid sportswoman-pilot, they found only a talkative older woman, a guest who didn't seem to know the party was over. A heroic reputation is too fragile to bear the weight of a rather eccentric, old, poor, but still energetic woman. People who wax enthusiastic, for those who do not, are a little irritating.

Interviewed about Marie, her priest-cousin Father Lhérault admitted to an unsaintly impatience with her interminable stories. "When I'd go home for the weekend and see her bike parked in front of the house, I couldn't help saying 'oh, merde!'"

Marie achieved legendary goals in her youth with her willpower, perseverance, and energy. Those same qualities in an older person looked like bossiness, stubbornness, and "attitude."

Jean Leclerc, who wrote about Marie's strength of character and pride, tells of the time in the 1950s when she took her bike to the Michenon shop to be repainted. Normally, a bike was fully dismantled there, sent to another shop to be re-coated, then fired and reassembled at Michenon's. The process took at least eight days. Marie wanted her bike done in one day, no dismantling, and repainted by hand with a brush. Monsieur Michenon, recognizing a superior force, did it her way.[17]

Good friend Jacques Coulet said, "When she climbed Mont Blanc, the whole world celebrated her, but afterward it was forgotten, and when she told you about it as an old woman, it was just another boring story, one that might not even have been true."[18]

The Coulet family were friends of Marie's for thirty years. Jacques Coulet said that in her later years she used to come into his pharmacy on the Place Jeanne d'Arc and drive

23-6.58 CARTE POSTALE

Chère Madame amie
Je rentre de voyage
et vous remercie
pour votre aimable
invitation
J'aurai le plaisir
d'être des vôtres demain
24. Avec mes bonnes
amitiés
M Marvingt

Madame Coulot
En Ville

Postcard to good friend Madame Coulot.

him crazy talking for an hour almost every day. Even as he said this, he was wiping his eyes.[19] Marie might have been annoying on occasion, but she was also loveable.[20] One friend wrote, "Whenever I encountered Marie, it was a great joy to me. I wish I could express what happiness it was to be with her."[21]

Notably outspoken (several people described her as "having a colossal nerve!"), she never hesitated to give a piece of her mind to people she deemed careless, ignorant, or rude. "She meddled in everything. If someone did something she didn't like, she let them know it!"[22]

Marie lived, appropriately enough, in a military neighborhood. Sharing the courtyard entrance with her was the 11th Military Division. She used to annoy the soldiers by leaving her old bike in the passageway so they had to walk around it.[23] It was often a case of "love me, love my bike," which prompted more than a few frosty looks when Marie insisted on bringing her bike into the most formal social gatherings. She propped it up in the entryway, in the coat room, or against the nearest wall. Her bicycle was not particularly attractive, and people had to skirt it coming and going.[24] But what were they to do—this was Marie Marvingt. Still, it was annoying.

Not everyone was unsympathetic. In 1958, Georges Gygax wrote in *L'Illustré*, "If she likes to recount what she's done to an attentive and sympathetic audience, she never seeks to impose her stories on anyone and she doesn't embroider on them. The truth is everything for her. Never any contradictions or imprecision."[25]

Marcel Cordier knew her when he was a schoolboy.[26] "I often ran across her. She was considered to be eccentric, an elderly 'original.' I found her likeable but I always thought, when she told me things she had done, that she was exaggerating or fantasizing. But then, when I started researching her for my books, I discovered that every single thing she had told me was true. In fact, she had only mentioned about a tenth of what she had actually done."[27]

Colonel Blanc said, "People thought she was exaggerating, but she wasn't."[28] Marie told him that during her travels in North Africa she met a certain illustrious general with whom she became friendly. With him and his troops, she visited some of the important *beys* (chieftains or leaders) in Tunisia and Morocco.

Although Blanc didn't like to doubt Marie's story, it seemed highly improbable that she had so deeply infiltrated both military and native circles in North Africa. "Later," he said, "I was in Morocco and ran across the fellow, and it was all true! And she hadn't even told me the half of it!" That lesson apparently wasn't enough for him. When she told him about her experiences with the new jet-engined helicopter, he didn't believe for one moment that she'd actually piloted one. He was, however, again mistaken.

A journalist wrote, "She spoke of Foch, of Glaoul, the powerful ruler of the South, of the Sultan Mohamed V who used to kneel to her. There is certainly a lot of fabrication in all that, but Marie Marvingt, 'the fiancée of danger,' is still a delightful woman."[29] Actually, there was no fabrication at all in that. Everything she said was true, and verifiable. Mohamed V knelt to her as a joke between them, but he did kneel to her.

A friend said, "Marie never minimized her exploits, but she didn't exaggerate them either."[30]

Her life was so improbable, and there were so many stories, that it was difficult for anyone, even old friends, to sort out what might have been true. Not only did people think sometimes that she was exaggerating or making up stories, but as one friend said, "If she liked to brag a little, she had good reason to talk like that."[31] Others wouldn't hear a word against her, saying that when she spoke about her accomplishments, "it was without any fanfare, as though what she'd done was just an ordinary thing."[32]

When a Parisian journalist came to interview her, Marie asked, "What do you want to know?"

"Everything!" he replied.

"Hmmm," said Marie. "In that case, you'd better extend your hotel reservation for a few days."[33]

Although the remark sounds self-aggrandizing, it would, in fact, take days to thoroughly interview Marie and see all her memorabilia.

Another journalist asserted, "In her, there is not a trace of pride or smugness—because she has never tried to become 'someone.' It has been sufficient for her to act."[34] Marie was only explaining who she was and where she had been. It's true, she was the main character in her stories, but she told them without any particular self-centeredness. She was simply sure you would find the things she had seen as fascinating as she did. It was innocent, naive, and completely age-appropriate. Many elders enjoy passing on their stories to families and friends. The difference was that Marie's tales seemed more like fairy tales. And with no close relatives, she told them to neighbors, acquaintances, and even strangers, who might have lacked the forbearance of family.

The first time people encounter Marie Marvingt's life, they suspect a hoax. They assume some embroidering. It is all just "too much," embarrassingly so. Several people attempted to diminish Marie's accomplishments under the understandable but mistaken idea that no one could have done everything she did.

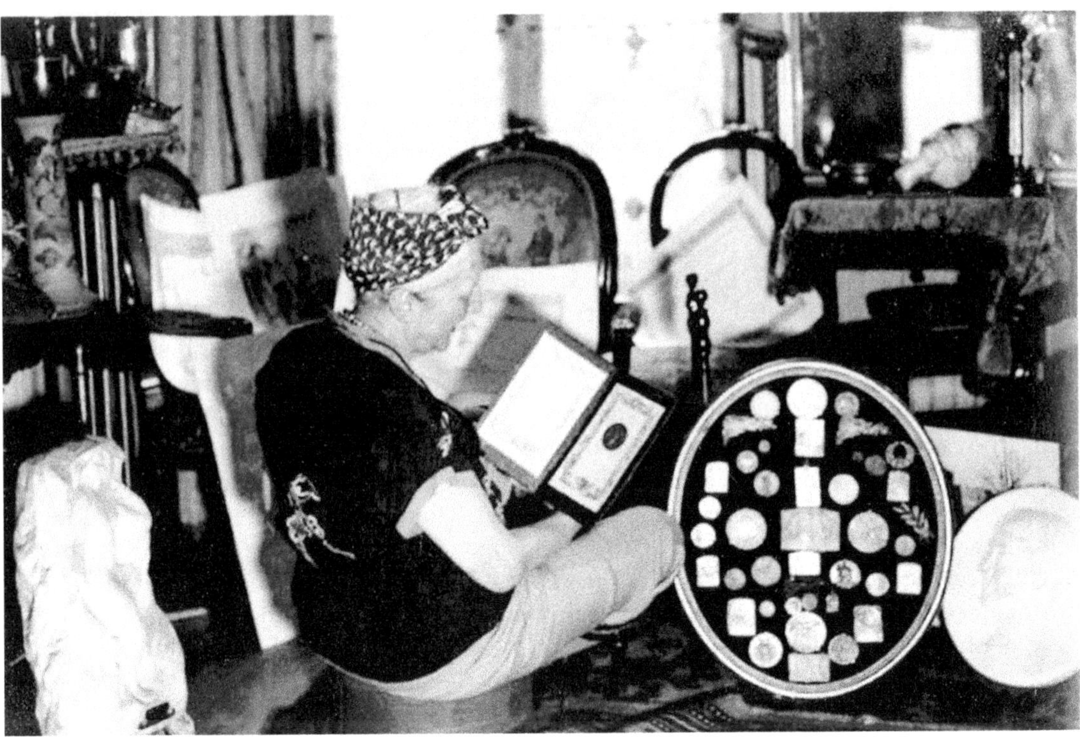

Marie with some of her medal and awards. Had they not disappeared after her death, there would be no question of what she had done.

In 1984, Marie-Josèphe de Beauregard, president and founder of the European Federation of Female Pilots, wrote that one shouldn't take Marie's life too seriously:

> She left a seductive legend, contrived by the media and by authors lacking academic rigor. According to her contemporaries, Marie Marvingt was an eccentric person, overly imaginative, and had a certain talent for publicity. The feats she is credited with are largely unverifiable, if not doubtful, to the point that, strictly historically speaking, one hardly knows what to make of them.
>
> Her aeronautic career can be summed up by three facts:
>
> She was a balloonist (license # 145) and, as such, the first woman to cross the North Sea in 1909. She made of this trip a rather colorful story, but given her mythomaniacal character, one doesn't know what part of the story is truth and what part is imagination.
>
> She obtained a pilot's license #281 in 1910.
>
> "She contributed, at the beginning of the first world war, to the launching, in France, of air rescue.
>
> All the rest is nothing but storytelling, which even so continues to support what I will call "the nebulous Marvingt."[35] (Madame de Beauregard added, "But perhaps your research will be able to authenticate one of the achievements of this cloudy personage. I hope so.")

Despite the research results that appeared in the 1991 *Marie Marvingt: Femme d'Un Siècle*, de Beauregard described Marie in her 1993 book as "a picturesque person whose vitality makes you forget her homely appearance."[36] She referred to Marie's "claims" and "presumed exploits," to Marie as someone who "cultivated" her Fiancée of Danger nickname, and said she was "supposedly a legend, based on a kernel of truth." Although she mentioned Marie's presence in the front lines of World War I, she wrote that it was only for "three little weeks."[37] Weeks are weeks, n'est-ce pas? The word "little" reveals a great deal about de Beauregard's determination to belittle Marie.

In the face of de Beauregard's well-trod disdain for Marie, it would probably be unproductive to reintroduce the medals, trophies, photographs, thousands of newspaper and magazine articles or to defend the authenticity of French government records, the French Air Club, friends like Foch, Weygand, and several presidents of the Republic. In addition, the facts of Marie's life can be taken straight from the headlines of the day, written by her contemporaries, recorded by on-site witnesses.

Throwing doubt on her crossing of the North Sea would probably annoy Marie the most. They had witnesses on both sides of the water, they had a verifiable tempest that night, and Major Garnier supported her on every point.[38]

Another great French pilot, Élisabeth Boselli, was much more complimentary of Marie, but even she said, "A certain circumspection is necessary to evaluate the sports performances that she added to her *palmares*. Some are recognized on the national level, others not. Her 'challenge' lacks precision. For example, she indicates that during the war of 1914–1918 she was a voluntary pilot. However, the French army never admitted her as a pilot, and there is a difference between asking to be a pilot, and really being a pilot."[39]

Sarcasm is not an argument. Government documents are more persuasive: Marie's Legion of Honor certificates and her Croix de Guerre testify that she piloted a plane and bombed an enemy air base twice in the first world war.

Boselli finished her letter by saying, "Whatever it amounts to, and even reducing her résumé to its essentials, Marie Marvingt was a cultivated woman with an open mind who figured among the pioneers of aviation before 1914 and of air rescue after 1918. She played an appreciable role in aviation propaganda. Her talents as a swimmer and mountain climber are undeniable as is her role as a sports pioneer."

The words of de Beauregard and Boselli, and similar uninformed opinions, tarnished Marie's reputation after her death, making it simpler to ignore Marie than to look for the truth. Her immoderate achievements underwrote the easy explanations of "storytelling" and "mythomania." The lengthy bibliography at the end of this book is an attempt to decertify "the Marvingt myth," and to replace it with accounts that can be read for themselves.

Had Marie been a certain Michel Marvingt, would it have been different? The disappearance of women after their deaths from the history, art, science, and record books is a generally acknowledged phenomenon. Great progress has been made in reversing this once-common practice, but Marie's activities pre-dated these advances. Her being a woman does not entirely account for her absence from the pantheons, but it is certainly a factor.

On a practical level, records were poorly kept in the newest fields—aviation, automobiles, winter sports. Women, in particular, literally didn't count. In an international aviation annual published in 1921, not one woman's name appeared even though women had been involved in aviation for years.

Marie may not have appeared in the record books, but the newspapers were another story. So many thousands of articles were written about Marie that errors crept in. News reports might mistake one date for another, one competition for another, even one female pilot for another. Foreign articles carried errors of translation. It was wrongly reported that she spoke seven languages (more like four or five), that she was a surgeon (she was a surgical assistant), that there was a statue of her in the Alps (no), that she was invited to Princess Grace's wedding (that was a different royal wedding), that she traveled to

Brazil (no proof of it and Marie certainly never claimed that). In Marie's own handwriting, she made several lists of her accomplishments. Her lists are accurate, but mistakes that appeared in articles by others tended to reflect poorly on her.

Marie had no family, no children to carry their mother's name forward, no close relatives with the energy and time (most were older) to take up the banner.

Some of Marie's longtime acquaintances felt that General de Gaulle didn't get along with Marie (jealousy, they suggest) and deliberately kept her out of the limelight.[40] This rather unlikely claim simply underlines the bewilderment of Marie's admirers, and their attempts to make some sense of her absence from the collective memory.

Slowly, the winds began to change. In 1982, mobilized by the possibility of Marie being removed from her place in the Préville Cemetery, Marcel Cordier and the International Marie Marvingt Committee, with the help of a cousin from Paris, gathered the funds to extend the lease on her burial plot until 2028. (Cordier notes somewhat bitterly that the mayor's office didn't contribute a *centime.*[41]) In 1983, Cordier published a book on Lorraine personalities, *Leurs Demeures en Lorraine: Tome II,* in which a chapter was devoted to Marie.

On April 1, 1984, the first stone was laid for the new "Marie Marvingt Residence," an apartment complex on the Boulevard de Scarpone in Nancy.[42] In the same year, on June 7, Cordier's International Marie Marvingt Committee installed a plaque on the building at 8, Place de la Carrière: "Here lived the Fiancée of Danger Marie Marvingt (1875–1963). Universal sportswoman, creator and pioneer of air rescue. Homage from her friends."[43] Townspeople contributed the cost of the plaque and gathered to honor Marie in a short dedication ceremony. That same day in nearby Maxéville, a new sports complex was named for her.[44] Again, that same day, a street in Nancy was named after Marie with a sign: "Marie Marvingt 1875–1963 Air Rescue Pioneer."[45]

In May 1987, a high school in nearby Tomblaine was officially named Marie Marvingt.[46] In September, Marie was posthumously named to the International Women's Sports Hall of Fame.[47] Introducing her to the many guests attending the award dinner at the Plaza Hotel in New York City was Jeana Yeager, co-pilot of the first nonstop, non-refueled flight around the world, the woman who spent nine days in a small, unpressurized cabin 3½' × 7'.[48] In 1990, the Zonta Club of Nancy sponsored an exhibit, "Marie Marvingt, Les Femmes, L'Aviation, l'Espace," at the Hôtel de Ville.[49] And in 1991, the first published biography appeared: *Marie Marvingt: Femme d'Un Siècle.*[50]

In 2004, a French postage stamp was issued honoring Marie, after years of lobbying by Marcel Cordier and his committee, as well as by the Soroptimist Club of Aurillac and many others. In blue, ocher, white, maroon, and beige, the stamp shows Marie in her flying outfit along with a sketch of her Antoinette.[51] Christophe Drochon, the artist responsible for the stamp said, "I wanted to show Marie Marvingt against the sky because she was one of those individuals who left their mark on it forever."

Since 2005, inspired by Colonel David M. Lam's introduction of Marie to the Aerospace Medical Association, the group has annually awarded an international prize called the Marie Marvingt Award to honor outstanding achievement in aerospace medicine. In 2015, the winner was Dr. Susan Ip Jewell for pioneering work in integrating technologies towards extreme environment survival skills necessary for deep space missions such as that planned for Mars. Mary F. Foley was given the award in 2009, Harriet Lester in 2013; the other winners have been equally deserving men.

In 2006, a fifty-two-minute documentary, *La Fiancée du Danger*, by Michèle Larue and Noël Burch, was aired on France 3.[52] That same year, the Museum of Air outside Paris held an exhibit of French air women that included one of the planes Marie flew.[53]

In October 2010, Benoît Pelard, Laurent Lajoye, and Sébastien Rolland, three balloonists from Lorraine, re-enacted Marie's 1909 balloon crossing of the North Sea in a balloon named "Marie Marvingt." They had planned to celebrate the centennial of her flight in 2009, but their first flight had to be aborted when the wind opened their valve and they lost hydrogen. In all, it took three men and five years of preparation to duplicate Marie's crossing of the North Sea.[54]

In 2013, Françoise Baron Boilley published *Marie Marvingt: A l'Aventure du Sport*, a comprehensive collection of information on Marie.[55]

Marie Marvingt's name can now also be found: on streets in Albi, Angers, Auneau, Aurillac, Bulgnéville, Cébazat, Dieulouard, Epinal, Fleuris-Mérogis, Heillecourt, Jarville-la-Malgrange, Metz, Migennes, Montpellier, Nantes, Reims, Saint-Dié, Saint-Etienne, Saint-Geneviève-des-Bois, Saint-Nicolas-de-Port, Saint-Parres-aux-Tertres, Strasbourg, Tarb, Toul, Ville-aux-Dames, and Yutz; and on alleys, intersections, or cul-de-sacs in Aussone, Laxou, Saulxures-lès-Nancy, Toulouse, and Vannes. Schools are named after her in Issy-lès-Moulineaux, Saint-Nicolas-de-Port, Tallard, and Vêzelise. Buildings carry her name in Essey-lès-Nancy and Pont-à-Mousson, sports complexes in Ludres and Villiers-lès-Nancy. Commemorative plaques are found in Dombasle-sur-Meurthe and Sainte-Alvère, in addition to the one in the Pierre de Coubertin stadium in Paris.[56] Aurillac named their air club after her, Rennes has a "Place Marie Marvingt," and the Metz-Nancy-Lorraine airport has an arrival hall named after her.[57]

Marie's life and accomplishments are thus beginning to re-emerge. With the admiration and efforts of many people, Marie seems poised to overcome the last, and most annihilating danger of all: the danger of being forgotten.

Afterword: Marie and Me

In 1980, I read a paragraph about Marie Marvingt in a now-forgotten book. The incredible accomplishments of this woman staggered me. But I recovered. "No, no, I have a degree in French. I certainly would have heard of her if this were true. It must be a hoax."

After a bit, it occurred to me that a hoax was a good story, too, so I did a little research. I found her obituary in the dignified, we-don't-know-from-jokes *New York Times*. In fewer than 200 words they highlighted the astonishing career of the "fiancée of danger." It was enough to keep me intrigued for nearly four decades.

Oddly enough, Marie and I missed each other by one year and nine months. She departed this world, and Nancy, in December 1963. I arrived in Nancy in September 1965. I was a French major, and had spent two summers as an administrative assistant for a National Defense and Education Act summer language institute in Rennes. I enrolled at the University of Nancy for the school year 1965–1966, earning several art-history certificates.

By the time I "discovered" Marie years later, my French was fluent, I was familiar with Nancy, and I had good friends there. I returned, in 1982, and with the help of Jacqueline Maire, a social worker and longtime friend, began interviewing people who had known Marie, visiting the departmental archives, and following threads here and there that might lead to more information about her.

My first surprise was that in Nancy, where Marie had lived for 75 years, only a handful of people remembered her. That intrigued me as much as her extraordinary life: How could such a woman be forgotten?

Jacqueline wanted the newspaper to interview me as the American writer who had come to Nancy to "resurrect" Marie. I'm bashful and didn't like the idea, but when she pointed out that we could request details about Marie in a newspaper that was read throughout the region, I saw her point. The last paragraph of the article gave my address in St. Paul, Minnesota, and asked for help.

Two good things came of that article. People sent me their memories of Marie. And, since the year I was there was the twentieth and final year of Marie's cemetery concession, many readers were upset to learn about its upcoming expiration and the plan to move Marie to a common grave. Some civic indignation inspired renewed interest in Marie.

That same year, Marcel Cordier, prolific and award-winning writer from Nancy, was writing a book about intriguing people who lived in Lorraine. One of his chapters would be devoted to Marie. He immediately established the International Marie Marvingt Committee and, with the help of concerned friends and a cousin of Marie's from Paris, secured her place in the cemetery until 2028.

I spent the next seven or eight years researching and writing a French-language biography of Marie while simultaneously trying to find a French publisher. I had several books in print by then, and while working on the biography, I had bits of Marie's story published in *Women's Sports & Fitness*, *Jack and Jill*, and an award-winning three-part series in *Cricket*. As for an English-language biography for U.S. readers, American publishers were less than intrigued. Much less.

"But she's French." Yes, I know. "We've never heard of her." Yes, I know. "Good luck."

France wasn't any easier. Maggie Doyle of the Paris literary agency La Nouvelle Agence was as enthused as I was about the book—and as helpless to sell the idea to a publisher.

France is a centralized country. You have Paris. And then you have everywhere else. I wanted a major, central, Parisian publisher. It wasn't going to happen.

After a few years, I turned to Marie's part of the country. And there I found the delightful and talented Jeannie Pierron Jung. Her father, a writer and artist (I have copies of five of his paintings in my living room), had founded the Éditions Pierron in nearby Sarreguemines, which published marvelous books: high quality, beautiful, significant. He had retired, and Jeannie was head of the company when I met her.

Jeannie read the manuscript and scheduled a meeting. She was blunt. "You're unknown. Marie is unknown. I like the manuscript but I can't sell two unknowns." However, Jeannie had an idea and his name was Marcel Cordier. "If you let us put Marcel's name on the book, we can sell it. He's well known in the area."

I would have to think about that. The book was completely written. Add someone else's name as author?

Back in Nancy, a group of intelligent and loyal supporters met in Marie-José Lionel-Pellerin's salon. A senator and physician, Marie-José was also the wife of a former mayor of Nancy. My friends (all women) were indignant. "A woman writes a biography about a woman, and they want to put a man's name on it? No!" They were adamant.

"But I live thousands of miles away," I told them. "I've been working on this for years, going back and forth to France on my own dime, and trying to sell it for most of that time. I have children at home and other books to do. But, most of all, I want Marie's story in print *now*. I've been to every publisher three and four times. This is our only choice."

Although I have had twinges when I read the foreign press that refers to Marcel Cordier as "Marie's biographer," the facts were that he had written about her in 1983, and in the years since then, no one had done more to restore her memory than he. I'm in the United States. Marcel is in France, giving hundreds and hundreds of two-hour talks about Marie, publishing countless articles in scores of publications about her. He's devoted. On her account, he will go anywhere, any time, speak to any group—even if it's only five people (I'm not saying he *has*, I'm saying he *would*). If I had to have a post-facto co-author, Marcel was perfect.

The book was published in 1991 as *Marie Marvingt: La Femme d'une Siècle*. I've never asked Jeannie how many copies the Éditions Pierron eventually sold, but I used to get a check for $40 now and then (Marcel got the other $40), so I can't imagine it did much more than earn back its costs. If that.

I've loved working on Marie's biography more than almost any other of my books, but it has not only earned just pennies, it has cost me a small fortune. And I'm fine with that. She is worth it.

I never meant to be a biographer. That's not where my interests lie. However, Marie is a restless, energetic woman and once she got hold of me (I can only think she couldn't find anyone else), she was focused and determined. More than once, I put "the Marie project" in the bottom drawer, only to find it back on my desk again.

After the French-language biography was published, I felt I'd done right by Marie Marvingt. But then came personal computers and digitized materials from the old days. Marie must have noticed because she became, well, persistent again. I saw that her story was richer, more detailed, and more intriguing than it had been when the only information I had was word-of-mouth, or a few articles in the local paper. To find a copy of a magazine from, say, 1910, I used to spend hours wandering among the booksellers along the Seine. Can you imagine doing research in that happenstance way?

Marie and I have been hanging out together for a long time now, and my best hope is to get the word out. Perhaps kind readers will pass along her story. Once her reputation regains the luster it once possessed, I feel sure she will allow me to work on a few other projects.

Chronology

1875: Marie Félicie Élisabeth Marvingt born in Aurillac (Cantal), France (February 20)

1878: Brother Félix Eugène Marvingt born

1880–1888: Swims; mountain climbs; rides tricycle, big bicycle, bicyclette; begins circus training, dance and fencing lessons; plays billiards

1889: Mother Elisabeth Brusquin Pallez Marvingt dies
Family moves to Nancy, 8 place Carrière
Competitive bicycling, fencing, drives automobiles

1890: Canoes from Nancy to Coblenz, Germany
Horseback riding, gymnastics, riflery, acting

1892: First prize in canoeing at Étretat

1897: Death of brother Eugène

1899: Obtains chauffeur's license

1900: Participates in all known sports

1901: First balloon ascension, as passenger

1903: Climbs Dent du Géant
Travels in balloon from Rouen to Lamballe
Travels in balloon from Nancy to Grabennendort
First referred to as "fiancée of danger"

1904: Nancy-Bordeaux bicycle race
Climbs Aiguille de l'M in the Mont Blanc range

1905: Climbs Trélaporte mountain peak in a culotte
Climbs Grépon and Grands Charmoz mountains in one day
Nancy–Naples bicycle race
First prize in freestanding canoe competition at Étretat
Drives automobile from Bussang to Aurillac in one day
Begins fifty years of writing for newspapers

1906: First Frenchwoman to swim across Paris (July 7)
Nancy–Toulouse bike race
Winter sports: skiing, skating
Wins swimming meet in English Channel
Opens first civilian ski school in France

1906: Climbs number of difficult peaks
Skis up several mountains
Ranks fifth best female mountain climber in world (1906–1910)

1907: First pilots a balloon
Wins Toulouse swim competition
Top prize at International Rifle Competition in 300-meter rifle division and in Flobert Carbine division; earns rank of first-class shot from Minister of War
Perfect score in clay pigeon shoot
Begins ski jumping

1908: Completes personal Tour de France
Wins and places in numerous winter sports competitions in Chamonix
Swims from Pallanza to the Iles Boromées at night
Swims across Lake Gérardmer
Swims 10 miles in Gulf of Naples

1909: First airplane ride, with Roger Sommer
Solo balloon flight, from France to Germany
Pilots *The Shooting Star* across the North Sea (October 26)
Wins and places in numerous winter sports competitions in Gérardmer
Begins flying lessons with Latham
Solos on the Antoinette

1910: Wins Léon Auscher Cup, international bobsled championship (January 26)
Awarded French Academy of Sports' gold medal for excellence in all sports (March 15)
First place in balloon race from Nancy to Neufchâteau
Receives balloon license no. 245 from French Air Club (June 10)
First woman to solo in Grand Prix of Aéro-Club de France distance competition
Obtains pilot's license no. 281, third woman in world, second in France to do so (November 8)
Sets first official women's world records in aviation (November 27)
Wins and places in numerous winter sports competitions in Ballon d'Alsace
Approaches Army with her design for ambulance-airplane

1911: Wins and places in skiing, ice skating, luging
Wins the Femina Cup
Participates in air meeting at Champirol, plane lands in acacia tree
Sightseeing in Dalmatia, Herzogovina, Montenegro
Developing ambulance-airplane
First prize in Eastern Air Club balloon competition
Begins giving conferences that will continue for 50 years

1912: Wins and places skiing, luging
Participates in Nancy-Jarville aviation meeting, in Toul aviation meeting
Paul Maurice Echeman dies in air crash (May 14)
Visits Ministry of War with plans for ambulance-airplane

1912: French Air Club balloon race, Paris-Irish Sea balloon race, Madame Salmageane balloon race, Paris-Brussells balloon race; a dozen other balloon ascensions
Orders airplane ambulance from Deperdussin company
Excursion boat capsizes, swims for 12 hours before rescue

1913: Balloon races in Lille, Châlons-sur-Saône
"Distinguishes herself" at European Rifle Championship
Crashes in her plane

1914–1918: Serves as Red Cross nurse and surgical assistant
"Lieutenant Beaulieu" fights in trenches, staffs listening post, fires and is fired upon
First woman combat pilot, bombs German airbase twice
Evacuates wounded soldiers from the Dolomites by means of skis and ropes
Receives safe conduct for Italian front, spends six months as war correspondent
Awarded the Croix de Guerre 1914–1918 *avec palmes* for her service as a pilot, soldier, nurse
Continues conferences to promote ambulance-airplane and air rescue

1916: Death of father Félix-Constant Marvingt

1917: Renews automobile license

1920: Hikes 35 miles in Maritime Alps, including difficult rock ascensions

1922–1927: Invited by President Millerand on fact-finding tour of North Africa
Ultimately travels 35,000 miles
Gives hundreds of conferences to promote aviation and air rescue
Visits most of North Africa, Malta, Sicily
Invents metal skis
Accompanies French, Italian, and Spanish troops as war correspondent and surgical assistant
Teaches health and hygiene

1922: Is issued safe conduct to travel for unknown reasons to Bonn, Cologne, Mayence, Coblenz, and Wiesbaden
Issues challenge to anyone in world to match her record

1923: Drives a Fiat 3549 across Sahara to In Salah
Founds ski school for Muslims
Experiments with metal skis on Saharan sands

1927–1928: On behalf of Republic of France, bestows medal on Charles Lindbergh following his historic air crossing of the Atlantic
Attends international aviation conference in London
Gives hundreds of talks throughout France

1929: Three-month lecture tour of North Africa
First international conference on flying ambulances
Establishes, with Robert Charlet, Friends of the Ambulance-Airplane

1931: Establishes Captain Echeman award

1932: Begins second tour of North Africa, which will last almost three years

1934: Writes, directs, and appears in two documentaries
Establishes civilian air rescue in Morocco

1935: Named Chevalier of the Legion of Honor
Receives first air nurse certificate
Autographs her wings at Aviators' Chapel, Mission Inn, Riverside, California
Tours the United States, gives talks with Amelia Earhart

1936: Inaugurates aviation nursing courses
First nurses graduate from her air-nurse program
Announces her challenge for the second time

1937: Named Chevalier of the Order of Public Health
Second lecture tour of the United States

1939: Establishes convalescent center for wounded aviators in Saint-Alvère
Works as surgical nurse
Invents new scar-minimizing surgical suture
Continues lecture circuit, radio talks, school conferences

1948: Issues her challenge for the third time
First prize in literary competition of International Women's Aeronautical Association

1949: Promoted to Officer of the Legion of Honor
First prize in literary competition of International Women's Aeronautical Association

1950: Awarded Silver Medal of Nancy

1951–1953: Travels 45,000 miles on third tour of North Africa

1954: Awarded Deutsch de la Meurthe Prize (January 30)

1955: Honored at Sorbonne by Fédération Nationale d'Aéronautique de France et d'Outre-Mer
Flies supersonic jet fighter, flies jet-engined helicopter

1957: Awarded Gold Medal in Physical Education
Awarded Silver Medal from Service of Air Safety

1958: Civic reception in honor of her birthday

1960: Obtains helicopter license, pilots first French jet helicopter, the Djinn 21

1961: Rides bicycle from Nancy to Paris
Solos in jet-engined helicopter

1963: Dies at age 88 (December 14)

Chapter Notes

Introduction

1. Leblond, J.-J., "Pour le 52e Anniversaire de Sa Première Ascension en Ballon Marie Marvingt a Piloté un Avion Sanitaire Entre Bordeaux et Marseille," *Dauphiné*, November 1952, p. 3.

2. Marvingt, Marie, letter to Frantz Reichel, in *Le Figaro*, "La Fiancée du Danger," January 3, 1914, p. 1; *L'Aéro*, "Petites Nouvelles," December 13, 1913, p. 2; *L'Industrie Vélocipédique*, "Chute d'une Aviatrice," December 20, 1913, p. 803; *L'Industrie Vélocipédique*, "L'Accident de Mlle Marvingt," December 27, 1913, p. 816; *Journal des Débats Politiques et Littéraires*, "Chute de Mlle Marvingt," December 14, 1913, p. 2; *Le Temps*, "Chute d'Une Aviatrice," August 8, 1911, p. 5; *London Evening News*, "Abroad," December 13, 1913, p. 1; Grahame-White, Claude, in collaboration with Harry Harper, *With the Airmen* (H. Frowde, Hodder and Stoughton, 1913), p. 47; *Gil Blas*, "Aéronautique et Aviation: Les Oiselles Tombent," December 13, 1913, p. 5; *The [New York] Sun*, "Mlle. Marvingt Has Record for Pluck: French Woman Aviation Showed Remarkable Courage in Recent Accident," February 1, 1914, p. 45; *Le Journal [Paris]*, "Mlle Marvingt Blessée Par Une Chute d'Aéroplane," December 13, 1913, p. 4; *The [London]Observer*, "Woman Aviator's Accident," December 14, 1913, p. 14.

3. Talbott, Earl G., "Marie Marvingt, 'Fiancée of Danger': Flirted With Death 88 Years," *New York Herald Tribune*, December 16, 1963.

4. *The Times of India*, "Marie Marvingt Dead," December 16, 1963, p. 7.

5. Saindizié, J.-M., "C'est Marie Marvingt, 85 Ans, la Plus Extraordinaire Femme du Siècle," *Pilote*, November 17, 1960, no. 56, p. 10; Barthélemy, Général, "La Femme la Plus Extraordinaire du Siècle: Marie Marvingt, la 'Fiancée du Danger,'" *L'Étrange Race des Hommes Volants* (Éditions France-Empire, 1979), pp. 241–254.

6. Porath, Jason, *Rejected Princesses: Tales of History's Boldest Heroines, Hellions, and Heretics* (2016).

7. *La Liberté*, "Marie Marvingt Est Morte à 88 Ans," December 16, 1963, no. 6164, pp. 1, 8.

8. A.W., *Tunisie France*, "La Femme la Plus Extraordinaire du Siècle Est à Tunis," February 2, 1952; Pelot, Paul, "Marie Marvingt Fut une des Femmes les Plus Extraordinaires du Siècle," *L'Équipe*, December 16, 1963; Barthélemy, Général, op. cit.; et alia.

9. Conte, Arthur, *Le Premier Janvier 1920* (Pion, 1979), p. 277; Daurat, Michel, "A Quatre-Vingts Ans, Marie Marvingt Demeure Toujours la Fiancée du Danger," *Lorraine Magazine*, no. 9, February 1955, pp. 10–13; Boetsch, Jacques, "Une Vie en Équilibre" *Anciens Combattants du Monde*, May 1955, pp. 11–14; *Les Ailes Brisées*, "Nos Aviatrices, les Véritables Pionniers de l'Aviation: Marie Marvingt," December 1959, pp. 21–22; Marchand, Robert, *Le Ciel n'a Pas de Toit* (Éditions Berger Levrault, 1962), pp. 258–262.

10. *La Sfaxienne*, "Mademoiselle Marvingt à Sfax," April 15, 1924, p. 1; Estrada de Tourniel, Jérôme, *Les Grands Événments de Meurthe-et-Moselle: De 1900 à Nos Jours* (De Borée, 2017), pp. 228–230.

11. Saindizié, J.-M., op. cit.

12. J.G., "L'Avion et le Vélo: Deux Amours de Marie Marvingt, 'la Fiancée du Danger,'" *L'Est Républicain*, February 16, 1975, p. 3.

13. "Marie Marvingt," French aviation publication, 1955, pp. 41–42.

14. Branchu, Marc, "Les Cent Vies de Marie Marvingt," *Magazine Air France*, no. 99, July 2005, pp. 42–44, 46.

15. Nicolle, Jean-Loup, "Marie Marvingt Racontée aux Enfants," Prix Jeanne Goury, 1961, Musée des Beaux Arts, Nancy.

16. Chevalier, Frédérique, "Marie Marvingt, la Fiancée des Airs (1875–1963)," *Les Grandes Aventurières* (City Editions, 2007), pp. 85–95.

Chapter 1

1. Michel-Royer, Dr. Jean, recorded interview with MM, January 1963.

2. Birth certificate, Civil Registry, Aurillac (Cantal), France.

3. Estrada de Tourniel, Jérôme, "La Fiancée du Danger," *Passions Grand Est*, no. 26, December 2003, pp. 58–62; Plancard, Frédéric, "La Fiancée du Danger," *L'Est Républicain*, September 6, 2013.

4. *Bulletin Mensuel des Postes et Télégraphes*, March 1879, p. 103.

5. *Ibid.*

6. Floret, Robert, "Grand-Mère 'Casse-Cou,'" *Détective*, November 18, 1960, no. 751, pp. 10–11.

7. Baron Boilley, Françoise, *Marie Marvingt: A l'Aventure du Sport* (L'Harmattan, 2013), p. 38.

8. R.P., "Mademoiselle Marvingt," *Le Cri de Constantin*, June 28, 1924; M.A., "Marie Marvingt, 'Fiancée du Danger,' Attend Encore Que Soit Relevé Son 'Défi Mondial,'" *Samedi Soir*, March 3, 1955, no. 505, pp. 1, 3; Kernel, Hélène, "Marie Marvingt, Doyenne des Aviatrices," *France Aviation*, August 1957, p. 8.

9. Rio, Armand, "La Fiancée du Danger, *Lecture Pour Tous*, April 1, 1913, vol. 5, no. 7, pp. 63–71; R.P., "Mademoiselle Marvingt," *Le Cri de Constantin*," June 28, 1924.
10. *Bulletin Mensuel des Postes et Télégraphes*, March 1879, p. 103.
11. *Bulletin Mensuel des Postes et Télégraphes*, November 1879, p. 680.
12. Floret, Robert, op. cit.
13. Rio, Armand, op. cit.
14. *Alrededor del Mundo*, "Biografía Extranjera Contemporánea: Mlle Marvingt," November 2, 1913, no. 753, p. 361.
15. Barthélemy, Général, "La Femme la Plus Extraordinaire du Siècle: Marie Marvingt, la 'Fiancée du Danger,'" *L'Étrange Race des Hommes Volants* (Éditions France-Empire, 1979), pp. 241–254.
16. *Paris-Match*, "Mlle Marvingt: Depuis 60 Ans 'Fiancée du Danger,'" May 7, 1949, no. 7, pp. 24–25.
17. Arnaud, Pierre, et Thierry Terret, *Histoire du Sport Féminin: Sport Masculin-Sport Féminin: Éducation et Société* (L'Harmattan, 1996).
18. Floret, Robert, op. cit.
19. Mauchaussée, Jean, "Marie Marvingt, Reine de l'Air," *Prestige*, October 1, 1959, no. 3, p. 1.
20. Michel-Royer, op. cit.
21. X.B., "D'Ultimes Témoignages Viennent Encore Préciser Quelques Traits Attachants de Marie Marvingt, Illustre Figure Lorraine," *Républicain Lorrain*, January 24, 1964, p. 2.
22. *Ibid.*
23. Cajelot, Maurice, "Marie Marvingt, 'la Fiancée du Diable,' 100 ans," *Républicain Lorrain*, February 13, 1975, p. 4.
24. *Ibid.*; *Le Lorrain*, "Deux Anciennes du Pensionnat Sainte-Chrétienne: 'La Fiancée du Danger' and la Soeur Blanche," May 11, 1950; Benoît, Michèle, "Marie Casse-Cou: Marie Marvingt," *Bulletin des Anciennes de Sainte-Chrétienne de Metz*, pp. 18–20.
25. Cajelot, Maurice, op. cit.
26. Ackerman, Gordon, "Fiancée of Danger," *Sports Illustrated*, June 26, 1961, pages 61–64.
27. *Républicain Lorrain*,"Les Anciennes Élèves de Sainte-Chrétienne Ont Retouvé, Avec Plaisir l'Ambiance de Leur Établissement," May 14, 1955.
28. Rio, Armand, op.cit.
29. *The Strand Magazine*, "'The Bride of Danger,' An Interview with Mlle. Marie Marvingt," vol. 46, September 1913, pp. 187–194.
30. Rio, Armand, op. cit.
31. Frémy, Dominique et Michèle, *Quid* (Robert Laffont, 1988), p. 1166; Marck, Bernard, *Dictionnaire Universel de l'Aviation* (Tallendier, 2005), pp. 688–689.
32. Mauchaussée, Jean, "Marie Marvingt, Reine de l'Air," *Prestige*, no. 3, October 1, 1959, p. 1; Jansen, Marianne, "Marie Marvingt: 86 Ans d'Âge, 51 Ans d'Aviation," *Écho de la Mode*, no. 47, November 19, 1961, pp. 18–19.
33. Dinan, R., "Les Enfants Prodiges: Mlle Marvingt Lance un Défi Mondial," *L'Ordre*, July 2, 1948.
34. *Le Petit Parisien*, "La revue de Printemps à Nancy," March 15, 1912, p. 5; *Journal des Débats Politiques et Littéraires*, "Manifestation Patriotique," March 16, 1912, p. 3.
35. Nightingale, Florence, *Notes on Nursing: What It Is, and What It Is Not* (1860), p. v.
36. Gygax, Georges, "La 'Fiancée du Danger': La Vie Prodigieuse de Marie Marvingt," *L'Illustré*, April 24, 1958, pp. 34–36; Floret, Robert, op. cit.
37. *Ibid.*; Ackerman, Gordon, op. cit.; Villard, Henry Serrano, *Contact! The Story of the Early Birds* (Crowell, 1969), p. 248.
38. Cordier, Marcel, "Les Escales de Marie Marvingt," *Leurs Demeures en Lorraine: Tome II* (Éditions Pierron, 1983), pp. 162–177.
39. Oudin, René, "Marie Marvingt Succède à Gyp," *L'Est Républicain*, July 17, 1968.
40. Liégeois, Achille, "Les Nancéiennes Loin de Chez Elles: Mlle Marie Marvingt Rentre d'un Long Voyage: Elle Nous Raconte Ses Impressions," *L'Est Républicain*, August 23, 1926, p. 1.
41. Aldrich, Mildred, *A Hilltop on the Marne: Being Letters Written June 3-September 8, 1914* (The Atlantic Monthly Co., 1915), p. 46.
42. *C'est la vie!*, "Marie Marvingt: Femme Universelle," April 28, 1950, no. 25, p. 3; Ackerman, Gordon, op. cit.; Mazenod, Lucienne, et Ghislaine Schoeller, *Dictionnaire des Femmes Célèbres* (1992), p. 579; Cordier, Marcel, op. cit.
43. "The World's Greatest Sportswoman," *Forest and Stream*, September 13, 1913, vol. 81, p. 323.
44. Gygax, George, op. cit.

Chapter 2

1. "The World's Greatest Sportswoman," *Forest and Stream*, vol. 81, 1913, p. 323.
2. Archdeacon, Ernest, "Mademoiselle Marvingt: Une Sportwoman Extraordinaire," *La Revue Aérienne*, December 25, 1910, pp. 702–706.
3. Marck, Bernard, "Marie Marvingt," *Aéroports Magazine*, 1992.
4. Floret, Robert, "Grand-Mère 'Casse-Cou,'" *Détective*, November 18, 1960, no. 751, pp. 10–11; Ackerman, Gordon, "Fiancée of Danger," *Sports Illustrated*, June 26, 1961, pages 61–64; Frémy, Dominique et Michèle, *Quid* (Robert Laffont, 1988), p. 1166; Barthélemy, Général, "La Femme la Plus Extraordinaire du Siècle: Marie Marvingt, la 'Fiancée du Danger,'" *L'Étrange Race des Hommes Volants* (Éditions France-Empire, 1979), pp. 241–254; Dinago, Georges, "Une Aviatrice Lorraine," *L'Éclair de l'Est*, January 7, 1911; Dinan, R., "Les Enfants Prodiges: Mlle Marvingt Lance un Défi Mondial," *L'Ordre*, July 2, 1948; Gygax, Georges, "La 'Fiancée du Danger': La Vie Prodigieuse de Marie Marvingt," *L'Illustré*, April 24, 1958, pp. 34–36.
5. Rio, Armand, "La Fiancée du Danger," *Lecture Pour Tous*, April 1, 1913, vol. 15, no. 7, pp. 63–71.
6. *The Strand Magazine*, "'The Bride of Danger,' An Interview with Mlle. Marie Marvingt," vol. 46, September 1913, pp. 187–194.
7. *Hastings and St. Leonards Observer*, "The Finest Sportswoman in the World," August 2, 1913, pp. 7–8; *Vestkusten [Sweden]*, "Marie Marvingt," March 26, 1914, p. 1; *Honolulu Star-Bulletin*, "Activities of Women," 30 June 1915, p. 3; *Lincolnshire [England] Echo*, "The First Sportswoman in the World," 13 August 13, 1913, p. 4; *Albert Lea Freeborn County [MN] Standard*, "Is Best Sportswoman," October 29, 1913, p. 3; et alia.
8. Seigel, Jessica, "10 Greatest Female Athletes Ever," *Glamour*, August 2008, p. 118.
9. Rio, Armand, op. cit.
10. Floret, Robert, op. cit.; Cordier, Marcel, *Lorraine, Secrète et Insolite* (Éditions du Sapin d'Or, 2011), p. 32; Barthélemy, Général, op. cit.; Bans, Georges, "Aéronautes Contemporains: Mademoiselle Marie Marvingt," *L'Aérophile*, November 15, 1909, vol. 17, no. 22, p. 505; Dinago, Georges, op. cit.

11. *L'Accessoire de l'Automobile* (1930), in Laget, Françoise and Serge, and Jean-Paul Mazot, with the collaboration of Elizabeth Foch, *Le Grand Livre du Sport Féminin* (FMT Éditions, 1982), p. 285.
12. Barthélemy, Général, op. cit.
13. Myriel, "Impressions d'une Française sur Guillaume II à Urville," *L'Est Républicain*, May 22, 1905, p. 1.
14. Cordier, Marcel, "Les Escales de Marie Marvingt," *Leurs Demeures en Lorraine: Tome II* (Éditions Pierron, 1983), pp. 162–177.
15. Floret, Robert, op. cit.
16. Dollet, Christophe, "Marie Marvingt Sans Visa," *L'Est Républicain*, 11 juillet 2011; Huguenot, Vianney, "Notre Incroyable Amnésie," *L'Estrade*, November 2015, no. 57.
17. Baron Boilley, Françoise, *Marie Marvingt: A l'Aventure du Sport* (L'Harmattan, 2013), p. 32.
18. Clemitson, Suze, ed., *Ride the Revolution: The Inside Stories from Women in Cycling* (2015), p. xi; Huguenot, Vianney, "Notre Incroyable Amnésie," *L'Estrade*, November 2015, no. 57.
19. Healy, Graham, *The Shattered Peloton: The Devastating Impact of World War I on the Tour de France* (2014), pp. 136–137; Poirier, Jean-Pierre, *La Véritable Jacqueline Auriol* (2005), pp. 51–52.
20. *Nancy Sportif*, "La Vie Sportive de Mlle Marvingt," June 4, 1914; Dinago, Georges, op. cit.; *The Strand Magazine*, op. cit.; Zwang, Annie, *100 Femmes Qui Ont Fait l'Histoire de France* (Éllipses, 2010), pp. 158–159; Lebow, Eileen F., *Before Amelia* (Brassey's, 2002), p. 37; Poirier, Jean-Pierre, *La Véritable Jacqueline Auriol* (2005), pp. 51–52; Cordier, Marcel, "Les Escales de Marie Marvingt," op. cit.; Clemitsen, Suze, and Mark Fairhurst, *P Is for Peloton: The A-Z of Cycling* (2015), p. 82; Thompson, Christopher S., *Tour de France: A Cultural History* (2008), p. 129; Thompson, Christopher, "Un Troisième Sexe? Les Bourgeoises et la Bicyclette Dans la France Fin de Siècle," *Le Mouvement Social*, July-September 2000, p. 32.
21. Laget, op. cit.
22. *L'Auto*, letter to editor (1909), in Suze Clemitsen and Mark Fairhurst, *P Is for Peloton: The A-Z of Cycling* (2015), p. 82.
23. Rio, Armand, op. cit.
24. Jansen, Marianne, "Marie Marvingt: 86 Ans d'Âge, 51 Ans d'Aviation," *Écho de la Mode*, no. 47, November 19, 1961, pp. 18–19.
25. Perren, Monsieur, in Jacqueline Maire, letter, July 5, 1982.
26. Maire, Jacqueline, letter, July 5, 1982.
27. *Marie France*, "La Fiancée du Danger' Lance un Défi aux Femmes du Monde Entier," June 22, 1948, p. 7.
28. "La Conférence de Mlle Marvingt à Notre-Dame de Sion," newspaper article, January 25, 1952.
29. Féral, Roger, "Marie Marvingt (86 ans), 'la Fiancée du Danger,' a Passé la Toussaint Dans un Hélicoptère à Réaction," *Télé-Paris*, November 1960.
30. Ackerman, Gordon, op. cit.
31. Cited in Alain Peyrefitte, ed., *L'Aventure du XXe Siècle* (1989), pp. 66–67.
32. *The [NY] Sun*, "Mlle. Marvingt Has Record for Pluck: French Woman Aviation Showed Remarkable Courage in Recent Accident," February 1, 1914, p. 45.
33. Cousin, Daniel, "La Femme Automobiliste," *La Presse*, January 10, 1914, p. 3.
34. Droz, Stanislas, op. cit.; J.P., "Marie Marvingt, Pilote d'Avion, Est Morte à 88 Ans: Championne en Tous Genres," *Le Monde*, December 17, 1963, p. 13; Barthélemy, Général, op. cit.
35. Droz, Stanislas, op. cit.
36. *L'Aéro*, "Propos en l'Air," August 31, 1912, p. 1; Gygax, Georges, "La Vie Prodigieuse de Marie Marvingt, la 'Fiancée du Danger,'" *L'Illustré*, May 1, 1958, pp. 40–42.
37. *L'Est Républicain*, "Une Course Militaire Internationale de Ski," February 13, 1911, p. 1; *La Musette [Massif Central]*, "Nos Echos," March 1911, p. 15.
38. Giroux, Jean, "Le 47e Salon de l'Automobile," February 1961, article annotated in Marie Marvingt's handwriting: "of which I've been a member since 1929."
39. *Le Radical*, "Le Grand Prix de la Corse," April 22, 1921, p. 4.
40. Ackerman, Gordon, op. cit.; Marck, Bernard, *Women Aviators* (Flammarion, 2013), pp. 14–21, 27, 39, 40, 43; *L'Est Républicain*, November 18, 1952, June 6, 1958, *L'Est Républicain*, November 18, 1952, June 6, 1958, or November 8, 1958; Pelot, Paul, "Marie Marvingt Fut Une des Femmes les Plus Extraordinaires du Siècle," *L'Équipe*, December 16, 1963; et alia.
41. Dinan, R., op. cit.; et alia.
42. Rio, Armand, op. cit.; Dinago, Georges, op. cit.
43. Rio, Armand, op. cit.
44. *L'Echo de Bougie*, "Mademoiselle Marie Marvingt à Bougie," December 31, 1922, p. 2; Barthélemy, Général, op. cit.; et alia.
45. *L'Est Républicain*, "Concours de Tir," June 19, 1906, p. 2; *Le Tir National*, "XIIIe Concours National et International de Tir," August 4, 1906, p. 365.
46. *Archdeacon, Ernest*, op. cit.; Almanach Hachette (Hachette, 1913), p. 195.
47. *L'Est Républicain*, "Concours de Tir," June 19, 1906, p. 2; *L'Est Républicain*, "Concours de Tir," June 26, 1906, p. 2; *L'Est Républicain*, "Concours de Tir," June 27, 1906, p. 2; *L'Est Républicain*, "Concours de Tir," June 30, 1906, p. 2; *L'Est Républicain*, "Concours de Tir: Journée de Clôture," July 3, 1906, p. 2; *L'Est Républicain*, "La Fête de Tir de Malzéville," August 5, 1906, p. 1; *L'Est Républicain*, "Concours de Tir à Malzéville," August 12, 1906, p. 1; *Le Tir National*, "XIIIe Concours National et International de Tir," August 11, 1906, p. 376; *Le Tir National*, "XIIIe Concours National et International de Tir," September 15, 1906, p. 425;
48. *Almanach Hachette*, op. cit.; Barthélemy, Général, op. cit.; Marchand, Robert, *Le Ciel n'a Pas de Toit* (Éditions Berger Levrault, 1962), pp. 258–262.
49. *Le Matin*, "Tir: Mlle Marvingt se Distingue au Championnat d'Europe de Dames," September 8, 1913, p. 5.
50. *Républicain Lorrain*, "A Nancy et à Metz, Marie Marvingt, la 'Fiancée du Danger' Dira Quels Sont les Buts Humanitaires de Sa Mission," October 19, 1949, p. 2; Féral, Roger, "Marie Marvingt (86 ans), 'la Fiancée du Danger,' a Passé la Toussaint Dans un Hélicoptère à Réaction," *Télé-Paris*, November 1960.
51. Barthélemy, Général, op. cit.
52. *L'Avenir de l'Est* [Algérie], "Mlle Marvingt," June 21, 1924, p. 2.
53. Ackerman, Gordon, op. cit.
54. Rio, Armand.
55. "La Conférence de Mlle Marvingt à Notre-Dame de Sion," newspaper clipping, January 25, 1952.
56. *L'Est Républicain*, "La Traversée de Paris à la Nage," July 30, 1906, p. 1.
57. *Journal de la Jeunesse*, "Athlétisme," August 18, 1906, no. 1759; *Journal des Débats Politiques et Littéraires*, "La Traversée de Paris à la Nage," July 30, 1906, p. 3. *L'Humanité*, "Paris à la Nage," July 30, 1906, p. 2; *Le Matin*, "Paris à la Nage no. 2," July 30, 1906, p. 4.

58. *Journal des Débats Politiques et Littéraires*, "La Traversée de Paris à la Nage," July 30, 1906, p. 3; Le XIXème siècle, "La Traversée de Paris à la Nage," July 31, 1906, p. 2; *L'Éclair*, "Paris à la Nage," July 30, 1906.
59. *Le Radical*, "Paris à la nage," July 30, 1906, p. 2.
60. *L'Echo de Paris*, "La Traversée de Paris à la Nage," July 30, 1906.
61. *La Justice*, "Nageuses," August 4, 1906, p. 1; *Le Temps*, "Paris à la Nage," July 30, 1906, p. 1; *L'Aurore*, "La Traversée de Paris à la Nage," July 30, 1906, p. 1; *Gil Blas*, "Natation," July 12, 1907, p. 4; *Sportif*, "Les Ondines Nagent Dans la Marne, à Joinville," 1906; *Le Radical*, "Paris à la Nage," July 30, 1906, p. 2; *Messidor*, "La Traversée de Paris à la Nage," July 9, 11, 17, 1907, p. 3; *Le Journal [Paris]*, "La Traversée de Paris à la Nage," July 29, 1906, p. 1; *Le Figaro*, "La Traversée de Paris," July 30, 1906, p. 6.
62. *L'Est Républicain*, "La Traversée de Paris à la Nage," July 31, 1906, p. 1.
63. *Journal de la Jeunesse*, "Athlétisme," August 18, 1906, no. 1759.
64. *Ibid.*
65. Ackerman, Gordon, op. cit.
66. A.C., "La Traversée de Toulouse à la Nage," *Express du Midi*, September 2, 1907; *Gil Blas*, "La Traversée de Toulouse," September 3, 1907, p. 4; *La Presse*, "La Traversée de Toulouse à la Nage," September 3, 1907, p. 3; *Le Petit Parisien*, "La Traversée de Toulouse à la Nage," September 2, 1907, p. 5.
67. A.C., op. cit.
68. *Le Journal [Paris]*, "La Fête de la Natation à la Sorbonne, December 10, 1906, p. 5.
69. Daurat, Michel, "A Quatre-Vingts Ans, Marie Marvingt Demeure Toujours la Fiancée du Danger," *Lorraine Magazine*, no. 9, February 1955, pp. 10–13.
70. *The Strand Magazine*, op. cit.
71. *Ibid.*
72. Estrada de Tourniel, Jérôme, "La Fiancée du Danger," *Passions Grand Est*, no. 26, December 2003, pp. 58–62; Grandidier, Gilbert, "Marie Marvingt (1875–1963), 'la Fiancée du Danger' Ouvre aux Femmes la Voie des Airs," newspaper article, May 1987.
73. Ackerman, Gordon, op. cit.
74. Desormière, Yves, "A 80 Ans Marie Marvingt, la 'Fiancée du Danger,' Est la Seule Femme au Monde à Être Titulaire de Quatre Brevets de Pilotage: Ballon, Avion, Hydravion, Secourisme de l'Air," *La Croix*, March 4, 1955, pp. 1, 4.
75. M.A., "Marie Marvingt, 'Fiancée du Danger,' Attend Encore Que Soit Relevé Son 'Défi Mondial,'" *Samedi Soir*, March 3, 1955, no. 505, pp. 1, 3.
76. Action Catholique des Femmes, *Femmes Remarquables: Je Broderai Vos Noms* (ACGF, 2001), pp. 212–215.
77. Burlingham, H., "Les Grimpeuses de Cimes," *Fémina*, no. 299, September 1, 1911, pp. 465–466, 470–471.
78. Boetsch, Jacques, "Une Vie en Équilibre" *Anciens Combattants du Monde*, May 1955, pp. 11–14.
79. Rio, Armand, op. cit.
80. Myriel, "Mes Débuts d'Alpiniste: 21 juillet 1903," *Éclair de l'Est*, December 6, 1908, p. 2.
81. *Ibid.*
82. *Ibid.*
83. *Ibid.*
84. *Ibid.*
85. Rio, Armand, op. cit.
86. Ottogalli-Mazzacavallo, Cécile, *Femmes et Alpinisme (1874–1919)*: *Un Genre de Compromis* (L'Harmattan, 2006), pp. 162–163; *Revue Mensuelle du Touring-Club de France*, November 1905, p. 516; *La Revue du Mont-Blanc et de Chamonix*, "Les Prouesses d'Une Française à Chamonix," August 1905.
87. *Dundee Evening Telegraph*, "Girl Climber's Record," September 7, 1905, p. 3; *The [New Orleans] Times-Democrat*, "Climbing Season of 1905: Record of Alpine Bravery," October 15, 1905, p. 35; et alia.
88. *La Revue du Mont-Blanc et de Chamonix*, "Les Prouesses d'Une Française à Chamonix," August 1905.
89. "La Conférence de Mlle Marvingt à Notre-Dame de Sion," newspaper clipping, January 25, 1952.
90. *La Revue du Mont-Blanc et de Chamonix*, op. cit.
91. Daurat, Michel, op. cit.
92. *La Revue du Mont-Blanc et de Chamonix*, op. cit.
93. *Ogden [UT] Standard*, "The Alpine Climbing Season," October 16, 1905, p. 1.
94. *La Montagne*, "Nouvelles Alpines," vol. 3, Septembre 1907, p. 418.
95. *The [Hobart, Tasmania] Mercury*, "Gleanings from the World's Press," 4 October 4, 1907, p. 2; *The Los Angeles Times*, "A Girl's Alpine Record," September 15, 1907, p. 17.
96. *The [Rockhampton, Australia] Capricornian*, "Gossip," October 19, 1907, p. 5.
97. M.A., op. cit.
98. Rio, Armand, op. cit.
99. *Ibid.*
100. Myriel, "Mes Débuts d'Alpiniste," op. cit.
101. Ackerman, Gordon, "Fiancée of Danger," *Sports Illustrated*, June 26, 1961, pages 61–64.
102. *La Tribune Républicaine*, "Débuts de Mlle Marvingt," August 4, 1911.
103. Codron, Charles, letter, circa 1905 (Lhérault collection).
104. Burlingham, H., op. cit.
105. *Ibid.*
106. *Ibid.*
107. Marvingt, Marie, "Sur les Alpes l'Hiver: Au Dessus de Mt Blanc en Avion—La Saison à Chamonix," *Lectures Pour Tous*, January 1929, pp. 46–53.
108. *Ogden [UT] Standard*, "The Alpine Climbing Season," op. cit.; Ottogalli-Mazzacavallo, op. cit.
109. *Le Figaro*, "Les Femmes au Mont-Blanc 1808–1921," October 18, 1921, p. 4.
110. Rio, Armand, op. cit.
111. *La Grande Encyclopédie de La Montagne* "Ski (Débuts en France)," (Éditions Atlas, 1978), pp. 2179–2180.
112. *Le Figaro*, "Les Sports d'Hiver," January 9, 1908, p. 5.
113. Dalbanne, J., "Le Concours de Ski des Vosges," *Armée et Marine*, March 20, 1909, pp. 95–96.
114. *Le Figaro*, "Sports d'Hiver," January 26, 1910, p. 7.
115. *L'Est Républicain*, "La Grand Semaine de Gérardmer," February 15, 1910, p. 2; *L'Est Républicain*, "Journée d'Excursion," February 16, 1910, p. 3; *Almanach Hachette* (Hachette, 1913), p. 195; Barthélemy, Général, "La Femme la Plus Extraordinaire du Siècle: Marie Marvingt, la 'Fiancée du Danger,'" *L'Étrange Race des Hommes Volants* (Éditions France-Empire, 1979), pp. 241–254; Boetsch, Jacques, "Une Vie en Équilibre" *Anciens Combattants du Monde*, May 1955, pp. 11–14; *Gil Blas*, "Les Sports à Gérardmer," February 14, 1910, p. 3; Dinago, Georges, "Une Aviatrice Lorraine," *L'Éclair de l'Est*, January 7, 1911; *The Strand Magazine*, "'The Bride of Danger,' An Interview with Mlle. Marie Marvingt," vol. 46, September 1913, pp. 187–194; *Le Rappel*, "La Vie Sportive: Le Ski," January 31, 1909, p. 3.

116. *Journal des Débats Politiques et Littéraires*, "Alpinisme," January 10, 1908, p. 3; *Le Figaro*, "Les Sports d'Hiver," January 9, 1908, p. 5.
117. *Le XIXème Siècle*, "Le Ski," January 31, 1909, p. 3; *La Lanterne*, "Sports d'Hiver: A Chamonix," January 30, 1909, p. 3; *Le Matin*, "La Vie Sportive: Patinage," January 29, 1909, p. 5; *Le Petit Parisien*, "La Grande Semaine du Touring-Club," January 29, 1909, p. 4; *La Presse*, "La Grande Semaine du T.C.F.," January 29, 1909, p. 1; Casella, George, "Les Sports d'Hiver," *Le Sport Universel Illustré*, February 14, 1909, p. 111.
118. Baron Boilley, Françoise, *Marie Marvingt: A l'Aventure du Sport* (L'Harmattan, 2013), p. 48.
119. *Journal des Débats Politiques et Littéraires*, "Alpinisme," January 29, 1909, p. 3; *L'Est Républicain*, "Au Concours de Skis de Chamonix," January 31, 1909, p. 2.
120. Casella, Georges, "'Sportswomen' d'Hiver à Chamonix," *La Culture Physique*, March 15, 1909, pp. 201–202; *L'Est Républicain*, "Le Concours de Ski à Gérardmer," February 23, 1909, p. 1; Gil Blas, "Patinage," January 30, 1909, p. 4.
121. *L'Est Républicain*, "Les Fêtes de Skis à Gérardmer," Feburary 24, 1909, p. 2.
122. *Le Matin*, "Mlle Marvingt, Aviatrice, Gagne une Course de Skis," January 24, 1911, p. 3; *Le Figaro*, "Sports Divers à Chamonix," January 25, 1911, p. 7.
123. Gil Blas, "Les Sports d'Hiver," February 16, 1911, p. 5; Drigny, Georges, "La Grande Semaine d'Hiver du T.C.F.," *Le Sport Universel Illustré*, January 9, 1910, p. 139; *Le Matin*, "Course de Luges à Plat Ventre," January 25, 1911, p. 3; *Le Journal [Paris]*, "Sports d'Hiver," January 23, 26, 27, 1911, p. 6; *La Presse*, "Le Championnat de Bobsleighs," January 26, 1911, p. 3.
124. *L'Est Républicain*, "Une Course Militaire Internationale de Ski," February 13, 1911, p. 1; Dalbanne, J., "Le Ve Concours International de Ski du 'Club Alpin Français,'" *Armée et Marine*, February 28, 1911, p. 246; *Le Matin*, "L'Aviatrice Mlle Marvingt a Gagné la Course de Dames," February 12, 1911, p. 3; *La Musette*, "Nos Echos," March 1911, p. 15; Le Journal [Paris], "La Course Militaire International de Skis du Lioran," February 12, 1911, p. 4.
125. *L'Avenir du Cantal*, "Mlle Marvingt à Aurillac," February 10, 1911; *La Musette* (Massif Central), "Nos Echos," March 1911, p. 15.
126. *Ibid.*
127. Casella, Georges, op. cit.
128. *Le Sport*, "La Grande Semaine à Gerardmer," February 19, 1910.
129. *Alrededor del Mundo*, "Biografía Extranjera Contemporánea: Mlle Marvingt," November 2, 1913, no. 753, p. 361; Baron Boilley, Françoise, op. cit.; *The Strand Magazine*, op. cit.
130. Marvingt, Marie, "Les Femmes et le Ski," in Louis Magnus et R. de la Fregeolière, *Les Sports d'Hiver* (Éditions Pierre Lafitte et Cie, 1911; Éditions Slatkine, 1979), pp. 176–181.
131. Casella, George, op. cit.; *Le Petit Parisien*, "Exploit d'Aviatrice: Mlle Marvingt Vole 53 Minutes," November 28, 1910, p. 2.
132. *L'Est Républicain*, "Roller Skating," February 26, 1911, p. 2.
133. *L'Illustration*, "Les Sports d'Hiver," February 6, 1909, no. 3441, p. 95; *L'Illustration*, "Course de Dames en Patin-Ski à Chamonix," February 9, 1935, no. 4797, p. 170.
134. *L'Est Républicain*, "Le Rallye Ballon-Automobile," September 13, 1924, p. 4.
135. Marvingt, Marie, postcard to Marthe Roumier, February 12, 1909.
136. *La Sfaxienne*, "Mademoiselle Marvingt à Sfax," April 15, 1924, p. 1; Chevalier, Frédérique, "Marie Marvingt, la Fiancée des Airs (1875–1963)," *Les Grandes Aventurières* (City Editions, 2007), pp. 85–95.
137. Chevalier, Frédérique, op. cit.; Valode, Philippe, "Marie Marvingt, l'Inégalable Casse-Cou," *Les Grandes Scandaleuses* (First Editions, 2015), pp. 227–235.
138. Myriel, "Comment J'ai Gagné la Coupe Léon Auscher," *Le Sport*, March 5, 1910.
139. *Le Petit Parisien*, "La Grande Semaine du Touring-Club," January 29, 1909, p. 4.
140. Laget, op. cit.
141. *Le Radical*, "Le Championnat de France des Bobsleighs," January 26, 1911, p. 5; *Le Matin*, "Sports d'Hiver," January 26, 1913, p. 7; *Le Journal [Paris]*, "Le Patinage en Savoie," January 20, 1913, p. 4.
142. *Le Journal [Paris]*, "La Coupe de France de Ski," January 29, 1928, p. 5.
143. *La Vie au Grand Air*, "Ecole de Ski," March 15, 1913, p. 187; Ackerman, Gordon, "Fiancée of Danger," *Sports Illustrated*, June 26, 1961, pages 61–64; Allen, E. John B., *The Culture and Sport of Skiing: From Antiquity to World War II* (University of Massachusetts Press, 2007), p. 177; Marchand, Robert, op. cit.
144. *L'Aero*, "L'Ecole Marvingt à Megève," January 31, 1914, p. 6.
145. Marvingt, Marie, "Sur les Alpes l'Hiver: Au Dessus de Mt Blanc en Avion—La Saison à Chamonix," *Lectures Pour Tous*, January 1929, pp. 46–53.
146. Ackerman, Gordon, op. cit.
147. Allen, E. John B., *Historical Dictionary of Skiing* (2011), pp. 125–126.
148. Pfeiffer, Doug, "Musée Dauphinois: La Grande Histoire du Ski," *Skiing Heritage Journal*, March 2004, pp. 33–37.
149. Myriel, "Le Ski," *Éclair de l'Est*, March 21, 1909.
150. Marvingt, Marie, "Winter Sports at Chamonix," *The Times [London)]*, February 2, 1909, p. 17; Marvingt, Marie, "The Alps in Winter, I," *The Times [London]*, December 14, 1909; Marvingt, Marie, "The Alps in Winter, II," *The Times [London]*, December 28, 1909.
151. *L'Est Républicain*, "Journée d'Excursion," February 16, 1910, p. 3.
152. *Ibid.*
153. Denning, Andrew, *Skiing Into Modernity: A Cultural and Environmental History* (2014), p. 46.
154. Marvingt, Marie, "Les Femmes et le Ski," in Louis Magnus et R. de la Fregeolière, *Les Sports d'Hiver* (Éditions Pierre Lafitte et Cie, 1911; Éditions Slatkine, 1979), pp. 176–181.
155. Rio, Armand, op. cit.
156. Rio, Armand, op. cit.
157. *L'Aéro*, "Académie des Sports," November 6, 1911, p. 6; *Journal des Débats Politiques et Littéraires*, "Sports Divers," November 12, 1911, p. 3; *Le Temps*, "L'Académie des Sports," 7 novembre 1911, p. 5; *Le Gaulois*, "L'Académie des Sports," November 11, 1911, p. 4; *Gil Blas*, "L'Académie des Sports," November 11, 1911, p. 4; *L'Echo Sportif du Centre et de l'Ouest*, "Une Carrière Féminine Sportive," 1914; et alia.
158. Balitrand, Suzanne, "Le Beau Voyage de Mademoiselle Marvingt," *Eve*, no. 307, August 15, 1926, p. 7; *La Nouvelle République*, "Marie Marvingt la 'Fiancée du Danger' Est Morte," no. 5855, December 16, 1963, p. 1, C; *L'Echo de Bougie*, "Mademoiselle Marie Marvingt à Bougie," December 31, 1922, p. 2; Petit, Edmond, "Hommage à Marie Marvingt," *Forces Aériennes Françaises*, no. 200, February 1964, pp. 268–270; et alia.
159. *Chicago Star Publications*, "Mistress of All Sports," October 31, 1913, p. 11.

160. de Coubertin, Pierre, in *Revue Olympique*, 1912.

161. Fédération des Internationaux du Sport Français (FISF), "Promotion 1995 des 'Gloires du Sport,'" 1995.

162. Ackerman, Gordon, op. cit.

163. *Républicain Lorrain*, "A Nancy et à Metz, Marie Marvingt, la 'Fiancée du Danger' Dira Quels Sont les Buts Humanitaires de Sa Mission," October 19, 1949, p. 2.

164. Cajelot, Maurice, "Marie Marvingt, 'la Fiancée du Diable,' 100 ans," *Républicain Lorrain*, February 13, 1975, p. 4; J.P., "Marie Marvingt, Pilote d'Avion, Est Morte à 88 Ans: Championne en Tous Genres," *Le Monde*, December 17, 1963, p. 13; *L'Est Républicain*, "Il y a Vingt Ans Mourait Marie Marvingt," December 14, 1983; Cordier, Marcel, "Les Escales de Marie Marvingt," op. cit.; Féral, Roger, op. cit.

165. Codron, Charles, op. cit.

166. Rio, Armand, op. cit.; *The Strand Magazine*, op. cit.

167. Gygax, Georges, "La Vie Prodigieuse de Marie Marvingt, la 'Fiancée du Danger,'" *L'Illustrë*, May 1, 1958, pp. 40–42.

168. R.P., "Mademoiselle Marvingt," *Le Cri de Constantin*, June 28, 1924; *Annales Africaines*, "300 Conférences, 4.000 Kilomètres en Afrique du Nord par Mlle Marie Marvingt," May 15, 1925, pp. 309, 312–313.

169. Gygax, Georges, May 1, 1958, op. cit.

170. Faure, Odile, "Marie Marvingt, La Fiancée du Danger," *Massif Central Magazine*, March-April 1996, pp. 68–71; Jacquemart, Claude, *Le Figaro*, "L'Épopée des Aventurières du Ciel," December 26, p. 22.

171. *Le Petit Parisien*, "Le Miroir des Sports," October 27, 1920, p. 3.

172. Action Catholique des Femmes, op. cit.; "Marie Marvingt, Inventeur de l'Aviation Sanitaire, Est Morte à Nancy: Une Grande Sportive, Une Grande Française," newspaper clipping, December 16, 1963.

173. Saladin, Raymond, "Figures et Événements: Marie Marvingt," *Aviation Magazine*, March 24, 1955, pp. 12–13.

174. Boetsch, Jacques, op. cit.

175. Boselli, Elisabeth, "Marie Marvingt 1875–1963 'la Fiancée du Danger,'" *Pionniers*, July 1984, pp. 18–22.

176. Violette, Marcel, *La Vie au Grand Air*, in *Républicain Lorrain*, "A Nancy et à Metz, Marie Marvingt, la 'Fiancée du Danger' Dira Quels Sont les Buts Humanitaires de Sa Mission," October 19, 1949, p. 2.

177. Arnaud, Pierre, et Thierry Terret, *Histoire du Sport Féminin: Sport Masculin-Sport Féminin: Éducation et Société*, Tome 2 (L'Harmattan, 1996), p. 158.

178. Baily, Paul J., *Gender and Education in China: Gender Discourses and Women's Schooling in the Early Twentieth Century* (2007), p. 71.

179. Stewart, Mary Lynn, *For Health and Beauty: Physical Culture for Frenchwoman 1880s-1930s* (The Johns Hopkins University Press, 2001), pp. 170–171.

180. Rozet, George, "L'Avènement du Sport Féminin," *Lectures Pour Tous*, April 1, 1919, p. 1565.

181. Gayraud, Amélie, *Les Jeunes Filles d'Aujourd'hui* (1914), pp. 110–111.

182. France, Geo, "Marie Marvingt Espère Fêter Ses 90 Ans ... Dans la Lune," *Ouest-France*, February 23, 1955.

Chapter 3

1. J.G., "L'Avion et le Vélo: Deux Amours de Marie Marvingt, 'la Fiancée du Danger,'" *L'Est Républicain*, February 16, 1975, p. 3.

2. Rio, Armand, "La Fiancée du Danger," *Lecture Pour Tous*, April 1, 1913, vol. 15, no. 7, pp. 63–71; Micelli, Corinne, "Marie Marvingt: Pionnière des Ailes Qui Sauvent," *Air Actualités*, April 2006, pp. 58–61.

3. Myriel, "Mes Débuts d'Alpiniste," *Éclair de l'Est*, December 6, 1908, p. 2; Erbstein, Roland, with Jean Matt, "Cartes Postales d'Hier et d'Aujourd'hui: "En 1909, 'la Fiancée du Danger' Ouvrait aux Femmes la Voie des Airs," *L'Est Républicain*," March 23, 1980; Faure, Odile, "Marie Marvingt, La Fiancée du Danger," *Massif Central Magazine*, March-April 1996, pp. 68–71; Branchu, Marc, "Les Cent Vies de Marie Marvingt," *Magazine Air France*, no. 99, July 2005, pp. 42–44, 46.

4. Ackerman, Gordon, "Fiancée of Danger," *Sports Illustrated*, June 26, 1961, pages 61–64.

5. Dollfus, Charles, *The Orion Book of Balloons* (The Orion Press, 1960).

6. Beaubois, Henry, "Ceux Qui Ouvrirent la Route au Ciel: Marie Marvingt," *Revue Aéronautique de France*, August 13, 1937; Gygax, Georges, "La 'Fiancée du Danger': La Vie Prodigieuse de Marie Marvingt," *L'Illustré*, April 24, 1958, pp. 34–36.

7. *Le Radical*, "Une Femme Aéronaute," November 27, 1909, p. 3; Bans, Georges, "Aéronautes Contemporains: Mademoiselle Marie Marvingt," *L'Aérophile*, November 15, 1909, vol. 17, no. 22, p. 505

8. Ackerman, Gordon, op. cit.

9. *L'Aérophile*, "Bulletin des Ascensions," September 1907, p. 263; *L'Aérophile*, "Ascensions au Parc de L'Aéro-Club de France," May 15, 1909, p. 238; *L'Aérophile*, "Les Ascensions au Parc de l'Aéro-Club de France," August 15, 1910, p. 383; *L'Est Républicain*, "Pendant le Congrès d'Aéronautique," September 21, 1909, p. 2.

10. Myriel, "Ma Première Ascension Comme Pilote," *Éclair de l'Est*, September 29, 1909.

11. *L'Est Républicain*, "Le Concours de Ballons Libres," September 25, 1909, p. 1.

12. Daniel, Florence, "Une Esquisse Inédite de Friant Ainsi Dédicacée: 'Temps Héroïque de l'Aviation, Somer de Passage à Nancy, à Marie Marvingt, Hommage d'E. Friant,'" *La Gazette Lorraine*, December 2005, no 44.

13. *Le Temps*, "Le Congrès Aéronautique International," September 25, 1909, p. 2.

14. Myriel, op. cit.

15. Méchelle, Ginette, interview, June 23, 1982.

16. *L'Est Républicain*, "Après le Concours de Ballons Libres: Ce Que Dit Mlle Marvingt," September 26, 1909, p. 2.

17. *Ibid.*

18. Dollfus, Charles, op. cit. p. 71.

19. Myriel, op. cit.

20. Tessier, Roland, *Femmes de l'Air* (Flammarion, 1948), p. 21.

21. *La Culture Physique*, "Une Belle École d'Énergie pour la Femme: La 'Stella,'" August 15, 1911, p. 510.

22. Baron Boilley, Françoise, *Marie Marvingt: A l'Aventure du Sport* (L'Harmattan, 2013), p. 61.

23. *L'Est Républicain*, "Les Deux Ascensions de Dimanche au Parc Aéronautique," October 10, 1909, p. 2; *L'Est Républicain*, "L'Ascension de Mlle Marvingt," October 12, 1909, p. 2.

24. *L'Est Républicain*, "Ascension au Ballon Libre," October 19, 1909, p. 2.

25. *L'Est Républicain*, "Au Parc d'Aéronautique," October 11, 1909, p. 1; *L'Est Républicain*, "Grande Fête Aéronautique du 15 Mai," May 14, 1910, p. 2; *L'Est Républicain*, "Prochaine Ascension de Mlle Marvingt," June 21, 1910, p. 2.

26. Bans, Georges, "Aéronautes Contemporains:

Mademoiselle Marie Marvingt," *L'Aérophile*, November 15, 1909, vol. 17, no. 22, p. 505.

27. Laurent, Michel-Yves, "Marie Marvingt, Surnommée 'Marie Casse-Cou,' Traversa au Début du Siècle, la Mer du Nord en Ballon; Elle Pilote Aujourd'hui à 86 Ans un Hélicoptère à Réaction," *L'Aurore* or *Paris Press*, November 2, 1960, no. 16.

28. *L'Est Républicain*, "Ascension à la Chiennerie," October 27, 1909, p. 3.

29. *The Times [London]*, "Across the North Sea in a Balloon," October 28, 1909, p. 7; *Le Figaro*, "Un Voyage Mouvementé," October 28, 1909, p. 1; *Yorkshire Evening Post*, "Lady's Adventurous Voyage: Journey of a Thousand Miles Ends in Suffolk," October 28, 1909, p. 3; *Illustration*, "Une Femme Traverse la Manche en Ballon," November 6, 1909, no. 3480, p. 335; *Western Gazette [Somerset, England]*, "Balloonists Cross Channel: Lady's Remarkable Adventure," October 29, 1909, p. 7; *Journal des Débats Politiques et Littéraires*, "Aérostation," October 29, 1909, p. 3; *Le Temps*, "Une Femme Traverse la Manche en Ballon," October 29, 1909, pp. 3–4; Chambe, René, *Histoire de l'Aviation* (Flammarion, 1980), p. 256; *L'Est Républicain*, "Une Ascension Mouvementée de Mlle Marvingt," October 28, 1909, p. 1; *L'Est Républicain*, "Une Ascension Mouvementée de Mlle Marvingt (suite)," October 29, 1909, p. 2; Association de Documentation Aéronautique, "Marie Marvingt (1875–1963), Issy-les-Moulineaux, École Marie Marvingt, December 12, 1988; *L'Aurore*, "La Manche en Ballon," October 29, 1909, p. 3; *Bibliography of Aeronautics* (1921), p. 846; *Le Gaulois*, "Ascension Aventureuse, de Nancy en Angleterre," October 28, 1909, p. 3; *Gil Blas*, "Nouvelles de Partout: Londres," October 28, 1909, p. 3; Brion, Hélène, "Aviation," *L'Action Féministe*, March-April 1915, p. 3; *Burnley [England] Express*, "Perilous Balloon Journey," October 30, 1909, p. 8; *Le XIXème Siècle*, "Aéronautique," October 30, 1909, p. 4; *Dublin Daily Express*, "Remarkable Flight," October 28, 1909, p. 5; "Eine Sturmfahrt im Ballon über die Norsee wehrend der Nacht von Nancy nach Southwold," *Deutsche Zietschr. Luftsch*, September 21, 1910; *The [Perth] Daily News*, "Woman Aeronaut Blown Out to Sea, Travels From France to England and Has Narrow Escape, Leap from the Car," November 29, 1909, p. 2; Shayler, David J., and Ian A. Moule, *Women in Space—Following Valentina* (Springer/Praxis, 2006), p. 11; Lecornu, Joseph Louis, *La Navigation Aérienne: Histoire Documentaire et Anecdotique* (1913), p. 306; *La Vie au Grand Air*, "Par-dessus la Mer du Nord," November 6, 1909, p. 7; *Le Radical*, "De Lorraine en Angleterre Par-Dessus la Manche," October 30,1909, p. 3; *The [Adelaide, Australia] Register*, "Remarkable Adventure," December 23, 1909, p. 12; Marck, Bernard, *Dictionnaire Universel de l'Aviation* (Tallendier, 2005), pp. 688–689; Mortane, Jacques, *La Femme Dans le Sport et l'Aviation* (Éditions J. Dupuis Fils et Cie, 1937), p. 23; Planck, Charles E., *Women With Wings* (1942), p. 302; Chambe, René, *Histoire de l'Aviation* (Flammarion, 1980), p. 256; *The Guardian*, "Balloon in a Gale, Perilous Voyage From France to Suffolk; Woman Aeronaut's Adventure," October 28, 1909, p. 7; et alia.

30. Marck, Bernard, *Dictionnaire Universel de l'Aviation* (Tallendier, 2005), pp. 688–689; Mauchaussée, Jean, "Marie Marvingt, Reine de l'Air," *Prestige*, October 1, 1959, no. 3, p. 1.

31. *La Vie au Grand Air*, "Par-Dessus la Mer du Nord," November 6, 1909, p. 7.

32. *The Times [London]*, "Across the North Sea in a Balloon," October 28, 1909, p. 7; *London Daily News*, "Lady's Thrilling Story," October 28, 1909, p. 5. Marck, Bernard, *Dictionnaire Universel de l'Aviation* (Tallendier, 2005), pp. 688–689.

33. *The [Melbourne] Australasian*, "Australians Abroad," December 4, 1909, p. 39.

34. *L'Est Républicain*, "Une Ascension Mouvementée de Mlle Marvingt (suite)," October 29, 1909, p. 2.

35. *Ibid.*

36. G.M., "Audacieux Voyage en Ballon: Interview de Mlle Marvingt," *L'Éclair de l'Est*, October 30, 1909.

37. *Ibid.*

38. *Ibid.*

39. Marvingt, Marie, "La Mer du Nord Traversée en Ballon en Pleine Tempête au Milieu de la Nuit," *Le Petit Journal*, November 22, 1909, no. 17132, pp. 1–2; Marvingt, Marie, "Traversée Nocturne de la Mer du Nord en Ballon," *L'Aérophile*, January 1, 1910, vol. 18, no. 1, pp. 17–20.

40. The *[Rockhampton, Australia] Capricornian*, "Balloonist's Thrilling Voyage," January 1, 1910, p. 28to come.

41. Ackerman, Gordon, op. cit.; Talbott, Earl G., "Marie Marvingt, 'Fiancée of Danger': Flirted With Death 88 Years," *New York Herald Tribune*, December 16, 1963.

42. *London Daily News*, "Lady's Thrilling Story," October 28, 1909, p. 5.

43. *London Evening News*, "Aerial Adventure," October 28, 1909, p. 16.

44. *London Evening News*, "Aerial Adventure," October 28, 1909, p. 16.

45. Ackerman, Gordon, op. cit.

46. *Ibid.*

47. Gautier, Émile, ed., *L'Année Scientifique et Industrielle 1909* (1910), p. 92.

48. *La Vie au Grand Air*, "Par-dessus la Mer du Nord," November 6, 1909, p. 7.

49. Bans, Georges, "Aéronautes Contemporains: Mademoiselle Marie Marvingt," *L'Aérophile*, vol. 17, no. 22, November 15, 1909, p. 505.

50. *L'Est Républicain*, "Ce Nouvel Accident ne Décourage pas Mlle Marvingt," May 15, 1910, p. 1.

51. *Ibid.*

52. *L'Est Républicain*, "La Fête Aéronautique de Dimanche," May 16–17, 1910, pp. 1–2.

53. Marvingt, Marie, "Trois Morts Mettent en Deuil les Ailes de Lorraine: Hommage d'une 'Vieille Tige,'" *Dimanche Éclair*, September 17, 1950, p. 3.

54. *L'Est Républicain*, "La Fête Aéronautique de Dimanche," op. cit.

55. *Le Petit Parisien*, "Petites Nouvelles Sportives," June 18, 1910, p. 4; *L'Aérophile*, "Brevet de Pilote Aéronaute," July 15, 1910, p. 334; *La Presse*, "Aviation," June 18, 1910, p. 3; *L'Aérophile*, "Brevets de Pilote d'Aérostats," August 15, 1910, p. 383; Aéro-Club de France, Certificat Attestant au Brevet, March 14, 1914.(The certificate says that license number 145 was awarded June 16, 1910. It is unclear why the certificate is dated later.)

56. *La Revue Aérienne*, "Nos Prix," July 10, 1910, p. 412.

57. *L'Est Républicain*, "La Fête Aéronautique de Dimanche," op. cit.; Marvingt, Marie, "Trois Morts Mettent en Deuil les Ailes de Lorraine: Hommage d'Une 'Vieille Tige,'" *Dimanche Éclair*, September 17, 1950, p. 3.

58. Bergeret, Tonin, "Les Ballons en Lorraine: L'Ascension de Mlle Marvingt," *Le Sport*, May 21, 1910, no. 85, pp. 1, 2.

59. *Ibid.*

60. *Ibid.*

61. *L'Est Républicain*, "Nouvelle Ascension de Mlle Marvingt," May 18, 1910, p. 2.

62. *Ibid.*

63. *Le Journal Amusant*, "Le Grand Prix de l'Aéro-Club de France," July 2, 1910, p. 14; *L'Aérophile*, "Le Sixième Grand Prix de l'Aéro-Club de France," July 15, 1910, pp. 330–332.; *Le Temps*, "Le Grand-Prix de l'Aéro-Club de France," June 28, 1910, p. 3; *Le Temps*, "Le Grand-Prix de l'Aéro Club de France," June 29, 1910, p. 4; Tessier, Roland, *Femmes de l'Air* (Flammarion, 1948), p. 22; *L'Aurore*, "Le Grand-Prix de l'Aéro-Club de France," June 25, 1910, p. 3; *Gil Blas*, "Aéronautique," June 25, 1910, p. 4; *Le XIXème Siècle*, "Le Grand Prix de l'Aéro-Club," June 30, 1910, p. 4; *Le Radical*, "Le Grand Prix de l'Aéro-Club," June 23, 25, 26, 1910, p. 5; *Le Rappel*, "Aéronautique: Le Grand-Prix de l'Aéro-Club," June 30, 1910, p. 4; *Le Matin*, "Le Grand Prix de l'Aéro-Club de France," June 27, 1910, p. 5.

64. *Le Journal [Paris]*, "Les Grandes Manifestations Sportives: Le Concours de Distance de l'Aéro-Club," June 26, 1910, p. 5.

65. *Le Journal [Paris]*, "Le Concours de l'Aéro-Club," June 27, 1910, p. 7.

66. *L'Est Républicain*, "Les Ascensions de Mlle Marvingt," June 28, 1910, p, 2; *La Lanterne*, "Le Grand Prix de l'Aéro-Club de France," June 30, 1910, p. 4.

67. *L'Est Républicain*, "Le Concours de Ballons Libres," September 25, 1909, p. 1.

68. *Le Journal [Paris]*, "Les Grandes Manifestations Sportives: Le Concours de Distance de l'Aéro-Club," June 26, 1910, p. 5.

69. *L'Aéro*, "A l'Aéro-Club de France: Journée des Sociétés Affiliées," July 7, 1912, p. 3; *L'Aéro*, "La Journée des Sociétés Affiliées—les Atterrisages," July 9, 1912, p. 2; *L'Aéro*, "Paris-Bruxelles en Sphérique," July 31, 1912, p. 1; *L'Aero*, "Le Grand Prix de l'Aero-Club de France," September 23, 1912, p. 1; *L'Aéro*, "Le Prix de Mme de Salmegeane," September 30, 1912, p. 2; *L'Aéro*, "Les Sphériques," October 19, 1912, p. 2; *L'Aéro*, "Propos en l'Air," December 13, 1912, p. 1; *L'Aérophile*, "Concours d'atterrissage," August 1, 1912, p. 358; *L'Aérophile*, "Ascensions au Parc de l'Aéro-Club de France," August 15, 1912, p. 383; *L'Aérophile*, "Le 8e Grand Prix de l'Aéro-Club de France," October 1, 1912, p. 450–451; *L'Aérophile*, "Ascensions au Parc de l'Aéro-Club de France," October 1, 1912, p. 455; *L'Aérophile*, "Le Prix de 'Mme de Salmegean,'" October 15, 1912, p. 476; *L'Aérophile*, "Ascensions au Parc de l'Aéro-Club de France," November 1, 1912, p. 503; *L'Aérophile*, "Ascensions au Parc de l'Aéro-Club de France," 1 janvier 1913, pp. 20–21; *Le Journal Amusant*, "Mlle Marvingt, Aéronaute, Va en Sphérique de Paris à Bruxelles," August 10, 1912, p. 14; *Le Temps*, "Les Voyages en Sphériques," July 30, 1912, p. 5; *Le Temps*, "Les Voyages en Sphériques," August 1, 1912, p. 5; *Le Temps*, "Le Prix Salmagean," September 30, 1912, p. 5; Le Gaulois, "Les Ballons de l'Aéro-Club," July 31, 1912, p. 6; Le Gaulois, "La Conquête de l'Air," September 30, 1912, p. 3; *Gil Blas*, "Aéronautique: le Prix de Distance," October 1, 1912, p. 5; Le XIXème Siècle, "Aéronautique: Le Prix Salmagean," October 1, 1912, p. 3; *Le Radical*, "Raid d'une Aviatrice en Ballon Sphérique," July 29, 1912, p. 3; *Le Radical*, "Mlle Marvingt, Avec Deux Passagers, s'Élève à 3,000 Mètres," September 19, 1912, p. 6; Marck, Bernard, *Dictionnaire Universel de l'Aviation* (Tallendier, 2005), pp. 688–689; *Le Figaro*, "De Paris à Bruxelles, les Ballons de l'Aéro-Club de France," July 31, 1912, p. 5.

70. Rio, Armand, op. cit.

71. *Le Matin*, "Mlle Marvingt Dans les Nuages," June 14, 1913, p. 6; *L'Aéro*, "Les Sphériques," August 6, 1913, p. 4.

72. *L'Aéro*, "La Coupe des Sociétés Affiliées," July 19, 1913, p. 2; *La Vie Sportive du Nord et du Pas-de-Calais*, "Les Fêtes Aérostatiques de Lille," July 19, 1913, p. 5; *La Vie Sportive du Nord et du Pas-de-Calais*, "La Fête Aérostatique de Lille," July 26, 1913, p. 2.

73. *L'Aéro*, "Les Sphériques," August 21, 1913, p. 2.

74. Florentin, Mademoiselle, in Jacqueline Maire, letter, July 23, 1982.

75. *L'Aéro*, "Le Grand Prix de l'Aéro Club," July 6, 1914, p. 1; *L'Aéro*, "Le Grand Prix de l'Aero-Club," July 9, 1914, p. 5; *L'Aéro*, "Les Grandes Épreuves de Sphériques: Le Xème Grand Prix de l'Aéro Club de France," July 20, 1914, p. 3. *L'Aérophile*, "Le 10e Grand Prix de l'Aéro-Club de France," July 15, 1914, p. 334; *L'Aérophile*, "Le Dixième Grand Prix de l'Aéro-Club de France," August 1, 1914, pp. 352–357; *Gil Blas*, "Le Grand Prix de l'Aéro-Club de France," July 19, 1914, p. 5; *Le XIXème Siècle*, "Le Grand Prix des Ballons, July 19, 1914, p. 3; *Le Radical*, "Le Grand Prix de l'Aéro Club," July 9, 11, 19, 21, 1914, p. 5; *Le Matin*, "Vingt-Deux Ballons Sphériques s'Élèvent des Tuileries pour le Grand Prix de Distance de l'Aéro-Club," July 20, 1914, p. 5; *Le Journal [Paris]*, "Vingt-Quatre Ballons Partent Aujourd'hui des Tuileries," July 19, 1914, p. 3; *Le Figaro*, "Le Grand Prix de l'Aéro-Club de France," July 19, 1914, p. 5.

76. *Gil Blas*, "Le Grand Prix de l'Aéro-Club de France," July 20, 1914, p. 6.

77. *Journal des Débats Politiques et Littéraires*, "Le Grand Prix de l'Aéro Club," July 21, 1914, p. 4; *The [Meadville, PA] Evening Republican*, "Serious Accident Mars Balloon Race," July 20, 1914, p. 4.

78. *L'Ouest Éclair [Rennes]*, "Le Grand Prix de l'Aéro-Club," July 21, 1914, p. 3; Ackerman, Gordon, "Fiancée of Danger," *Sports Illustrated*, June 26, 1961, pages 61–64; *Journal des Débats Politiques et Littéraires*, "Le Grand Prix de l'Aéro Club," July 21, 1914, p. 4; D'Aubigny, Eugène, "La premiéreTraversée de la Manche par une Aéronaute," *L'Aéro*, July 24, 1914, p. 1; *Le Journal [Paris]*, "Le Grand-Prix de l'Aéro-Club," July 21, 1914, p. 9.

79. Butler, Frank Hedges, *Fifty Years of Travel by Land, Water, and Air* (1920), p. 411.

80. *Le Temps*, "Le Grand-Prix de l'Aéro-Club de France," July 21, 1914, pp. 5, 6; *Le Gaulois*, "Le 10e Grand Prix de l'Aéro-Club de France," July 19, 1914, p. 4; *Le Gaulois*, "Le 10e Grand Prix de l'Aéro-Club de France," July 21, 1914, p. 5.

81. Marvingt, Marie, "En Ballon, de Paris à la Mer d'Irlande: Mlle Marvingt Raconte à *L'Éclair de l'Est* Son Intrépide Voyage Aérien," *L'Éclair de l'Est*, July 22, 1914.

82. *The Guardian*, "French Balloon Race, Descents Near Cardigan Bay," July 21, 1914, p. 10; *L'Homme Libre*, "Le Grand Prix de l'Aéro-Club: Mlle Marvingt Atterrit Sans Encombre," July 21, 1914, p. 2; D'Aubigny, Eugène, "La Premiére Traversée de la Manche par une Aéronaute," *L'Aéro*, July 24, 1914, p. 1; *The Leader [Orange, Australia]*, "Balloonists in a Gale, 60 Miles an Hour," September 11, 1914, p. 1.

83. D'Aubigny, Eugène, "La Premiére Traversée de la Manche par une Aéronaute," *L'Aéro*, July 24, t 1914, p. 1.

84. Marvingt, Marie, "En Ballon, de Paris à la Mer d'Irlande: Mlle Marvingt Raconte à *L'Éclair de l'Est* Son Intrépide Voyage Aérien," *L'Éclair de l'Est*, July 22, 1914.

85. *L'Aérophile*, "Le 10e Grand Prix de l'Aéro-Club de France," November 1–15, 1914, pp. 409–410.

86. Marvingt, Marie, "En Ballon, de Paris à la Mer d'Irlande: Mlle Marvingt Raconte à *L'Éclair de l'Est* Son Intrépide Voyage Aérien," *L'Éclair de l'Est*, July 22, 1914.

87. Marvingt, Marie, "En Ballon, de Paris à la Mer d'Irlande: Mlle Marvingt Raconte à *L'Éclair de l'Est* Son Intrépide Voyage Aérien," *L'Éclair de l'Est*, July 22, 1914.

88. Marvingt, Marie, "En ballon, de Paris à la mer d'Irlande: Mlle Marvingt raconte à *L'Éclair de l'Est* son intrépide voyage aérien," *L'Éclair de l'Est*, 22 juillet 1914; Cordier, Marcel, "Les Escales de Marie Marvingt," *Leurs Demeures en Lorraine: Tome II* (Éditions Pierron, 1983), pp. 162–177.

Chapter 4

1. Lilienthal, Otto (1896), in *The American Magazine*, vol. 63, 1906, p. 618.
2. Mauchaussée, Jean, "Marie Marvingt, Reine de l'Air," *Prestige*, October 1, 1959, no. 3, p. 1.
3. Marck, Bernard, "Ces Merveilleux Fous Volants," *Le Figaro Histoire*, August-September 2015, no. 21, p. 48.
4. Myriel, "Un Nouveau Monoplan à Champigneulles," *Éclair de l'Est*, August 4, 1909.
5. Myriel, "La Grande Semaine d'Aviation," *L'Éclair de l'Est*, August 27, 1909.
6. Marck, Bernard, op. cit.
7. Pelletier, George E., "Wings for the Wounded," *Oakland Tribune*, September 17, 1939, p. 78; *Lincoln, [NE] Sunday Journal and Star*, September 17, p. 33, and others.
8. Myriel, "La Grande Semaine d'Aviation," *L'Éclair de l'Est*, August 27, 1909.
9. *Ibid.*
10. *Ibid.*
11. *Ibid.*
12. *Ibid.*
13. *Ibid.*
14. *Ibid.*
15. *L'Aérophile*, "Roger Sommer à Nancy," October 1, 1909, p. 436; *L'Est Républicain*, "A Jarville Aviation," September 11, 1909, p. 2.
16. *L'Est Républicain*, "A Jarville Aviation," op. cit.
17. *Ibid.*
18. *Ibid.*
19. *L'Est Républicain*, "Mlle Marvingt—Apprenti-Pilote," December 1, 1909, p. 2; *Kalgoorlie [Australia] Miner*, "Two Hundred Flying Men; Growing Army of Aviators; 'Stars' and Their Pupils," May 5, 1910, p. 8; *L'Est Républicain*, "Mlle Marvingt, Aviatrice," December 28, 1909, p. 2; *L'Est Républicain*, "Latham et Mlle Marvingt," January 9, 1910, p. 1; *L'Aurore*, "A Mourmelon," November 24, 1909, p. 3; "*Bilder aus aller Welt*," Vienna newspaper, 1910.
20. *L'Aéro*, "Les Conférences de Mlle Marvingt," October 10, 1912, p. 1.
21. *Le Matin*, "Sous le Ciel d'Egypte, un Aviateur Tombe de 45 Mètres," February 2, 1910, p. 1; *L'Aurore*, "Aéronautique," February 3, 1910, p. 3.
22. Grahame-White, Claude, in collaboration with Harry Harper, *With the Airmen* (H. Frowde, Hodder and Stoughton, 1913), p. 47.
23. *Indépendant*, "Mlle Marie Marvingt à Pau: Deux Causeries Littéraires et Sportives," June 16, 1927.
24. *L'Aérophile*, "Au Camp de Châlons," December 15, 1909, p. 554; *[Adelaide] Chronicle*, "Notes by 'Aerial,'" May 14, 1910, p. 20; *L'Aérophile*, "Les Aéroplanes," September 1, 1910, p. 398; *L'Ouest Éclair [Rennes]*, "La Première Aviatrice Brevetée," October 23, 1910, p. 1; *Le Journal Amusant*, "Commission d'Aviation," December 3, 1910, p. 14; Thirion, M., "Hommage de l'Aéro-Club de l'Est à Marie Marvingt," *L'Est Républicain*, December 18, 1963; *L'Est Républicain*, "Mlle Marvingt, Pilote Aéronaute," June 18, 1910, p. 2; *L'Est Républicain*, "Mlle Marvingt Passe Brillamment Son Brevet de Pilote," October 23, 1910, p. 2; Le Matin, "Chez les Hommes Oiseaux," July 29, 1910, p. 4; *Le Matin*, "Mlle Marvingt, Pilote Brevetée," October 23, 1910, p. 5; *La Revue Aérienne*, "A Châlons," December 25, 1909, pp. 775–776; *Le Figaro*, "Les Promeneuses de l'Air," September 21, 1907, p. 1.
25. *Oakland Tribune*, "Daring Birdwomen Invade Cloudland," 23 July 1911, p. 12; Hoyt, Helen, "Helen Hoyt Has Instances of Daring of Women Aviators," *Gazette and Bulletin [Williamsport, PA]*, August 8, 1911, p. 7.
26. Chevalier, Frédérique, "Marie Marvingt, la Fiancée des Airs (1875–1963)," *Les Grandes Aventurières* (City Editions, 2007), pp. 85–95.
27. *L'Est Républicain*, "Latham et Mlle Marvingt," January 9, 1910, p. 1; *Rochester [NY] Democrat*, "Daring Aviatresses Whose Exploits Are Winning Fame," July 16, 1911, p. 14.
28. Archdeacon, Ernest, "Mademoiselle Marvingt: Une Sportwoman Extraordinaire," *La Revue Aérienne*, December 25, 1910, pp. 702–706.
29. The Motor, *The Aero Manual* (1909).
30. Chancel, Jules, "A la Conquête du Brevet d'Aviateur," *Lecture Pour Tous*, January 1911, pp. 355–366.
31. Myriel, "Comment On Devient Pilote-Aviateur," *Le Sport*, November 24, 1910, no. 102, p. 1.
32. *Le Journal [Paris]*, "Dernières Nouvelles Sportives," August 7, 1911, p. 4.
33. Myriel, "Comment On Devient Pilote-Aviateur," op. cit.
34. *L'Année Mondiale Illustrée*, "Aéronautique," 1914, p. 521.
35. Daurat, Michel, "A Quatre-Vingts Ans, Marie Marvingt Demeure Toujours la Fiancée du Danger," *Lorraine Magazine*, no. 9, February 1955, pp. 10–13; Grandidier, Gilbert, "Marie Marvingt (1875–1963), 'La Fiancée du Danger' Ouvre aux Femmes la Voie des Airs," newspaper article, May 1987; Erbstein, Roland, in collaboration with Jean Matt, "Cartes Postales d'Hier et d'Aujourd'hui: "En 1909, 'la Fiancée du Danger' Ouvrait aux Femmes la Voie des Airs," *L'Est Républicain*," March 23, 1980.
36. *Fémina*, "Au Pays de l'Aviation," August 1, 1910, p. 402.
37. *Le Matin*, "Deux Aviateurs se Tuent," December 29, 1910, pp. 1–2.
38. *Ibid.*
39. *Ibid.*
40. *Le Matin*, "Deux Aviateurs Voulaient Vaincre le Vent; Ils Sont Morts d'une Mort Affreuse," May 19, 1911, pp. 1–2.
41. *Oakland Tribune*, "Daring Birdwomen Invade Cloudland," July 23, 1911, p. 12; *Rochester [NY] Democrat*, "Daring Aviatresses Whose Exploits Are Winning Fame," July 16, 1911, p. 14.
42. Zeyons, Serge, "Marie Marvingt, Une femme d'Exception," *Timbres Magazine*, no. 151, December 2013, pp. 80–82.
43. Woodhouse, Henry, "Women of the Air," *Outdoor World & Recreation*, May 1913, vol. 48, pp. 305–309.
44. Earhart, Amelia, *The Fun of It* (1932), p. 138.
45. Fleming, Candace, *Amelia Lost: The Life and Disappearance of Amelia Earhart* (2012), p. 109
46. *L'Aérophile*, "Hélène Boucher," vol. 42, no. 12, December 1934. p. 366.
47. *The Catholic Press [Sydney, Australia]*, "Flying Women: Their Fearlessness and Nerve," May 26, 1910, p. 18.
48. Marvingt, Marie, postcard, no date.
49. *Association Général Aéronautique*, "Une Conférence de Mlle Marvingt," March 1913, p. 38.

50. Chancel, Jules, op. cit.
51. *Le Petit Parisien*, "Mlle Marvingt, Détentrice de la Coupe d'Aviation Fémina," November 29, 1910.
52. *La Revue Aérienne*, "A Reims," December 25, 1909, pp. 776–777; *L'Aérophile*, "Les Aéroplanes," January 1, 1910, p. 6.
53. *La Presse*, "Ca et Là," August 11, 1910, p. 3; *L'Aérophile*, "Les Aéroplanes," October 1, 1910, p. 436; *L'Aérophile*, "Les Aéroplanes," October 15, 1910, pp. 458–463; *Le Petit Parisien*, "A Mourmelon," September 6, 1910, p. 4.
54. *Le Matin*, "Chez les Hommes Oiseaux," July 29, 1910, p. 4.
55. Marvingt, Marie, "Mon Vol le Plus Émouvent," *La Vie au Grand Air*, no. 638, December 10, 1910, p. 897.
56. *L'Aérophile*, "A Mourmelon," November 10, 1910, p. 486; Beaubois, Henry, "Ceux Qui Ouvrirent la Route au Ciel: Marie Marvingt," *Revue Aéronautique de France*, August 13, 1937; Le Gaulois, "Nouveaux Pilotes," November 13, 1910, p. 4.
57. Beaubois, Henry, "Ceux Qui Ouvrirent la Route au Ciel: Marie Marvingt," *Revue Aéronautique de France*, August 13, 1937; Daurat, Michel, op. cit.; et alia.
58. Charvet, Louis, *Petite Histoire de l'Aviation* (1942).
59. *Ibid.*
60. *L'Aérophile*, "Les Records Féminins d'Aviation," December 15, 1910, no. 24, p. 556;
Journal des Débats Politiques et Littéraires, "Aviation," October 24, 1910, p. 3.
61. Marvingt, Marie, "Mon Vol le Plus Émouvent," op. cit.
62. *L'Aérophile*, "Brevets de Pilotes-Aviateurs," December 1, 1910, p. 549; *L'Aérophile*, "Les Records Féminins d'Aviation," December 15, 1910, no. 24, p. 556; *L'Aérophile*, "Liste Alphabétique des Pilotes-Aviateurs Titulaires du Brevet de l'Aéro-Club de France," January 15, 1911, pp. 36, 40; Lassalle, Émile J., *Les Cent Premiers Aviateurs Brevetés au Monde et la Naissance de l'Aviation* (Nauticaero, 1960), p. 21; et alia.
63. Petit, Edmond, "Hommage à Marie Marvingt," *Forces Aériennes Françaises*, February 1964, no. 200, pp. 268–270.
64. Custis, John Trevor, "Women Make Real Advent Into Aerial Navigation," *The Washington Post*, December 18, 1910, p. S4.
65. *Ibid.*
66. *Ibid.*
67. *[Sydney] Sunday Times*, "World's Famous Aviators Write of Their Experiences in Learning to Fly," November 12, 1911, p. 27.
68. *Collier's*, "Sky Women," September 30, 1911, pp. 15, 17, 34.
69. Marvingt, Marie, "The Intoxication of Flight," *Collier's*, September 30, 1911, p. 15.
70. *Ibid.*; Pound, Richard, ed., *Quotations for the Fast Lane* (2013), p. 199.
71. Marck, Bernard, *Women Aviators* (Flammarion, 2013), pp. 14–21, 27, 39, 40, 43.
72. *L'Est Républicain*, "La Conférence de Mlle Marvingt: Deux Heures Dans les Airs," October 13, 1929, p. 3.
73. *L'Accessoire de l'Automobile* (1910), in Serge Laget and Françoise Laget, and Jean-Paul Mazot, with the collaboration of Elizabeth Foch, *Le Grand Livre du Sport Féminin* (FMT Éditions, 1982), p. 269.
74. Marchand, Robert, "Les Conquérants du Ciel," *Le Petit Écho de la Mode Magazine*, no. 17, April 25, 1964, pp. 46–49.
75. de Saint-Fégor, L., *Le Royaume de l'Air* (Société d'Édition et de Publications, 1910), pp. 338, 341.
76. *The [Perth] Daily News*, "Adventurous Voyages," December 19, 1910, p. 8.
77. *L'Équipe*, "Mlle Marvingt, Détentrice des Records Féminins en Aéroplane," octobre 1910 ; Heinmuller, John P.V., *Man's Fight to Fly* (Aero Print Co., 1945), p. 276; Frémy, Dominique et Michèle, *Quid* (Robert Laffont, 1988), p. 1166; Houard, Georges, "Marie Marvingt, Pilote et Propagandiste Exemplaire," *Les Ailes*, January 1, 1964, p. 21; *Evening News* [Sault Sainte Marie, Michigan], "Another Contender for Aviation Honors: Mme Marvingt," January 10, 1911, p. 6; *Aircraft*, "Foreign News," vol. 1, 1910, p. 393; *Astronautics and Aeronautics, 1963: Chronology on Science, Technology, and Policy* (1964), p. 479; Drigny, Georges, "La Fin de la Saison," *Le Sport Universel Illustré*, December 4, 1910, p. 783; *Stanberry [Missouri] Owl-Headlight*, "Breaks Airwomen's Records," January 31, 1911, p. 9; et alia.
78. *Le Journal [Paris]*, "Mlle Marvingt, Détentrice de la Coupe d'Aviation Fémina," November 29, 1910, p. 6; *Le Figaro*, "Aviation," November 29, 1910, p. 6.
79. *Indianapolis Star*, "Hopes to Win Prize for Sustained Flight," December 22, 1910, p. 2; *Gil Blas*, "Aviation: Une Aviatrice Concourt," November 28, 1910, p. 3; *Fémina*, "La Coupe d'Aviation 'Fémina,'" December 15, 1910, p. 680.
80. *Historische*, "Die Französische Fliegerin Mlle. Marvingt," 1910.
81. Welch, Roseanne, *Encyclopedia of Women in Aviation and Space* (1998), p. 138; *L'Est Républicain*, "Mlle Marvingt Aviatrice," November 28, 1910, p. 1; *Gil Blas*, "Mlle Marvingt vole 53 minutes," November 29, 1910, p. 3, 4; *Cairns [Australia] Post*, "Lady Sky Pilot," December 1, 1910, p. 3; *Aberdeen Journal*, "At Châlons," December 1, 1910, p. 11; *Sunderland Daily Echo and Shipping Gazette*, November 30, 1910, p. 1, 2; *Yorkshire Evening Post*, "Long Flight by a Woman," November 29, 1910, p. 4; Journal des Débats Politiques et Littéraires, "Aviation," November 29, 1910, p. 3.; *Le Temps*, "Une Aviatrice Vole Cinquante-Trois Minutes," November 29, 1910; *Cairns [Australia] Post*, "Lady Sky Pilot," December 1, 1910, p. 3; *The [Brisbane] Telegraph*, "Marie Marvingt," April 10, 1911, p. 12; Villard, Henry Serrano, *Contact! The Story of the Early Birds* (Crowell, 1969), p. 248; et alia.
82. *La Vie au Grand Air*, "Mlle Marvingt Portée en Triomphe," December 3, 1910, cover; de Laroche, Raymonde, "Recordwomen," *Je Sais Tout*, January 15. 1911, vol. 6, no. 72, pp. 801–809; Laget, Françoise and Serge, and Jean-Paul Mazot, with the collaboration of Elizabeth Foch, *Le Grand Livre du Sport Féminin* (FMT Éditions, 1982), p. 275.
83. *The Illustrated London News*, "Portraits and World's News," December 3, 1910, p. 862; *Chicago Star Publications*, "Breaks Airwomen's Record," February 2, 1911, p. 4; *La Lanterne*, "Une Aviatrice Vole 53 Minutes," November 30, 1910, p. 4; et alia.
84. *L'Aérophile*, "Les Records Féminins d'Aviation," no. 24, December 15, 1910, p. 556.
85. Myriel, "Mon Premier Vol de Durée: Comment J'ai Établi le Premier Record de la Coupe Fémina," *Le Sport*, December 8, 1910.
86. Dollfus, Charles, Henry Beaubois, Camille Rougeron, *L'Homme, l'Air et l'Espace* (Éditions de l'Illustration, 1965), pp. 130, 144.
87. *Kingston [Jamaica] Gleaner*, "Aerial Flights," December 24, 1910, p. 6; *Hammond Lake Country [IN] Times*, "Fair French Women Rivals for Long Distance Aeroplane Flight Prize Offered by Paris Paper," December 27, 1910, p. 4; *Brandon [Manitoba] Weekly Sun*, "Fair Aviators," January 12, 1911, p. 23; *The Salt Lake [UT] Trib-

une, "Feminine Aviators Vie With Each Other for Title of Queen of the Air," December 11, 1910, p. 2; "The World of Women,"*[Mansfield, OH] News-Journal*, January 18, 1911, p. 4; et alia.

88. *L'Aviation Illustrée*, "Nos Aviatrices et la Coupe Fémina," December 15, 1910, no. 61, cover.

89. *Los Angeles Herald*, "Midwinter Aviators," December 26, 1910, p. 4.

90. *The Tacoma [WA] Times*, "Four Little Bird Women, All in a Row, Soaring For Prizes, Hi-lee, Oh, Hi-low," January 3, 1911, p. 5.

91. *L'Est Républicain*, "Mlle Marvingt Va s'Attaquer au Record de Mlle Dutrieu," December 9, 1910, p. 2.

92. *L'Est Républicain*, "Mlle Marvingt," January 1–2, 1911, p. 1; *Le Temps*, "Tentatives/Portateurs des Brevets d'Aviation," January 1, 1911, p. 3; *La Presse*, "Au Camp de Châlons: Une Tentative de Mlle Marvingt," December 31, 1910, p. 11.

93. Myriel, "Pourquoi Je n'Ai Pas Pu Reprendre la Coupe Fémina," *Courrier de l'Est*, January 4, 1911, p. 2.

94. *Ibid.*

95. *L'Aviation Illustrée*, "Nos Aviatrices et la Coupe Fémina," December 15, 1910, no. 61, cover.

96. *Fémina*, "La Coupe Fémina," January 1, 1911, pp. 14, 18.

97. Marck, Bernard, *Elles Ont Conquis le Ciel* (Arthaud, 2009), pp. 14–21; Marck, Bernard, *Women Aviators* (Flammarion, 2013), pp. 14–21; Cordier, Marcel, *Lorraine, Secrète et Insolite* (Éditions du Sapin d'Or, 2011), p. 83; Chevalier, Frédérique, op. cit.; *La Liberté*, "Marie Marvingt Est Morte à 88 ans," December 16, 1963, no. 6164, pp. 1, 8.

98. *Journal des Débats Politiques et Littéraires*, "L'Aviation," October 3, 1913, p. 4.

99. *Le Petit Parisien*, "Le Meeting de Saint Etienne," July 28, 1911, p. 4.

100. *La Tribune Républicaine*, "Débuts de Mlle Marvingt," August 4, 1911.

101. *Gil Blas*, "Deux aviateurs Blessés 'en Automobile,'" August 4, 1911, p. 44; *Le Temps*, "Aviateurs Blessés en Automobile," August 4, 1911, p. 5; *Le Journal [Paris]*, "Un Meeting d'Aviation Fertile en Incidents," August 15, 1911, p. 4; *L'Est Républicain*, "Mlle Marvingt Blessée en Automobile," August 4, 1911, p. 1; *Le Journal [Paris]*, "Aviateurs Blessés en Auto," August 3, 1911, p. 4.

102. *La Tribune Républicaine*, "Débuts de Mlle Marvingt," August 4, 1911.

103. *Ibid.*

104. Beaubois, Henry, "Ceux Qui Ouvrirent la Route au Ciel: Marie Marvingt," *Revue Aéronautique de France*, August 13, 1937.

105. Woodhouse, Henry, "Women of the Air," *Outdoor World & Recreation*, May 1913, vol. 48, pp. 305–309.

106. Beaubois, Henry, "Ceux Qui Ouvrirent la Route au Ciel: Marie Marvingt," *Revue Aéronautique de France*, August 13, 1937.

107. *La Vie au Grand Air*, "Une Aviatrice Tombe Dans un Jeu de Boules," August 19, 1911, cover; *Journal des Débats Politiques et Littéraires*, "Une Aviatrice Tombe Dans un Jeu de Boules," August 8, 1911, p. 3; Le Gaulois, "Petites Nouvelles de la Nuit," August 7, 1911, p. 33; *La Gazette du Village*, "Chute d'une Aviatrice Dans un Café," August 13, 1911, p. 515; Dollfus, Charles, and Henri Bouché, *L'Histoire de l'Aéronautique* (Éditions de l'Illustration, 1938), pp. 233, 235.

108. *Illustration*, "Un aéroplane Dans un Jeu de Boules," August 12, 1911, no. 3572, p. 128.

109. *L'Echo de Sanflorain*, "Une Aviatrice Tombe Dans un Café," August 12, 1911.

110. *Le Journal [Paris]*, "Dernières Nouvelles Sportives," August 7, 1911, p. 4.

111. Morgane, Yves, "La Fiancée du Danger: Marie Marvingt Fut l'Une des Premières Femmes Aviatrices Égales des Hommes Dans la Conquête de l'Air," *Miroir de l'Histoire*, November-December 1978, no. 307, pp. 80–84.

112. *Ric et Rac*, "Prouesses d'Il y a Vingt Ans," December 6, 1930, p. 2.

113. J.P., "Marie Marvingt, Pilote d'Avion, Est Morte à 88 Ans: Championne en Tous Genres," *Le Monde*, December 17, 1963, p. 13; Barthélemy, Général, "La Femme la Plus Extraordinaire du Siècle: Marie Marvingt, la 'Fiancée du Danger,'" *L'Étrange Race des Hommes Volants* (Éditions France-Empire, 1979), pp. 241–254; *Le Parisien*, "Marie Marvingt, Surnommée 'la Fiancée du Danger,' Est Décédée Samedi à Nancy," December 16, 1963. et alia.

114. *Le Rappel*, "Mlle Marvingt tombe encore," August 17, 1911, p. 3.

115. Reichel, Frantz, "La Fiancée du Danger," *Le Figaro*, January 3, 1914, p. 1.

116. *Ibid.*

117. *Le Matin*, "Être Blessé n'Est Rien," December 17, 1913, p. 5.

118. *Le Figaro*, "Les Femmes Pilotes Affrontent le Danger," January 20, 1914, cited in Alain Peyrefitte, ed., *L'Aventure du XXe Siècle* (1989), p. 180.

119. Marck, Bernard, "Ces Merveilleux Fous Volants," *Le Figaro Histoire*, August-September 2015, no. 21, p. 48.

120. *L'Aéro*, "Sur les Aérodromes," July 11, 1912, p. 2; *L'Aéro*, "Sur les Aérodromes," July 19, 1912, p. 3; *L'Aéro*, "Sur les Aérodromes," August 13, 1912, p. 2; *L'Aéro*, "Sur les Aérodromes," October 1, 1912, p. 2; *L'Aéro*, "Sur les Aérodomes: Bétheny," October 11, 1912, p. 2; *L'Aéro*, "Sur les Aérodromes," August 8, 1913, p. 2; *L'Aéro*, "Un Voyage de Mlle Marvingt," October 3, 1913, p. 2; *L'Aéro*, "Petites Nouvelles," October 29, 1913, p. 2; *L'Ouest Éclair [Caen]* and *L'Ouest Éclair [Rennes]*, "Prouesse d'Aviatrice: Elle s'Élève à Trois Mille Mètres Avec Deux Passagers," September 19, 1912, p. 5; *L'Ouest Éclair [Caen]* and *L'Ouest Éclair [Rennes]*, "Le Temps, "L'Aviation en Province," 14 août 1911, p. 5Une aviatrice," October 3, 1913, p. 3; *Le Temps*, "L'Aviation en Province," 16 août 1911, p. 6; *L'Aurore*, "Chez Deperdussin à Courcy," July 28, 1912, p. 3; *L'Aurore*, "Aux Ecoles Deperdussin à Reims," July 30, 1912, p. 3; L'Aurore, "L'Entraînement aux écoles de l'aviation," 2 août 1912, p. 3; L'Aurore, "Aux Ecoles Deperdussin," 6 août 1912, p. 3; L'Aurore, "Aux Ecoles Deperdussin—Aérodrome de la Champagne," 8 août 1912, p. 3; L'Aurore, "Aux Ecoles Deperdussin," 13 août 1912, p. 3; L'Aurore, "Aux Ecoles Deperdussin," 8 septembre 1912, p. 3; L'Aurore, "L'Entraînement à l'Ecole Deperdussin de Reims," 9 septembre 1912, p. 3; L'Aurore, "Aux Ecoles Deperdussin à Betheny," 17 septembre 1912, p. 3; L'Aurore, "Un beau vol de Mlle Marvingt," 19 septembre 1912, p. 3; L'Aurore, "Aux Ecoles Deperdussin," 20 septembre 1912, p. 3; L'Aurore, "A l'Ecole Deperdussin de Bétheny," 22 septembre 1912, p. 3; L'Aurore, "Aux Ecoles Deperdussin à Betheny," 26 septembre 1912, p. 3; L'Aurore, "Aux Ecoles Deperdussin à Betheny," 28 septembre 1912, p. 3; L'Aurore, "Chez Deperdussin à Estampes," 11 octobre 1912, p. 3; L'Aurore, "Au Salon de l'Aéronautique," 31 octobre 1912, p. 2; L'Aurore, "A l'Ecole Deperdussin à Reims," 23 novembre 1912, p. 3; L'Aurore, "A l'Ecole Deperdussin à Reims," 1 décembre 1912, p. 3; L'Aurore, "Dans les écoles d'aviation," 2 décembre 1912, p. 3; L'Aurore, "A l'Ecole Deperdussin, aérodrome de la Cham-

pagne," 6 décembre 1912, p. 3; L'Aurore, "A Betheny: Ecole Deperdussin," 12 décembre 1912, p. 3; L'Aurore, "Mlle Marvingt vole de Reims à Mourmelon," 3 octobre 1913, p. 3; L'Aurore, "Un beau vol de Mlle Marvingt," 26 octobre 1913, p. 3; *Le Gaulois*, "Chez Deperdussin à Courcy," July 28, 1912, p. 4; Le Gaulois, "Aux Ecole Deperdussin à Reims," July 30, 1912, p. 3; Le Gaulois, "Aux Ecoles Deperdussin à Betheny," August 4, 1912, p. 4; Le Gaulois, "Aux Ecoles Deperdussin," August 6, 1912, p. 3; *Le Gaulois*, "Aux Ecoles Deperdussin—Aérodrome de la Champagne," August 8, 1912, p. 4; *Le Gaulois*, "Aux Ecoles Deperdussin," August 13, 1912, p. 3; *Le Gaulois*, "L'Ecole Deperdussin à Bétheny," August 22, 1912, p. 4; *Le Gaulois*, "Aux Ecoles Deperdussin," September 8, 1912, p. 3; *Le Gaulois*, "Aux Ecoles Deperdussin à Bétheny," September 17, 1912, p. 3; *Le Gaulois*, "Aux Ecoles Deperdussin à Bétheny," September 21, 1912; *Le Radical*, "A l'Aérodrome de la Champagne," December 7, 1912, p. 5; Le Gaulois, "Aux Ecoles Deperdussin à Bétheny," 23 septembre 1912, p. 4 Le Gaulois, "Aux Ecoles Deperdussin à Bétheny," 26 septembre 1912, p. 4 Le Gaulois, "Aux Ecoles Deperdussin à Bétheny," 28 septembre 1912, p. 3 Le Gaulois, "Aux Ecoles Deperdussin à Bétheny," 2 octobre 1912, p. 4; Le Gaulois, "A Bétheny—Ecole Deperdussin militaire," 11 octobre 1912, p. 3 Le Gaulois, "Aérodrome de la Champagne—Ecoles Deperdussin," 31 octobre 1912, p. 4 Le Gaulois, "Chez Deperdussin," 9 novembre 1912, p. 4 Le Gaulois, "Chez Deperdussin à Reims," 23 novembre 1912, p. 4; Le Gaulois, "Aérodrome de la Champagne—Ecoles Deperdussin," 6 décembre 1912, p. 4 Le Gaulois, "A l'Ecole Deperdussin, aérodrome de la Champagne," 10 août 1913, p. 4; *Gil Blas*, "Promenade d'Aviatrice," September 19, 1912, p. 5; *L'Echo d'Alger*, "Mlle Marvingt Emmène Deux Passagers à Trois Mille Mètres de Hauteur," September 19, 1912, p. 2; *L'Est Républicain*, "Les Fétes de Toul: Ce Fut une Journée Splendide: On Applaudit de Beaux Vols," April 15, 1912, p. 1; *Le Radical*, "Aviation: Envolées Diverses," July 30, 1912, p. 5; Rastelli, E., "La Semaine d'Aviation de Turin," *L'Aérophile*," August 1, 1911, p. 357; *La Lanterne*, "Les Vols de Mlle Marvingt," December 23, 1912, p. 4; *La Lanterne*, "Un Vol de Mlle Marvingt," October 3, 1913, p. 4; *La Lanterne*, "Un Raid de l'Aviateur Marvingt," October 26, 1913, p. 3; *Le Matin*, "L'Éclectique Aviatrice s'Entraîne," October 30, 1913, p. 5; *LaRevue Aérienne*, "La Semaine d'Aviation à Turin," July 10, 1911, p. 359; *Le Journal [Paris]*, "Aviation," August 14, 1911, p. 6; *Le Journal [Paris]*, "Nos Aviateurs Volent à Toul," April 15, 1912, p. 4; *Le Journal [Paris]*, "Aviation," December 1, 1912, p. 8; *Le Journal [Paris]*, "A l'Aérodrome de la Champagne," August 9, 1913, p. 5; *Le Figaro*, "Aviation," August 6, 19, September 17, October 1, December 10, 1912; *Le Petit Parisien*, "Une Aviatrice Ammène Deux Passagers," September 19, 1912, p. 5; *La Presse*, "Aviation," August 7, 1912, p. 3; *La Presse*, "A l'École Deperdussin," August 8, 1912, p. 3; *La Presse*, "Série de Beaux vols," August 14, 1912, p. 3; *London Evening News*, "Ladies' Flying Meeting: First International Contest Among Women at the Hendon Aerodrome," July 1, 1912, p. 7; *London Standard*, "Women Aviators' Meeting," June 29, 1912, p. 11; *London Standard*, "Ladies in the Air: First International Contest for Women," July 2, 1912, p. 10; *Journal des Débats Politiques et Littéraires*, "Prouesse d'Aviatrice," September 19, 1912, p. 4; et alia.

121. Grandidier, Gilbert, op. cit.

122. Marvingt, Marie, "Le Meeting d'Aviation de Reims,"*Le Temps*, September 29, 1913, p. 4.

123. *L'Aurore*, "A l'Ecole Deperdussin de Bétheny," September 22, 1912, p. 3; *L'Aurore*, "A l'Ëcole Deperdussin de Reims," September 23, 1912, p. 3; *Dundee Evening Telegraph*, "Women Conquerors of the Air," October 30, 1912, p. 6; *La Presse*, "Une Conférence de Mlle Marvingt," June 14, 1912, p. 3; *La Presse*, "Les Exploits de Mlle Marvingt," October 6, 1912, p. 3.

124. Lallier, Roger, "Par-Dessus les Moulins," *VU: La Journal de la Semaine*, May 15, 1929, no. 61, pp. 390–391.

125. *Les Ailes Brisées*, "Nos Aviatrices, les Véritables Pionniers de l'Aviation: Marie Marvingt," December 1959, pp. 21–22; Boase, Wendy, *The Sky's the Limit: Women Pioneers in Aviation* (Macmillan, 1979), pp. 11–12.

126. Sumner, Ian, *Kings of the Air: French Aces and Airmen of the Great War* (2015), pp. 23–24.

127. Robertston, Patrick, *Robertson's Book of Firsts* (2011), p. 546.

128. https://www.womenofaviationweek.org/women-aviation-week-aussies/

129. *The Tacoma [WA] Times*, "Four Little Bird Women, All in a Row, Soaring For Prizes, Hi-lee, Oh, Hi-low," January 3, 1911, p. 5.

130. *Huntingdon [PA] Daily News*, "Aviation Is New Field for Women of Europe," May 3, 1926, p. 8.

131. *San Antonio Light And Gazette*, "Daring French "Aviatresses," January 1, 1911, p. 49; *Rochester [NY] Democrat*, "Daring Aviatresses Whose Exploits Are Winning Fame," July 16, 1911, p. 14.

132. *Hammond Lake County [IN] Times*, "Lady-Birds Are Going to Take Wing at Squantum Aviation Meet Next Month; They'll Ask No Odds From Men Contestants," July 15, 1911, p. 4.

133. *Oakland Tribune*, "Flyerettes," August 7, 1911, p. 6.

134. Marvingt, Marie, in Roland Erbstein, in collaboration with Jean Matt, "Cartes Postales d'Hier et d'Aujourd'hui: "En 1909, 'la Fiancée du Danger' Ouvrait aux Femmes la Voie des Airs," *L'Est Républicain*," March 23, 1980.

135. Ainsworth, Ed, "Along El Camino Real, *The Los Angeles Times*, October 27, 1935, p. 35.

136. *Le Temps*, "De l'aéroplane au Ballon Sphérique," March 4, 1911, p. 3; *L'Est Républicain*, "La 'Fiancée du Danger' Est Morte, Hier, à 88 Ans," December 15, 1963, p. 1; Barthélemy, Général, "La Femme la Plus Extraordinaire du Siècle: Marie Marvingt, la 'Fiancée du Danger,'" *L'Étrange Race des Hommes Volants* (Éditions France-Empire, 1979), pp. 241–254; *L'Auvergnat de Paris*, "Marie Marvingt n'Est Plus," December 21, 1963, no. 51, pp. 1, 3; Boetsch, Jacques, "Une Vie en Équilibre" *Anciens Combattants du Monde*, May 1955, pp. 11–14; Estrada de Tourniel, Jérôme, "La Fiancée du Danger," *Passions Grand Est*, December 2003, no. 26, pp. 58–62; Desormière, Yves, "A 80 ans Marie Marvingt, la 'Fiancée du Danger,' est la Seule Femme au Monde à Être Titulaire de Quatre Brevets de Pilotage: Ballon, Avion, Hydravion, Secourisme de l'Air," *La Croix*, March 4, 1955, pp. 1, 4; Laurent, Michel-Yves, "Marie Marvingt, Surnommée 'Marie Casse-Cou,' Traversa au Début du Siècle, la Mer du Nord en Ballon; Elle Pilote Aujourd'hui à 86 Ans un Hélicoptère à Réaction," *L'Aurore* or *Paris Press*, November 2, 1960, no. 16; Riverain, Jean, *Dictionnaire des Aéronautes Célèbres* (Larousse, 1970), p. 106; Mauchaussée, Jean, "Marie Marvingt, Reine de l'Air," *Prestige*, October 1, 1959, no. 3, p. 1; *La Liberté*, "Marie Marvingt Est Morte à 88 ans," December 16, 1963, no. 6164, pp. 1 & 8; Desormière, Yves, "A 80 Ans Marie Marvingt, la 'Fiancée du Danger,' Est la Seule Femme au Monde à Être Titulaire de Quatre Brevets de Pilotage: Ballon, Avion, Hydravion, Secourisme de l'Air," *La Croix*, March 4, 1955, pp. 1, 4; *La Liberté*, "Marie Marvingt Est Morte à 88 Ans," December 16, 1963, no. 6164, pp. 1 & 8; Gygax,

Georges, "La 'Fiancée du Danger': La Vie Prodigieuse de Marie Marvingt," *L'Illustré*, April 24, 1958, pp. 34–36.

137. Gygax, Georges, "La Vie Prodigieuse de Marie Marvingt, la 'Fiancée du Danger'," *L'Illustré*, May 1, 1958, pp. 40–42.

138. *Les Ailes*, "Sur les Terrains de Vol à Voile," September 15, 1932, p. 13; *Les Ailes*, "Faisons Confiance à la Banne d'Ordanche," September 21, 1933, p. 12; Dinan, R., "Les Enfants Prodiges: Mlle Marvingt Lance un Défi Mondial," *L'Ordre*, July 2, 1948; *L'Echo d'Alger*, "Le Vol à Voile: Le Meeting à Biskra Commence Aujourd'hui," January 26, 1923, p. 2; *Le Petit Parisien*, "Le Meeting de Vol à Voile de Biskra," January 27, 1923, p. 3; Féral, Roger, "Marie Marvingt (86 ans), 'la Fiancée du Danger,' a Passé la Toussaint Dans un Hélicoptère à Réaction," *Télé-Paris*, November 1960.

139. Mathis, Suzy, "Aviatrices d'Hier et d'Aujourd'hui: Marie Marvingt," *Aviation Française*, May 8, 1946, p. 9.

140. Myriel, "L'Aviation," *Éclair de l'Est*, May 4, 1909; Myriel, "Le 'Ville de Nancy': Il est Arrivé!" *Le Sport*, July 24, 1909, p. 2; *L'Aérophile*, "Le 'Ville-de-Nancy' à Nancy," August 1, 1909, p. 353; Myriel, "La Grande Semaine d'Aviation," *L'Éclair de l'Est*, August 27, 1909; Myriel, "La Grande Semaine d'Aviation," *L'Éclair de l'Est*, September 1, 1909, p. 2.

141. Laurent, Michel-Yves, "Marie Marvingt, Surnommée 'Marie Casse-Cou,' Traversa au Début du Siècle, la Mer du Nord en Ballon; Elle Pilote Aujourd'hui à 86 ans un Hélicoptère à Réaction," *L'Aurore* or *Paris Press*, November 2, 1960, no. 16.

142. *Long Beach [CA] Independent*, "Still Learning at 80," April 5, 1955, p. 7; *Burlington [NC] Daily Times News*, "Aviation Pioneer Learns Something New," 6 April 6, 1955, p. 5; *Carbondale Southern Illinoisan*, "Woman Takes Up Helicopter at 80," April 5, 1955, p. 16; *Sikeston [MO] Daily Standard*, "80-Year-Old Woman Tries 'Copter," April 5, 1955, p. 1; *Cumberland [MD] Evening Times*, "Never Too Old to Learn, She Says," April 4, 1955, p. 1; Laurent, Michel-Yves, "Marie Marvingt, Surnommée 'Marie Casse-Cou,' Traversa au Début du Siècle, la Mer du Nord en Ballon; Elle Pilote Aujourd'hui à 86 Ans un Hélicoptère à Réaction," *L'Aurore* or *Paris Press*, November 2, 1960, no. 16; et alia.

143. Floret, Robert, "Grand-Mère 'Casse-Cou,'" *Détective*, November 18, 1960, no. 751, pp. 10–11.

144. *Pacific Stars and Stripes*, "Never Too Old," April 4, 1955, p. 1; et alia.

145. Bichet, Gabriel (G.B.), "Marie Marvingt, la Fiancée du Danger, Est Morte Hier à Nancy," *L'Est Républicain*, December 15, 1963.

146. Prévost, Marie-Eudes Lauriot et Adélaïde de Clermont-Tonnerre, "Les Chevalières du Ciel: Marie Marvingt, la Fiancée du Danger," *Point de Vue*, December 16, 2009, no. 3204, pp. 38–43.

147. Gygax, Georges, "La 'Fiancée du Danger': La Vie Prodigieuse de Marie Marvingt," *L'Illustré*, April 24, 1958, pp. 34–36.

148. Féral, Roger, "Marie Marvingt (86 ans), 'la Fiancée du Danger,' a Passé la Toussaint Dans un Hélicoptère à Réaction," *Télé-Paris*, November 1960; Laurent, Michel-Yves, "Marie Marvingt, Surnommée 'Marie Casse-Cou,' Traversa au Début du Siècle, la Mer du Nord en Ballon; Elle Pilote Aujourd'hui à 86 Ans un Hélicoptère à Réaction," *L'Aurore* or *Paris Press*, November 2, 1960, no. 16.

149. Bichet, Gabriel (G.B.), op. cit.

150. Kernel, Hélène, "Marie Marvingt, Doyenne des Aviatrices," *France Aviation*, August 1957, p. 8; *L'Humanité*, "La Grandmère (Volante) Fait (Aussi) du Vélo," no. 10, October 1961; "La Grandmère Volante Obtient, à Quatre-Vingt-Six Ans, le Renouvellement de Son Brevet de Pilote," October 1961; *Le Meusien*, "Toujours Pilote d'Avion à 86 Ans," October 20, 1961.

151. Saindizié, J.-M., "C'est Marie Marvingt, 85 ans, la Plus Extraordinaire Femme du Siècle," *Pilote*, November 17 1960, no. 56, p. 10.

152. Bichet, Gabriel (G.B.), op. cit.

153. *Ibid.*

154. Bichet, Gabriel (G.B.), op. cit.

155. Enselme-Trichard, Renée, letter, July 3, 1982.

Chapter 5

1. Palmer, Svetlana, and Sarah Wallis, *Intimate Voices From the First World War* (William Morrow, 2003), jacket copy.

2. Aldrich, Mildred, *A Hilltop on the Marne: Being Letters Written June 3-September 8, 1914* (The Atlantic Monthly Co., 1915), p. 46.

3. *Ibid.*

4. Clodfelter, Micheal, *Warfare and Armed Conflicts: A Statistical Encyclopedia of Casualty and Other Figures, 1492–2015*, 4th ed. (McFarland, 2017), p. 430.

5. Middlebrook, Martin, *The First Day on the Somme* (Allen Lane, 1971; rev. ed. Pen & Sword Books, Ltd., 2016), p. 264.

6. Leland, Anne, *American War and Military Operations Casualties: Lists and Statistics* (Congressional Research Service, February 26, 2010, pp. 2–4.

7. Strachan, Hew, ed., *The Oxford Illustrated History of the First World War* (Oxford University Press, 2014), ch. 3, unpaged.

8. Bowen, Ezra, *Knights of the Air* (Time-Life Books, 1980), p. 27.

9. Palmer, Svetlana, and Sarah Wallis, *A War in Words* (HarperCollins, 2003), p. ix.

10. Floret, Robert, "Grand-Mère 'Casse-Cou,'" *Détective*, November 18, 1960, no. 751, pp. 10–11.

11. Marvingt, Marie, letter to Marthe Roumier, August 10, 1914.

12. *L'Aérophile*, "Dirigeable et Avions Allemands sur Nancy," January 1–15, 1915, p. 20.

13. *Ibid.*

14. Marvingt, Marie, "Dans Nancy Bombardée," *Samedi Soir*, January 29, 1916.

15. Cordier, Marcel, "Marie Marvingt: Une Lorraine Qui Volait au Secours des Blessés," *La Vie la Santé*, July 1998, no. 6, pp. 28–29.

16. *Le Gaulois*, "Parmi Ceux Qui Sont Partis," August 7, 1914, p. 2.

17. Darrow, Margaret H., *French Women and the First World War: War Stories of the Home Front* (2000), p. 240; Proctor, Tammy M., *Civilians in a World at War, 1914–1918* (2010), p. 97.

18. Foch, Ferdinand, cited in Clifford A. Pickover, *Time : A Traveler's Guide* (1998), p. 249.

19. Sumner, Ian, *Kings of the Air: French Aces and Airmen of the Great War* (2015), pp. 23–24.

20. to come

21. Franks, Norman L.R., *Over the Front: A Complete Record of the Fighter Aces and Units of the United States and French Air Services, 1914–1918* (1992).

22. Driggs, Laurence La Tourette, *Heroes of Aviation* (1918), p. 32.

23. Longstreet, Stephen, *The Canvas Falcons: The Men and the Planes of World War I* (Ballantine, 1970), p. 43.

24. *Daily Mirror*, "Joan of Arc of the Clouds," May 20, 1915, p. 10.
25. *L'Est Républicain*, "Mlle Marvingt Veut S'engager," May 21, 1915, p. 2.
26. *Ibid.*
27. *Ibid.*
28. Bere, Emile, "Nos Aviatrices S'ennuient," *Le Figaro*, September 27, 1915, p. 3; Darrow, Margaret H., *French Women and the First World War: War Stories of the Home Front* (2000), p. 229.
29. *Indianapolis Star*, "The Woman With War Wings," October 11, 1914, p. 55.
30. *Van Nuys [California] and The Van Nuys Call*, "Women Serve As Air Scouts," August 14, 1914, p. 5; *L'Echo d'Alger*, "Quelques Réservistes de Marque," August 7, 1914, p. 2.
31. *Boston Sunday Post*, "Women Airbirds Eager to Serve in War," August 23, 1914, p. 47.
32. *Ibid.*
33. *Ibid.*
34. *Ibid.*
35. *The [Spokane, Washington] Spokesman-Review*, "Women Aviators for the Allied Army?," November 8, 1914, p. 2; *The Montana Standard*, "Women Aviators for the Allied Army?," November 8, 1914, p. 25; et alia.
36. *Ibid.*
37. *Ibid.*
38. *Automobile Journal*, "Women to Be War Chauffeurs," vol. 39, 1915, p. 49; *Arizona Republican*, "War Chauffeurs," July 4, 1915, p. 4.
39. *The [Chicago] Sun*, "Women Seek War Jobs: Want Positions As Ambulance Drivers in France," April 14, 1915, p. 2.
40. Darrow, Margaret H., op. cit., pp. 240–241.
41. *Indépendant*, "Mlle Marie Marvingt à Pau: Deux Causeries Littéraires et Sportives," June 16, 1927; Beaubois, Henry, "Ceux Qui Ouvrirent la Route au Ciel: Marie Marvingt," *Revue Aéronautique de France*, August 13, 1937.
42. *Norwalk Reflector Herald*, "Three Frenchwomen Volunteer to Serve in Aviation Corps," September 11, 1914, p. 4; *Racine [Wisconsin] Journal News*, "What Women Are Doing," January 2, 1915, p. 6; *Brian [Denton, TX] Daily Eagle and Pilot*, "Activities of Women," September 1, 1914, p. 4; *Chillicothe [Missouri] Constitution*, "Activities of Women," October 12, 1914, p. 3; et alia.
43. R.G., "Une Femme Qui Pratique Tous les Sports: Mlle Marvingt," *Le Miroir des Sports*, October 28, 1920, pp. 260–261.
44. Shipton, Elisabeth, *Female Tommies: The Frontline Women of the First World War* (The History Press, 2014), pp. 83–85, 128, 237.
45. Robertston, Patrick, *Robertson's Book of Firsts* (2011), p. 546; *L'Afrique du Nord Illustré*, "Un Rare Exemple d'Énergie Féminine," July 15, 1922, p. 7; Leblond, J.-J., "Pour le 52e Anniversaire de Sa Première Ascension en Ballon Marie Marvingt a Piloté un Avion Sanitaire Entre Bordeaux et Marseille," *Dauphiné*, November 1952, p. 3; Godot, J.-H., "La Vie de Marie Marvingt" et "Témoignage: A Nancy, Sur les Traces de Marie Marvingt: Rosalie, Écrivain Américain Veut Faire Revivre 'la Fiancée du Danger,'" *Est Républicain*, June 25, 1982; Grandidier, Gilbert, "Marie Marvingt (1875–1963), 'la Fiancée du Danger' Ouvre aux Femmes la Voie des Airs," newspaper, May 1987; et alia.
46. Sumner, Ian, op. cit.
47. Shayler, David J., and Ian A. Moule, *Women in Space—Following Valentina* (Springer/Praxis, 2006), p. 16; Master, Nancy Robinson, *Airplanes* (2008), p. 12.
48. Chevalier, Frédérique, "Marie Marvingt, la Fiancée des Airs (1875–1963)," *Les Grandes Aventurières* (City Editions, 2007), pp. 85–95.
49. Shipton, Elisabeth, *Female Tommies: The Frontline Women of the First World War* (The History Press, 2014), pp. 83–85.
50. Bichet, Gabriel, "Marie Marvingt, la Fiancée du Danger, est Morte Hier à Nancy," *Est Républicain*, December 15, 1963, p. 1.
51. Pelot, Paul, "Marie Marvingt Fut une des Femmes les Plus Extraordinaires du Siècle," *L'Équipe*, December 16, 1963; Shipton, Elisabeth, op. cit.
52. Clara Barton, "The Women Who Went to the Field," *The Red Cross in Peace and War* (1899), p. 510.
53. Desormière, Yves, "A 80 Ans Marie Marvingt, la 'Fiancée du Danger,' est la Seule Femme au Monde à Être Titulaire de Quatre Brevets de Pilotage: Ballon, Avion, Hydravion, Secourisme de l'Air," *La Croix*, March 4, 1955, pp. 1, 4.
54. Ackerman, Gordon, "Fiancée of Danger," *Sports Illustrated*, June 26, 1961, pages 61–64; J.P., "Marie Marvingt, Pilote d'Avion, Est Morte à 88 Ans: Championne en Tous Genres," *Le Monde*, December 17, 1963, p. 13.
55. *Les Ailes Brisées*, "Nos Aviatrices, les Véritables Pionniers de l'Aviation: Marie Marvingt," December 1959, pp. 21–22.
56. Boetsch, Jacques, "Une Vie en Équilibre" *Anciens Combattants du Monde*, May 1955, pp. 11–14; Scott, R.J., *Albuquerque Journal*, "Scott's Scrap Book," July 10, 1940, p. 11.
57. R.G., "Une Femme Qui Pratique Tous les Sports: Mlle Marvingt," *Le Miroir des Sports*, October 28, 1920, pp. 260–261.
58. *The Washington Post*, "France's Pioneer Woman Flier Wins Acclaim in a Dozen Lines," January 15, 1936, p. 13.
59. Bichet, Gabriel, "Marie Marvingt, la Fiancée du Danger, est Morte Hier à Nancy," *Est Républicain*, December 15, 1963, p. 1.
60. R.P., "Mademoiselle Marvingt," *Le Cri de Constantin*, June 28, 1924.
61. Micromegas, "Une Femme," *L'Echo d'Alger*, July 6, 1922, p. 2.
62. R.G., "Une Femme Qui Pratique Tous les Sports: Mlle Marvingt," *Le Miroir des Sports*, October 28, 1920, pp. 260–261; Baron Boilley, Françoise, *Marie Marvingt: A l'Aventure du Sport* (L'Harmattan, 2013), p. 78; R.P., op. cit.
63. Bichet, Gabriel, op. cit.
64. *New York Times*, "Early Aviator, 88," December 16, 1963, p. 33.
65. Légion d'Honneur, La Grande Chancellerie de la Légion d'Honneur: documents supporting and confirming her grades of Chevalier and then Officer of the Légion d'Honneur, 1935 and 1950.
66. Droz, Stanislas, *Les Femmes Dans la Grande Guerre 1914–1918* (Vent d'Est, 2014), pp. 74, 75–78; R.G., "Une Femme Qui Pratique Tous les Sports: Mlle Marvingt," *Le Miroir des Sports*, October 28, 1920, pp. 260–261.
67. Marvingt, Marie, postcard to Mr. And Mrs. Roumier, December 23, 1916.
68. Thompson, Mark, *The White War: Life and Death on the Italian Front 1915–1919* (2009), p. 204.
69. Boetsch, Jacques, "Une Vie en Équilibre" *Anciens Combattants du Monde*, May 1955, pp. 11–14.
70. R.G., op. cit.
71. Mazaud, François, cited in Michel Daurat, "A Quatre-Vingts Ans, Marie Marvingt Demeure Toujours

la Fiancée du Danger," *Lorraine Magazine*, no. 9, February 1955, pp. 10–13.
72. Bouchet, Madame, in Jacqueline Maire, letter, July 4, 1982.
73. to come
74. Bichet, Gabriel, "C'était la Fiancée du Danger...," *Est Républicain*, December 15, 1963.
75. Légion d'Honneur, La Grande Chancellerie de la Légion d'Honneur: documents supporting and confirming her grades of Chevalier and then Officer of the Légion d'Honneur, 1935 and 1950;
Nancy-Ochey Flash, "Marie Marvingt, 'La Fiancée du Danger,' July 2003, no. 171, p. 17.
76. Florentin, Mademoiselle, cited in Jacqueline Maire, letter, July 23, 1982.
77. to come
78. *L'Est Républicain*, "Décès du 10 Mars: Félix-Constant Marvingt," March 13, 1916, p. 2.
79. *Ibid.*; *Journal de la Meurthe et des Vosges*, "Etat-Civil Quotidien de Nancy du 10 Mars," March 12, 1916, p. 2.
80. Besançon, Madame, letter, June 28, 1982.
81. *Bulletin de la Meurthe et Moselle*, "Mlle Marvingt," August 12, 1917, p. 2.
82. Marvingt, Marie, visa for safe conduct, issued August 24, 1921; Gygax, Georges, "La Vie Prodigieuse de Marie Marvingt, la 'Fiancée du Danger,'" *L'Illustrë*, May 1, 1958, pp. 40–42.
83. Telegrams between the Prefect of Lyon and the Prefect of Nancy, and letter from the Special Commissioner in Nancy to the Prefect of Meurthe-et-Moselle, on the subject of an Italian visa for Marie Marvingt, October 26, 1917; Dollet, Christophe, "Marie Marvingt Sans Visa," *L'Est Républicain*, July 11, 2011.
84. *Ibid.*
85. *Ibid.*
86. Marvingt, Marie, letter to Mr. And Mrs. Roumier, December 26, 1916.
87. *Bulletin de la Meurthe et Moselle*, "Mlle Marvingt," August 12, 1917, p. 2.
88. Marvingt, Marie, letter to "Marie-Thérèse and Monsieur," November 18, 1917.
89. Mechelle, Ginette, interview, June 23, 1982.
90. *Ibid.*
91. *Ibid.*
92. Coulet, Jacques, interview, June 24, 1982.
93. Balitrand, Suzanne, "Le Beau Voyage de Mademoiselle Marvingt," *Eve*, August 15, 1926, no. 307, p. 7; *L'Echo d'Alger*, "Une Conférence de Mlle Marvingt à Mascara," July 8, 1925, p. 6; Légion d'Honneur, La Grande Chancellerie de la Légion d'Honneur: documents supporting and confirming her grades of Chevalier and then Officer of the Légion d'Honneur, 1935 and 1950.
94. Cordier, Marcel, "Les Combats de Marie Marvingt," *La Nouvelle Revue Lorraine*, no. 27, August-September 2014, p. 20–23.
95. *Ibid.*
96. *Indépendant*, "Mlle Marie Marvingt à Pau: Deux Causeries Littéraires et Sportives," June 16, 1927; *L'Echo d'Alger*, "Huit Cours et Conférences à Alger de Mlle Marie Marvingt," May 15, 1925, p. 3; Renoux-Barès, E., "Une Héroïne Parle à Pau de l'Aviation, École d'Énergie," *Revue Aéronautique de France*, October 1927, p. 11.
97. *L'Echo d'Alger*, "Ceryville: Conférence," July 28, 1925, p. 4; Renoux-Barès, E., op. cit.
98. *Indépendant*, op. cit.
99. Marvingt, Marie, signed autograph statement, Paris, June 5, 1931.
100. Floret, Robert, "Grand-Mère 'Casse-Cou,'" *Détective*, November 18, 1960, no. 751, pp. 10–11.
101. Laget, Françoise and Serge, and Jean-Paul Mazot, with the collaboration of Elizabeth Foch, *Le Grand Livre du Sport Féminin* (FMT Éditions, 1982), p. 275; Cordier, Marcel, "Marie Marvingt: Une Lorraine Qui Volait au Secours des Blessés," *La Vie la Santé*, July 1998, no. 6, pp. 28–29.
102. Cordier, Marcel, "Les Combats de Marie Marvingt," *La Nouvelle Revue Lorraine*, no. 27, August-September 2014, p. 20–23.
103. *L'Aérophile*, "Chronique des Vieilles Tiges," October 1943, p. iii; Coulet, Jacques, and Madame Coulet, interview, June 24, 1982; Mechelle, Ginette, interview, June 23, 1982.
104. Cordier, Marcel, "Les Combats de Marie Marvingt," *La Nouvelle Revue Lorraine*, no. 27, August-September 2014, p. 20–23; Marvingt, Marie, documents discovered in April 2012, maintained by the International Marie Marvingt Committee.
105. Martin-Auberdon, Astrid, "Marie Marvingt, une Femme Résistante Célébrée," *Sud Ouest*, May 20, 2016.
106. Bitsch, Maître, cited in Jacqueline Maire, letter, August 5, 1982.

Chapter 6

1. Pétain, Marshal Philippe, autograph signed December 8, 1931, in Cordier, Marcel, "Les Escales de Marie Marvingt," *Leurs Demeures en Lorraine: Tome II* (Éditions Pierron, 1983), p. 172.
2. *L'Est Républicain*, "Mlle Marie Marvingt, Chevalier de la Santé Publique," November 5, 1937.
3. J.P., "Marie Marvingt, Pilote d'Avion, Est Morte à 88 Ans: Championne en Tous Genres," *Le Monde*, December 17, 1963, p. 13; Davis, Jeffrey R., Robert Johnson, Jan Stepanek, Jennifer A. Fogarty, eds., *Fundamentals of Aerospace Medicine*, 4th ed. (2008), pp. 9, 14; *La Sfaxienne*, "Mademoiselle Marvingt à Sfax," April 15, 1924, p. 1; Stanley, Autumn, *Mothers and Daughters of Invention: Notes for a Revised History of Technology* (Scarecrow Press, 1993; Rutgers University Press, 1995), p. 201; *La Liberté*, "Marie Marvingt Est Morte à 88 ans," December 16, 1963, no. 6164, pp. 1 & 8.
4. *Le Matin*, "L'Aviation Sanitaire," October 24, 1921, p. 5; Reynolds, Siân, *France Between the Wars: Gender and Politics* (Routledge, 1996), p. 67; Ladame, Albert, "Petite Histoire de l'Aviation Sanitaire," *Air ANSORAA*, no. 133, April-May-June 2013, pp. 15–18; Probst, Ernst, "Marie Marvingt, Die Mütter der Luftambulanz," *Königinnen der Lufte von A bis Z* (2010), pp. 249–252, 602; *L'Illustration*, "Les Voyages Aeriens," February 15, 1919, no. 3963, p. 172; et alia.
5. *L'Air*, "Informations: Le Challenge du Capitaine Écheman," November 1, 1931; *Comoedia*, "L'Oeuvre de Mlle Marvingt Pour l'Aviation Sanitaire," November 18, 1931, p. 5.
6. Bichet, Gabriel (G.B.), "Marie Marvingt, la Fiancée du Danger, Est Morte Hier à Nancy," *L'Est Républicain*, December 15, 1963.
7. *Gil Blas*, "L'Ève Moderne," August 4, 1906, p. 1; Marchand, Robert, *Le Ciel N'a Pas de Toit* (Editions Berger Levrault, 1962), pp. 258–262.
8. *L'Est Républicain*, "Mlle Marie Marvingt Reçoit Deux Nouveaux Témoignages Officiels de Gratitude," February 21, 1958, p. 4; Action Catholique des Femmes, *Femmes Remarquables: Je Broderai Vos Noms* (ACGF, 2001), pp. 212–215.

9. *Républicain Lorrain*, "A Nancy et à Metz, Marie Marvingt, la 'Fiancée du Danger' Dira Quels Sont les Buts Humanitaires de Sa Mission," October 19, 1949, p. 2.

10. M.A., "Marie Marvingt, 'Fiancée du Danger,' Attend Encore Que Soit Relevé Son 'Défi Mondial,'" *Samedi Soir*, March 3, 1955, no. 505, pp. 1, 3; *La Sfaxienne*, "Mademoiselle Marvingt à Sfax," April 15, 1924, p. 1.

11. Floret, Robert, "Grand-Mère 'Casse-Cou,'" *Détective*, November 18, 1960, no. 751, pp. 10–11.

12. Proctor, Tammy M., *Civilians in a World at War, 1914–1918* (2010), p. 97.

13. Droz, Stanislas, *Les Femmes dans la Grande Guerre 1914–1918* (Vent d'Est, 2014), pp. 74, 75–78.

14. Dinan, R., "Les Enfants Prodiges: Mlle Marvingt Lance un Défi Mondial," *L'Ordre*, July 2, 1948; *Nancy-Ochey Flash*, "Marie Marvingt, 'La Fiancée du Danger,' no. 171, July 2003, p. 17.

15. *L'Aérophile*, "Chronique des Vieilles Tiges," October 1943, p. iii.

16. Brun, Mario, "Riviera Gazette," *Nice Matin*, November 22, 1956.

17. Gygax, Georges, "La Vie Prodigieuse de Marie Marvingt, la 'Fiancée du Danger'," *L'Illustré*, May 1, 1958, p. 42.

18. Gygax, Georges, "La 'Fiancée du Danger': La Vie Prodigieuse de Marie Marvingt," *L'Illustré*, April 24, 1958, pp. 34–36; L'Action Catholique Générale Féminine, op. cit.

19. Leblond, J.-J., "Pour le 52e Anniversaire de Sa Première Ascension en Ballon Marie Marvingt a Piloté un Avion Sanitaire Entre Bordeaux et Marseille," *Dauphiné*, November 1952, p. 3.

20. Maire, Jacqueline, letter, July 23, 1982.

21. Gygax, Georges, May 1, 1958, op. cit.

22. Boetsch, Jacques, "Une Vie en Équilibre" *Anciens Combattants du Monde*, May 1955, pp. 11–14.

23. Gygax, Georges, April 24, 1958, op. cit.

24. Bichet, Gabriel (G.B.), op. cit.

25. Lawson, Eric, and Jane Lawson, *The First Air Campaign: August 1914-November 1918* (2007), p. 57; Moore, Alison, ed., *Sexing Political Culture in the History of France* (2012).

26. *L'Aéro*, "L'avion-Ambulance," August 18, 1913, p. 2; Bell, Ryan Corbett, *The Ambulance: A History* (2009), pp, 121, 122, 124; Davis, Jeffrey R., Robert Johnson, Jan Stepanek, Jennifer A. Fogarty, eds., *Fundamentals of Aerospace Medicine*, 4th ed. (2008), pp. 9, 14; De Pauw, Linda Grant, *Battle Cries and Lullabies: Women in War From Prehistory to the Present* (2014), p. 224; Stanley, Autumn, *Mothers and Daughters of Invention: Notes for a Revised History of Technology* (Scarecrow Press, 1993; Rutgers University Press, 1995), p. 201; *Revue de l'Union des Femmes de France*, "Les Journées d'Aviation Sanitaire à l'Exposition Coloniale," Cctober 1931, p. 402.

27. Moolman, Valerie, *Women Aloft* (Time/Life Books, 1981), p. 30/Moolman, Valerie, *Les Femmes Aviateurs* (Time/Life Books, 1982), p. 30; Lawson, Eric, and Jane Lawson, *The First Air Campaign: August 1914-November 1918* (2007), p. 57.

28. *L'Aéro*, "L'avion-Ambulance," August 18, 1913, p. 2; Laurent, Michel-Yves, "Marie Marvingt, Surnommée 'Marie Casse-Cou,' Traversa au Début du Siècle, la Mer du Nord en Ballon; Elle Pilote Aujourd'hui à 86 Ans un Hélicoptère à Réaction," *L'Aurore* or *Paris Press*, November 2, 1960, no. 16.

29. Marvingt, Marie, "La Femme et l'Aviation au Service de la Charité: les Infirmières-Convoyeuses de l'Aviation Sanitaire," *Les Ailes*, December 21, 1933, p. 11; Marvingt, Marie, "Comment J'ai Conçu le Premier Avion Sanitaire," *Revue de la Société Scientifique et Historique de Documentation Aérienne*, war edition, no. 6, January-February 1940.

30. Rio, Armand, "La Fiancée du Danger," *Lecture Pour Tous*, April 1, 1913, vol. 15, no. 7, pp. 63–71; *The Strand Magazine*, "'The Bride of Danger,' An Interview With Mlle. Marie Marvingt," vol. 46, September 1913, pp. 187–194; D'Orliac, J., "L'Avion-Ambulance," *Le Journal [Paris]*, June 23, 1914, p. 6.

31. Beaupré, Norman, *The Man With the Easel of Horn: The Life and Works of Emile Friant* (Lumina Press, 2010), pp. 339–340.

32. Bichet, Gabriel (G.B.), "Marie Marvingt Lance Aujourd'hui l'Aviation Sanitaire Mondiale," *L'Est Républicain*, July 5, 1956, p. 2.

33. *L'Est Républicain*, "Conférence de Mlle Marvingt," October 16, 1929, p. 3.

34. *L'Est Républicain*, "La Remise du Challenge de Mlle Marvingt," November 17, 1931, p. 2.

35. Julliot, Ch.-L., "Une Ligue pour la Propagande de l'Aviation Sanitaire," *Les Ailes*, February 13, 1930, p. 7; *L'Aéro*, "Le Challenge Capitaine Echeman," June 14, 1935, p. 3; *Les Cahiers de la Santé Publique*, "Journées de l'Aviation Sanitaire Coloniale: Réception à L'Aéro-Club de France," June 25, 1931, p. 1043; *Les Ailes*, "L'Aviation Sanitaire à Orly," August 6, 1931, p. 14; *Les Ailes*, "Petites Nouvelles de France," October 15, 1931, p. 15; *Les Ailes*, "L'Aviation Sanitaire," July 7, 1938, p. 4; *L'Air*, "Informations: Le Challenge du Capitaine Écheman," November 1, 1931.

36. *L'Aérophile*, "Le IIe Congrès International de l'Aviation Sanitaire," September 1933, p. 265.

37. *Journal Officiel de la République Française*, "Ministère de l'Air: Dons et Legs., Officiers d'Académie," July 2, 1932, p. 7200.

38. D'Orliac, J., "L'Avion-Ambulance," *Le Journal [Paris]*, June 23, 1914, p. 6.

39. Noetinger, Jacques, "Reflets du Ciel: La Première de Toutes," *Vie et Bonté: Revue Officielle de la Croix Rouge Française*, no. 160, November 1965.

40. Powell, E. Alexander, *Vive la France!* (Charles Scribner's Sons, 1915), p. 74.

41. *L'Est Républicain*, "Nancy: La Journée," July 1, 1912, p. 2; *Le Petit Parisien*, "Mlle Marvingt: Conférence en Faveur de l'Aviation Militaire, June 13, 1912, p. 4; *L'Est Républicain*, "Féte Patriotique," November 7, 1912, p. 3; *L'Est Républicain*, "Au Theatre," November 12, 1912, p. 3; *Journal des Débats Politiques et Littéraires*, "L'Aviation au jour le jour," August 20, 1912, p. 2; *Journal de la Meurthe et des Vosges*, "Une Conference Sur l'Aviation," March 20, 1914, p. 3; *Journal de la Meurthe et des Vosges*, "Deux Heures Dans les Airs," March 28, 1914, p. 3; *Journal de la Meurthe et des Vosges*, "Ambulance de la Place St-Georges," December 30, 1915, p. 2; *Le Temps*, "Conférences d'Aviatrice," June 13, 1912, p. 5; *L'Homme Libre*, "Une Intéressante Conférence," December 19, 1938, p. 4; *Ligue Aéronautique de France*, "Les Conférences de Mlle Marvingt," May 1914, p. 75; *Les Ailes*, "Coups d'Ailes," October 6, 1921, p. 2; *Les Ailes*, "Aujourd'hui ...," November 3, 1932, p. 5; *Les Ailes*, "Coups d'Ailes," January 19, 1933, p. 15; *Les Ailes*, "Ici et Là," December 22, 1938, p. 4; *Association Général Aéronautique*, "Une Conférence de Mlle Marvingt," March 1913, p. 38; *Gazette des Eaux*, "Vittel: Aviation," January 6, 1912, p. 762; *L'Est Républicain*, "L'Aéroplane-Ambulance," 14 juin 1912, p. 3; *Le Petit Matin* [Tunis], "La Conférence de Mlle Marvingt au Cercle de l'Automobile-Club," March 10, 1932; *Le Petit Matin* [Tunis], "La Fiancée du Danger à l'Alliance Française,"

February 5, 1952; Verviers, Pierre, "Une Femme Veut Doter la France d'Avions-Ambulances," *La Vie Sportive du Nord et du Pas-de-Calais*, November 16, 1912, p. 5 ; *Républicain Lorrain*, "A Nancy et à Metz, Marie Marvingt, la 'Fiancée du Danger' Dira Quels Sont les Buts Humanitaires de Sa Mission," October 19, 1949, p. 2; *Le Matin*, "Aéronautique," June 13, 1912, p. 5; *Le Matin*, "Conférencière," August 19, 1912, p. 4; *Revue Aéronautique de France*, "Sur la Côte d'Azur: Conférences de Mlle Marvingt," May 1928, p. 12; *La Presse*, "L'Oevure de Mlle Marvingt: L'Avion-Ambulance," August 18, 1913, p. 3; *Revue Aéronautique de France*, "Une Belle Tournée de Conférences," October 1932, p. 5; *Revue Aéronautique de France*, "Mlle Marvingt au Maroc," February-March 1934, p. 4; *Le Progrès*, "Mlle Marvingt à Orléansville," July 2, 1925, p. 2; *Comoedia*, "Les Conférences Scolaires de Mlle Marvingt Sur l'Activité de Notre Aviation," November 26, 1931, p. 4; and many hundreds more.

42. *Indianapolis Star*, "The Woman With War Wings," October 11, 1914, p. 55.

43. *L'Aéro*, "Les Conférences de Mlle Marvingt," October 10, 1912, p. 1; *L'Aéro*, "Propos en l'Air," November 11, 1912, p. 1; *L'Aéro*, "L'avion-Ambulance," August 18, 1913, p. 2; *L'Aéro*, "Mlle Marvingt Conférence au Havre," November 19, 1913, p. 2; *L'Aéro*, "Les Conférences de Mlle Marvingt," March 20, 1914, p. 6; *L'Aéro*, "Les Conférences de Mlle Marvingt," May 2, 1914, p. 6; *L'Aéro*, "Mlle Marvingt Conférencie à Bourges," May 14, 1914, p. 4; *L'Aéro*, "Une Conférence de Mlle Marvingt à Nevers," May 15, 1914, p. 3; *L'Ouest Éclair [Caen]*, "Anciens de l'Aéronautique," 2 juin 1928, p. 6; *L'Oued-Sahel*, "Conférence de Mlle Marie Marvingt," January 4, 1923, p. 2; Barthélemy, Général, "La Femme la Plus Extraordinaire du Siècle: Marie Marvingt, la 'Fiancée du Danger,'" *L'Étrange Race des Hommes Volants* (Éditions France-Empire, 1979), pp. 241–254; D'Orliac, J., "L'Avion-Ambulance," *Le Journal [Paris]*, June 23, 1914, p. 6; *L'Echo d'Alger*, "Mlle Marvingt Conférence: Elle Veut Réunir les Fonds Nécessaires à l'Achat d'un Avion-Ambulance," May 23, 1914, p. 2; *Revue Aéronautique de France*, "Mademoiselle Marvingt: Une Grande Propagandiste," October 1929, pp. 7, 15.

44. Saladin, Raymond, "Figures et Événements: Marie Marvingt," *Aviation Magazine*, March 24, 1955, pp. 12–13; *Le Petit Parisien*, "Les Principales Émissions Françaises," January 21, 1933, p. 10.

45. Saladin, Raymond,op. cit.

46. Beaubois, Henry, "Ceux Qui Ouvrirent la Route au Ciel: Marie Marvingt," *Revue Aéronautique de France*, August 13, 1937.

47. Bowen, Barry D., "Marie Marvingt: The Premier Sportswoman of France and Pioneer in the Development of Aeromedical Evacuation," special to AirAmbulanceServices.com, March 6, 2013.

48. *L'Oued-Sahel*, "Conférence de Mlle Marie Marvingt," January 4, 1923, p. 2; Journal des Débats Politiques et Littéraires, "Conférences de Mlle Marvingt," June 13, 1912, p. 2.;; *L'Echo d'Alger*, "Une Conférence de Mlle Marvingt à Mascara," July 8, 1925, p. 6.

49. *Revue Aéronautique de France*, "Mlle Marvingt au Mans," July 1928, p. 10.

50. *L'Echo d'Alger*, "Initiation Aux Sciences Psychiques," May 6, 1925, p. 2; *L'Echo d'Alger*, "Huit Cours et Conférences à Alger de Mlle Marie Marvingt," May 15, 1925, p. 3.

51. J.P., "Marie Marvingt, Pilote d'Avion, Est Morte à 88 Ans: Championne en Tous Genres," *Le Monde*, December 17, 1963, p. 13.

52. Marvingt, Marie, postcard to Marthe Roumier, December 29, 1937.

53. J.B.L., "On Peut Fort Bien 'Être et Avoir Été,' Démontre Marie Marvingt," *La Presse*, February 3, 1952.

54. Marvingt, Marie, postcard to Dr. and Mrs. Thévenin, January 19, 1952.

55. *Les Annales Coloniales*, "Conférences," May 7, 1925, p. 2; *L'Echo d'Alger*, "Une Conférence de Mlle Marvingt à Mascara," July 8, 1925, p. 6.

56. *L'Ouest Éclair [Caen]*, "Ligue Aéronautique de France," May 23, 1928, p. 6; *L'Ouest Éclair [Caen]*, "Le Gala de la Ligue Aéronautique de France," April 16, 1939, p. 4; *L'Ouest Éclair [Nantes]* and *L'Ouest Éclair [Rennes]*, "Une Exploratrice Vient de Parcourir 56.114 Kilomètres," 16 juin 1926, p. 3; *Indépendant*, "Mlle Marie Marvingt à Pau: Deux Causeries Littéraires et Sportives," June 16, 1927.

57. Bowen, Barry D., op. cit.

58. *Indépendant*, "Mlle Marie Marvingt à Pau: Deux Causeries Littéraires et Sportives," June 16, 1927.

59. *Républicain Lorrain*, "Marie Marvingt, la 'Fiancée du Danger,' N'est Plus," December 15, 1963, pp. 1, 4.

60. Olivaint, Maurice, "Causerie d'un Algérois," *Annales Africaines*, May 22, 1925, pp. 324–325; *Indépendant*, "Mlle Marie Marvingt à Pau: Deux Causeries Littéraires et Sportives," June 16, 1927.

61. *L'Ouest Éclair [Caen]*, "Pour l'Essor de l'Aviation Française," March 25, 1932, p. 6.

62. *La Croix*, Mlle Marvingt Conference en Faveur de l'Aviation Militaire," no. 8972, June 16–17, 1912, p. 6; *L'Aéro*, "Mlle Marvingt Conférencie à Bourges," May 14, 1914, p. 4; *L'Ouest Éclair [Caen]*, "Le Gala de l'Aéro-Club de Caen et du Calvados," March 10, 1933, p. 4; *L'Echo d'Alger*, "Département d'Alger: Teniet-el-Haad," July 2, 1925, p. 3; *L'Echo d'Alger*, "Constantine: Ain-M'lila," July 16, 1932, p. 7; *Revue Aéronautique de France*, "Mademoiselle Marvingt: Une Grande Propagandiste," October 1929, pp. 7, 15; *Revue Aéronautique de France*, "Une Conférence de Mlle Marvingt à Nancy," November 1929, pp. 12–13; *Revue Aéronautique de France*, "Mademoiselle Marvingt," January 1931, pp. 11, 12; Renoux-Barès, E., "Une Héroïne Parle à Pau de l'Aviation, École d'Énergie," *Revue Aéronautique de France*, October 1927, p. 11; *Républicain Lorrain*, "Marie Marvingt, la 'Fiancée du Danger,' n'Est Plus," December 15, 1963, pp. 1, 4; *L'Echo d'Alger*, "Département d'Alger: Teniet-el-Haad," July 2, 1925, p. 3; et alia.

63. *The Washington Post*, "France's Pioneer Woman Flier Wins Acclaim in a Dozen Lines," January 15, 1936, p. 13.

64. *L'Echo de Bougie*, "Mademoiselle Marvingt à Bougie," October 18, 1924, p. 3.

65. *L'Est Républicain*, "Mlle Marvingt Parle de la Tunisie," January 26, 1932, p. 3.

66. *Ibid.*

67. Marvingt, Marie, "Invitation a Une Causerie," handwritten by Marie Marvingt and reprinted by L'Imprimerie Berger-Levrault, Nancy and Paris.

68. *Journal de la Meurthe et des Vosges*, "Deux Heures Dans les Airs," March 28, 1914, p. 3; et alia.

69. Marvingt, Marie, "Invitation a Une Causerie," op. cit.

70. "Marie Marvingt: 56.000 kilomètres en Afrique du Nord," newspaper article, April 16, 1928.

71. Marvingt, Marie, "La Femme et l'Aviation au Service de la Charité." op. cit.

72. Marvingt, Marie, "La Femme et l'Aviation au Service de la Charité: Les Infirmières-Convoyeuses de l'Aviation Sanitaire," *Les Ailes*, December 21, 1933, p. 11.

73. D'Orliac, J., "L'Avion-Ambulance," *Le Journal [Paris]*, June 23, 1914, p. 6.

74. *L'Est Républicain*, "Causerie Scolaire sur l'Aviation," July 24, 1929, p. 3.
75. Revue Aéronautique de France, "La Ligue Aéronautique en Province," May 1927, pp. 8–9, 13.
76. *L'Est Républicain*, "Le Buste de Mlle Marvingt," July 9, 1932, p. 3.
77. Leblond, J.-J., op. cit.
78. *Ibid.*
79. *Revue Aéronautique de France*, "Distinctions Méritées," July 1926, p. 47; Julliot, Ch.-L., "Les Onze Voeux de l'Aviation Sanitaire," *Les Ailes*, September 17, 1931, p. 7; Mathis, Suzy, "Aviatrices d'Hier et d'Aujourd'hui: Marie Marvingt," *Aviation Française*, May 8, 1946, p. 9; *L'Est Républicain*, "Le Suprême Adieu de Nancy à Marie Marvingt," December 18, 1963; Petit, Edmond, *Histoire mondiale de l'aviation* (Hachette, 1967), p. 73; et alia.
80. Lam, David M., "Marie Marvingt and the Development of Aeromedical Evacuation," *Aviation, Space, and Environmental Medicine*, vol. 74, no. 8, August 2003, pp. 863–868.
81. *Indianapolis Star*, "The Woman With War Wings," October 11, 1914, p. 55.
82. Marvingt, Marie, "Comment J'ai Crée l'A.S.C.A. (Aviation Sanitaire Civile au Maroc)," *L'Echo des Ailes*, June 12, 1935, p. 5.
83. *Ibid.*
84. Marvingt, Marie, "La Femme et l'Aviation au Service de la Charité: Les Infirmières-Convoyeuses de l'Aviation Sanitaire," *Les Ailes*, December 21, 1933, p. 11.
85. Lassalle, Émile J., *Les Cent Premiers Aviateurs Brevetés au Monde et la Naissance de l'Aviation* (Nauticaero, 1960), p. 21.
86. Marvingt, Marie, "La Femme et l'Aviation au Service de la Charité," op. cit.; Holleran, Renee S., *ASTNA Patient Transport: Principles and Practice* (2009), p. 5.
87. *Ibid.*
88. *Ibid.*
89. Bichet, Gabriel (G.B.), op. cit.; Davis, Jeffrey R., Robert Johnson, Jan Stepanek, Jennifer A. Fogarty, eds., *Fundamentals of Aerospace Medicine*, 4th ed. (2008), pp. 9, 14; Stewart, Mary Lynn, *For Health and Beauty: Physical Culture for Frenchwoman 1880s-1930s* (The Johns Hopkins University Press, 2001), pp. 168–169; Ladame, Albert, "Mémoires de Lorraine: Marie Marvingt," *Air ANSORAA*, no. 137, April-May-June, 2014; *Revue Aéronautique de France*, "Présentation au Bourget de la Ppremière Promotion des Infirmières de l'Air," June 1935, pp. 5, 7.
90. *Les Ailes*, "Ici et Là," June 13, 1935, p. 19; Holleran, Renee S., *ASTNA Patient Transport: Principles and Practice* (2009), p. 5.
91. Marvingt, Marie, "Comment J'ai Conçu le Premier Avion Sanitaire," op. cit.
92. *Ibid.*
93. Pelletier, George E., "Wings for the Wounded," *Oakland Tribune*, September 17, 1939, p. 78; also published in the *Nebraska Sunday Journal and Star*, September 17, 1939, p. 33.
94. *Ibid.*
95. *The Sydney Morning Herald*, "Airwomen's Flying Ambulance Corps: The Work of the Pioneer Frenchwoman," November 6, 1939, p. 2.
96. Marvingt, Marie, "Anniversaire de Fondation de l'A.S.C.A.M.," *La Vigie Marocaine*, January 30, 1935.
97. *Les Annales Coloniales*, "L'Aviation," April 13, 1935, p. 2; *Le Matin*, "Constitution au Maroc d'Une Ligue de l'Aviation Sanitaire Civile," April 11, 1935, p. 5.
98. Marvingt, Marie, "Anniversaire de Fondation de l'A.S.C.A.M.," op. cit.; Marvingt, Marie, "Comment J'ai Conçu le Premier Avion Sanitaire," op. cit.
99. *Ibid.*
100. Chevalier, Frédérique, "Marie Marvingt, la Fiancée des Airs (1875–1963)," *Les Grandes Aventurières* (City Editions, 2007), pp. 85–95; et alia.
101. Grey, C.G., *New York Times*, "The Utility of the Autogiro: A Debate," August 16, 1931.
102. Ackerman, Gordon, "Fiancée of Danger," *Sports Illustrated*, June 26, 1961, pages 61–64.
103. Enselme-Trichard, Renée, letter, July 2, 1982.
104. Marvingt, Marie, "Les Ailes Féminines au Service de la Charité," *La Croix*, no. 4956, November 28, 1931, pp. 1–2; Marvingt, Marie, "La Femme et l'Aviation au Service de la Charité," op. cit.
105. Lam, David M., op. cit.
106. Marvingt, Marie, "Comment J'ai Conçu le Premier Avion Sanitaire," op. cit.
107. *Ibid.*
108. *Ibid.*
109. *L'Illustration*, "Les Voyages Aeriens," February 15, 1919, no. 3963, p. 172.
110. *L'Est Républicain*, "Une Soeur Blanche Est Evacuée en Avion d'Alger à Paris," April 13, 1938, p. 5; *Les Ailes*, "Une Évacuation Aérienne d'Alger à Paris," April 21, 1938, p. 4; *Revue Aéronautique de France*, "Nouvelles," May-June 1938, p. 10.
111. Marvingt, Marie, "Un Avion de la Swissair Effectue au Havre une Évacuation Sanitaire," *Les Ailes*, August 8, 1935, p. 7.
112. *Ibid.*
113. *Ibid.*
114. *Revue Aéronautique de France*, "Une Héroïne Parle à Pau de l'Aviation, École d'Énergie," October 1927, p. 11.
115. Julliot, Ch.-L., "Une Ligue Pour la Propagande de l'Aviation Sanitaire," *Les Ailes*, February 13, 1930, p. 7. *L'Homme Libre*, "Les Amis de l'Aviation Sanitaire," July 10, 1929, p. 4; *Le Journal [Paris]*, "Le 1er Congrès de l'Aviation Sanitaire," May 16, 1929, p. 4.
116. *Le Monde*, "Marie Marvingt, Pionnière de l'Aviation," July 24, 2004.
117. *Journal des Débats Politiques et Littéraires*, "L'Aviation Sanitaire," July 7, 1929, p. 5; *Les Ailes*, "Petites Nouvelles de France," July 11, 1929, p. 14; *La Médecine Internationale*, "Les Amis de l'Aviation Sanitaire," August 1929, p. lxi.
118. *L'Association Médicale*, "Les Amis de l'Aviation Sanitaire," March 1930, pp. 115–116.
119. *L'Est Républicain*, "Au Pavillon de l'Aéronautique: Le Gala des Ailes Qui Sauvent," July 23, 1936, p. 4.
120. *L'Aérophile*, "Le IIe Congrès International de l'Aviation Sanitaire," September 1933, p. 265; *Le Temps*, "Au Congrès d'Aviation Sanitaire," June 1, 1933, p. 4; *L'Est Républicain*, "IIe Congrès International de l'Aviation Sanitaire," May 21, 1933, p. 1.
121. *Journal des Débats Politiques et Littéraires*, "La Vie Aérienne," June 9, 1935, p. 5; *L'Est Républicain*, "Le Challenge Capitaine Echeman," June 9, 1935, p. 4; *L'Européen*, "Ce Que Fut le 3e Congrès International de l'Aviation Sanitaire," June 28, 1935, p. 5; Richet, Charles, "L'Aviation Sanitaire C'est le Progrès Sans Larmes," *Le Matin*, July 6, 1935, p. 1.
122. *Ibid.*
123. *L'Européen*, "Ce Que Fut le 3e Congrès International de l'Aviation Sanitaire," June 28, 1935, p. 5.
124. Marvingt, Marie, "Comment j'ai conçu le premier avion sanitaire," op. cit.

125. Marvingt, Marie, "Anniversaire de Fondation de l'A.S.C.A.M.," op. cit.
126. Marvingt, Marie, "Comment j'ai conçu le premier avion sanitaire," *Revue de la Société Scientifique et Historique de Documentation Aérienne*, édition de guerre, no. 6, janvier-février 1940.
127. *Les Ailes*, "L'Aviation Sanitaire et l'Exposition de Lille," June 22, 1939, p. 5.
128. *L'Homme Libre*, "Le Challenge Capitaine-Echeman," October 7, 1931, p. 4.
129. *Journal des Débats Politiques et Littéraires*, "A l'Exposition Coloniale: Les Journées d'Aviation Sanitaire," July 26, 1931, p. 4; *Journal des Débats Politiques et Littéraires*, "Le Congrès d'Aviation Sanitaire Coloniale," August 1, 1931, p. 2.
130. Marvingt, Marie, "L'Aviation sanitaire coloniale aux derniers congrès d'aéronautique," *L'Aérophile*, vol. 26, no. 11, 15 novembre 1931, pp. 321, 345.
131. *Ibid.*
132. Marvingt, Marie, "L'Aviation Sanitaire Mondiale," report no. 1262, speech give at the Third Congrès National de L'Aviation Française, April 1–5, 1947.
133. *Ibid.*
134. Marvingt, Marie, "La Femme et l'Aviation au Service de la Charité," op. cit.
135. *Ibid.*
136. *L'Est Républicain*, "Mlle Marie Marvingt, Chevalier de la Santé Publique," November 5, 1937.
137. *L'Homme Libre*, "Nouvelles Aériennes: Mlle Marvingt à Argenteuil," January 10, 1933, p. 4.
138. Marvingt, Marie, "Comment J'ai Crée l'A.S.C.A. (Aviation Sanitaire Civile au Maroc)," *L'Echo des Ailes*, June 12, 1935, p. 5; R. de B., *L'Illustration*, "Les Ailes Qui Sauvent," November 17, 1934, no. 4785, p. 33; *L'Afrique du Nord Illustré*, "Les Ailes Qui Sauvent," July 28, 1934, p. 7; *Les Ailes*, "Les 'Ailes Qui Sauvent,'" November 18, 1934, p. 8.
139. *L'Afrique du Nord Illustré*, "Les Ailes Qui Sauvent," July 28, 1934, p. 7.
140. *L'Afrique du Nord Illustré*, op. cit.
141. *Ibid.*
142. *Ibid.*
143. Colin, Marie-Thérèse, "On a Retrouvé 'les Ailes de la Colombe,' *Est Républicain*, May 9, 1993.
144. *L'Afrique du Nord Illustré*, "Les Ailes Qui Sauvent," July 28, 1934, p. 7.
145. *L'Européen*, "Un Film sur l'Aviation," November 23, 1934, p. 5; *Les Ailes*, "Les 'Ailes Qui Sauvent,'" November 18, 1934, p. 8.
146. Marvingt, Marie, postcard to Madame Roumier, July 13, 1934.
147. *L'Afrique du Nord Illustré*, "Les Ailes Qui Sauvent," July 28, 1934, p. 7.
148. *L'Est Républicain*, "Mlle Marie Marvingt, Chevalier de la Santé Publique," November 5, 1937.
149. *Ibid.*
150. Bichet, Gabriel, op. cit.
151. Lam, David M., op. cit.

Chapter 7

1. Rio, Armand, "La Fiancée du Danger," *Lecture Pour Tous*, April 1, 1913, vol. 15, no. 7, pp. 63–71; Margot, Olivier, "L'Indomptable: Marie Marvingt," *L'Équipe*, August 17, 2013, pp. 60–63.
2. Atwood, Margaret, "The Whirlpool Rapids," *Bluebeard's Egg* (Houghton Mifflin, 1986), p. 120.
3. *Républicain Lorrain*, "A Nancy et à Metz, Marie Marvingt, La 'Fiancée du Danger' Dira Quels Sont les Buts Humanitaires de Sa Mission," October 19, 1949, p. 2; Féral, Roger, "Marie Marvingt (86 Ans), 'La Fiancée du Danger,' A Passé La Toussaint Dans un Hélicoptère à Réaction," *Télé-Paris*, November 1958; Estrada de Tourniel, Jérôme, "La Fiancée du Danger," *Passions Grand Est*, December 2003, no. 26, pp. 58–62.
4. *L'Echo Sportif du Centre et de l'Ouest*, "Une Carrière Féminine Sportive," 1914; Barthélemy, Général, "La Femme la Plus Extraordinaire du Siècle: Marie Marvingt, la 'Fiancée du Danger,'" *L'Étrange Race des Hommes Volants* (Éditions France-Empire, 1979), pp. 241–254.
5. Marvingt, Marie, "La Mer du Nord Traversée en Ballon en Pleine Tempête au Milieu de la Nuit," *Le Petit Journal*, November 22, 1909, no. 17132, pp. 1–2.
6. Garnier, Émile, in *L'Aéro*, "Souvenirs Aériens," October 4, 1912, p. 2.
7. Rio, Armand, op. cit.
8. Boetsch, Jacques, "Une Vie en Équilibre" *Anciens Combattants du Monde*, May 1955, pp. 11–14.
9. Blanc, Gaston, letter, May 23, 1983.
10. Marvingt, Marie, "Trois Morts Mettent en Deuil les Ailes de Lorraine: Hommage d'une 'Vieille Tige,'" *Dimanche Éclair*, September 17, 1950, p. 3.
11. H.O., "Mlle Marvingt, La 'Fiancée du Danger,' Nous Révèlera à Tunis les Sources de Son Extraordinaire Dynamisme," *Le Petit Matin* [Tunis], February 3, 1952.
12. Boulet, Jean, "De Passage au Maroc: Mlle Marie Marvingt, la Fiancée du Danger, Témoin ou Victime de Vingt Catastrophes, Elle Est la Femme la Plus Extraordinaire du Monde," Moroccan newspaper, 1951, pp. 1, 4; Leblond, J.-J., "Pour le 52e Anniversaire de Sa Première Ascension en Ballon Marie Marvingt a Piloté un Avion Sanitaire Entre Bordeaux et Marseille," *Dauphiné*, November 1952, p. 3.
13. Dall'Asta, Monica, "Pearl, the Swift One, or the Extraordinary Adventures of Pearl White in France," in Dahlquist, Marina, ed., *Exporting Perilous Pauline: Pearl White and the Serial Film Craze* (2013), p. 71.
14. White, Pearl, *Just Me* (1919), p. 160.
15. Dall'Asta, Monica, op. cit., pp. 73–74.
16. Dahlquist, Marina, ed., *Exporting Perilous Pauline: Pearl White and the Serial Film Craze* (2013), p. 3.
17. Dall'Asta, Monica, op. cit., pp. 71, 72.
18. White, Pearl, op. cit., pp. 51, 158, 162, 163.
19. *Ibid.*, p. 32.
20. *Ibid.*, p. 14.
21. *Ibid.*, p. 161.
22. *Ibid.*, pp. 162, 163.
23. *Ibid.*, pp. 173–174.
24. Daurat, Michel, "A Quatre-Vingts Ans, Marie Marvingt Demeure Toujours la Fiancée du Danger," *Lorraine Magazine*, no. 9, February 1955, pp. 10–13; Talbott, Earl G., "Marie Marvingt, 'Fiancée of Danger': Flirted With Death 88 Years," *New York Herald Tribune*, December 16, 1963; Ackerman, Gordon, "Fiancée of Danger," *Sports Illustrated*, June 26, 1961, pages 61–64; Estrada de Tourniel, Jérôme, op. cit.
25. *Marie France*, "La Fiancée du Danger' Lance un Défi aux Femmes du Monde Entier," June 22, 1948, p. 7; Lauwick, Hervé, *Heroines of the Sky (Conquérantes du Ciel)*, tr. from French by James Cleugh (Frederick Muller, 1960), pp. 25, 28; Branchu, Marc, "Les Cent Vies de Marie Marvingt," *Magazine Air France*, no. 99, July 2005, pp. 42–44, 46; *L'Est Républicain*, November 18, 1952, June 6, 1958, or November 8, 1958.
26. Coulet, Madame, interview, June 24, 1982.
27. Greenwell, Daisy, "The Top 10 Female Risk Takers, *The Times [London]*, March 3, 2011.
28. Besançon, Madame, letter, June 28, 1982.

29. Marvingt, Marie, "Les Ailes Féminines au Service de la Charité," *La Croix*, November 28, 1931, no. 4956, pp. 1–2; Marvingt, Marie, "Comment J'ai Conçu le Premier Avion Sanitaire," *Revue de la Société Scientifique et Historique de Documentation Aérienne*, war edition, no. 6, January-February 1940; Boetsch, Jacques, op. cit.; Talbott, Earl G., "Marie Marvingt, 'Fiancée of Danger': Flirted With Death 88 Years," *New York Herald Tribune*, December 16, 1963; Barthélemy, Général, op. cit.

30. Ackerman, Gordon, op. cit.; Action Catholique des Femmes, *Femmes Remarquables: Je Broderai Vos Noms* (ACGF, 2001), pp. 212–215.

31. Boetsch, Jacques, op. cit.

32. Barthélemy, Général, op. cit.

33. Marchand, Robert, *Le Ciel n'a Pas de Toit* (Éditions Berger Levrault, 1962), pp. 258–262.

34. Boetsch, Jacques, op. cit.

35. Saladin, Raymond, "Figures et Événements: Marie Marvingt," *Aviation Magazine*, March 24, 1955, pp. 12–13.

36. Marchand, Robert, op. cit.

37. Daurat, Michel, op. cit.; Estrada de Tourniel, Jérôme, op. cit.; Lam, David M., "Marie Marvingt and the Development of Aeromedical Evacuation," *Aviation, Space, and Environmental Medicine*, Vol. 74, No. 8, August 2003, pp. 863–868; Margot, Olivier, op. cit.; Erbstein, Roland, in collaboration with Jean Matt, "Cartes Postales d'Hier et d'Aujourd'hui: En 1909, 'la Fiancée du Danger' Ouvrait aux Femmes la Voie des Airs," *L'Est Républicain*," March 23, 1980.

38. *L'Aurore*, "La Coupe 'Fémina,'" November 29, 1910, p. 3.

39. *Gil Blas*, "L'Aviation en Province," August 16, 1911, p. 4; Boetsch, Jacques, op. cit.; Gygax, Georges, "La Vie Prodigieuse de Marie Marvingt, la 'Fiancée du Danger,'" *L'Illustré,* May 1, 1958, pp. 40–42.

40. Rio, Armand, op. cit.

41. *La Presse*, "Nouvelle Chute de Mlle Marvingt," August 16, 1911, p. 3; *Le XIXème Siècle*, "Mlle Marvingt Tombe Encore," August 17, 1911, p. 3; *Le Rappel*, "Mlle Marvingt Tombe Encore," August 17, 1911, p. 3.

42. Liégeois, Achille, "Les Nancéiennes Loin de Chez Elles: Mlle Marie Marvingt Rentre d'un Long Voyage: Elle Nous Raconte Ses Impressions," *L'Est Républicain*, 23 août 1926, p. 1.

43. Codron, Charles, letter, circa 1905 (Lhérault collection).

44. Daurat, Michel, op. cit.; Gygax, Georges, op. cit.

45. "La Conférence de Mlle Marvingt à Notre-Dame de Sion," newspaper clipping, January 25, 1952.

46. *Alrededor del Mundo*, "Biografía Extranjera Contemporánea: Mlle Marvingt," November 2, 1913, no. 753, p. 361.

47. Noetinger, Jacques, "Reflets du Ciel: La Première de Toutes," *Vie et Bonté: Revue Officielle de la Croix Rouge Française*, no. 160, November 1965.

48. Ackerman, Gordon, op. cit.

49. Rio, Armand, op. cit.

50. Boulet, Jean, op. cit.; Leblond, J.-J., op. cit.

51. *Le Monde*, "Nouveaux Attentats," May 20, 1952.

52. Leblond, J.-J., op. cit.

53. *Le Monde*, op. cit.

54. Desormière, Yves, "A 80 Ans Marie Marvingt, la 'Fiancée du Danger,' Est la Seule Femme au Monde à Être Titulaire de Quatre Brevets de Pilotage: Ballon, Avion, Hydravion, Secourisme de l'Air," *La Croix*, March 4, 1955, pp. 1, 4.

55. Marvingt, Marie, postcard to Dr. And Mrs. Thévenin, January 19, 1952.

56. *Le Pays Lorrain*, "Nouvelles Lorraines," no. 10, 1952, p. 142.

57. E.B., "La Fiancée du Danger: Une Célèbre Femme de l'Air Est à Limoges" *Notre Province [Limoges]*, June 1944, p. 160.

58. *Ibid.*

59. Mechelle, Ginette, interview, June 23, 1982.

60. Marvingt, Marie, "Souvenirs de Chalcidique Tirés de Mon Carnet de Route: A Propos du Tremblement de Terre de Grèce," *L'Éclair de l'Est*, October 1, 1932, p. 1.

61. *Revue Aéronautique de France*, "Mademoiselle Marvingt," January 1931, pp. 11, 12.

62. X.B., "D'Ultimes Témoignages Viennent Encore Préciser Quelques Traits Attachants de Marie Marvingt, Illustre Figure Lorraine," *Républicain Lorrain*, January 24, 1964, p. 2.

63. Michel-Royer, Dr. Jean, recorded interview with Marie Marvingt, January 1963.

64. Rio, Armand, op. cit.; *Alrededor del Mundo*, op. cit.

65. Rio, Armand, op. cit.

66. H.O., op. cit.

67. Michel-Royer, Dr. Jean, op. cit.

Chapter 8

1. Ackerman, Gordon, "Fiancée of Danger," *Sports Illustrated*, June 26, 1961, pages 61–64.

2. Bethem-Edwards, Matilda, *Home Life in France* (1905), p. 98.

3. *France Aviation*, "Nos Lecteurs Nous Écrivent: Marie Marvingt," September 1957, p. C.

4. *L'Illustration*, "Les Obsèques du Capitaine Echeman," no. 3613, May 25, 1912, p. 444.

5. *Ibid.*

6. *Le Petit Parisien*, "A Mourmelon," October 18, 1910, p. 5.

7. Pelletier, George E., "Wings for the Wounded," *Oakland Tribune*, September 17, 1939, p. 78.

8. Dalbanne, J., "Le Ve Concours International de Ski du 'Club Alpin Français,'" *Armée et Marine*, February 28, 1911, cover, p. 246.

9. *La Vie au Grand Air*, "Les Aviateurs aux Sports d'Hiver," February 17, 1912, p. 3; "Ski (Débuts en France)," *La Grande Encyclopédie de La Montagne* (Éditions Atlas, 1978), no. 109, pp. 2179–2180.

10. R.G., "Une Femme Qui Pratique Tous les Sports: Mlle Marvingt," *Le Miroir des Sports*, October 28, 1920, pp. 260–261.

11. *L'Illustration*, "La Chute Mortelle du Capitaine Echeman," May 18, 1912, pp. 420–421;
L'Illustration, "Les Obsèques du Capitaine Echeman," May 25, 1912, no. 3613, p. 444.

12. *Ibid.*

13. *L'Illustration*, "Les Obsèques du Capitaine Echeman," May 25, 1912, no. 3613, p. 444.

14. to come

15. *Nord-Touriste*, "Nos Aviateurs Militaires: Lettre de Paul-Maurice Echeman à Sa Soeur," October 1912, pp. 231–232.

16. *L'Aérophile*, "Réception en l'Honneur du Maréchal Lyautey et des Membres du Congrès Aéronautique Coloniale," November 15, 1931, pp. 349–351.

17. *Ibid.*; *Les Ailes*, "Coups d'Ailes," November 5, 1931, p. 15; *L'Est Républicain*, "La Remise du Challenge de Mlle Marvingt," November 17, 1931, p. 2; *Comoedia*, "L'Oeuvre de Mlle Marvingt Pour l'Aviation Sanitaire," November 18, 1931, p. 5; et alia.

18. *L'Est Républicain*, "Il y a Vingt Ans Mourait Marie Marvingt," December 14, 1983.
19. Blanc, Colonel Gaston, and Madame, interview, June 24, 1982.
20. Coulet, Jacques, and Madame Coulet, interview, June 24, 1982; Méchelle, Ginette, interview, June 23, 1982.
21. Coulet, op. cit.
22. Gerbaud, A. Lorac, letter to Phillipe Bayart, June 28, 1988.
23. M.A., "Marie Marvingt, 'Fiancée du Danger,' Attend Encore Que Soit Relevé Son 'Défi Mondial,'" *Samedi Soir*, March 3, 1955, no. 505, pp. 1, 3; Faure, Odile, "Marie Marvingt, La Fiancée du Danger," *Massif Central Magazine*, March-April 1996, pp. 68–71.
24. *L'Aérophile*, "Le Banquet de l'Aéro-Club de France en l'Honneur de M. Laurent-Eynac, Premier Ministre de l'Air," December 1–15, 1928, p. 374.
25. Coulet, op. cit.; *L'Est Républicain*, "Il y a Vingt Ans Mourait Marie Marvingt," December 14, 1983.
26. Méchelle, Ginette, op. cit.
27. Bitsch, Monsieur, in Jacqueline Maire, letter, August 5, 1982.
28. Bouchet, Madame, in Jacqueline Maire, letter, July 5, 1982.
29. Marvingt, Marie, "Les Femmes et le Ski," in Louis Magnus and R. de la Fregeolière, *Les Sports d'Hiver* (Éditions Pierre Lafitte et Cie, 1911; Éditions Slatkine, 1979), pp. 176–181.
30. Codron, Charles, letter, circa 1905 (Lhérault collection).
31. Cordier, Marcel, "Marie Marvingt: Une Femme de Foi," *Chrétiens Dans la Ville*, no 73, June 1992, p. 6.
32. Curral-Couttet, Gaby, "Marie Marvingt, la Fiancée du Danger," *Les Folles Années de Chamonix* (Éditions France-Empire, 1984), pp. 75–78.
33. Gygax, Georges, "La Vie Prodigieuse de Marie Marvingt, la 'Fiancée du Danger,'" *L'Illustré*, May 1, 1958, pp. 40–42.
34. Gygax, Georges, "La 'Fiancée du Danger': La Vie Prodigieuse de Marie Marvingt," *L'Illustré*, April 24, 1958, pp. 34–36.
35. Didier, Gény, in Jacqueline Maire, letter, July 5, 1982.
36. "Éleve-pilote à 80 ans," newspaper clipping, 1955.
37. Plauche-Gillon, Jéhanne, in Jacqueline Maire, letter, July 29, 1982.
38. Conte, Arthur, *Le Premier Janvier 1920* (Pion, 1979), p. 277; Cordier, Marcel, "Les Escales de Marie Marvingt," *Leurs Demeures en Lorraine: Tome II* (Éditions Pierron, 1983), pp. 163; Yelnick, Claude, "Adieu à Marie Marvingt," *Icare*, no. 29, March 1964, pp. 60–65; Marchal, Mireille, "Du Ciel à la Carrière Pour la 'Fiancée du Danger," *Le Republicain Lorrain*, June 8, 1984; Frémy, Dominique et Michèle Frémy, *Quid* (Robert Laffont, 1988), p. 1166.
39. Boetsch, Jacques, "Une Vie en Équilibre" *Anciens Combattants du Monde*, May 1955, pp. 11–14, with additions and underlining in MM's handwriting.
40. Gygax, Georges, April 24, 1958, op. cit.
41. *Revue Aéronautique de France*, "Mlle Marvingt au Mans," July 1928, p. 10; et alia.
42. Plauche-Gillon, Jéhanne, op. cit.
43. Didier, Gény, op. cit.
44. Monsieur Y, in Jacqueline Maire, letter, July 5, 1982.
45. *Revue Aéronautique de France*, "Une Belle Tournée de Conférences," October 1932, p. 5.
46. *Ibid.*
47. Diehl, Nicole, letter, July 8, 1982.
48. *Ibid.*
49. Perren, Monsieur, in Jacqueline Maire, letter, July 5, 1982.
50. X.B., "D'Ultimes Témoignages Viennent Encore Préciser Quelques Traits Attachants de Marie Marvingt, Illustre Figure Lorraine," *Républicain Lorrain*, January 24, 1964, p. 2.
51. Marvingt, Marie, letter to Jeanne Roumier, 1915.
52. *L'Est Républicain*, "La Fiancée du Danger Marie Marvingt Donne Son Nom au Gymnase de Maxéville," January 8, 1984.
53. Curral-Couttet, Gaby, op. cit.
54. R.P., "Mademoiselle Marvingt," *Le Cri de Constantin*, June 28, 1924.
55. Blanc, op cit.; Coulet, op. cit.
56. Gerbaud, A. Lorac, letter to Phillipe Bayart, June 28, 1988.
57. Codron, Charles, op. cit.
58. Reboul, Madame, cited in Jacqueline Maire, letter, July 5, 1982.
59. Coulet, op. cit.; Blanc, op. cit.; et alia.

Chapter 9

1. R.P., "Mademoiselle Marvingt," *Le Cri de Constantin*, June 28, 1924.
2. *L'Est Républicain*, "Au Pavillon de l'Aéronautique: Le Gala des Ailes Qui Sauvent," July 23, 1936, p. 4; *L'Est Républicain*, "A l'Exposition de Nancy, Le Succès de la Fête Enfantine," July 24, 1936, p. 4; *L'Est Républicain*, "Au Stand de l'Aéronautique," July 26, 1936, p. 4; *L'Est Républicain*, "Nancy, la Ville de Stanislas, Fait au Général Rydz-Smigly une Enthousiaste Accueil," September 4, 1936, p. 3; *L'Est Républicain*, "La Réception du Colonel Weiss," October 19, 1936, p. 4; *L'Est Républicain*, "Le Meeting au Vélodrome du Rassemblement National Lorrain," October 26, 1936, p. 4; *L'Est Républicain*, "Ligue des Amis de l'Aviation Sanitaire," December 2, 1936, p. 4; *L'Est Républicain*, "Henry de Monfreid Chez les Sauvages," December 7, 1936, p. 3; *L'Est Républicain*, "La Séance Recreative de l'A.M.C.," March 22, 1937, p. 3; *L'Est Républicain*, "La Conférence du Docteur Benech," May 21, 1937, p. 5; et alia.
3. Rio, Armand, "La Fiancée du Danger," *Lecture Pour Tous*, vol. 15, no. 7, April 1, 1913, pp. 63–71.
4. *L'Est Républicain*, "M. Jean Beziaud Parle de l'Aviation Française," April 24, 1938, p. 4; *L'Est Républicain*, "M. Jacques Mortane a Parlé de Guynemer et de Mermoz," May 22, 1938, p. 4; *L'Est Républicain*, "Le Salon de la Société Lorraine des Amis des Arts s'Ouvre Aujourd'hui," October 9, 1938; *L'Est Républicain*, "M. Robert Leurquin a Raconté Ce Qu'il a Vu en Chine," November 8, 1938, p. 5; *L'Est Républicain*, "La Soirée de l'Aviation Populaire," November 11, 1938, p. 4; *L'Est Républicain*, "Les Fêtes du 11 Novembre Dans la Capitale Lorraine Pavoisée," November 12, 1938, p. 4; *L'Est Républicain*, "Matinee de la Jeunesse de l'Empire Français," 13 novembre 1939, p. 3; *L'Est Républicain*, "La Matinee Artistique de la Salle Poirel," February 5, 1940, p. 3; *L'Est Républicain*, "La Jeunesse de l'Empire Français a Reçu les Gars du 170e," March 18, 1940, p. 3; *L'Est Républicain*, "La Matinee de la Salle Poirel," April 22, 1940, p. 3; *L'Est Républicain*, "Avec les Félicitations de Marie Marvingt," July 31, 1953; Les Ailes, "Le Tour de France des Avions Légers," September 20, 1928, p. 14; et alia.
5. *L'Est Républicain*, "Une Soirée Sensationnelle Mercredi 13 Courant," January 6, 1932, p. 3; *L'Est Républicain*, "Mlle Marvingt à la Salle Poirel, Mercredi

Prochain," January 10, 1932, p. 3; *L'Est Républicain*, "La Soirée Exceptionnelle de Demain Mercredi," January 12, 1932, p. 3; *L'Est Républicain*, "Aujourd'hui, Salle Poirel, la Société de Géographie Reçoit Mlle Marvingt," January 13 janvier 1932, p. 3.

6. *L'Est Républicain*, "Une Bonne Conférence de Mlle Marvingt," January 15, 1932, p. 2.

7. *L'Est Républicain*, "Conférence," January 20, 1932, p. 3; *L'Est Républicain*, "La Soirée de la Faculté de Lettres," January 22, 1932, p. 3; *L'Est Républicain*, "Les Explorations de Mlle Marvingt," January 23, 1932, p. 3; *L'Est Républicain*, "La Vie Musulmane en Afrique en Tunisie et aux Colonies Italiennes," January 25, 1932, p. 3; *L'Est Républicain*, "Mlle Marvingt Parle de la Tunisie," January 26, 1932, p. 3; *L'Est Républicain*, "Quand Mlle Marvingt Fait une Conférence," January 27, 1932, p. 3.

8. For example, *L'Echo d'Alger* ran articles on March 9, 11, 12, 14, 16, and 18, 192, for a conference of hers.

9. *L'Est Républicain*, "Visions d'Orient," October 4, 6, 7, 10, 1932, p. 3.

10. *L'Echo d'Alger*, "Département d'Alger: Blida," December 4, 1922, p. 4.

11. *L'Est Républicain*, "Ligues Aéronautique et d'Aviation Sanitaire," January 17, 1932, p. 3.

12. *L'Est Républicain*, "L'Émulation Parmi les Groupes Scolaires de la Ligue Aéronautique," December 5, 1929, p. 3.

13. *L'Est Républicain*, "Conférence de Mlle Marvingt," October 16, 1929, p. 3.

14. *L'Est Républicain*, "Mademoiselle Marvingt," June 18, 1924, p. 5.

15. *The [Hobart, Tasmania] Mercury*, "Gleanings From the World's Press," October 4, 1907 p. 2.

16. *Brandon [Manitoba] Weekly Sun*, "The Aviatress," December 8, 1910, p. 8; *Kingston [Jamaica] Gleaner*, "Aerial Flights," December 24, 1910, p. 6; *The [Perth] West Australian*, "A Lady Aeronaut," 30 November 1910, p. 7.

17. *Pocahontas County [IA] Sun*, "Breaks Airwomen's Record," January 26, 1911, p. 3.

18. *The Strand Magazine*, "'The Bride of Danger,' An Interview With Mlle. Marie Marvingt," vol. 46, September 1913, pp. 187–194; Gygax, Georges, "La 'Fiancée du Danger': La Vie Prodigieuse de Marie Marvingt," *L'Illustré*, April 24, 1958, pp. 34–36, and May 1, 1958, pp. 40–42; *Forest and Stream*, "The World's Greatest Sportswoman," September 13, 1913, vol. 81, p. 323; *Vestkusten*, "Marie Marvingt," March 26, 1914, p. 1; *Marie France*, "La Fiancée du Danger' Lance un Défi aux Femmes du Monde Entier," June 22, 1948, p. 7; Saladin, Raymond, "Figures et Événements: Marie Marvingt," *Aviation Magazine*, March 24, 1955, pp. 12–13; Save, Colette, "L'Âge n'est Plus une Question d'Années," *Point de Vue Images du Monde*, March 3, 1955, p. 20; Desormière, Yves, "A 80 Ans Marie Marvingt, la 'Fiancée du Danger,' est la Seule Femme au Monde à Être Titulaire de Quatre Brevets de Pilotage: Ballon, Avion, Hydravion, Sécourisme de l'Air," *La Croix*, March 4, 1955, pp. 1, 4; Boetsch, Jacques, "Une Vie en Équilibre" *Anciens Combattants du Monde*, May 1955, pp. 11–14; M.A., *Samedi Soir*, "Marie Marvingt, 'Fiancée du Danger,' Attend Encore Que Soit Relevé Son Défi Mondial'," March 3, 1955, no. 505, pp. 1, 3; Cordier, Marcel, "Les Escales de Marie Marvingt," *Leurs Demeures en Lorraine: Tome II* (Éditions Pierron, 1983), p. 175.

19. Gygax, Georges, April 24, 1958, op. cit.

20. *Hattiesburg [MS] American*, "French Woman Flier Predicts Air Travel Will Predominate," November 16, 1928, p. 2; *Appleton [WI] Post Crescent*, "French Aviatrix Sees Big Flying Growth," December 8, 1928, p. 5; *Fairbanks [AK] Daily News Miner*, "Veteran French Aviatrix Sees Big Flying Growth," December 1, 1928, p. 5; *Ogden [UT] Standard Examiner*, "Vet Aviatrix Sees Big Flying Growth," November 25, 1928, p. 20; *The Gettysburg [PA] Times*, "Veteran French Aviatrice Claims Big Flying Growth," November 21, 1928, p. 2; et alia.

21. *Albuquerque Journal*, "First French Woman Pilot Here, Gave Lindy Medal for Paris Hop," October 10, 1935, pp. 1, 12.

22. *L'Aéro*, "Marie Marvingt on Her Antoinette," April 15, 1911; *L'Aviation Illustrée*, "Nos Aviatrices et la Coupe Fémina," December 15, 1910, no. 61; *La Vie au Grand Air*, "Mlle Marvingt Portée en Triomphe," December 3, 1910.

23. Palmer, Svetlana, and Sarah Wallis, *Intimate Voices From the First World War* (William Morrow, 2003).

24. Rio, Armand, op. cit.

25. R.P., op. cit.

26. *Ibid.*

27. Marvingt, Marie, autograph book, no date, annoted by MM: "My colleague from *Le Figaro*."

28. *New York Times*, "Early Aviator, 88," December 16, 1963, p. 33.

29. Myriel, "Comment On Devient Pilote-Aviateur," *Le Sport*, no. 102, November 24, 1910, p. 1.

30. *Albuquerque Journal*, "First French Woman Pilot Here, Gave Lindy Medal for Paris Hop," October 10, 1935, pp. 1, 12.

31. Myriel, "L'Aviation," *Éclair de l'Est*, May 4, 1909; Myriel, "Clôture de la Grande Semaine des Vosges, Organisée Par le Touring-Club de France," *Le Sport*, April 28, 1910, supplement to no. 73; Myriel, "Pourquoi Je n'Ai Pas Pu Reprendre la Coupe Fémina," *Courrier de l'Est*, January 4, 1911; etc.

32. *L'Est Républicain*, "Au Pavillon de l'Aéronautique: Le Gala des Ailes Qui Sauvent," July 23, 1936, p. 4; Laurent, Michel-Yves, "Marie Marvingt, Surnommée 'Marie Casse-Cou,' Traversa au Début du Siècle, la Mer du Nord en Ballon; Elle Pilote Aujourd'hui à 86 Ans un Hélicoptère à Réaction," *L'Aurore* or *Paris Press*, no. 16, November 2, 1960; et alia.

33. Marvingt, Marie, "Dans Nancy Bombardée," *Samedi Soir*, January 29, 1916.

34. Myriel, "Ma Première Ascension Comme Pilote," *Éclair de l'Est*, September 29, 1909.

35. Boetsch, Jacques, "Une Vie en Équilibre" *Anciens Combattants du Monde*, May 1955, pp. 11–14.

36. *L'Aérophile*, "Roger Sommer à Nancy," October 1, 1909, p. 436.

37. Myriel, "Un Nouveau Monoplan à Champigneulles," *Éclair de l'Est*, August 4, 1909.

38. *Ibid.*

39. Myriel, "La Grande Semaine d'Aviation," *L'Éclair de l'Est*, September 1, 1909, p. 2.

40. Marvingt, Marie, "Les Sports d'Hiver dans les Vosges," *Le Sport*, March 5, 1910, p. 1; Allen, E. John B., *The Culture and Sport of Skiing: From Antiquity to World War II* (University of Massachusetts Press, 2007), p. 84.

41. *La Vie au Grand Air*, "La Grande Semaine des Vosges," February 26, 1910, no. 597, pp. 136, 142.

42. Marvingt, Marie, "The Alps in Winter, I," *The Times [London]*, December 14, 1909.

43. *Ibid.*

44. *Ibid.*

45. Myriel, "Clôture de la Grande Semaine des Vosges, Organisée Par le Touring-Club de France," *Le Sport*, supplement to no. 73, April 28, 1910.

46. *L'Echo d'Alger*, "Une Femme," July 6, 1922, p. 2.

47. Marvingt, Marie, "The Alps in Winter, I," op. cit.

48. Myriel, "Les Sports d'Hiver," *Éclair de l'Est*, March 8, 1909.
49. *Ibid.*
50. Duvernay, Dr. L., "Le Mont Revard," *Le Sport Universel Illustré*, March 10, 1912, pp. 158–159; *Le Sport Universel Illustré*, "Le Mont Revard," January 12, 1912, pp. 157–158.
51. Marvingt, Marie, "Deux Grands Lorrains Réunis Dans la Mort," *Petit Marocain*, October 18, 1934.
52. *Ibid.*
53. Marvingt, Marie, "Leurs Femmes," *La Vigie Marocaine*, October 22, 1934.
54. Myriel, "Impressions d'une Française sur Guillaume II à Urville," *Est Républicain*, May 22, 1905, p. 1.
55. Marvingt, Marie, "A Propos de la Randonnée Aérienne de S. Em. le Cardinal Luçon," *La Croix*, no. 13897, June 22, 1928, p. 1; *Le Gaulois*, "Le Baptême de l'Air du Cardinal Luçon," June 22, 1928, p. 4.
56. *Revue Aéronautique de France*, "Une Initiative de Mlle Marvingt," August 1928, p. 13.
57. Marvingt, Marie, "A Propos de la Randonnée Aérienne de S. Em. le Cardinal Luçon," *La Croix*, no. 13897, June 22, 1928, p. 1; *Le Gaulois*, "Le Baptême de l'Air du Cardinal Luçon," June 22, 1928, p. 4.
58. Marvingt, Marie, "Souvenirs de Chalcidique Tirés de Mon Carnet de Route: A Propos du Tremblement de Terre de Grèce," *L'Éclair de l'Est*, October 1, 1932, p. 1.
59. Marvingt, Marie, "Trente Minutes Avec la Trinité du Banquet Royal à Versailles," *La Croix*, July 28, 1938, no. 17016, p. 3.
60. *Ibid.*
61. *Les Ailes Brisées*, "Nos Aviatrices, les Véritables Pionniers de l'Aviation: Marie Marvingt," December 1959, pp. 21–22.
62. Coulet, Jacques, and Madame Coulet, interview, June 24, 1982.
63. Marvingt, Marie, "Sagesse d'Une Vieille Messine: En Marge de la Croisade de l'Amabilité," *Le Lorrain*, May 11, 1950.
64. Marvingt, Marie, "Trois Morts Mettent en Deuil les Ailes de Lorraine: Hommage d'Une 'Vieille Tige,'" *Dimanche Éclair*, September 17, 1950, p. 3.
65. Marvingt, Marie, "Soeur Emile de la Doctrine Chrétienne: Une Vie au Service des Orphelines," newspaper clipping, c. 1950, initialed by MM.
66. Féral, Roger, "Marie Marvingt (86 ans), 'la Fiancée du Danger,' a Passé la Toussaint Dans un Hélicoptère à Réaction," *Télé-Paris*, November 1960.
67. Houard, Georges, "Marie Marvingt, Pilote et Propagandiste Exemplaire," *Les Ailes*, January 1, 1964, p. 21.
68. Acte de Notoriété de Marie Marvingt, September 29, 1964, Étude du Maître Jules Thomas, notaire, Nancy.

Chapter 10

1. *Berkeley Daily Gazette*, "French Woman Has Record for Traveling and Sports," September 13, 1926, p. 5.
2. *Revue Mensuelle du Touring-Club de France*, April 1896, p. 156.
3. *Républicain Lorrain*, "Marie Marvingt, la 'Fiancée du Danger,' n'Est Plus," December 15, 1963, pp. 1, 4; Ackerman, Gordon, "Fiancée of Danger," *Sports Illustrated*, June 26, 1961, pages 61–64; *Winnipeg [Manitoba] Free Press*, "Daring French Woman Retires at Eighty," July 5, 1955, p. 14.
4. *Le Matin*, "L'Aviatrice Fait du Tourisme," September 21, 1911, p. 5.
5. *Comoedia*, "Nos Aviatrices: Hommage à Mlle Marvingt," August 2, 1930, p. 4; *Le Figaro*, "Marie-Marvingt Parle et Vole," July 31, 1930, p. 8.
6. Marvingt, Marie, postcard dated June 20, 1954, in Cordier, Marcel, "Les Escales de Marie Marvingt," *Leurs Demeures en Lorraine: Tome II* (Éditions Pierron, 1983), p. 175.
7. Saindizié, J.-M., "C'est Marie Marvingt, 85 Ans, la Plus Extraordinaire Femme du Siècle," *Pilote*, November 17, 1960, no. 56, p. 10.
8. Londet, Maurice, "Les Sports d'Hiver," *Le Figaro*, January 9, 1908, p. 5.
9. *Berkeley Daily Gazette*, op. cit.
10. *L'Echo d'Alger*, "Une Conférence de Mlle Marvingt aux Beaux-Arts," March 11, 1929, p. 2.
11. Marchand, Robert, *Le Ciel N'a Pas de Toit* (Éditions Berger Levrault, 1962), pp. 258–262; "Éleve-Pilote à 80 ans," newspaper article, 1955; Marvingt, Marie, "Chronologie," Musée de l'Air, Paris.
12. "Marie Marvingt, Surnommée 'la Fiancée du Danger,' Parlera Vendredi à 18 Heures," Algerian journal, December 19, 1951.
13. *L'Est Républicain*, "Une Bonne Conference de Mlle Marvingt," January 15, 1932, p. 2.
14. *Bulletin de la Meurthe et Moselle*, "Mlle Marvingt," March 11, 1917, p. 1.
15. *Les Ailes*, "Petites Nouvelles de France," January 15, 1931, p. 15.
16. *Revue Aéronautique de France*, "Mademoiselle Marvingt," January 1931, pp. 11, 12.
17. Ackerman, Gordon, op. cit.
18. Marvingt, Marie, "L'Aviation et la Jeunesse des Écoles en Amérique," *Revue Aéronautique de France*, December 1936, pp. 2–3.
19. *Ibid.*
20. *Ibid.*
21. *Les Ailes*, "Coups d'Ailes," November 9, 1933, p. 15.
22. Marvingt, Marie, "L'Aviation et la Jeunesse des Écoles en Amérique," op. cit.
23. *Ibid.*
24. *Popular Aviation*, vol. 17, no. 6, December 1935, p. 374.
25. "Served in Trenches, Declares Aviatrix," *[Rochester, NY] Democrat and* Chronicle, *August 23, 1935, p. 7;* The [York, PA] Gazette and Daily, *"Announcements," October 11, 1935, p. 6.*
26. *East Liverpool [OH] Review*, "French Flyer Arrives," August 29, 1935, p. 14; *The [Moline, IL] Dispatch*, "French Flyer Arrives," September 2, 1935, p. 11.
27. *The Washington Post*, "France's Pioneer Woman Flier Wins Acclaim in a Dozen Lines," January 15, 1936, p. 13. Also: *The Washington Post*, "Alliance to Hear French Aviatrix," January 14, 1936, p. 10; *The Washington Post*, "Wings of the World," March 8, 1936, p. PY6.
28. *The Washington Post*, "France's Pioneer Woman Flier Wins Acclaim in a Dozen Lines," January 15, 1936, p. 13.
29. Marvingt, Marie, "L'Aviation et la Jeunesse des Écoles en Amérique," op. cit.
30. Marvingt, Marie, "Sky Women," *Collier's*, September 30, 1911, p. 15.
31. *Chicago Tribune*, "Two Groups to Have Reception for Aviatrix Here From France," October 20, 1935, p. 77; *Chicago Daily Tribune*, "Marie Marvingt, Noted French Flyer, Is Guest in City," October 19, 1935, p. 21.
32. *Chicago Tribune*, "First Licensed Aviatrix in World Visits Chicago," September 26, 1935, p. 2; *Chicago*

Tribune, October 22, 1935, p. 15: "Mlle. Marvingt, French Aviatrix, Will Speak Before the Cercle Francais," October 22, 1935, p. 15.

33. "Alliancistes Can't Be Bothered With Reservations," *Chicago Tribune*, October 20, 1935, p. 75.

34. Marvingt, Marie, "L'Aviation et la Jeunesse des Écoles en Amérique," op. cit.

35. *Los Angeles Times*, "Woman War Flyer Here," September 30, 1935, p. 19; *Los Angeles Times*, "Woman War Flyer in Los Angeles," October 1, 1935.

36. Young, Pam, Director of Collections, Mission Inn, Riverside, California, letter, November 24, 1982; Association de Documentation Aéronautique, "Marie Marvingt (1875–1963)," Issy-les-Moulineaux, École Marie Marvingt, December 12, 1988; Curral-Couttet, Gaby, "Marie Marvingt, La Fiancée du Danger," *Les Folles Années de Chamonix* (Èditions France-Empire, series "Si 1900 m'Était Conté," 1984), pp. 75–78 ; Dubreuil, Yvonne, "Marie Marvingt 'La Fiancée du Danger,' Extrait de la Brochure Éditée à l'Occasion du Congrès de Nancy, October 1998, pp. 23–25; Saladin, Raymond, "Figures et Événements: Marie Marvingt," *Aviation Magazine*, March 24, 1955, pp. 12–13.

37. *Arizona Republic*, September 30, 1935, p. 6.

38. Marvingt, Marie, "L'Aviation et la Jeunesse des Écoles en Amérique," *Revue Aéronautique de France*, December 1936, pp. 2–3.

39. Gardner, Major Lester D., "Women Soon Took to Air: Their 'Firsts' in Various Fields of Aviation Are Recorded," *New York Times*, July 21, 1940.

40. *The Central New Jersey Home News*, "Mlle. Marie Marvingt Says Planes to Take Place of Ambulances,"April 21, 1936, p. 9; Pittsburg Post-Gazette, "Famous French Flier Will Address Alliance Francaise," January 21, 1936, p. 14; Weiss, Anne, "Current Issues Hold Attention of Women: List of Speakers to Be Heard by Clubs Would Seem to Be Celebrities' 'Blue Book'," *The Pittsburgh Press*, January 12, 1936, p. 25; *Pittsburgh Post-Gazette*, "Famous French Aviatrix Speaks Under Auspices of Alliance Francaise," November 1, 1935, p. 12; *The [Wilkes-Barre, PA] Evening News*, "Do You Know," January 15, 1936, p. 8; *The Pittsburgh Press*, "French Aviatrix Speaks," November 1, 1935, p. 68; *Detroit Free Press*, "First Woman Pilot," January 23, 1936; *Pittsburgh Post-Gazette*, "Air Ambulance Is Latest Thing," August 8, 1931, p. 22; et alia.

41. Marvingt, Marie, "L'Aviation et la Jeunesse des Écoles en Amérique," op. cit.

42. *Ibid.*

43. *Ibid.*

44. Marvingt, Marie, letter to the Grand Chancellor of the Legion of Honor, 1937.

45. *Les Ailes Brisées*, "Nos Aviatrices, les Véritables Pionniers de l'Aviation: Marie Marvingt," December 1959, pp. 21–22.

46. *Les Ailes Brisées*, op. cit.; Marck, Bernard, *Women Aviators* (Flammarion, 2013), pp. 14–21, 27, 39, 40, 43; Saladin, Raymond, "Figures et Événements: Marie Marvingt," *Aviation Magazine*, March 24, 1955, pp. 12–13.

47. Marvingt, Marie, handwritten notes on an article by Jacques Boetsch, "Une Vie en Équilibre" *Anciens Combattants du Monde*, May 1955, pp. 11–14.

48. *L'Afrique du Nord Illustré*, "Conférence de Mlle Marvingt," May 2, 1925, p. 6; *Les Ailes*, "Coups d'Ailes," April 27, 1922, p. 2; A.L., "Infatigable Globe-Trotter Raconte ses 12 Mois d'Afrique," *L'Est Républicain*, August 25, 1927; Boetsch, Jacques, "Une Vie en Équilibre" *Anciens Combattants du Monde*, May 1955, pp. 11–14.

49. *Annales Africaines*, "Mlle Marvingt en Afrique du Nord," July 13, 1922, p. 958.

50. Boulet, Jean, "De Passage au Maroc: Mlle Marie Marvingt, la Fiancée du Danger, Témoin ou Victime de Vingt Catastrophes, Elle est la Femme la Plus Extraordinaire du Monde," Moroccan newspaper, 1951, pp. 1, 4.

51. Baron Boilley, Françoise, *Marie Marvingt: A l'Aventure du Sport* (L'Harmattan, 2013), p. 91.

52. *Annales Africaines*, "Les Prochaines Conférences de Mlle Marvingt," May 29, 1925, p. 338; Balitrand, Suzanne, "Le Beau Voyage de Mademoiselle Marvingt," *Eve*, August 15, 1926, no. 307, p. 7.

53. *Bulletin de la Société de Géographie et d'Etudes Coloniales de Marseille*, "50,000 Kilomètres en Afrique du Nord," par Marie Marvingt, Vice-Présidente des "Amis de l'Aviation Sanitaire," Déléguée de la "Ligue Aéronatique," vol. 51, 1930, pp. 93–94.

54. *La Sfaxienne*, "Mademoiselle Marvingt à Sfax," April 15, 1924, p. 1.

55. A.L., "Infatigable Globetrotter Raconte Ses 52 Mois d'Afrique," *L'Est Républicain*, August 25, 1927.

56. Liégeois, Achille, "Les Nancéiennes Loin de Chez Elles: Mlle Marie Marvingt Rentre d'un Long Voyage: Elle Nous Raconte Ses Impressions," *L'Est Républicain*, August 23, 1926, p. 1.

57. *Ibid.*

58. L'Est Républicain, "Une Soirée Sensationnelle Mercredi 13 Courant," January 6, 1932, p. 3; *L'Est Républicain*, "Mlle Marvingt à la Salle Poirel, Mercredi Prochain," January 10, 1932, p. 3; *L'Est Républicain*, "La Soirée Exceptionnelle de Demain Mercredi," January 12, 1932, p. 3; *L'Est Républicain*, "Aujourd'hui, Salle Poirel, la Société de Géographie Reçoit Mlle Marvingt," January 13 janvier 1932, p. 3; et alia.

59. *Annales Africaines*, "300 Conférences, 4.000 Kilomètres en Afrique du Nord par Mlle Marie Marvingt," May 15, 1925, pp. 309, 312–313.

60. Balitrand, Suzanne, "Le Beau Voyage de Mademoiselle Marvingt," *Eve*, August 15, 1926, no. 307, p. 7.

61. *L'Echo d'Alger*, "Le Congrès Afrique du Nord-Midi," August 4, 1927, p. 3; *Le Figaro*, "Le Congrès de l'Afrique du Nord," July 25, 1927, p. 2.

62. *Annales Africaines*, "Mlle Marvingt en Afrique du Nord," July 13, 1922, p. 960; *Annales Africaines*, "300 Conférences, 4.000 Kilomètres en Afrique du Nord par Mlle Marie Marvingt," May 15, 1925, pp. 309, 312–313.

63. *L'Est Républicain*, "Les Explorations de Mlle Marvingt," January 23, 1932, p. 3.

64. *Revue Aéronautique de France*, "Avec Mademoiselle Marvingt: Présidente Fondatrice de l'Aviation Sanitaire Civile au Maroc," July 1934, pp. 2, 15.

65. Balitrand, Suzanne, op. cit.

66. R.P., "Mademoiselle Marvingt," *Le Cri de Constantin*, June 28, 1924.

67. *Annales Africaines*, "300 Conférences, 4.000 Kilomètres en Afrique du Nord," op. cit.

68. Cordier, Marcel, "Marie Marvingt: Une Femme de Foi," *Chrétiens Dans la Ville*, no 73, June 1992, p. 6.

69. Marvingt, Marie, postcard to Marthe Roumier, April 29, 1922.

70. Marvingt, Marie, postcard to Madame Goupit, 1922.

71. Marvingt, Marie, postcard to the Misses Winsback, December 28, 1951.

72. Enselme-Trichard, Renée, letter, July 2, 1982.

73. Cordier, Marcel, "Les Combats de Marie Marvingt," *La Nouvelle Revue Lorraine*, no. 27, August-September 2014, p. 20–23.

74. J.B.L., "On Peut Fort Bien 'Être et Avoir Été,' Démontre Marie Marvingt," *La Presse*, February 3, 1952.

75. Blanc, Colonel Gaston, interview, June 24, 1982.
76. *Ibid.*
77. *L'Afrique du Nord Illustré*, "Un Rare Exemple d'Énergie Féminine," July 15, 1922, p. 7; *L'Afrique du Nord Illustré*, "Conférence de Mlle Marvingt," May 2, 1925, p. 6.
78. *Annales Africaines*, "300 Conférences, 4.000 Kilomètres en Afrique du Nord par Mlle Marie Marvingt," May 15, 1925, pp. 309, 312–313.
79. Petit, Edmond, "Hommage à Marie Marvingt," *Forces Aériennes Françaises*, no. 200, February 1964, pp. 268–270.
80. *L'Echo d'Alger*, "Une Conférence de Mlle Marvingt aux Beaux-Arts," March 11, 1929, p. 2.
81. A.L., "Infatigable Globetrotter Raconte Ses 52 Mois d'Afrique," *L'Est Républicain*, August 25, 1927; Liégeois, Achille, "Les Nancéiennes Loin de Chez Elles: Mlle Marie Marvingt Rentre d'un Long Voyage: Elle Nous Raconte Ses Impressions," *L'Est Républicain*, August 23, 1926, p. 1.
82. Liégeois, Achille, "Les Nancéiennes Loin de Chez Elles: Mlle Marie Marvingt Rentre d'un Long Voyage: Elle Nous Raconte Ses Impressions," *L'Est Républicain*, August 23, 1926, p. 1.
83. *L'Est Républicain*, "Une Bonne Conference de Mlle Marvingt," January 15, 1932, p. 2.
84. Marvingt, Marie, postcard to Madame Roumier, August 26, 1922.
85. "Marie Marvingt, Surnommée 'la Fiancée du Danger,' Parlera Vendredi à 18 heures," Algerian newspaper, December 19, 1951.
86. "Marie Marvingt, Surnommée 'la Fiancée du Danger,' Parlera Vendredi à 18 Heures," Algerian newspaper, December 19, 1951.
87. Enselme-Trichard, Renée, letter, October 5, 1982.
88. *Les Annales Coloniales*, "En Auto: A Travers le Sahara," February 20, 1923, p. 1.
89. Liégeois, Achille, "Les Nancéiennes Loin de Chez Elles: Mlle Marie Marvingt Rentre d'un Long Voyage: Elle Nous Raconte Ses Impressions," *L'Est Républicain*, August 23, 1926, p. 1.
90. Thétard, Henri, "Le Général Estienne et M. Citroën Sont Arrivés à In-Salah," *Le Petit Parisien* February 20, 1923, p. 1.
91. *Ibid.*
92. Renaud, Jean, "La Triomphale Arrivée des Autos-Chenilles Qui Ont Traversé le Sahara," *Le Journal [Paris]*, March 6, 1923, pp. 1–2.
93. Marvingt, Marie, letter to Marcel Knecht, January 1, 1962.
94. Leblond, J.-J. "Pour le 52e Anniversaire de Sa Première Ascension en Ballon Marie Marvingt a Piloté un Avion Sanitaire Entre Bordeaux et Marseille," *Dauphiné*, November 1952, p. 3.
95. Boulet, Jean, "De Passage au Maroc: Mlle Marie Marvingt, la Fiancée du Danger, Témoin ou Victime de Vingt Catastrophes, Elle est la Femme la Plus Extraordinaire du Monde," Moroccan newspaper, 1951, pp. 1, 4.
96. Walter, Don, "In the Spotlight for 50 Years," *European Stars and Stripes*, February 16, 1958, pp. 11–12.
97. Ackerman, Gordon, "Fiancée of Danger," *Sports Illustrated*, June 26, 1961, pages 61–64.
98. Jansen, Marianne, "Marie Marvingt: 86 Ans d'Âge, 51 Ans d'Aviation," *Écho de la Mode*, no. 47, November 19, 1961, pp. 18–19.
99. Desormière, Yves, "A 80 Ans Marie Marvingt, la 'Fiancée du Danger,' Est la Seule Femme au Monde à Être Titulaire de Quatre Brevets de Pilotage: Ballon, Avion, Hydravion, Sécourisme de l'Air," *La Croix*, March 4, 1955, pp. 1, 4.
100. de Mazières, M., "Mouvement Touristique au Maroc en 1934," *Revue de Géographie Marocaine*, January 1935, p. 107; *L'Illustration*, "A Ski et à Chameau au Coeur du Grand Erg Occidental," August 7, 1937, no. 4927, p. 456.
101. Liégeois, Achille, "Les Nancéiennes Loin de Chez Elles: Mlle Marie Marvingt Rentre d'un Long Voyage: Elle Nous Raconte Ses Impressions," *L'Est Républicain*, August 23, 1926, p. 1.
102. *Revue Aéronautique de France*, "Avec Mademoiselle Marvingt: Présidente Fondatrice de l'Aviation Sanitaire Civile au Maroc," July 1934, pp. 2, 15.
103. A.L., "Infatigable Globe-Trotter Raconte Ses 12 Mois d'Afrique," *L'Est Républicain*, August 25, 1927.
104. A.L., "Infatigable Globe-Trotter Raconte Ses 12 Mois d'Afrique," *L'Est Républicain*, August 25, 1927.
105. A.L., "Infatigable Globetrotter Raconte Ses 52 Mois d'Afrique," *L'Est Républicain*, August 25, 1927.
106. *L'Est Républicain*, "Une Bonne Conference de Mlle Marvingt," January 15, 1932, p. 2.
107. *L'Aérophile*, "Le Tour du Monde Aérien," July 1–15, 1926, p. 217; *L'Ouest Éclair [Nantes]* and *L'Ouest Éclair [Rennes]*, "Une Exploratrice Vient de Oarcourir 56.114 Kilomètres," June 16, 1926, p. 3.
108. Ackerman, Gordon, "Fiancée of Danger," *Sports Illustrated*, June 26, 1961, pages 61–64.
109. Albaret, Laurent, "Marie Marvingt, la Fiancée du Danger," *L'Echo de la Timbrology*," no. 1901, December 2015, pp. 68–72.
110. X.B., "D'Ultimes Témoignages Viennent Encore Préciser Quelques Traits Attachants de Marie Marvingt, Illustre Figure Lorraine," *Républicain Lorrain*, January 24, 1964, p. 2.Barthélemy, Général, "La femme la plus extraordinaire du siècle: Marie Marvingt, la 'Fiancée du Danger,'" *L'Étrange race des hommes volants* (Éditions France-Empire, 1979), pp. 241–254
111. "Marie Marvingt: 56.000 Kilomètres en Afrique du Nord," newspaper article, April 16, 1928.
112. Balitrand, Suzanne, "Le Beau Voyage de Mademoiselle Marvingt," *Eve*, no. 307, August 15, 1926, p. 7.
113. *Annales Africaines*, "300 Conférences, 4.000 Kilomètres en Afrique du Nord par Mlle Marie Marvingt," May 15, 1925, pp. 309, 312–313.
114. Boetsch, Jacques, "Une Vie en Équilibre" *Anciens Combattants du Monde*, May 1955, pp. 11–14.
115. *Ibid.*
116. A.L., "Infatigable Globetrotter Raconte Ses 52 Mois d'Afrique," *L'Est Républicain*, August 25, 1927; Liégeois, Achille, "Les Nancéiennes Loin de Chez Elles: Mlle Marie Marvingt Rentre d'un Long Voyage: Elle Nous Raconte Ses Impressions," *L'Est Républicain*, August 23, 1926, p. 1.
117. A.L., "Infatigable Globetrotter Raconte Ses 52 Mois d'Afrique," *L'Est Républicain*, August 25, 1927; Liégeois, Achille, "Les Nancéiennes Loin de Chez Elles: Mlle Marie Marvingt Rentre d'un Long Voyage: Elle Nous Raconte Ses Impressions," *L'Est Républicain*, August 23, 1926, p. 1.
118. *Ibid.* 7; Liégeois, Achille, "Les Nancéiennes Loin de Chez Elles: Mlle Marie Marvingt Rentre d'un Long Voyage: Elle Nous Raconte Ses Impressions," *L'Est Républicain*, August 23, 1926, p. 1.
119. Cordier, Marcel, "Les Escales de Marie Marvingt," *Leurs Demeures en Lorraine: Tome II* (Éditions Pierron, 1983), p. 170.
120. Boulet, Jean, "De Passage au Maroc: Mlle Marie Marvingt, la Fiancée du Danger, Témoin ou Victime de Vingt Catastrophes, Elle est la Femme la plus Extraordinaire du Monde," Moroccan newspaper, 1951, pp. 1, 4.

121. Marvingt, Marie, letter to Madame Roumier, July 13, 1934.

122. Marvingt, Marie, "Anniversaire de Fondation de l'A.S.C.A.M.," *La Vigie Marocaine*, January 30, 1935.

123. Boulet, Jean, "De Passage au Maroc: Mlle Marie Marvingt, la Fiancée du Danger, Témoin ou Victime de Vingt Catastrophes, Elle est la Femme la plus Extraordinaire du Monde," Moroccan newspaper, 1951, pp. 1, 4.

124. *Ibid.*

125. Marvingt, Marie, letter to Louis Lhérault, February 10, 1952, in Cordier, Marcel, "Les Escales de Marie Marvingt," *Leurs Demeures en Lorraine: Tome II* (Éditions Pierron, 1983), p. 174.

126. "Les Anciennes Élèves de Sainte-Chrétienne Ont Retouvé, Avec Plaisir l'Ambiance de Leur Établissement," *Républicain Lorrain*, May 14, 1955.

127. Cordier, Marcel, "Les Escales de Marie Marvingt," *Leurs Demeures en Lorraine: Tome II* (Éditions Pierron, 1983), p. 175.

128. *Ibid.*

129. Nicolle, Jean-Loup, "Marie Marvingt Racontée aux Enfants," Prix Jeanne Goury, 1961, Musée des Beaux Arts, Nancy.

130. Liégeois, Achille, "Les Nancéiennes Loin de Chez Elles: Mlle Marie Marvingt Rentre d'un Long Voyage: Elle Nous Raconte Ses Impressions," *L'Est Républicain*, August 23, 1926, p. 1.

131. Bichet, Gabriel (G.B.), "Marie Marvingt, la Fiancée du Danger, Est Morte Hier à Nancy," *Est Républicain*, December 15, 1963; Branchu, Marc, "Les Cent Vies de Marie Marvingt," *Magazine Air France*, no. 99, July 2005, pp. 42–44, 46; Daurat, Michel, "À Quatre-Vingts Ans, Marie Marvingt Demeure Toujours la Fiancée du Danger," *Lorraine Magazine*, no. 9, February 1955, pp. 10–13; Dollet, Christophe, "Marie Marvingt Sans Visa," *L'Est Républicain*, July 11, 2011; *L'Est Républicain*, "La 'Fiancée du Danger' Est Morte, Hier, à 88 Ans," December 15, 1963; *L'Est Républicain*, "Le Lycée Professionnel Dédié à la 'Fiancée du Danger,'" May 24, 1987; France, Geo, "Marie Marvingt Espère Fêter Ses 90 Ans ... Dans la Lune," *Ouest-France*, February 23, 1955; J.G., "L'Avion et le Vélo: Deux Amours de Marie Marvingt, 'la Fiancée du Danger,'" *L'Est Républicain*, February 16, 1975, p. 3; *L'Union*, "'La Fiancée du Danger' Est Décédée à Nancy, Samedi, à 88 Ans," December 16, 1963; Zeyons, Serge, "Marie Marvingt, Une Femme d'Exception," *Timbres Magazine*, no. 151, December 2013, pp. 80–82; France, Geo, "Marie Marvingt Espère Fêter Ses 90 Ans ... Dans la Lune," *Ouest-France*, February 23, 1955.

Chapter 11

1. Rio, Armand, "La Fiancée du Danger," *Lecture Pour Tous*, vol. 15, no. 7, April 1, 1913, pp. 63–71.

2. Marvingt, Marie, "Un Défi Mondial!" flyer distributed to the press, with corrections and additions in Marie Marvingt's handwriting (1922); *Marie France*, "La Fiancée du Danger' Lance un Défi aux Femmes du Monde Entier," June 22, 1948, p. 7; Morgane, Yves, "La Fiancée du Danger: Marie Marvingt Fut l'Une des Premières Femmes Aviatrices Égales des Hommes Dans la Conquête de l'Air," *Miroir de l'Histoire*, November-December 1978, no. 307, pp. 80–84; *C'est la Vie!*, "Marie Marvingt: Femme Universelle," April 28, 1950, no. 25, p. 3.

3. Boetsch, Jacques, "Une Vie en Équilibre" *Anciens Combattants du Monde*, May 1955, pp. 11–14; Branchu, Marc, "Les Cent Vies de Marie Marvingt," *Magazine Air France*, no. 99, July 2005, pp. 42–44, 46; Cordier, Marcel, "Les Escales de Marie Marvingt," *Leurs Demeures en Lorraine: Tome II* (Éditions Pierron, 1983), pp. 162–177; E.B., "La Fiancée du Danger: Une Célèbre Femme de l'Air Est à Limoges," *Notre Province (Limoges)*, June 1944, p. 160; Mauchaussée, Jean, "Marie Marvingt, Reine de l'Air," *Prestige*, October 1, 1959, no. 3, p. 1; Petit, Edmond, "Hommage à Marie Marvingt," *Forces Aériennes Françaises*, February 1964, no. 200, pp. 268–270; Save, Colette, "L'Âge n'est Plus une Question d'Années," *Point de Vue Images du Monde*, March 3, 1955, p. 20.

4. Floret, Robert, "Grand-Mère 'Casse-Cou,'" *Détective*, no. 751, November 1960, pp. 10–11.

5. *Marie France*, "La Fiancée du Danger' Lance un Défi aux Femmes du Monde Entier," June 22, 1948, p. 7; Dinan, R., "Les Enfants Prodiges: Mlle Marvingt Lance un Défi Mondial," *L'Ordre*, July 2, 1948.

6. *C'est la vie!*, op. cit.; Morgane, Yves, "La Fiancée du Danger: Marie Marvingt Fut l'Une des Premières Femmes Aviatrices Égales des Hommes Dans la Conquête de l'Air,"*Miroir de l'Histoire*, op. cit.; Dinan, R., op. cit.

7. M.A., "Marie Marvingt, 'Fiancée du Danger,' Attend Encore Que Soit Relevé Son 'Défi Mondial,'" *Samedi Soir*, March 3, 1955, no. 505, pp. 1, 3.

8. J.G., "L'Avion et le Vélo: Deux Amours de Marie Marvingt, 'la Fiancée du Danger,'" *L'Est Républicain*, February 16, 1975, p. 3.

9. Baron Boilley, Françoise, *Marie Marvingt: A l'Aventure du Sport* (L'Harmattan, 2013), p. 74.

10. *Paris-Match*, "Mlle Marvingt: Depuis 60 Ans 'Fiancée du Danger,'" no. 7, May 7, 1949, pp. 24–25; *Nancy-Ochey Flash*, "Marie Marvingt, 'La Fiancée du Danger,' no. 171, July 2003, p. 17; Zeyons, Serge, "Marie Marvingt, Une femme d'Exception," *Timbres Magazine*, no. 151, December 2013, pp. 80–82; *La Nouvelle République*, "Marie Marvingt la 'Fiancée du Danger' Est Morte," no. 5855, December 16, 1963, pp. 1, C; Saindizié, J.-M., "C'est Marie Marvingt, 85 Ans, la Plus Extraordinaire Femme du Siècle," *Pilote*, no. 56, November 17, 1960, p. 10; *Le Parisien*, "Marie Marvingt, Surnommée 'la Fiancée du Danger,' Est Décédée Samedi à Nancy," December 16, 1963; Pelot, Paul, "Marie Marvingt Fut Une des Demmes les Plus Extraordinaires du Siècle," *L'Équipe*, December 16, 1963; et alia.

11. *La Justice*, "Nageuses," August 4, 1906, p. 1; *La Sfaxienne*, "Mademoiselle Marvingt à Sfax," April 15, 1924, p. 1.

12. *Nord-Touriste*, "Une Conférence sur l'Aérostation et l'Aviation par Mlle Marvingt," November 1912, pp. 251–252; Gygax, Georges, "La 'Fiancée du Danger': La Vie Prodigieuse de Marie Marvingt," *L'Illustré*, April 24, 1958, pp. 34–36.

13. Cordier, Marcel, op. cit., p. 164.

14. Cordier, Marcel, "Marie Marvingt: Une Lorraine Qui Volait au Secours des Blessés," *La Vie la Santé*, no. 6, July 1998, pp. 28–29.

15. Mollier, Christian, and Jean-Pierre Gallay, *La Mémoire des Sports d'Hiver au Pied du Mont-Blanc* (Imprimerie Nouvelles Chamonix/Sallanches, 2000), p. 80.

16. Curral-Couttet, Gaby, "Marie Marvingt, la Fiancée du Danger," *Les Folles Années de Chamonix* (Éditions France-Empire, 1984), pp. 75–78.

17. Daurat, Michel, "A Quatre-Vingts Ans, Marie Marvingt Demeure Toujours la Fiancée du Danger," *Lorraine Magazine*, no. 9, February 1955, pp. 10–13.

18. Lexa-Chomard, Annette, *Lucien Cuénot, L'Intuition Naturaliste* (2004), p. 56–57.

19. *Ibid.*
20. Dinan, R., op. cit.
21. Geoffroy, J.M., letter, March 29, 1983.
22. M.A., "Marie Marvingt, 'Fiancée du Danger,' Attend Encore Que Soit Relevé Son 'Défi Mondial,'" *Samedi Soir*, March 3, 1955, no. 505, pp. 1, 3.
23. Blanc, Colonel Gaston, interview, June 24, 1982.
24. *Ibid.*
25. to come
26. *L'Aérophile*, "Aujourd'hui," Decembre 1934, p. 365; *L'Afrique du Nord Illustré*, "Les Ailes Qui Sauvent," July 28, 1934, p. 7.
27. Jansen, Marianne, "Marie Marvingt: 86 Ans d'Âge, 51 Ans d'Aviation," *Écho de la Mode*, no. 47, November 19, 1961, pp. 18–19.
28. Picot, Madame Augusta, letter, June 30, 1982.
29. Kernel, Hélène, "Marie Marvingt, Doyenne des Aviatrices," *France Aviation*, August 1957, p. 8.
30. M.A., op. cit.
31. *Paris-Esperanto*, "Nos Dîners, December 1910, p. 55; *Le Figaro*, "Banquet d'Espérantistes," November 13, 1910, p. 3.
32. J.B.L., "On Peut Fort Bien 'Être et Avoir Été,' Démontre Marie Marvingt," *La Presse*, February 3, 1952.
33. Jansen, Marianne, op. cit.; Rio, Armand, "La Fiancée du Danger," *Lecture Pour Tous*, vol. 15, no. 7, April 1, 1913, pp. 63–71; Floret, Robert, "Grand-Mère 'Casse-Cou,'" *Détective*, no. 751, November 18, 1960, pp. 10–11.
34. *Républicain Lorrain*, "A Nancy et à Metz, Marie Marvingt, la 'Fiancée du Danger' Dira Quels Sont les Buts Humanitaires de Sa Mission," October 19, 1949, p. 2.
35. Boetsch, Jacques, "Une Vie en Équilibre" *Anciens Combattants du Monde*, May 1955, pp. 11–14; Gygax, Georges, "La Vie Prodigieuse de Marie Marvingt, la 'Fiancée du Danger,'" *L'Illustré*, May 1, 1958, pp. 40–42; Walter, Don, "In the Spotlight for 50 Years," *European Stars and Stripes*, February 16, 1958, pp. 11–12.
36. *La Presse*, "Les Grandes Emotions du Sport," August 12, 1913, p. 1.
37. *La Sfaxienne*, "Mademoiselle Marvingt à Sfax," April 15, 1924, p. 1.
38. *C'est la vie!*, op. cit.
39. Codron, Charles, letter, circa 1905 (Lhërault collection).
40. Gygax, Georges, May 1, 1958, op. cit.; Faure, Odile, "Marie Marvingt, La Fiancée du Danger," *Massif Central Magazine*, March-April 1996, pp. 68–71.
41. *Le Petit Matin* [Tunis], "Mlle Marvingt, la 'Fiancée du Danger,' Nous Révélera à Tunis les Sources de Son Extraordinaire Dynamisme," February 3, 1952; Dollet, Christophe, "Marie Marvingt Sans Visa," *L'Est Républicain*, July 11, 2011.
42. Pfeiffer, Doug, "Musée Dauphinois: La Grande Histoire du Ski," *Skiing Heritage Journal*, March 2004, pp. 33–37.
43. *Républicain Lorrain*, "A Nancy et à Metz, Marie Marvingt, la 'Fiancée du Danger' dira quels sont les buts humanitaires de sa mission," 19 octobre 1949, p. 2.
44. Ackerman, Gordon, "Fiancée of Danger," *Sports Illustrated*, June 26, 1961, pages 61–64.
45. Gygax, Georges, May 1, 1958, op. cit.
46. *Paris-Match*, "Mlle Marvingt: Depuis 60 Ans 'Fiancée du Danger,'" no. 7, May 7, 1949, pp. 24–25.
47. *L'Echo d'Alger*, "Huit Cours et Conférences à Alger de Mlle Marie Marvingt," May 15, 1925, p. 3.
48. *Annales Africaines*, "300 Conférences, 40.000 Kilomètres en Afrique du Nord par Mlle Marie Marvingt," May 15, 1925, pp. 309, 312–313; *L'Avenir de l'Est* [Algérie], "Conférence," December 20, 1924, p. 2; *L'Echo de Tiaret*, "Conférences," July 11, 1925, p. 2.
49. *L'Afrique du Nord Illustré*, "Conférence de Mlle Marvingt," May 2, 1925, p. 6; et alia.
50. R.P., "Mademoiselle Marvingt," *Le Cri de Constantin*, June 28, 1924; Barthélemy, Général, "La Femme la Plus Extraordinaire du Siècle: Marie Marvingt, la 'Fiancée du Danger,'" *L'Étrange Race des Hommes Volants* (Éditions France-Empire, 1979), pp. 241–254.
51. Gygax, Georges, May 1, 1958, op. cit.
52. *L'Est Républicain*, "La Fin prévue," 20 juillet 1912, p. 4.
53. Barthélemy, Général, "La Femme la Plus Extraordinaire du Siècle: Marie Marvingt, la 'Fiancée du Danger,'" *L'Étrange Race des Hommes Volants* (Éditions France-Empire, 1979), pp. 241–254.
54. *Le Matin*, "Le Tourisme Aérien," October 17, 1913, p. 2; *Indépendant*, "Mlle Marie Marvingt à Pau: Deux Causeries Littéraires et Sportives," June 16, 1927.
55. *Honolulu Star-Bulletin*, "Aviatrix Sees Big Growth in Flying," July 1, 1928, p. 21.
56. *L'Immeuble et la Construction Dans l'Est*, "L'Aviation et la Future Construction," January 30, 1910, p. 651.
57. *La Presse*, "La Fiancée du Danger à l'Alliance Française," February 5, 1952.
58. *La Sfaxienne*, "Mademoiselle Marvingt à Sfax," April 15, 1924, p. 1.
59. *Ibid.*
60. Barthélemy, Général, op. cit.
61. *Ibid.*
62. *Ibid.*
63. *La Sfaxienne*, op. cit.
64. J.P., "Marie Marvingt, Pilote d'Avion, Est Morte à 88 Ans: Championne en Tous Genres," *Le Monde*, December 17, 1963, p. 13; Bichet, Gabriel, "C'était la Fiancée du Danger...," *L'Est Républicain*, December 15, 1963.
65. Coulet, Madame, interview, June 24, 1982.
66. *L'Echo de Bougie*, "Mademoiselle Marvingt à Bougie," October 18, 1924, p. 3.
67. *L'Echo d'Alger*, "Une Conférence de Mlle Marvingt à Mascara," July 8, 1925, p. 6; *L'Echo de Bougie*, "Mademoiselle Marvingt à Bougie," October 18, 1924, p. 3.
68. Scott, Sarah, *A Description of Millenium Hall* (1762), p. 6.
69. Winterson, Jeanette, *Oranges Are Not the Only Fruit* (1985), p. 94.

Chapter 12

1. Bergson, Henri, *Time and Free Will* (1910), p. 172.
2. Arikha, Noga, *Passions and Tempers: A History of the Humours* (2007).
3. *L'Est Républicain*, "Le Concours de Skis à Gérardmer," February 23, 1909, p. 1; *Nancy Sportif*, "La Vie Sportive de Mlle Marvingt," June 4, 1914; *Indépendant*, "Mlle Marie Marvingt à Pau: Deux Causeries Littéraires et Sportives," June 16, 1927; Saladin, Raymond, "Figures et Événements: Marie Marvingt," *Aviation Magazine*, March 24, 1955, pp. 12–13; Krysaniac, C., *Le Cartophile de Meurthe-&-Moselle*, no. 1, February 1983; et alia.
4. Pearson, Carol, *The Hero Within: Six Archetypes We Live By* (1986).
5. *Ibid.*, p. 51.
6. Friedman, Meyer, and Rosenman, R.H., *Type A Behavior and Your Heart* (1974).
7. *Ibid.*, p. 92.

8. J.B.L., "On Peut Fort Bien 'Être et Avoir Été,' Démontre Marie Marvingt," *La Presse*, February 3, 1952.
9. Marvingt, Marie, letter to Marcel Knecht, October 30, 1954.
10. Parti, Rajiv, *The Soul of Wellness: 12 Holistic Principles for Achieving a Healthy Body, Mind, Heart, and Spirit* (2012).
11. Saladin, Raymond, op. cit.
12. Sand, George, and Gustave Flaubert, *The George Sand-Gustave Flaubert Letters* (1921).
13. Wentworth, Patricia, *Out of the Past* (1953), p. 150.
14. Crouter, Fred, *Body Types: What Mesomorph, Endomorph And Ectomorph Tell You About Character* (2013).
15. Gosling, Sam, cited in Zuckerman, Marvin, "Are You a Risk-Taker?," *Psychology Today*, November 1, 2000.
16. Breuning, Loretta Graziano, *Meet Your Happy Chemicals: Dopamine, Endorphin, Oxytocin, Serotonin* (2012).
17. Gygax, Georges, "La Vie Prodigieuse de Marie Marvingt, la 'Fiancée du Danger'," *L'Illustré*, May 1, 1958, pp. 40–42.
18. *Ibid.*
19. Rio, Armand, "La Fiancée du Danger," *Lecture Pour Tous*, April 1, 1913, vol. 15, no. 7, pp. 63–71.
20. Zuckerman, Marvin, "Are You a Risk-Taker?," November 1, 2000.
21. *Ibid.*
22. *Marie France*, "La Fiancée du Danger' Lance un Défi aux Femmes du Monde Entier," June 22, 1948, p. 7; *L'Est Républicain*, November 18, 1952, June 6, 1958, or November 8, 1958; Branchu, Marc, "Les Cent Vies de Marie Marvingt," *Magazine Air France*, no. 99, July 2005, pp. 42–44, 46.
23. Briggs, Katharine Cook, and Isabel Briggs Myers, *Briggs Myers Type Indicator Handbook* (1944).
24. Keirsey, David, and Marilyn Bates, *Please Understand Me: Character and Temperament Types* (1984).
25. *Ibid.*
26. Kuder, George Frederic, *Kuder Preference Record: Examiner's Manual* (1953).
27. Keirsey and Bates, op. cit.
28. Barthélemy, Général, "La Femme la Plus Extraordinaire du Siècle: Marie Marvingt, la 'Fiancée du Danger,'" *L'Étrange Race des Hommes Volants* (Éditions France-Empire, 1979), pp. 241–254.
29. Marvingt, Marie, postcard to her "dear cousins," June 28, 1941.
30. Saladin, Raymond, op. cit.
31. Woodhouse, Henry, "Women of the Air," *Outdoor World & Recreation*, May 1913, vol. 48, pp. 305–309.
32. James, William, "Is Life Worth Living?" *The Will to Believe And Other Essays in Popular Philosophy* (1896), p. 59.
33. M.A., "Marie Marvingt, 'Fiancée du Danger,' Attend Encore Que Soit Relevé Son 'Défi Mondial,'" *Samedi Soir*, March 3, 1955, no. 505, pp. 1, 3.
34. *Ibid.*

Chapter 13

1. *L'Est Républicain*, November 18, 1952; June 6, 1958; November 8, 1958.
2. Liégeois, Achille, "Les Nancéiennes Loin de Chez Elles: Mlle Marie Marvingt Rentre d'un Long Voyage: Elle Nous Raconte Ses Impressions," *L'Est Républicain*, August 23, 1926, p. 1.
3. Save, Colette, "LÂge n'Est Plus une Question d'Années," *Point de Vue Images du Monde*, March 3, 1955, p. 20; X.B., "D'Ultimes Témoignages Viennent Encore Préciser Quelques Traits Attachants de Marie Marvingt, Illustre Figure Lorraine," *Républicain Lorrain*, January 24, 1964, p. 2; Barthélemy, Général, "La Femme la Plus Extraordinaire du Siècle: Marie Marvingt, la 'Fiancée du Danger,'" *L'Étrange Race des Hommes Volants* (Éditions France-Empire, 1979), pp. 241–254.
4. *Le Parisien*, "Marie Marvingt, Surnommée 'la Fiancée du Danger,' Est Décédée Samedi à Nancy," December 16, 1963; *L'Est Républicain*, "Champ-le-Boeuf: Hommage à la Fiancée du Danger," April 19, 1993; *La Dépêche Meusienne*, "Marie Marvingt (1875–1963)," November 15, 2003; Larue, Michèle, "Marie Marvingt: Reine du Ciel et Aventurière des Sables/Queen of the Skies and Desert Adventuress," *RAM (Royal Air Maroc) Magazine*, March-April 2006, no. 136, pp. 96–104; Margot, Olivier, "L'Indomptable: Marie Marvingt," *L'Équipe*, August 17, 2013, pp. 60–63.
5. *La Sfaxienne*, "Mademoiselle Marvingt à Sfax," April 15, 1924, p. 1; de Champclos, G. Davin, "Mlle Marie Marvingt et Ses 56.000 Kilomètres en Afrique du Nord: La Première Femme de Sport du Monde," *L'Éclaireur du Soir*, January 3, 1928; Gygax, Georges, "La Vie Prodigieuse de Marie Marvingt, la 'Fiancée du Danger,'" *L'Illustré*, May 1, 1958, pp. 40–42; Krysaniac, C., *Le Cartophile de Meurthe-&-Moselle*, bulletin de liaison du Cercle Cartophile de Meurthe & Moselle, no. 1, February 1983; *La Revue Lorraine Populaire*, "Personnages de Chez Nous: Marie Marvingt," December 1, 2000, p. 45; Marck, Bernard, *Dictionnaire Universel de l'Aviation* (Tallendier, 2005), pp. 688–689; Poirier, Jean-Pierre, *La Véritable Jacqueline Auriol* (2005), pp. 51–52; Cordier, Marcel, "Marie Marvingt (1875–1963)," *Revue Lorraine Populaire*, June 2004, p. 12.
6. Valode, Philippe, "Marie Marvingt, l'Inégalable Casse-Cou," *Les Grandes Scandaleuses* (First Editions, 2015), pp. 227–235.
7. Eiermann, J., Grand Chancellor, Chief of the Bureau of the Legion of Honor, letter confirming the two promotions of Marie Marvingt, to Jacqueline Maire, December 31, 1982; Archives Nationales, Certificat de l'Admission de Marie Marvingt à la Légion d'Honneur, Signé par le Ministre de l'Air, January 19, 1935; *L'Aérophile*, "Dans la Légion d'Honneur," February 1935, pp. xxvi, 57; *Journal des Débats Politiques et Littéraires*, "Légion-d'Honneur," January 27, 1935, p. 2; *Journal Officiel de la République Française*, "Légion d'Honneur," January 26, 1935, p. 871; *Le Temps*, "Légion d'Honneur," January 27, 1935, p. 4; *L'Homme Libre*, "Légion d'Honneur," January 26, 1935, p. 3.; *L'Est Républicain*, "L'Aviatrice Nancéïenne Marie Marvingt Chevalier de la Légion d'Honneur," January 21, 1935, p. 4; *Les Ailes*, "La Légion d'Honneur: La Promotion Civile du Ministère de l'Air," January 31, 1935, p. 9; *Revue Aéronautique de France*, "Distinctions Honorifiques: Légion d'Honneur," February 1935, p. 2.
8. Shayler, David J., and Ian A. Moule, *Women in Space—Following Valentina* (Springer/Praxis, 2006), p. 16.
9. Le Ministère des Travaux Publics, des Transports et du Tourisme, report, October 3,1949.
10. Légion d'Honneur, La Grande Chancellerie de la Légion d'Honneur: documents supporting and confirming her grades of Chevalier and then Officer of the Légion d'Honneur, 1935 and 1949; Le Ministère des Travaux Publics, des Transports et du Tourisme, report, October 3, 1949: Renseignements Produits à l'Appui d'un Project de Décret Portant Promotion au Grade d'OF-

FICIER dans l'Ordre National de la Légion d'Honneur, for Mlle MARVINGT Marie; Archives Nationales, "Procès-Verbal de Réception d'un Officer de la Légion d'Honneur," May 14, 1950.

11. *Ibid.*

12. *Ibid.*

13. *C'est La Vie!*, "Marie Marvingt: Femme Universelle," April 28, 1950, no. 25, p. 3.

14. *Le Figaro*, "Le Dîner de l'Académie des Sports," December 15, 1911, p. 3.

15. Thirion, M., "Hommage de l'Aéro-Club de l'Est à Marie Marvingt," *L'Est Républicain*, December 18, 1963; Association de Documentation Aéronautique, "Marie Marvingt (1875–1963), Issy-les-Moulineaux, École Marie Marvingt, December 12, 1988; Albaret, Laurent, "Marie Marvingt, la Fiancée du Danger," *L'Echo de la Timbrology*," December 2015, no. 1901, pp. 68–72.

16. *L'Est Républicain*, "La 'Victoire de Marie Marvingt," February 2, 1955; Gygax, Georges, "La Vie Prodigieuse de Marie Marvingt, La 'Fiancée du Danger'," *L'Illustré*, May 1, 1958, pp. 40–42.

17. Fédération Nationale Aéronautique, award program, and granting of diplomas for 1954 graduates, January 30, 1955.

18. "Marie Marvingt," French aviation publication, 1955, pp. 41–42.

19. Fédération Nationale Aéronautique, Programme de la Remise des Prix et Récompenses aux Lauréats 1954, January 30, 1955.

20. *L'Est Républicain*, "La 'Fiancée' de Nancy," June 4, 1955; *L'Avenir à Argenteuil*, "La Célèbre Aviatrice Marie Marvingt S'est Intéressée aux Journées de Propagande Aéronautique," May 13, 1955; Gygax, Georges, "La 'Fiancée du Danger': La Vie Prodigieuse de Marie Marvingt," *L'Illustré*, April 24, 1958, pp. 34–36; Berlioux, Monique, *Gloires de Sport* (Atlantica, 2009), p. 180; Bowen, Barry D., "Marie Marvingt: The Premier Sportswoman of France and Pioneer in the Development of Aeromedical Evacuation," special to AirAmbulanceServices.com, March 6, 2013.

21. *L'Est Républicain*, "Mlle Marie Marvingt, Chevalier de la Santé Publique," November 5, 1937; Kernel, Hélène, "Marie Marvingt, Doyenne des Aviatrices," *France Aviation*, August 1957, p. 8.

22. *Journal Officiel de la République Française*, "Officiers d'Académie," November 19, 1911, p. 9207.

23. *L'Est Républicain*, "Mlle Marie Marvingt Reçoit Deux Nouveaux Témoignages Officiels de Gratitude," February 21, 1958, p. 4.

24. *Ibid.*

25. *Ibid.*

26. Cordier, Marcel, "Les Escales de Marie Marvingt," *Leurs Demeures en Lorraine: Tome II* (Éditions Pierron, 1983), p. 176; Gygax, Georges, op. cit.

27. *L'Est Républicain*, "Jeanne d'Arc, la Libératrice et le Cinquième Anniversaire de la Capitulation Allemande Ont Été Fêtés Ensemble au Cours de Manifestations Sobres, Mais Imposantes," May 15, 1950.

28. *Revue Aéronautique de France*, "Les Distinctions de la Ligue Aéronautique de France," January 1926, p. 13.

29. *Revue Aéronautique de France*, "Distinctions Méritées," July 1926, p. 47.

30. Liégeois, Achille, "Les Nancéiennes Loin de Chez Elles: Mlle Marie Marvingt Rentre d'un Long Voyage: Elle Nous Raconte Ses Impressions," *L'Est Républicain*, August 23, 1926, p. 1; A.L., "Infatigable Globe-Trotter Raconte Ses 12 Mois d'Afrique," *L'Est Républicain*, August 25, 1927.

31. Young, Pam, Director of Collections, Mission Inn, Riverside, California, letter, November 24, 1982; Association de Documentation Aéronautique, "Marie Marvingt (1875–1963)," Issy-les-Moulineaux, École Marie Marvingt, December 12, 1988; Curral-Couttet, Gaby, "Marie Marvingt, La Fiancée du Danger," *Les Folles Années de Chamonix* (Èditions France-Empire, series "Si 1900 M'était Conté," 1984), pp. 75–78 ; Dubreuil, Yvonne, "Marie Marvingt 'La Fiancée du Danger,' Extrait de la Brochure Éditée à l'Occasion du Congrès de Nancy, October 1998, pp. 23–25; Saladin, Raymond, "Figures et Événements: Marie Marvingt," *Aviation Magazine*, March 24, 1955, pp. 12–13.

32. Marvingt, Marie, "L'Aviation et la Jeunesse des Écoles en Amérique," *Revue Aéronautique de France*, December 1936, pp. 2–3.

33. McQueen, Elizabeth Lippincott (Mrs. Ulysses Grant), Women's International Association of Aeronautics, letter, September 25, 1949.

34. Cordier, Marcel, "Les Escales de Marie Marvingt,"op. cit.; Jansen, Marianne, "Marie Marvingt: 86 Ans d'Âge, 51 Ans d'Aviation," *Écho de la Mode*, no. 47, November 19, 1961, pp. 18–19; Valode, Philippe, op. cit.

35. *Paris-Match*, "Mlle Marvingt: Depuis 60 Ans 'Fiancée du Danger,'" May 7, 1949, no. 7, pp. 24–25.

Chapter 14

1. Cordier, Marcel, "Les Escales de Marie Marvingt," *Leurs Demeures en Lorraine: Tome II* (Éditions Pierron, 1983), p. 175.

2. Blanc, Colonel Gaston, interview, June 24, 1982; Coulet, Jacques, and Madame Coulet, interview, June 24, 1982.

3. Gygax, Georges, "La Vie Prodigieuse de Marie Marvingt, la 'Fiancée du Danger'," *L'Illustré*, May 1, 1958, p. 41.

4. *L'Est Républicain*, November 18, 1952.

5. *L'Est Républicain*, "La Conférence de Mlle Marvingt: Deux Heures Dans les Airs," October 13, 1929, p. 3; *L'Est Républicain*, "La Remise du Challenge de Mlle Marvingt," November 17, 1931, p. 2; *L'Est Républicain*, "Mlle Marvingt Parle de la Tunisie," January 26, 1932, p. 3; *L'Est Républicain*, "Au Stand de l'Aéronautique," July 26, 1936, p. 4; et alia.

6. Labourel, Monsieur, in Jacqueline Maire, letter, July 5, 1982.

7. Myriel, "La Grande Semaine d'Aviation," *L'Éclair de l'Est*, August 27, 1909.

8. *L'Aéro*, "Le Grand Prix de l'Aéro-Club de France," September 23, 1912, p. 1.

9. Méchelle, Ginette, interview, June 23, 1982; Friry, Jean, in Jacqueline Maire, letter, July 5, 1982.

10. Renoux-Barès, E., "Une Héroïne Parle à Pau de l'Aviation, École d'Énergie," *Revue Aéronautique de France*, October 1927, p. 11.

11. *Ibid.*

12. Raffalovich, G.D., "Souvenirs des Temps Héroïques," *Le Monde Illustré*, November 17, 1934, pp. 964, 966; Liégeois, Achille, "Les Nancéiennes Loin de Chez Elles: Mlle Marie Marvingt Rentre d'un Long Voyage: Elle Nous Raconte Ses Impressions," *L'Est Républicain*, August 23, 1926, p. 1.

13. *L'Aéro*, "Les Obsèques de Latham," January 17, 1914, p. 3; *L'Aérophile*, "Obsèques d'Hubert Latham," February 1, 1914, p. 62.; *Journal des Débats Politiques et Littéraires*, "Obsèques d'Hubert Latham," January 18, 1914, pp. 2–3.

14. *L'Aéro*, "Les Sphériques," August 6, 1913, p. 4.
15. *Le Gaulois*, "Le Service du Duc de Chevreuse," February 5, 1918, p. 2.
16. *Journal des Débats Politiques et Littéraires*, "Les Obsèques du Capitaine Féquant," September 11, 1915, p. 3.
17. Marvingt, Marie, "Mon Filleul, le Général Féquant," *Excelsior*, December 29, 1938.
18. Féquant, General Philippe, October 8, 1931.
19. Marvingt, Marie, "Mon Filleul, le Général Féquant," op.cit.
20. *Ibid.*
21. H.O., "Mlle Marvingt, la 'Fiancée du Danger,' Nous Révélera à Tunis les Sources de Son Extraordinaire Dynamisme," *Le Petit Matin [Tunis]*, February 3, 1952.
22. *L'Avenir à Argenteuil*, "La Célèbre Aviatrice Marie Marvingt S'est Intéressée aux Journées de Propagande Aéronautique," May 13, 1955.
23. Blanc, op. cit.
24. *Journal des Débats Politiques et Littéraires*, "Les Obsèques du Commandant Félix," June 22, 1914, p. 4; *L'Homme Libre*, "Obsèques du Commandant Félix," June 21, 1914, p. 2.
25. *L'Est Républicain*, "Ce Nouvel Accident ne Décourage pas Mlle Marvingt," May 15, 1910, p. 1.
26. *Ibid.*
27. Myriel, "L'Aviation," *Éclair de l'Est*, May 4, 1909.
28. *Ibid.*
29. *Paris-Soir*, "Avec les 'As,' en Attendant les Concurrents de la Coupe Deutsch de la Meurthe," May 30, 1933, p. 5; Marvingt, Marie, "Les Ailes Féminines au Service de la Charité," *La Croix*, November 28, 1931, no. 4956, pp. 1–2.
30. Archdeacon, Ernest, "Mademoiselle Marvingt: Une Sportwoman Extraordinaire," *La Revue Aérienne*, December 25, 1910, pp. 702–706.
31. *Ibid.*
32. *La Vie au Grand Air*, "Un Tour à Mourmelon Aviation," January 1, 1910, pp. 8–9.
33. *Ibid.*
34. *La Revue Aérienne*, "Nos Prix," January 10, 1910, p. 28.
35. *Le Figaro*, "Banquet d'Espérantistes," November 13, 1910, p. 3.
36. Albaret, Laurent, "Marie Marvingt, la Fiancée du Danger," *L'Echo de la Timbrology*," December 2015, no. 1901, pp. 68–72.
37. *La Presse*, "Pour Aller Faire une Conférence," April 18, 1913, p. 3.
38. *Gil Blas*, "Au Meeting de Monaco," April 17, 1913, p. 6.
39. *L'Echo d'Alger*, "Constantine: Ain-M'lila," July 16, 1932, p. 7.
40. Earhart, Amelia, *For the Fun of It* (1932), p. 180.
41. Marvingt, Marie, "Maryse Bastié, la Grande Championne des Ailes Françaises, *Revue Aéronautique de France*, February-March 1937, pp. 2–3.
42. *Ibid.*
43. *Albuquerque Journal*, "First French Woman Pilot Here, Gave Lindy Medal for Paris Hop," October 10, 1935, pp. 1, 12; *France Aviation*, "Nos Lecteurs Nous Écrivent: Marie Marvingt," September 1957, p. C.
44. Foch, Ferdinand, attributed in Raymond Recouly, *Foch: Le Vainqueur de la Guerre* (1919).
45. Foch, Ferdinand, cited in Charles Bugnet, *Foch Speaks* (1929), p. 171.
46. *Revue Aéronautique de France*, "Mademoiselle Marvingt: Une Grande Propagandiste," October 1929, pp. 7, 15.
47. Mauchaussée, Jean, "Marie Marvingt, Reine de l'Air," *Prestige*, October 1, 1959, no. 3, p. 1.
48. Bichet, Gabriel, "Marie Marvingt, la Fiancée du Danger, est Morte Hier à Nancy," *Est Républicain*, December 15, 1963, p. 1; Daurat, Michel, "À Quatre-Vingts Ans, Marie Marvingt Demeure Toujours la Fiancée du Danger," *Lorraine Magazine*, no. 9, February 1955, pp. 10–13; Saindizié, J.-M., "C'est Marie Marvingt, 85 Ans, la Plus Extraordinaire Femme du Siècle," *Pilote*, November 17 1960, no. 56, p. 10; et alia.
49. Walter, Don, "In the Spotlight for 50 Years," *European Stars and Stripes*, February 16, 1958, pp. 11–12.
50. Pétain, Philippe, autograph signed December 8, 1931, in Marcel Cordier, op. cit., p. 172.
51. Bichet, Gabriel (G.B.), op. cit.; Saindizié, J.-M., op. cit.
52. Marvingt, Marie, "Nos Maréchaux de France Vivants," *Petit Marocain*, September 26, 1934.
53. *Ibid.*
54. Marvingt, Marie, "Deux Grands Lorrains Réunis Dans la Mort," *Petit Marocain*, October 18, 1934; *Gil Blas*, "La Lorraine à Paris," December 9, 1912, p. 3.
55. *Ibid.*
56. *Ibid.*
57. Lyautey, Hubert, letter to his sister, in André Maurois, *Marshal Lyautey* (1931), p. 71.
58. Hoffmann, Eleanor, *Realm of the Evening Star: A History of Morocco and the Lands of the Moors* (1965), p. 231.
59. Marvingt, Marie, "Deux Grands Lorrains Réunis Dans la Mort," op.cit.
60. *Ibid.*
61. *L'Aérophile*, "Réception en l'Honneur du Maréchal Lyautey et des Membres du Congrès Aéronautique Coloniale," November 15, 1931, pp. 349–351.
62. Marvingt, Marie, "Deux Grands Lorrains Réunis Dans la Mort," op. cit.
63. Coulet, Jacques, and Madame Coulet, interview, June 24, 1982.
64. Marvingt, Marie, "Nos Maréchaux de France Vivants," *Petit Marocain*, September 26, 1934.
65. Mauchaussée, Jean, "Marie Marvingt, Reine de l'Air," *Prestige*, October 1, 1959, no. 3, p. 1.
66. Susset, Bruno, *L'Est Républicain*, "Un des Livres de Marie Marvingt Retrouvé à Nancy," July 1, 1993.
67. Bichet, Gabriel, op. cit.; Saindizié, J.-M., op. cit.
68. Weygand, Maxime, autograph signed December 21, 1958, cited in Marcel Cordier, op. cit.
69. Marvingt, Marie, "Nos Maréchaux de France Vivants," op. cit.
70. *Ibid.*
71. Méchelle, Ginette, interview, June 23, 1982.
72. Marvingt, Marie, "Les Obsèques du Commandant du Plessis à Palerme," *La Dépêche Tunisienne*, January 1924.
73. *Annales Africaines*, "300 Conférences, 4.000 Kilomètres en Afrique du Nord par Mlle Marie Marvingt," May 15, 1925, pp. 309, 312–313.
74. Liégeois, Achille, op. cit.
75. *Giornale di Sicilia*, "Una Conferenza dell'Aviatrice Marvingt al Circolo di Cultura," December 30, 1923; A.L., "Infatigable Globe-Trotter Raconte Ses 12 Mois d'Afrique," *L'Est Républicain*, August 25, 1927.
76. A.L., "Infatigable Globe-Trotter Raconte Ses 12 Mois d'Afrique," *L'Est Républicain*, August 25, 1927.
77. *L'Ouest Éclair*, "Le Général Marty Félicite Mlle Marvingt," June 1, 1928, p. 5.
78. *Ibid.*; *Revue Aéronautique de France*, "Mlle Marvingt au Maroc," February-March 1934, p. 4.

79. Marvingt, Marie, "Nos Maréchaux de France Vivants," op. cit.

80. Marvingt, Marie, "Deux Grands Lorrains Réunis Dans la Mort," op. cit.; *Le Radical*, "Ministres en Voyage," July 1, 1912, p. 2; *Le Rappel*, "MM. Poincaré et Lebrun à Bar-le-Duc," July 2, 1912, p. 3; *La Lanterne*, "Un Nouveau Champ d'Aviation," July 2, 1912, p. 2; *La Revue Aérienne*, "Les Fêtes du 30 Juin à Bar-le-Duc," July 25, 1912, pp. 398, 401; *Le Journal [Paris]*, "M. Raymond Poincaré Inaugure le Champ d'Aviation de Bar-le-Duc," July 1, 1912, p. 4.

81. "Une Délégation de Lorraine Offre un Object d'Art au Président," newspaper item, October 20, 1913.

82. Marvingt, Marie, "Deux Grands Lorrains Réunis Dans la Mort," op. cit.

83. *Ibid.*

84. *Ibid.*

85. *Ibid.*

86. Bidelman, Patrick Kay, *Pariahs Stand Up! The Founding of the Liberal Feminist Movement in France, 1858–1889* (1982), p. 39.

87. *Journal des Débats Politiques et Littéraires*, "A l'Elysée," January 13, 1939, p. 3; *Le Journal [Paris)]*, "A l'Elysee," January 12, 1939, p. 4.

88. Liégeois, Achille, op. cit.

89. Légion d'Honneur, La Grande Chancellerie de la Légion d'Honneur: documents supporting and confirming her grades of Chevalier and then Officer of the Légion d'Honneur, 1935 and 1949; Le Ministère des Travaux Publics, des Transports et du Tourisme, report, October 3, 1949: Renseignements Produits à l'Appui d'un Project de Décret Portant Promotion au Grade d'OFFICIER dans l'Ordre National de la Légion d'Honneur, for Mlle MARVINGT Marie; Archives Nationales, "Procès-Verbal de Réception d'un Officer de la Légion d'Honneur," May 14, 1950.o come.

90. Cajelot, Maurice, "Marie Marvingt, 'la Fiancée du Diable,' 100 ans," *Républicain Lorrain*, February 13, 1975, p. 4.

91. Plauche-Gillon, Jéhanne, in Jacqueline Maire, letter, July 29, 1982.

92. Brun, Mario, "Riviera Gazette," *Nice Matin*, November 22, 1956.

93. *Ibid.*

94. *L'Est Républicain*, numerous articles, May 1–10, 1951.

95. Reichardt, Gisèle, "'Marie Casse-Cou' un siècle d'exploits," *Républicain Lorrain*, 26 avril 1992, p. 7.

96. Cordier, Marcel, "Les Escales de Marie Marvingt," *Leurs Demeures en Lorraine: Tome II* (Éditions Pierron, 1983), p. 175.

97. Florentin, Mademoiselle, cited in Jacqueline Maire, letter, July 23, 1982.

98. Marchand, Robert, *Le Ciel N'a Pas de Toit* (Éditions Berger Levrault, 1962), pp. 258–262; Curral-Couttet, Gaby, "Marie Marvingt, la fiancée du danger," *Les Folles Années de Chamonix* (Éditions France-Empire, series "si 1900 m'était conté," 1984), pp. 75–78; Action Catholique des Femmes, *Femmes Remarquables: Je Broderai Vos Noms* (ACGF, 2001), pp. 212–215; Margot, Olivier, "L'Indomptable: Marie Marvingt," *L'Équipe*, August 17, 2013, pp. 60–63.

99. M.A., *Samedi Soir*, "Marie Marvingt, 'Fiancée du Danger,' Attend Encore Que Soit Relevé Son 'Défi Mondial,'" March 3, 1955, no. 505, pp. 1, 3.

100. *Ibid.*

101. Boulet, Jean, "De Passage au Maroc: Mlle Marie Marvingt, la Fiancée du Danger, Témoin ou Victime de Vingt Catastrophes, Elle est la Femme la plus Extraordinaire du Monde," Moroccan newspaper, 1951, pp. 1, 4; Leblond, J.-J. "Pour le 52e Anniversaire de Sa Première Ascension en Ballon Marie Marvingt a Piloté un Avion Sanitaire Entre Bordeaux et Marseille," *Dauphiné*, November 1952, p. 3.

102. *Ibid.*

103. Cordier, Marcel, op. cit., p. 175.

104. *L'Est Républicain*, "L'Aviatrice Nancéïenne Marie Marvingt Chevalier de la Légion d'honneur," January 21, 1935, p. 4.

105. *L'Est Républicain*, "L'Aviatrice Nancéïenne Marie Marvingt Chevalier de la Légion d'Honneur," January 21, 1935, p. 4.

106. *La Dépêche Meusienne*, "Marie Marvingt (1875–1963)," November 15, 2003.

107. Gygax, Georges, "La 'Fiancée du Danger': La Vie Prodigieuse de Marie Marvingt," *L'Illustré*, April 24, 1958, pp. 34–36.

108. Marvingt, Marie, "Trois Morts Mettent en Deuil les Ailes de Lorraine: Hommage d'Une 'Vieille Tige,'" *Dimanche Éclair*, September 17, 1950, p. 3.

109. Ballu, Yves, "Les Femmes et le Ski ou la Découverte des Sports d'Hiver," *L'Épopée du Ski* (Éditions Arthaud, 1981), p. 194.

110. Morlay, Gaby, autograph signed December 2, 1961, cited in Marcel Cordier, "Les Escales de Marie Marvingt," *Leurs Demeures en Lorraine: Tome II* (Éditions Pierron, 1983), p. 176.

111. Leblond, J.-J., op. cit.

112. *L'Est Républicain*, "Le Suprême Adieu de Nancy à Marie Marvingt," December 18, 1963; J.G., "L'Avion et le Vélo: Deux Amours de Marie Marvingt, 'la Fiancée du Danger,'" *L'Est Républicain*, February 16, 1975, p. 33.

113. Blanc, Colonel Gaston, and Madame, interview, June 24, 1982.

114. *Le Figaro*, "La Saison à Vittel," July 28, 1914, p. 10; Le Gaulois, "La Saison à Vittel," July 28, 1914, p. 5.

115. *Le Temps*, "Les Fêtes d'Aviation de Nancy," April 10, 1912, p. 4; *L'Est Républicain*, "Gala de L'Aéro-Club de L'Est," June 5, 1937, p. 4; et alia.

116. *L'Homme Libre*, "A l'Union des Oeuvres de Bienfaisance de l'Aéronautique," June 30, 1938, p. 4; *L'Est Républicain*, "Matinee de la Jeunesse de l'Empire Français," November 13, 1939, p. 3.

117. *L'Est Républicain*, "Le Gala de Bienfaisance du 26e," December 25, 1939, p. 3.

118. *L'Est Républicain*, "Le Gala des Lions," May 3, 1940, p. 3.

119. Union des Femmes de France, "Les Journées d'Aviation Sanitaire à l'Exposition," *Bulletin Mensuel*, October 1931, p. 402.

120. *L'Est Républicain*, "Le Bal de la Presse," March 2, 1938, p. 4.

121. *L'Est Républicain*, "La Soirée de la Faculté de Lettres," January 22, 1932, p. 3.

122. M.A., "Marie Marvingt, 'Fiancée du Danger,' Attend Encore Que Soit Relevé Son 'Défi Mondial,'" *Samedi Soir*, March 3, 1955, no. 505, pp. 1, 3; Mechelle, op. cit.; Blanc, op. cit.

123. *L'Est Républicain*, "Matinée Dansante du Pelican Club," November 19, 1929, p. 3.

124. *L'Est Républicain*, "Au Stand de l'Aéronautique," July 26, 1936, p. 4; *L'Est Républicain*, "Le Bal de la Presse," March 2, 1938, p. 4; et alia.

125. *L'Est Républicain*, "M. Jean Beziaud Parle de l'Aviation Française," April 24, 1938, p. 4; *L'Est Républicain*, "Nancy, la Ville de Stanislas, Fait au Général Rydz-Smigly une Enthousiaste Accueil," September 4, 1936, p. 3; *L'Est Républicain*, "Henry de Monfreid Chez les

Sauvages," December 7, 1936, p. 3; *L'Est Républicain*, "La conférence du docteur Benech," May 21, 1937, p. 5; L'Est Républicain, "M. Jean Beziaud parle de l'aviation française," April 24, 1938, p. 4; *L'Est Républicain*, "M. Jacques Mortane a parlé de Guynemer et de Mermoz," May 22, 1938, p. 4; et alia.

126. *L'Est Républicain*, "Obsèques: Madame Humblot," September 9, 1937, p. 4; *L'Est Républicain*, "Necrologie: M. Kleber Saudry," October 12, 1939, p. 3; et alia.

127. *L'Est Républicain*, "Les Obsèques du Professeur Gaston Michel," March 17, 1937, p. 4.

128. *L'Est Républicain*, "Au Pavillon de l'Aéronautique: Le Gala des Ailes Qui Sauvent," July 23, 1936, p. 4.

129. Boetsch, Jacques, "Une Vie en Équilibre" *Anciens Combattants du Monde*, May 1955, pp. 11–14.

130. Cordier, Marcel, op. cit., p. 173.

131. Michel-Royer, Dr. Jean, interview with Marie Marvingt, January 1963.

132. Gygax, Georges, April 24, 1958, op. cit.

133. Boetsch, Jacques, "Une Vie en Équilibre" *Anciens Combattants du Monde*, May 1955, pp. 11–14; Saindizié, J.-M., op. cit.; *La Nouvelle République*, "Marie Marvingt la 'Fiancée du Danger' Est Morte," December 16, 1963, no. 5855, pp. 1, C; et alia.

134. Aurenche, Dr. Henry, "Notre Doyenne," *La Presse Scientifique*, 1961; A.W., *Tunisie France*, "La Femme la Plus Extraordinaire du Siècle Est à Tunis," February 2, 1952; Baron Boilley, Françoise, *Marie Marvingt: A l'Aventure du Sport* (L'Harmattan, 2013), p. 33; Bichet, Gabriel, "Marie Marvingt Lance Aujourd'hui l'Aviation Sanitaire Mondiale," *L'Est Républicain*, July 5, 1956, p. 2; *Gil Blas*, "La Lorraine à Paris," December 9, 1912, p. 3; Walter, Don, "In the Spotlight for 50 Years," *European Stars and Stripes*, February 16, 1958, pp. 11–12; *Revue Mensuelle du Touring-Club de France*, "Liste des Candidats," April 1896, p. 156; *Les Vieilles Tiges: Annuaire*, "Membres Titulaires," 1924, p. 52; Rozet, George, "L'Avènement du Sport Féminin," *Lectures Pour Tous*, April 1, 1919, p. 1565; et alia.

135. *L'Est Républicain*, "Banquet en l'Honneur de l'Aviateur Sommer," September 10, 1909, p. 2;

Les Ailes, "L'Aéro-Club d'Auvergne à Paris," December 23, 1926, p. 4; *Les Ailes*, "Les Ailes des T.O.E.," March 23, 1933, p. 5; *Les Ailes*, "L'Aviation Médicale à Berck," June 27, 1935, p. 13; *Les Ailes*, "Ici et Là," February 3, 1938, p. 4; A.L., "Infatigable Globe-Trotter Raconte Ses 12 Mois d'Afrique," *L'Est Républicain*, August 25, 1927; *Gil Blas*, "Le Dîner de l'Académie des Sports," December 16, 1911, p. 6; *L'Est Républicain*, "Fêtes d'Aviation à Nancy," April 8–9, 1912, p. 2; *L'Est Républicain*, "Les Bals de Samedi: A l'Union des Femmes de France," November 28, 1926, p. 3; *L'Est Républicain*, "La Soirée de la Faculté de Lettres," January 22, 1932, p. 3; *Revue Aéronautique de France*, "Notre Fête du 21 Décembre," January 1931, p. 14; *Le Matin*, "L'Assemblée Générale des Anciens des Régiments d'Aviation du Maroc et de la Syrie," March 25, 1933, p. 10; *Le Matin*, "Le Ministre de l'Air Chez les Vieilles Tiges de l'Aviation," November 18, 1931, p. 3; *Le Figaro*, "Banquet d'Espérantistes," November 13, 1910, p. 3; *Le Figaro*, "Le Dîner de l'Académie des Sports," December 15, 1911, p. 3; et alia.

136. *L'Aérophile*, "Le Banquet de l'Aéro-Club de France en l'Honneur de M. Laurent-Eynac, Premier Ministre de l'Air," December 1–15, 1928, p. 374; *Les Ailes*, "L'Aéro-Club de France a Reçu M. Laurent Eynac," November 29, 1928, p. 4.

137. *L'Aérophile*, "Le Banquet Annuel des Vieilles Tiges," December 15, 1931, p. 377; *Le Journal [Paris]*, "M. J.-L. Dumesnil Expose aux 'Vieilles Tiges' l'Ensemble du Programme du Ministère de l'Air," November 18, 1931, p. 3.

138. *L'Echo d'Alger*, "Le Banquet de la Betterave," January 20, 1930, p. 4.

139. *Le Pays Lorrain*, "L'Entente Cordiale et la Lorraine," no. 7, 1907, pp. 350–351.

140. *Le Matin*, "Les Fêtes d'Aviation de Nancy," April 9, 1912, p. 3; *Les Ailes*, "L'Aéro-Club de France a Reçu M. Laurent Eynac," November 29, 1928, p. 4; *La Revue Aérienne*, "Nos Prix," July 10, 1910, p. 412; *Le Matin*, "La Féte de l'Air le 10 Juillet à Villacoublay," June 30, 1938, p. 8; *Le Journal [Paris]*, "M. Pierre Cot Est Venu à La Baule Accueillir les Concurrents du Tour de France des Avions de Tourisme Partis le Matin de Biarritz," July 28, 1933, pp. 1, 4; *Le Petit Parisien*, "Les Fêtes de Nancy; Le Banquet," July 2, 1906, p. 2; *Les Cahiers de la Santé Publique*, "Journées de l'Aviation Sanitaire Coloniale: Réception à L'Aéro-Club de France," June 25, 1931, p. 1043; *Comoedia*, "L'Oeuvre de Mlle Marvingt Pour l'Aviation Sanitaire," November 18, 1931, p. 5.

141. *Les Ailes*, "L'Aéro-Club de France a Reçu M. Laurent Eynac," November 29, 1928, p. 4.

142. *Ibid.*

143. *L'Aéro*, "Académie des Sports," November 6, 1911, p. 6; *Journal des Débats Politiques et Littéraires*, "Sports Divers," November 12, 1911, p. 3; *Le Temps*, "L'Académie des Sports," 7 novembre 1911, p. 5; *Le Gaulois*, "L'Académie des Sports," November 11, 1911, p. 4; *Gil Blas*, "L'Académie des Sports," November 11, 1911, p. 4; *L'Echo Sportif du Centre et de l'Ouest*, "Une Carrière Féminine Sportive," 1914.

144. *Le Journal [Paris]*, "Au Club du Faubourg," December 16, 1936, p. 7.

145. *Le Journal [Paris]*, "Au Club du Faubourg," January 9, 1937, p. 7.

146. *Le Matin*, "La Première de Ce Soir," October 27, 1937, p. 10; *Le Journal [Paris]*, "Au Club du Faubourg," October 27, 1937, p. 7.

147. Nocher, Jean, *En Direct Avec Vous*, radio program, January 1964.

148. *Les Ailes*, "Coups d'Ailes," December 15, 1932, p. 15.

149. *Ibid.*

Chapter 15

1. Daurat, Michel, "A Quatre-Vingts Ans, Marie Marvingt Demeure Toujours la Fiancée du Danger," *Lorraine Magazine*, no. 9, February 1955, pp. 10–13.

2. Pilâtre, Philippe Buron, *Lorraine, Fille de l'Air* (Éditions Serpenoise, 2010), pp. 52–58; Favier, Claude (1960), in Vianney Huguenot, "Notre Incroyable Amnésie," *L'Estrade*, November 2015, no. 57.

3. Didier, Gény, in Jacqueline Maire, letter, July 5, 1982.

4. *Le Matin*, "Reims-Mourmelon-Reims en Aéroplane Avec Six Passagers," October 19, 1911, p. 2.

5. Codron, Charles, letter, circa 1905 (Lhérault collection).

6. *Ibid.*

7. Barthélemy, Général, "La Femme la Plus Extraordinaire du Siècle: Marie Marvingt, la 'Fiancée du Danger,'" *L'Étrange Race des Hommes Volants* (Éditions France-Empire, 1979), pp. 241–254.

8. Saladin, Raymond, "Figures et Événements: Marie Marvingt," *Aviation Magazine*, March 24, 1955, pp. 12–13.

9. Bitsch, Maître, cited in Jacqueline Maire, letter, August 5, 1982.
10. *Brandon [Manitoba] Weekly Sun*, "Fair Aviators," January 12, 1911, p. 23.
11. Berlioux, Monique, *Gloires de Sport* (Atlantica, 2009), p. 180.
12. Colombani, Dr. Jules, cited in Jean Boulet, "De Passage au Maroc: Mlle Marie Marvingt, la Fiancée du Danger, Témoin ou Victime de Vingt Catastrophes, Elle est la Femme la Plus Extraordinaire du Monde," Moroccan newspaper, 1951, pp. 1, 4.
13. Michel-Royer, Dr. Jean, recorded interview with MM, January 1963.
14. Gygax, Georges, "La 'Fiancée du Danger': La Vie Prodigieuse de Marie Marvingt," *L'Illustré*, April 24, 1958, pp. 34–36.
15. *Nord-Touriste*, "La Conférence de Mlle Marvingt sur l'Aérostation et l'Aviation," December 1912, pp. 269, 274–275.
16. Curral-Couttet, Gaby, "Marie Marvingt, la Fiancée du Danger," *Les Folles Années de Chamonix* (Éditions France-Empire, 1984), pp. 75–78.
17. Brun, Madame Emmanuel, interview, June 22, 1982.
18. Daurat, Michel, op.cit.; Floret, Robert, "Grand-Mère 'Casse-Cou,'" *Détective*, November 18, 1960, no. 751, pp. 10–11.
19. R.P., "Mademoiselle Marvingt," *Le Cri de Constantin*, June 28, 1924.
20. Lionel-Pellerin, Marie-José, letter, July 6, 1992.
21. Liégeois, Achille, "Les Nancéiennes Loin de Chez Elles: Mlle Marie Marvingt Rentre d'un Long Voyage: Elle Nous Raconte Ses Impressions," *L'Est Républicain*, August 23, 1926, p. 1.
22. *Nord-Touriste*, op. cit., *L'Ouest Éclair [Caen]*, "Le Général Marty Félicite Mlle Marvingt," June 1, 1928, p. 5; *L'Echo d'Alger*, "Huit Cours et Conférences à Alger de Mlle Marie Marvingt," May 15, 1925, p. 3; *Ligue Aéronautique de France*, "Les Conférences de Mlle Marvingt," May 1914, p. 75; et alia.
23. Thirion, M., "Hommage de l'Aéro-Club de l'Est à Marie Marvingt," *L'Est Républicain*, December 18, 1963.
24. Brun, Emmanuel, interview, June 22, 1982.
25. *Ibid.*
26. Prudhomme, Monsieur, cited in Jacqueline Maire, letter, August 5, 1982.
27. Didier, Gény, in Jacqueline Maire, letter, July 5, 1982.
28. Coulet, Jacques, interview, June 24, 1982.
29. J.B.L., "On Peut Fort Bien 'Être et Avoir Été,' Démontre Marie Marvingt," *La Presse*, February 3, 1952.
30. Rio, Armand, "La Fiancée du Danger," *Lecture Pour Tous*, April 1, 1913, vol. 15, no. 7, pp. 63–71.
31. Thirion, M., "Hommage de l'Aéro-Club de l'Est à Marie Marvingt," *L'Est Républicain*, December 18, 1963.
32. *L'Est Républicain*, "L'Aéro-Club de l'Est a Tenu son Assemblée Générale Annuelle," "Le Cinquantenaire du Comité de Nancy de l'Union des Femmes de France," "Mariage de Paul-Cavallier et de Suyrot," March 6, 1938, p. 4.
33. Gil Blas, "Les Sports d'Hiver," February 16, 1911, p. 5; Drigny, Georges, "La Grande Semaine d'Hiver du T.C.F.," *Le Sport Universel Illustré*, January 9, 1910, p. 139; *Le Matin*, "Course de Luges à Plat Ventre," January 25, 1911, p. 3; *Le Journal [Paris]*, "Sports d'Hiver," January 23, 26, 27, 1911, p. 6; *La Presse*, "Le Championnat de Bobsleighs," January 26, 1911, p. 3.
34. *Le Matin*, "Une Aviatrice Très Occupée," December 10, 1912, p. 6; *L'Aéro*, "Propos en l'Air," December 13, 1912, p. 1.
35. *L'Aéro*, "Propos en l'Air," August 31, 1912, p. 1.
36. *Le Petit Parisien*, "Les Voyages d'une Aviatrice," January 4, 1913, p. 5.
37. Liégeois, Achille, op. cit.
38. Leblond, J.-J., "Pour le 52e Anniversaire de Sa Première Ascension en Ballon Marie Marvingt a Piloté un Avion Sanitaire Entre Bordeaux et Marseille," *Dauphiné*, November 1952, p. 3.
39. Marck, Bernard, *Women Aviators* (Flammarion, 2013), pp. 14–21, 27, 39, 40, 43 / Marck, Bernard, *Elles Ont Conquis le Ciel* (Arthaud, 2009), pp. 14–21.
40. *[Washington, DC] Evening Star*, "A Female Admirable Crichton," October 18, 1913, p. 5; *Border Watch [Mount Gambier, Australia]*, "Female Admirable Crichton," November 29, 1913, p. 6; *Brunswick and Coburg [Australia] Leader*, "Female Admirable Crichton," March 6, 1914, p. 4; *Camperdown [Australia] Chronicle*, "The Admirable Crichton," November 27, 1913, p. 5; *Dimboola [Australia] Banner and Wimmera and Mallee Advertiser*, "Female Admirable Crichton," March 10, 1914, p. 4; *The Gundagai [Australia] Independent and Pastoral, Agricultural and Mining Advocate*, "Female Admirable Crichton," April 8, 1914, p. 2; *The South Eastern Times [Millicent, Australia]*, "Female Admirable Crichton," August 1, 1922, p. 4; *West Gippsland [Australia] Gazette*, "Female Admiral Crichton," December 2, 1913, p. 6; *Zeehan and Dundas [Tasmania] Herald*, "Female Admirable Crichton," November 19, 1913, p. 3; et alia.
41. Marvingt, Marie, postcard to Mesdemoiselles Winsbach, April 5, 1957; postcard to her cousin, February 19, 1961; et alia.
42. Cordier, Marcel, "Les Escales de Marie Marvingt," *Leurs Demeures en Lorraine: Tome II* (Éditions Pierron, 1983), p. 176.
43. Acte de Notoriété de Marie Marvingt, Étude du Maître Jules Thomas, notary, Nancy, September 29, 1964.
44. Méchelle, Ginette, interview, June 23, 1982.
45. Gygax, Georges, "La Vie Prodigieuse de Marie Marvingt, la 'Fiancée du Danger'," *L'Illustré*, May 1, 1958, pp. 40–42.
46. Wanss, Annie, "La Mode Chez les Aviatrices," *La Revue Aérienne*, March 10, 1912, no. 82, pp. 125–128.
47. *Ibid.*
48. *Ibid.*
49. Blanc, Colonel Gaston, letter, May 13, 1983.
50. Curral-Couttet, Gaby, op. cit.; Masson, Lucien, "Programme des Spectacles: Marie Marvingt," May 29, 1993; Marcacci, Philippe, "Marie Marvingt Redécolle," *Est Magazine*, June 27, 2004, pp. 10–11; *L'Est Républicain*, "Laxou: Honneur à Marie Marvingt, May 12, 2002; et alia.
51. Boetsch, Jacques, "Une Vie en Équilibre" *Anciens Combattants du Monde*, May 1955, pp. 11–14.
52. Peyrefitte, Alain, ed., "Les Femmes Pilotes Affrontent le Danger" (*Le Figaro*, January 20, 1914), *L'Aventure du XXe SiŠcle* (1989), p. 180.
53. Ballu, Yves, "Les Femmes et le Ski ou la Découverte des Sports d'Hiver," *L'Épopée du Ski* (Éditions Arthaud, 1981), p. 163.
54. Curral-Couttet, Gaby, op cit.
55. Friry, Jean, in Jacqueline Maire, letter, July 5, 1982.
56. Gygax, Georges, April 24, 1958, op. cit.
57. *L'Est Républicain*, "Mlle Marie Marvingt Reçoit Deux Nouveaux Témoignages Officiels de Gratitude," February 21, 1958, p. 4.
58. Ackerman, Gordon, "Fiancée of Danger," *Sports Illustrated*, June 26, 1961, pages 61–64.
59. Save, Colette, "L'Âge N'est Plus une Question

d'Années," *Point de Vue Images du Monde*, March 3, 1955, p. 20.
60. Gygax, Georges, April 24, 1958, op. cit.
61. Barthélemy, Général, "La Femme la Plus Extraordinaire du Siècle: Marie Marvingt, la 'Fiancée du Danger,'" *L'Étrange Race des Hommes Volants* (Éditions France-Empire, 1979), pp. 241–254.
62. *L'Est Républicain*, "Un Beau Trait de Camaraderie Sportive," July 9, 1911, p. 1.
63. *New Castle [Pennsylvania] News*, "Pilots Commended by French Aviatrix," January 9, 1936, p. 10; et alia.
64. *Indépendant*, "Mlle Marie Marvingt à Pau: Deux Causeries Littéraires et Sportives," June 16, 1927; Renoux-Barès, E., "Une Héroïne Parle à Pau de l'Aviation, École d'Énergie," *Revue Aéronautique de France*, October 1927, p. 11.
65. Marvingt, Marie, "L'Aviation et la Jeunesse des Écoles en Amérique," *Revue Aéronautique de France*, December 1936, pp. 2–3.
66. *Les Ailes*, "Coups d'Ailes," April 27, 1922, p. 2.
67. *Indépendant*, "Mlle Marie Marvingt à Pau: Deux Causeries Littéraires et Sportives," June 16, 1927.
68. Codron, Charles, op. cit.
69. Blanc, Colonel Gaston, interview, June 24, 1982.
70. Maire, Jacqueline, letter, February 12, 1983.
71. X.B., "D'Ultimes Témoignages Viennent Encore Préciser Quelques Traits Attachants de Marie Marvingt, Illustre Figure Lorraine," *Républicain Lorrain*, January 24, 1964, p. 2.
72. Marvingt, Marie, letter to Louis Lhérault, January 10, 1952, in Marcel Cordier, "Les Escales de Marie Marvingt," *Leurs Demeures en Lorraine: Tome II* (Éditions Pierron, 1983), p. 174.
73. Rio, Armand, op. cit.
74. Currel-Couttet, Gaby, op cit.
75. Marchand, Robert, *Le Ciel N'a Pas de Toit* (Éditions Berger Levrault, 1962), pp. 258–262.
76. J.G., "L'Avion et le Vélo: Deux Amours de Marie Marvingt, 'la Fiancée du Danger,'" *L'Est Républicain*, February 16, 1975, p. 3.
77. Valode, Philippe, "Marie Marvingt, l'Inégalable Casse-Sou," *Les Grandes Scandaleuses* (First Editions, 2015), pp. 227–235.
78. Micromegas, "Une Femme," *L'Echo d'Alger*, July 6, 1922, p. 2.
79. Cordier, Marcel, *Lorraine, Secrète et Insolite* (Éditions du Sapin d'Or, 2011), p. 83; Chevalier, Frédérique, "Marie Marvingt, la Fiancée des Airs (1875–1963)," *Les Grandes Aventurières* (City Editions, 2007), pp. 85–95; Comte, Marie-Gaëtane, "Marie Marvingt," *Oxygen*, February 2006, no. 4, p. 24; Dollet, Christophe, "Marie Marvingt Sans Visa," *L'Est Républicain*, July 11, 2011.
80. J.P., "Marie Marvingt, Pilote d'Avion, Est Norte à 88 ans: Championne en Tous Genres," *Le Monde*, December 17, 1963, p. 13.
81. Fisher, Élise, *Le Roman de la Place Stanislas* (Éditions Place Stanislas, 2007), pp. 51–53.
82. Flobert, Laure-Paul, "La Femme et le Costume Masculin," *Le Vieux Papier* (Bulletin de la Société Archéologique, Historique et Artistique), July 1, 1911, pp. 349–366.
83. Yelnick, Claude, "Adieu à Marie Marvingt," *Icare*, March 1964, no. 29, pp. 60–65.
84. *L'Est Républicain*, "Il y a Vingt Ans Mourait Marie Marvingt," December 14, 1983.
85. Baron Boilley, Françoise, *Marie Marvingt: A l'Aventure du Sport* (L'Harmattan, 2013), pp. 16, 23.
86. "Marie Marvingt," www.wehuntedthemammoth.com, December 8, 2016.
87. Marvingt, Marie, "L'Aviation Sanitaire et les Femmes," *Revue Aéronautique de France*, September-October 1931, pp. 4, 5–6; Marvingt, Marie, "Les Ailes Féminines au Service de la Charité," *La Croix*, November 28, 1931, no. 4956, pp. 1–2.
88. Davis, Jeffrey R., Robert Johnson, Jan Stepanek, Jennifer A. Fogarty, eds., *Fundamentals of Aerospace Medicine*, 4th ed. (2008), pp. 9, 14.
89. Larue, Michèle, in Monique Ayoun, "La Fille de l'Air," *Le Nouvel Observateur*, September 7, 2012.
90. *Brandon [Manitoba] Weekly Sun*, "The Aviatress," 8 December 1910, p. 8.
91. *West Gippsland [Australia] Gazette*, "Flights of Women: Steering Aeroplanes As Though They Were Perambulators," March 14, 1911, p. 6.
92. *Winnipeg [Manitoba] Free Press*, "Over the World With Women," January 7, 1911, p. 20.
93. Barthélemy, Général, op. cit.
94. Lahausse, Jean-Bernard, and Romain Sertelet, "Biographie du Mois: Marie Marvingt, 'La Femme la Plus Décorée du Monde," March 2013, http://verdun-meuse.fr.
95. Baron Boilley, Françoise, in *L'Est Républicain*, October 10, 2013.
96. Shipton, Elisabeth, *Female Tommies: The Frontline Women of the First World War* (The History Press, 2014), pp. 83–85, 128, 237.
97. *Le Figaro*, "Un Voyage Mouvementé," October 28, 1909, p. 1.
98. Dahlquist, Marina, ed., *Exporting Perilous Pauline: Pearl White and the Serial Film Craze* (2013), p. 6.
99. Gayraud, Amélie, cited in Robertson, Priscilla, *An Experience of Women: Pattern and Change in Nineteenth-Century Europe* (Temple University Press, 1982), pp. 341–342.
100. Baily, Paul J., *Gender and Education in China: Gender Discourses and Women's Schooling in the Early Twentieth Century* (2007), p. 71.
101. Combeau-Mari, Evelyne, et Valérie Boulain, "Transgression des Normes Sexuées et Violence Symbolique: Itinéraire de Jeunesse d'une Sportive, Marie Marvingt (1875–1963)," *International Review on Sport and Violence* (2014), no. 8, pp. 48–61.
102. Heilbrun, Carolyn, *Writing a Woman's Life* (1988).
103. *Comoedia*, "Nos Aviatrices: Hommage à Mlle Marvingt," August 2, 1930, p. 4; *Le Figaro*, "Marie-Marvingt Parle et Vole," July 31, 1930, p. 8.
104. Pétain, Maréchal Philippe, signed receipt to Marie Marvingt, August 28, 1941.
105. *Revue Aéronautique de France*, "Partie Officielle," August-September 1932, pp. 3–4.
106. Marvingt, Marie, "Maryse Bastié, la Grande Championne des Ailes Françaises, *Revue Aéronautique de France*, February-March 1937, pp. 2–3.
107. Marvingt, Marie, "L'Aviation Sanitaire et les Femmes," *Revue Aéronautique de France*, September-October 1931, pp. 4, 5–6; Marvingt, Marie, "Les Ailes Féminines au Service de la Charité," *La Croix*, November 28, 1931, no. 4956, pp. 1–2.
108. Marvingt, Marie, "Leurs Femmes," *La Vigie Marocaine*, October 22, 1934.
109. *Ibid.*
110. *Ibid.*
111. *Ibid.*
112. *Ibid.*
113. Grandidier, Gilbert, "Marie Marvingt (1875–1963), 'la Fiancée du Danger' Ouvre aux Femmes la Voie des Airs," newspaper, May 1987.

114. *The Washington Herald*, "Women Not Good Flyers," November 29, 1910, p. 7; *El Paso Herald*, "Flying Not the Sport for Women," 6 December 1910, p. 11; *Trenton Evening Times*, "Says Women Will Not Excel in Air," 29 November 1910, p. 7; "Not Suitable for Women," *New Castle [Pennsylvania] Herald*, Dec. 6, 1910.
115. Rudorff, Raymond, *The Belle Epoque—Paris in the Nineties* (1972), p. 13.
116. Ackerman, Gordon, op. cit.
117. *Républicain Lorrain*, "Marie Marvingt, la 'Fiancée du Danger,' N'est Plus," December 15, 1963, pp. 1, 4.
118. Godot, J.-H., interview, June 24, 1982.
119. *The Times [London]*, "Across the North Sea in a Balloon," letter to the editor, December 18, 1909, p. 12.
120. *Le Journal [Paris]*, "A Propos d'Une Ascension," November 12, 1909.
121. *Geelong [Australia] Advertiser*, "Lady Aviator Runs Into Trouble," August 18, 1911, p. 4.
122. James, P.D., *A Taste for Death* (1986), p. 183.
123. Lessing, Doris, in *Sunday Times* (1992).

Chapter 16

1. Daurat, Michel, "A Quatre-Vingts Ans, Marie Marvingt Demeure Toujours la Fiancée du Danger," *Lorraine Magazine*, no. 9, February 1955, pp. 10–13.
2. Ackerman, Gordon, "Fiancée of Danger," *Sports Illustrated*, June 26, 1961, pages 61–64.
3. *Ibid.*
4. Bichet, Gabriel (G.B.), "Marie Marvingt Lance Aujourd'hui l'Aviation Sanitaire Mondiale," *L'Est Républicain*, July 5, 1956, p. 2.
5. Maire, Jacqueline, letter, July 12, 1983.
6. Gygax, Georges, "La 'Fiancée du Danger': La Vie Prodigieuse de Marie Marvingt," *L'Illustré*, April 24, 1958, pp. 34–36.
7. Boetsch, Jacques, "Une Vie en Équilibre" *Anciens Combattants du Monde*, May 1955, pp. 11–14.
8. Poirier, Marcelle, "Danger's Sweetheart, at 80, Still Astonishes Paris," *Yorkshire Post and Leeds Intelligencer*, March 10, 1955, p. 5.
9. Daurat, Michel, op. cit."
10. Liégeois, Achille, "Les Nancéiennes Loin de Chez Elles: Mlle Marie Marvingt Rentre d'un Long Voyage: Elle Nous Raconte Ses Impressions," *L'Est Républicain*, August 23, 1926, p. 1.
11. Féral, Roger, "Marie Marvingt (86 ans), 'La Fiancée du Danger,' a Passé La Toussaint Dans un Hélicoptère à Réaction," *Télé-Paris*, November 1958.
12. Gygax, Georges, "La Vie Prodigieuse de Marie Marvingt, La 'Fiancée du Danger'," *L'Illustré*, May 1, 1958, pp. 40–42.
13. Blanc, Colonel Gaston, letter, May 13, 1983.
14. Maire, Jacqueline, letter, July 5, 1982.
15. Coulet, Jacques, and Madame Coulet, interview, June 24, 1982; Monsieur Remy, cited in Jacqueline Maire, letter, August 5, 1982; Bichet, Gabriel, "C'était la Fiancée du Danger...," *L'Est Républicain*, December 15, 1963; J.G., "L'Avion et le Vélo: Deux Amours de Marie Marvingt, 'la Fiancée du Danger,'" *L'Est Républicain*, February 16, 1975, p. 3.
16. Coulet, op. cit.
17. Prévote. Louis-Philippe, in Jacqueline Maire, letter, July 5, 1982; J.G., op. cit.
18. Jansen, Marianne, "Marie Marvingt: 86 Ans d'Âge, 51 Ans d'Aviation," *Écho de la Mode*, no. 47, November 19, 1961, pp. 18–19.
19. Blanc, op. cit.
20. Cordier, Marcel, "Les Escales de Marie Marvingt," *Leurs Demeures en Lorraine: Tome II* (Éditions Pierron, 1983), p. 171.
21. Bouchet, Madame, in Jacqueline Maire, letter, July 5, 1982.
22. Coulet, op. cit.
23. Oudin, René (July 17, 1968), in letter from J.M. Geoffroy, March 29, 1983; Cordier, Marcel, op. cit.; Estrada de Tourniel, Jérôme, "La Fiancée du Danger," *Passions Grand Est*, December 2003, numéro 26, pp. 58–62.
24. Bichet, Gabriel, op. cit.
25. Remy, Jean-Louis, in Jacqueline Maire, letter, August 5, 1982.
26. Légion d'Honneur, La Grande Chancellerie de la Légion d'Honneur: documents supporting and confirming her grades of Chevalier and then Officer of the Légion d'Honneur, 1935 and 1950.
27. Marchard, Robert, *Le Ciel N'a Pas de Toit* (1962), p. 244; Normand, Suzanne, "Les Aviatrices," *Marianne*, March 4, 1936, p. 9.
28. *Le Figaro*, "Le Prix de l'Héroïsme," October 21, 1910, p. 4.
29. *Ibid.*
30. *Anaconda (Montana) Standard*, "Aviators' Earnings: Record of the European Fliers up to Sept. 10 of This Year," November 27, 1910, p. 35.
31. *L'Est Républicain*, "La 'Fiancée' de Nancy," June 4, 1955; *L'Avenir à Argenteuil*, "La Célèbre Aviatrice Marie Marvingt S'est Intéressée aux Journées de Propagande Aéronautique," May 13, 1955; Gygax, Georges, April 24, 1958, op. cit.; Berlioux, Monique, *Gloires de Sport* (Atlantica, 2009), p. 180; Bowen, Barry D., "Marie Marvingt: The Premier Sportswoman of France and Pioneer in the Development of Aeromedical Evacuation," special to AirAmbulanceServices.com, March 6, 2013.
32. Boselli, Elisabeth, "Marie Marvingt 1875–1963 'la Fiancée du Danger,'" *Pionniers*, July 1984, pp. 18–22.
33. Masson, Lucien, "Programme des Spectacles: Marie Marvingt," May 29, 1993.
34. *La Revue du Touring Club de France*, "Petites Annonces," no. 387, February 1927, p. 37.
35. Coulet, op. cit.
36. Blanc, op. cit.
37. Florentin, Mademoiselle, in Jacqueline Marie, letter, July 23, 1982.
38. *L'Est Républicain*, "Il y a Vingt Ans Mourait Marie Marvingt," December 14, 1983.
39. Didier, Gény, in Jacqueline Maire, letter, July 5, 1982.
40. Bitsch, Monsieur X., conversation with Jacqueline Maire, 1984.
41. de Lalanne, Mademoiselle, in J.M. Geoffroy, letter, March 29, 1983.
42. Blanc, op. cit.
43. Blanc, op. cit.
44. Coulet, op. cit.
45. Bouchet, op. cit.
46. Jansen, Marianne, "Marie Marvingt: 86 Ans d'Âge, 51 Ans d'Aviation," *Écho de la Mode*, no. 47, November 19, 1961, pp. 18–19.
47. Krysaniac, C., *Le Cartophile de Meurthe-&-Moselle*, no. 1, February 1983; Oudin, René, newspaper article, July 17, 1968.
48. Willotte, Jeanne, letter, August 6, 1982.
49. de Lalanne, Mademoiselle, in J.M. Geoffroy, letter, March 29, 1983.
50. Brun, Emmanuel, and Madame Brun, interview, June 22, 1982.

51. Bichet, Gabriel, "C'était La Fiancée du Danger...," *Est Républicain*, December 15, 1963.
52. Blanc, op. cit.
53. Coulet, op. cit.
54. *Ibid.*
55. Godot, J.-H., interview, June 24, 1982.
56. Plauche-Gillon, Jehanne, in Jacqueline Maire, letter, July 29, 1982.
57. Monsieur Hulster, Gény Didier, and Monsieur Hodt, in Jacqueline Maire, letter, July 5, 1982.
58. Grandidier, Gilbert, "Marie Marvingt (1875–1963), 'La Fiancée du Danger' Ouvre aux Femmes la Voie des Airs," newspaper article, May 1987; Krysaniac, C., *Le Cartophile de Meurthe-&-Moselle*, no. 1, February 1983.
59. Méchelle, Ginette, interview, June 23, 1982.
60. Plauche-Gillon, op. cit.
61. Maire, Jacqueline, letter, April 4, 1984.
62. Grandidier, Gilbert, op. cit.
63. *L'Est Républicain*, "Il y a Vingt Ans Mourait Marie Marvingt," December 14, 1983.
64. Plauche-Gillon, op. cit.
65. Maire, Jacqueline, letter, July 29, 1983.
66. Bichet, op. cit.
67. Michel-Royer, Dr. Jean, recorded interview with MM, January 1963.
68. Plauche-Gillon, Jehanne, letter, July 29, 1982.
69. *Ibid.*
70. *Ibid.*
71. Enselme-Trichard, Renée, letter, July 3, 1982.
72. Centre Psychothérapique de Nancy, "Constation de Décès," and medical notes, 14 December 1963.
73. *Ibid.*
74. Blanc, op. cit.
75. Thévenin, Madame, in Marcel Cordier, "Marie Marvingt: une Femme de Foi," *Chrétiens Dans la Ville*, no 73, June 1992, p. 6.
76. Willotte, op. cit.
77. *Ibid.*
78. Talbott, Earl G., "Marie Marvingt, 'Fiancée of Danger': Flirted With Death 88 Years," *New York Herald Tribune*, December 16, 1963.
79. *Républicain Lorrain*, "Marie Marvingt, la 'Fiancée du Danger,' N'est Plus," December 15, 1963, pp. 1, 4.
80. Pelot, Paul, "Marie Marvingt Fut une des Femmes les Plus Extraordinaires du Siècle," *L'Équipe*, December 16, 1963.
81. Thirion, M., "Hommage de l'Aéro-Club de l'Est à Marie Marvingt," *L'Est Républicain*, December 18, 1963.
82. Bichet, Gabriel (G.B.), "Marie Marvingt, la Fiancée du Danger, Est Morte Hier à Nancy," *Est Républicain*, December 15, 1963.
83. J.P., "Marie Marvingt, Pilote d'Avion, Est Morte à 88 Ans: Championne en Tous Genres," *Le Monde*, December 17, 1963, p. 13.
84. *The Times of India*, "Marie Marvingt Dead," December 16, 1963, p. 7.
85. Conte, Arthur, *op. cit.;* Boetsch, Jacques, op. cit.; Petit, Edmond, "Hommage à Marie Marvingt," *Forces Aériennes Françaises*, no. 200, February 1964, pp. 268–270; Houard, Georges, "Marie Marvingt, Pilote et Propagandiste Exemplaire," *Les Ailes*, January 1, 1964, p. 21; J.G., op. cit.; Lederlé, Olivier, "Marie Marvingt, Femme d'un Siècle," *Républicain Lorrain*, November 2, 1991; Droz, Stanislas, *Les Femmes Dans la Grande Guerre 1914–1918* (Vent d'Est, 2014), pp. 74, 75–78; Erbstein, Roland, op. cit.; Estrada de Tourniel, Jérôme, op. cit.; et alia.
86. Bichet, Gabriel, "C'était La Fiancée du Danger...," *Est Républicain*, Décembre 15, 1963.
87. *L'Est Républicain*, "Le Suprême Adieu de Nancy à Marie Marvingt," December 18, 1963.
88. Thirion, M., "Hommage de l'Aéro-Club de l'Est à Marie Marvingt," *L'Est Républicain*, December 18, 1963.
89. *Républicain Lorrain*, "Les Ailes Françaises Ont Rendu un Ultime et Émouvant Hommage à Marie Marvingt, 'La Colombe Qui Sauve,'" Decemberl8, 1963, p. 2.
90. Cordier, Marcel, letter, January 16, 1984.
91. Erbstein, Marguerite, "Marie Marvingt, 'La Fiancée du Danger,' Entre à l'Olympe des Plus Grandes Sportives," *L'Est Républicain*, September 8, 1987.
92. Acte de Notoriété de Marie Marvingt, Étude du Maître Jules Thomas, notary, September 29, 1964.
93. *Ibid.*

Chapter 17

1. Rogers, Will, in Donald Day, ed., *The Autobiography of Will Rogers* (Houghton Mifflin Company, 1926), p. 107.
2. Nocher, Jean, *En Direct Avec Vous*, radio program, January 1964.
3. *L'Est Républicain*, "Il y a Vingt Ans Mourait Marie Marvingt," December 14, 1983.
4. J.G., "Trois Plaques Pour la 'Fiancée du Danger': Marie Marvingt Honorée Dans Sa Ville," *Est Républicain*, June 8, 1984; Estrada de Tourniel, Jérôme, "La Fiancée du Danger," *Passions Grand Est*, no. 26, December 2003, pp. 58–62.
5. Marvingt, Marie, "The Alps in Winter, I," *The [London] Times*, December 14, 1909.
6. Rogers, Will, op. cit.
7. de Sévigné, Madame (1675), *Letters of Madame de Sévigné to Her Daughter and Her Friends*, vol. 3 (1811), p. 104.
8. *L'Aérophile*, "Hélène Boucher," vol. 42, no. 12, December 1934, p. 366; *Flight*, "Mlle. Boucher Killed," vol. 26, no. 1334, December 6, 1934. p. 1298.
9. Glasgow, Ellen, *Letters of Ellen Glasgow* (1958), p. 267.
10. James, Alice (1891), in Anna Robeson Burr, *Alice James* (1934), p. 249.
11. Willotte, Jeanne, letter, August 6, 1982.
12. Cordier, Marcel, *Hommes et Lieux de Mémoire en Lorraine* (Éditions Pierron, 1991), p. 94.
13. Emerson, Ralph Waldo, *Representative Men: Seven Lectures* (1856), p. 20.
14. France, Geo, "Marie Marvingt Espère Fêter Ses 90 Ans ... Dans la Lune," *Ouest-France*, February 23, 1955.
15. Michel-Royer, Dr. Jean, recorded interview, January 1963.
16. Pessel, Jeanne-Georgette, letter, June 27, 1982.
17. Leclerc, Jean, "Le Vélo de Marie Marvingt," *CAIN*, March 10, 2016.
18. Coulet, Jacques, interview, June 24, 1982.
19. *Ibid.*
20. Dr. Bitsch, in Jacqueline Maire, letter, August 5, 1982; Knecht, Marcel, letters, 1984; Saladin, Raymond, "Figures et Événements: Marie Marvingt," *Aviation Magazine*, March 24, 1955, pp. 12–13.
21. Besançon, Madame, letter, June 28, 1982.
22. Coulet, Jacques, and Madame Coulet, interview, June 24, 1982.
23. Brun, Emmanuel, and Madame Brun, interview, June 22, 1982; Godot, J.-H., interview, June 24, 1982.
24. Laprévote, Louis-Philippe, in Jacqueline Maire, letter, July 5, 1982.

25. Gygax, Georges, "La Vie Prodigieuse de Marie Marvingt, La 'Fiancée du Danger'," *L'Illustré*, May 1, 1958 pp. 40–42.
26. Charut, Gérard, "Les Cent Vies de Marie Marvingt," *L'Est Républicain*, March 4, 1990.
27. Marcacci, Philippe, "Marie Marvingt redécolle," *Est Magazine*, 27 juin 2004, pp. 10–11
28. Blanc, Colonel Gaston, interview, June 24, 1982.
29. Cajelot, Maurice, "Marie Marvingt, 'la Fiancée du Diable,' 100 Ans," *Républicain Lorrain*, February 13, 1975, p. 4.
30. Mechelle, Ginette, interview, June 23, 1982.
31. Coulet, Jacques, interview, June 24, 1982.
32. Mechelle, Ginette, op cit.
33. Gygax, Georges, "La 'Fiancée du Danger': La Vie Prodigieuse de Marie Marvingt," *L'Illustré*, April 24, 1958, pp. 34–36.
34. "Marie Marvingt," French aviation publication, 1955, pp. 41–42.
35. de Beauregard, Marie-Josèphe, letter, April 12, 1984.
36. de Beauregard, Marie-Josèphe, *Femmes de l'Air: Chronique d'Une Conquête* (1993), p. 55.
37. *Ibid.*, pp. 55–58.
38. *L'Aéro*, "Souvenirs Aériens," October 4, 1912, p. 2.
39. Boselli, Élisabeth, letter, October 28, 1982.
40. Enselme-Trichard, Renée, letter, July 2, 1982.
41. Cordier, Marcel, letter, January 16, 1984.
42. *L'Est Républicain*, "La Résidence Marie Marvingt a Déjà Une Pierre," April 3, 1984.
43. *L'Est Républicain*, "A la Mémoire de Marie Marvingt," December 1, 1983; *L'Est Républicain*, "Marie Marvingt Aura Sa Plaque," May 4, 1984; Marchal, Mireille, "Du Ciel à la Carrière Pour la 'Fiancée du Danger,'" *Le Republicain Lorrain*, June 8, 1984.
44. *L'Est Républicain*, "Marie Marvingt Donnera Son Nom au Complexe Sportif," December 28, 1983; *L'Est Républicain*, "Tout est Prêt pour l'Inauguration du Complèxe Sportif Marie Marvingt," January 6, 1984; *L'Est Républicain*, "La Fiancée du Danger Marie Marvingt Donne Son Nom au Gymnase de Maxéville," January 8, 1984.
45. *L'Est Républicain*, "Nancy: en Mémoire de Marie Marvingt, 'la Fiancée du Danger,'" June 4, 1984; Robaux, P. et D., *Les Rues de Nancy* (1984), p. 208.
46. *L'Est Républicain*, "Samedi à Tomblaine: Baptême du LEP Marie Marvingt," May 21, 1987; *L'Est Républicain*, "Le Lycée Professionnel Dédié à la 'Fiancée du Danger,'" May 24, 1987.
47. International Women's Sports Hall of Fame, program, September 21, 1987; *Women's Sports and Fitness*, "1987 Hall of Fame: The Winners' Circle," October 1987, p. 58; Erbstein, Marguerite, "Marie Marvingt, 'la Fiancée du Danger,' Etre à l'Olympe des Plus Grandes Sportives," *L'Est Républicain*, September 8, 1987.
48. *Women's Sports and Fitness*, "1987 Hall of Fame: The Winners' Circle," October 1987, p. 58.
49. *Républicain Lorrain*, "Hommage à Marie Marvingt," March 2, 1990; Charut, Gérard, "Les Cent Vies de Marie Marvingt," *L'Est Républicain*, March 4, 1990.
50. Maggio, Rosalie, and Marcel Cordier, *Marie Marvingt: Femme d'un Siècle* (Éditions Pierron, 1991); *L'Est Républicain*, "Les Cent Vies de Marie Marvingt," September 22, 1991.
51. Marcacci, Philippe, "Marie Marvingt Redécolle," *Est Magazine*, June 27, 2004, pp. 10–11; *Républicain Lorrain*, "Marie Marvingt Dans le Panthéon Postal," July 1, 2004; *Le Monde*, "Marie Marvingt, Pionnière de l'Aviation," July 24, 2004.
52. Ayoun, Monique, "La Fille de l'Air," *Le Nouvel Observateur*, September 7, 2012.
53. *Le Figaro*, "Le Musée de l'Air Fête Ses Héroïnes," March 7, 2006.
54. Nicolas, Eric, *L'Est Républicain*, "Marie Marvingt Va Revoler," April 4, 2010.
55. Baron Boilley, Françoise, *Marie Marvingt: A l'Aventure du Sport* (L'Harmattan, 2013).
56. Fédération des Internationaux du Sport Français (FISF), "Promotion 1995 des 'Gloires du Sport,'"1995; Cordier, Marcel, "Marie Marvingt: Gloire du Sport," *La Dépêche Meusienne*, 1995; Cordier, Marcel, "Hommage à Marie Marvingt (1875–1963)," *L'Ami Hebdo*, January 1, 2014.
57. J.G., *Est Républicain*, "Trois Plaques Pour la 'Fiancée du Danger': Marie Marvingt Honorée Dans Sa Ville," June 8, 1984; Baudais, Pierrick, "Marie Marvingt, 'la Fiancée du Danger' Oubliée," *Ouest-France*, January 5, 1918; Marcel Cordier and the International Marie Marvingt Committee keep a running list of places honoring Marie.

Bibliography

This bibliography contains only items that refer to Marie Marvingt; materials consulted on the two world wars, early aviation, French history, the ambulance airplane, sports, and other topics are not included. Unless otherwise indicated, all letters were written to the author.

Aberdeen Journal, "At Châlons," December 1, 1910, p. 11; *Sunderland Daily Echo and Shipping Gazette*, November 30, 1910, p. 1, 2; *Yorkshire Evening Post*, "Long Flight by a Woman," November 29, 1910, p. 4.

A.C., "La traversée de Toulouse à la nage," *Express du Midi*, September 2, 1907.

Ackerman, Gordon, "Fiancée of Danger," *Sports Illustrated*, June 26, 1961, pages 61–64.

Acte de Notoriété de Marie Marvingt, September 1964, Étude du Maître Jules Thomas, notary, Nancy.

Action Catholique des Femmes, *Femmes remarquables: je broderai vos noms* (ACGF, 2001), pp. 212–215.

[Adelaide] Chronicle, "Notes by 'Aerial,'" May 14, 1910, p. 20.

L'Aéro, "Académie des sports," November 6, 1911, p. 6.

L'Aéro, "A l'Aéro-Club de France: Journée des société affiliées," July 7, 1912, p. 3.

L'Aéro, "La journée des société affiliées—les atterrisages," July 9, 1912, p. 2.

L'Aéro, "Sur les Aérodromes," July 11, 1912, p. 2.

L'Aéro, "Sur les Aérodromes," July 19, 1912, p. 3.

L'Aéro, "Paris-Bruxelles en sphérique," July 31, 1912, p. 1.

L'Aéro, "Sur les Aérodromes," August 13, 1912, p. 2.

L'Aéro, "Propos en l'Air," August 31, 1912, p. 1.

L'Aéro, "Le Grand Prix de l'Aero-Club de France," September 23, 1912, p. 1.

L'Aéro, "Le Prix de Mme de Salmegeane," September 30, 1912, p. 2.

L'Aéro, "Sur les Aérodromes," October 1, 1912, p. 2.

L'Aéro, "Souvenirs aériens," October 4, 1912, p. 2.

L'Aéro, "Les conférences de Mlle Marvingt," October 10, 1912, p. 1.

L'Aéro, "Propos en l'Air," October 10, 1912, p. 1.

L'Aéro, "Sur les Aérodomes: Bétheny," October 11, 1912, p. 2.

L'Aéro, "Les Sphériques," October 19, 1912, p. 2.

L'Aéro, "Propos en l'Air," November 11, 1912, p. 1.

L'Aéro, "Propos en l'Air," December 13, 1912, p. 1.

L'Aéro, "La Coupe des Sociétés Affiliées," July 19, 1913, p. 2.

L'Aéro, "Les Sphériques," August 6, 1913, p. 4.

L'Aéro, "Sur les aérodromes," August 8, 1913, p. 2.

L'Aéro, "L'avion-ambulance," August 18, 1913, p. 2.

L'Aéro, "Les Sphériques," August 21, 1913, p. 2.

L'Aéro, "Les concurrents accomplissent les épreuves de longuer et hauteur," September 28, 1913, p. 2.

L'Aéro, "En piste pour la Coupe," September 30, 1913, p. 1.

L'Aéro, "Un Voyage de Mlle Marvingt," October 3, 1913, p. 2.

L'Aéro, "Petites Nouvelles," October 29, 1913, p. 2.

L'Aéro, "Mlle Marvingt conférence au Havre," November 19, 1913, p. 2.

L'Aéro, "Petites Nouvelles," December 13, 1913, p. 2.

L'Aéro, "Les obsèques de Latham," January 17, 1914, p. 3.

L'Aero, "L'École Marvingt à Megève," January 31, 1914, p. 6.

L'Aéro, "Les Conférences de Mlle Marvingt," March 20, 1914, p. 6.

L'Aéro, "Les Conférences de Mlle Marvingt," May 2, 1914, p. 6.

L'Aéro, "Mlle Marvingt Conférencie à Bourges," May 14, 1914, p. 4.

L'Aéro, "Une Conférence de Mlle Marvingt à Nevers," May 15, 1914, p. 3.

L'Aéro, "Le Grand Prix de l'Aéro Club," July 6, 1914, p. 1.

L'Aéro, "Le Grand Prix de l'Aéro-Club," July 9, 1914, p. 5.

L'Aéro, "Les Grandes Épreuves de sphériques: Le Xème Grand Prix de l'Aéro Club de France," July 20, 1914, p. 3.

L'Aéro, "Le Challenge Capitaine Echeman," June 14, 1935, p. 3.

Aéro-Club de France, certificate attesting to her balloon license, March 14, 1914.

L'Aérophile, "Bulletin des ascensions," September 1907, p. 263.

L'Aérophile, "Ascensions au Parc de L'Aéro-Club de France," May 15, 1909, p. 238.

L'Aérophile, "Le 'Ville-de-Nancy' à Nancy," August 1, 1909, p. 353.
L'Aérophile, "Roger Sommer à Nancy," October 1, 1909, p. 436.
L'Aérophile, "Au Camp de Châlons," December 15, 1909, p. 554.
L'Aérophile, "Les Aéroplanes," January 1, 1910, p. 6.
L'Aérophile, "Le Sixième grand prix de l'Aéro-Club de France," July 15, 1910, pp. 330–332; "Brevet de pilote aéronaute," p. 334.
L'Aérophile, "Brevets de pilote d'aérostats" and "Les Ascensions au Parc de l'Aéro-Club de France," August 15, 1910, p. 383.
L'Aérophile, "Les Aéroplanes," September 1, 1910, p. 398.
L'Aérophile, "Les Aéroplanes," October 1, 1910, p. 436.
L'Aérophile, "Les Aéroplanes," October 15, 1910, pp. 458–463.
L'Aérophile, "A Mourmelon," November 1, 1910, p. 486.
L'Aérophile, "Brevets de pilotes-aviateurs," December 1, 1910, p. 549.
L'Aérophile, "Les Records féminins d'aviation," December 15, 1910, p. 556.
L'Aérophile, "Pour la Coupe 'Femina,'" January 1, 1911, p. 10.
L'Aérophile, "Liste alphabétique des pilotes-aviateurs titulaires du Brevet de l'Aéro-Club de France," January 15, 1911, pp. 36, 40.
L'Aérophile, "Concours d'atterrissage," August 1, 1912, p. 358.
L'Aérophile, "Ascensions au Parc de l'Aéro-Club de France," August 15, 1912, p. 383.
L'Aérophile, "Le 8e Grand Prix de l'Aéro-Club de France," October 1, 1912, p. 450–451; "Ascensions au Parc de l'Aéro-Club de France," p. 455.
L'Aérophile, "Le Prix de 'Mme de Salmegeane,'" October 15, 1912, p. 476.
L'Aérophile, "Ascensions au Parc de l'Aéro-Club de France," November 1, 1912, p. 503.
L'Aérophile, "Ascensions au Parc de l'Aéro-Club de France," January 1, 1913, pp. 20–21.
L'Aérophile, "Ascensions au Parc de l'Aéro-Club de France," September 1, 1913, p. 407.
L'Aérophile, "Obsèques d'Hubert Latham," February 1, 1914, p. 62.
L'Aérophile, "Principales épreuves d'aéroplanes en 1913," April 1, 1914, p. 161.
L'Aérophile, "Le 10e Grand Prix de l'Aéro-Club de France," July 15, 1914, p. 334.
L'Aérophile, "Le dixième Grand Prix de l'Aéro-Club de France," August 1, 1914, pp. 352–357.
L'Aérophile, "Le 10e Grand Prix de l'Aéro-Club de France," November 1–15, 1914, pp. 409–410.
L'Aérophile, "Dirigeable et avions allemands sur Nancy," January 1–15, 1915, p. 20.
L'Aérophile, "L'Année aéronautique 1914 et l'Aéro-Club de France," May 1–15, 1915, p. 101.
L'Aérophile, "Le Tour du monde aérien," July 1–15, 1926, p. 217.
L'Aérophile, "Autour du monde aérien," July 1–15, 1928, p. 209.
L'Aérophile, "Le Banquet de l'Aéro-Club de France en l'honneur de M. Laurent-Eynac, premier ministre de l'Air," December 1–15, 1928, p. 374.
L'Aérophile, "Les Journées d'aviation sanitaire coloniale," August 15, 1931, p. 231; "Aujourd'hui," p. 232.
L'Aérophile, "Réception en l'honneur du Maréchal Lyautey et des membres du Congrès aéronautique coloniale," November 15, 1931, pp. 349–351.
L'Aérophile, "Le banquet annuel des Vieilles Tiges," December 15, 1931, p. 377.
L'Aérophile, "Le IIe Congrès International de l'Aviation Sanitaire," September 1933, p. 265.
L'Aérophile, "Aujourd'hui," December 1934, p. 365.
L'Aérophile, "Dans la Légion d'Honneur," February 1935, pp. xxvi, 57.
L'Aérophile, "Les Foyers du soldats de l'air," March 1940, p. 71.
L'Aérophile, "Chronique des Vieilles Tiges," October 1943, p. iii.
L'Aérophile, "Concours des Jeunes Tiges 1943," June 1944, p. 74.
L'Afrique du Nord Illustré, "Un rare exemple d'énergie féminine," July 15, 1922, p. 7.
L'Afrique du Nord Illustré, "Conférence de Mlle Marvingt," May 2, 1925, p. 6.
L'Afrique du Nord Illustré, "Les ailes qui sauvent," July 28, 1934, p. 7.
Les Ailes, "Coups d'ailes," October 6, 1921, p. 2.
Les Ailes, "L'Aéro-Club d'Auvergne à Paris," December 23, 1926, p. 4.
Les Ailes, "L'an prochain se tiendra, à Paris, le Congrès d'Aviation Sanitaire," August 10, 1928, p. 8.
Les Ailes, "Le Tour de France des avions légers," September 20, 1928, p. 14.
Les Ailes, "L'Aéro-Club de France a reçu M. Laurent Eynac," November 29, 1928, p. 4.
Les Ailes, "Petites nouvelles de France," July 11, 1929, p. 14.
Les Ailes, "L'Aviation Sanitaire à Orly," August 6, 1931, p. 14.
Les Ailes, "Petites Nouvelles de France," January 15, 1931, p. 15.
Les Ailes, "Petites Nouvelles de France," October 15, 1931, p. 15.
Les Ailes, "Coups d'Ailes," November 5, 1931, p. 15.
Les Ailes, "Sur les terrains de Vol à Voile," September 15, 1932, p. 13.
Les Ailes, "Aujourd'hui ...," November 3, 1932, p. 5.
Les Ailes, "Coups d'Ailes," December 15, 1932, p. 15.
Les Ailes, "Coups d'Ailes," January 19, 1933, p. 15.
Les Ailes, "Chronique des Clubs," March 16, 1933, p. 7.
Les Ailes, "Les Ailes des T.O.E.," March 23, 1933, p. 5.
Les Ailes, "Faisons confiance à la Banne d'Ordanche," September 21, 1933, p. 12.
Les Ailes, "Coups d'Ailes," November 9, 1933, p. 15.
Les Ailes, "Les 'Ailes qui sauvent,'" November 18, 1934, p. 8.
Les Ailes, "La Légion d'Honneur: La promotion civile du Ministère de l'Air," January 31, 1935, p. 9.
Les Ailes, "Ici et Là," June 13, 1935, p. 19.
Les Ailes, "L'Aviation médicale à Berck," June 27, 1935, p. 13.
Les Ailes, "Ici et Là," August 6, 1936, p. 15.
Les Ailes, "Les premières aviatrices," June 24, 1937, p. 13.
Les Ailes, "Ici et Là," February 3, 1938, p. 4.
Les Ailes, "Une évacuation aérienne d'Alger à Paris," April 21, 1938, p. 4.
Les Ailes, "L'Aviation sanitaire," July 7, 1938, p. 4.

Les Ailes, "Ici et Là," December 22, 1938, p. 4.

Les Ailes, "L'Aviation Sanitaire et l'Exposition de Lille," June 22, 1939, p. 5.

Les Ailes, "Le trait d'union des gens de l'air," January 4, 1940, p. 6.

Les Ailes Brisées, "Nos aviatrices, les véritables pionniers de l'aviation: Marie Marvingt," December 1959, pp. 21–22.

"Les Ailes qui sauvent: la propagande par le film," unsourced, undated newspaper article.

L'Air, "Informations: Le Challenge du Capitaine Écheman," November 1, 1931.

Aircraft, "Foreign News," vol. 1, 1910, p. 393.

A.L., "Infatigable globe-trotter raconte ses 12 mois d'Afrique," *L'Est Républicain*, August 25, 1927.

Albaret, Laurent, "Marie Marvingt, la Fiancée du Danger," *L'Echo de la Timbrology*," no. 1901, December 2015, pp. 68–72.

Albert Lea Freeborn County [MN] Standard, "Is Best Sportswoman," October 29, 1913, p. 3; *Boynton [OK] Index*, "Is Best Sportswoman," October 31, 1913, p. 3; *Fair Play [Ste. Genevieve, MO]*, "Is Best Sportswoman," November 1, 1913, p. 4; *Fort Gibson [OK] New Era*, "Is Best Sportswoman," October 30, 1913, p. 3; *The Greenville [OH] Journal*, "Is Best Sportswoman," October 30, 1913, p. 7; *The Hays Free Press [KS]*, "Is Best Sportswoman," November 15, 1913; *Iron County [Ironton, MO] Register*, "Is Best Sportswoman," October 30, 1913, p. 3; *Lawrence [Lawrenceburg, TN] Democrat*, "Is Greatest Sportswoman," November 5, 1913, p. 7; *The Madison [Tallulah, LA] Journal*, "Is Best Sportswoman," November 1, 1913, p. 7; *Nocona [TX] News*, "Is Best Sports-woman," November 21, 1913, p. 8; *Oelwein [IA] Daily Register*, "Is Best Sportswoman," November 7, 1913, p. 3; *Tensas Gazette [St. Joseph, LA]*, "Is Best Sportswoman," November 21, 1913, p. 7.

Albuquerque Journal, "First French Woman Pilot Here, Gave Lindy Medal for Paris Hop," October 10, 1935, pp. 1, 12.

Albuquerque Journal, "Twenty Years Ago," October 10, 1955, p. 11.

Albuquerque Journal, "Pioneer Woman Pilot Succumbs," December 26, 1963, p. 8; *Abilene [TX] Reporter News*, "Aviatrix Succumbs," December 17, 1963, p. 33; *Altoona [PA] Mirror*, "Marie Marvingt," December 16, 1963, p. 49; *Anniston [AL] Daily Star*, "Miss Marie Marvingt," December 16, 1963, p. 9; *Arkansas City Daily Traveler*, "Marie Marvingt," December 16, 1963, p. 23; *Auburn [NY] Citizen Advertiser*, "Marie Marvingt," December 16, 1963, p. 18; *Austin [MN] Daily Herald*, "French Sportswoman Dies," December 16, 1963, p. 16; *Connellsville [PA] Daily Courier*, "Marie Marvingt," December 16, 1963, p. 2; *Delaware County [PA] Daily Times*, "Marie Marvingt," December 26, 1963, p. 4; *East Liverpool [OH] Review*, "Sportswoman Dies," December 16, 1963, p. 25; *Eureka Humboldt [CA] Standard*, "Marie Marvingt," December 16, 1963, p. 9; *Fairfield [IA] Daily Ledger*, "Died in Poverty," December 16, 1963, p. 3; *[Frederick, MD] News*, "Sportswoman Dies," December 16, 1963, p. 5; *Galesburg [IL] Register Mail*, "Marie Marvingt," December 16, 1963, p. 20; *Galveston [TX] Tribune*, "Marie Marvingt," December 16, 1963, p. 6; *Great Bend [KS] Tribune*, "Marie Marvingt," December 16, 1963, p. 8; *Greeley [CO] Daily Tribune*, "Deaths Around the World," December 16, 1963, p. 27; *Hattiesburg [MS] American*, "Famed Sportswoman Dies in Nursing Home," 17 December 1963, p. 10; *Hayward California Daily Review*, "Noted Aviatrix Marvingt Dies," December 16, 1963, p. 2; *Holland [MI] Evening Sentinel*, "French Pilot Dies," December 16, 1963, p. 8; *Idaho Falls Post Register*, "Marie Marvingt," December 16, 1963; *Idaho State Journal*, "Marie Marvingt," December 16, 1963, p. 11; *Ironwood [MI] Daily Globe*, "Deaths in the News," December 16, 1963, p. 2; *Keokuk [IA] Daily Gate City*, "Famed French Aviatrix Dead," December 16, 1963, p. 16; *The Kokomo [IN] Tribune*, "Sportswoman Dies," December 16, 1963, p. 17; *Logansport [IN] Pharos Tribune*, "Deaths in the News," December 16, 1963, p. 3; *[Long Beach, CA] Press Telegram*, "Marie Marvingt, 88, Early Aviator, Dies," December 16, 1963, p. 17; *Nevada State Journal*, "Flyer Succumbs," December 16, 1963, p. 8; *New Castle [PA] News*, "Marie Marvingt," December 16, 1963, p. 11; *Northwest Arkansas Times*, "Deaths in the News," December 16, 1963, p. 14; *Oxnard [CA] Press Courier*, "Marie Marvingt, Early Flier, Dies," December 16, 1963, p. 3; *Phoenix Arizona Republic*, "Famed French Aviatrix Dies," December 16, 1963, p. 15; *Portsmouth [OH] Times*, "Sportswoman Dies," December 16, 1963, p. 23; *Western Kansas Press*, "Marie Marvingt," December 17, 1963, p. 13; *Wichita [TX] Fall Times*, "She Pioneered Flying," December 20, 1963, p. 22.

Albury [Australia] Banner and Wodonga Express, "World's Greatest Lady Athlete," June 12, 1914, p. 38.

Allen, E. John B., *The Culture and Sport of Skiing: From Antiquity to World War II* (University of Massachusetts Press, 2007), pp. 84, 177.

Allen, E. John B., *Historical Dictionary of Skiing* (2011), pp. 125–126.

Almanach Hachette (Hachette, 1913), p. 195.

Alrededor del Mundo, "Biografía extranjera contemporánea: Mlle Marvingt," no. 753, November 2, 1913, p. 361.

Anaconda [MT] Standard, "Aviators' Earnings: Record of the European Fliers up to Sept. 10 of This Year," November 27, 1910, p. 35; *The Philadelphia Inquirer*, "Aviators' Earnings," March 30, 1911, p. 2.

Anaconda [MT] Standard, "Red Cross Aeroplanes," March 22, 1914, p. 86; *Fergus County [MT] Democrat*, "Red Cross Aeroplanes," March 31, 1914, p. 5.

Annales Africaines, "Mlle Marvingt en Afrique du Nord," July, 13, 1922, pp. 958, 960.

Annales Africaines, "300 Conférences, 4.000 kilomètres en Afrique du Nord par Mlle Marie Marvingt," May 15, 1925, pp. 309, 312–313.

Annales Africaines, "Les prochaines conférences de Mlle Marvingt," May 29, 1925, p. 338.

Les Annales Coloniales, "En Auto: À travers le Sahara," February 20, 1923, p. 1.

Les Annales Coloniales, "En auto: Á travers le Sahara," March 12, 1923, p. 1.

Les Annales Coloniales, "Les Evenements et les Hommes," March 19, 1923, p. 2.

Les Annales Coloniales, "Conférences," May 7, 1925, p. 2.
Les Annales Coloniales, "L'Aviation," April 13, 1935, p. 2.
L'Année Mondiale Illustrée, "Aéronautique," 1914, p. 521.
Annuaire de la noblesse de France, 1931–1933, p. 272.
Archdeacon, Ernest, "Mademoiselle Marvingt: Une sportwoman extraordinaire," *La Revue Aérienne*, December 25, 1910, pp. 702–706.
Arizona Republican, "War Chauffeurs," July 4, 1914, p. 4.
Arnaud, Pierre, et Thierry Terret, *Histoire du Sport Féminin: Sport masculin-sport féminin: éducation et société*, Tome 2 (L'Harmattan, 1996), p. 158.
Association de Documentation Aéronautique, "Marie Marvingt (1875–1963)," Issy-les-Moulineaux, École Marie Marvingt, December 12, 1988.
Association général aéronautique, "Pour un Poste de Secours à Soissons," July 1912, p. 129.
Association général aéronautique, "Une Conférence de Mlle Marvingt," March 1913, p. 38.
L'Association Médicale," Les Amis de l'aviation sanitaire," March 1930, pp. 115–116.
Astronautics and Aeronautics, 1963: Chronology on Science, Technology, and Policy (1964), p. 479.
L'Astrosophie: Revue Mensuelle de l'Astrologie ésotérique et exotérique, de Psychisme et des Sciences Occultes, "Mlle Marvingt," September 21, 1932, pp. 6–7.
Atlas Editions, "Marie Marvingt," *Les As de l'Aviation* (1986), pp. 53–54.
D'Aubigny, Eugène, "La premiére traversée de la Manche par une aéronaute," *L'Aéro*, July 24, 1914, p. 1.
Aurenche, Dr. Henry, "Notre Doyenne," *La Presse Scientifique* (Paris), 1961.
L'Aurore, "La traversée de Paris à la nage," July 30, 1906, p. 1.
L'Aurore, "La Manche en ballon," October 29, 1909, p. 3.
L'Aurore, "A Mourmelon," November 24, 1909, p. 3.
L'Aurore, "Aéronautique," February 3, 1910, p. 3.
L'Aurore, "Le Comité de Direction de l'Aéro-Club de France," June 20, 1910, p. 3.
L'Aurore, "Le Grand-Prix de l'Aéro-Club de France," June 25, 1910, p. 3.
L'Aurore, "Au Camp de Châlons," October 23, 1910, p. 3.
L'Aurore, "Mlle Marvingt s'engage pour la Coupe Femina," November 27, 1910, p. 3.
L'Aurore, "Une aviatrice concourt," November 28, 1910, p. 3.
L'Aurore, "La Coupe 'Femina,'" November 29, 1910, p. 3.
L'Aurore, "Aviation," December 6, 1910, p. 3.
L'Aurore, "Chez Deperdussin à Courcy," July 28, 1912, p. 3.
L'Aurore, "Aux Ecoles Deperdussin à Reims," July 30, 1912, p. 3.
L'Aurore, "L'Entraînement aux écoles de l'aviation," August 2, 1912, p. 3.
L'Aurore, "Aux Ecoles Deperdussin," August 6, 1912, p. 3.
L'Aurore, "Aux Ecoles Deperdussin—Aérodrome de la Champagne," August 8, 1912, p. 3.
L'Aurore, "Aux Ecoles Deperdussin," August 13, 1912, p. 3.
L'Aurore, "Aux Ecoles Deperdussin," September 8, 1912, p. 3.
L'Aurore, "L'Entraînement à l'Ecole Deperdussin de Reims," September 9, 1912, p. 3.
L'Aurore, "Aux Ecoles Deperdussin à Bétheny," September 17, 1912, p. 3.
L'Aurore, "Un beau vol de Mlle Marvingt," September 19, 1912, p. 3.
L'Aurore, "Aux Ecoles Deperdussin," September 20, 1912, p. 3.
L'Aurore, "A l'Ecole Deperdussin de Bétheny," September 22, 1912, p. 3.
L'Aurore, "A l'école Deperdussin de Reims," September 23, 1912, p. 3.
L'Aurore, "Aux Ecoles Deperdussin à Bétheny," September 26, 1912, p. 3.
L'Aurore, "Aux Ecoles Deperdussin à Bétheny," September 28, 1912, p. 3.
L'Aurore, "Chez Deperdussin à Estampes," October 11, 1912, p. 3.
L'Aurore, "Au Salon de l'Aéronautique," October 31, 1912, p. 2.
L'Aurore, "A l'Ecole Deperdussin à Reims," November 23, 1912, p. 3.
L'Aurore, "A l'Ecole Deperdussin à Reims," December 1, 1912, p. 3.
L'Aurore, "Dans les écoles d'aviation," December 2, 1912, p. 3.
L'Aurore, "A l'Ecole Deperdussin, aérodrome de la Champagne," December 6, 1912, p. 3.
L'Aurore, "A Bétheny: Ecole Deperdussin," December 12, 1912, p. 3.
L'Aurore, "Mlle Marvingt vole de Reims à Mourmelon," October 3, 1913, p. 3.
L'Aurore, "Un beau vol de Mlle Marvingt," October 26, 1913, p. 3.
L'Aurore, "Chute de Mlle Marvingt," December 13, 1913, p. 3.
L'Aurore, "Le Grand-Prix de l'Aéroclub," July 19, 1914, p. 3.
L'Aurore, "La 'Fiancée du danger' est morte à 88 ans," December 16, 1963.
Automobile Journal, "Women to Be War Chauffeurs," vol. 39, 1915, p. 49.
L'Auvergnat de Paris, "Marie Marvingt n'est plus," no. 51, December 21, 1963, pp. 1, 3.
L'Auvergne, "La fiancée du danger à Aurillac," October 29, 1942.
L'Avenir à Argenteuil, "La célèbre aviatrice Marie Marvingt s'est intéressée aux journées de propagande aéronautique," May 13, 1955.
L'Avenir de l'Est [Algérie], "Mlle Marvingt," June 21, 1924, p. 2.
L'Avenir de l'Est [Algérie], "Mlle Marvingt," July 5, 1924, p. 2.
L'Avenir de l'Est [Algérie], "Conférence," December 20, 1924, p. 2.
L'Avenir du Cantal, "Mlle Marvingt à Aurillac," February 10, 1911.
L'Aviation Illustrée, "Nos aviatrices et la Coupe Femina," no. 61, December 15, 1910.

A.W., "La femme la plus extraordinaire du siècle est à Tunis," *Tunisie France*, February 2, 1952.

Ayoun, Monique, "La fille de l'air," *Le Nouvel Observateur*, September 7, 2012.

Ayoun, Monique, "Des femmes au Panthéon?" *Le Nouvel Observateur*, October 10, 2013.

Baily, Paul J., *Gender and Education in China: Gender Discourses and Women's Schooling in the Early Twentieth Century* (2007), p. 71.

Balitrand, Suzanne, "Le beau voyage de Mademoiselle Marvingt," *Eve*, no. 307, August 15, 1926, p. 7.

Ballu, Yves, "Les femmes et le ski ou la découverte des sports d'hiver," *L'épopée du ski* (Éditions Arthaud, 1981), pp. 163, 184, 186, 191, 194–202.

Ballu, Yves, "Les femmes et le ski (suite)," *Le Pélérin*, no. 5166, December 6, 1981, p. 79.

Ballu, Yves, "Faits d'hivers: de glisse et de glace," *Télérama*, no. 2195, February 5, 1992, pp. 8–12.

Bans, Georges, "Aéronautes contemporains: Mademoiselle Marie Marvingt," *L'Aérophile*, vol. 17, no. 22, November 15, 1909, p. 505.

Baraton, Alain, ed., *Almanach des Terres de France 2013* (Presses de la Cité, 2012).

Baron Boilley, Françoise, *Marie Marvingt: À l'Aventure du Sport* (L'Harmattan, 2013).

Barthélemy, Général, "La femme la plus extraordinaire du siècle: Marie Marvingt, la 'Fiancée du Danger,'" *L'Étrange race des hommes volants* (Éditions France-Empire, 1979), pp. 241–254.

Baudais, Pierrick, "Marie Marvingt, 'la fiancée du danger' oubliée," *Ouest-France*, January 5, 1918.

Baudry, M., "Une touriste hors du commun au Pouliguen: Marie Marvingt," newspaper article, February 17, 1982.

Beaubois, Henry, "Ceux qui ouvrirent la route au ciel: Marie Marvingt," *Revue aéronautique de France*, August 13, 1937.

Beaupré, Norman, *The Man With the Easel of Horn: The Life and Works of Emile Friant* (Lumina Press, 2010), pp. 339–340.

Bell, Ryan Corbett, *The Ambulance: A History* (McFarland, 2009), pp. 121, 122, 124.

Benoît, Michèle, "Marie casse-cou: Marie Marvingt," *Bulletin des Anciennes de Sainte-Chrétienne de Metz*, pp. 18–20.

Béré, Emile, "Nos aviatrices s'ennuient," *Le Figaro*, September 27, 1915, p. 3.

Bergeret, Tonin, "Les ballons en Lorraine: L'ascension de Mlle Marvingt," *Le Sport*, no. 85, May 21, 1910, pp. 1, 2.

Berkeley Daily Gazette, "French Woman Has Record for Traveling and Sports," September 13, 1926, p. 5.

Berlioux, Monique, *Gloires de Sport* (Atlantica, 2009), p. 180.

Besançon, Madame, letter, June 28, 1982.

Beverley and East Riding Recorder, "Balloonists' Long Journey," October 30, 1909, p. 3.

Bibliographie nationale française (1992), pp. 171, 210.

Bibliography of Aeronautics, 1921, p. 846.

Bichet, Gabriel (G.B.), "Marie Marvingt lance aujourd'hui l'aviation sanitaire mondiale," *L'Est Républicain*, July 5, 1956, p. 2.

Bichet, Gabriel (G.B.), "Marie Marvingt, la fiancée du danger, est morte hier à Nancy," *L'Est Républicain*, December 15, 1963.

Bichet, Gabriel, "Marie Marvingt, la 'fiancée du danger,' n'est plus, *L'Est Républicain*, December 15, 1963.

Bichet, Gabriel, "C'était la fiancée du danger...," *L'Est Républicain*, December 15, 1963.

"Bilder aus aller Welt," Vienna newspaper, 1910.

Birckel, Véronique, "Il y a 100 ans dans l'Est: La prouesse de Mlle Marvingt," *L'Est Républicain*, November 8, 2009, p. 16.

Birckel, Véronique, "Il y a 100 ans dans l'Est: Marie Marvingt apprentie pilote," *L'Est Républicain*, December 13, 2009, p. 16.

Birckel, Véronique, "Rien ne décourage Mlle Marvingt!" *L'Est Républicain*, May 16, 2010, p. 16.

Birckel, Véronique, "Nouvelle ascension de Marie Marvingt," *L'Est Républicain*, May 23, 2010, p. 16.

Blanc, Colonel Gaston, and Madame Blanc, interview, June 24, 1982; letter, May 13, 1983.

Boase, Wendy, *The Sky's the Limit: Women Pioneers in Aviation* (Macmillan, 1979), pp. 11–12.

Boetsch, Jacques, "Une vie en équilibre" *Anciens combattants du monde*, May 1955, pp. 11–14.

Boetsch, Jacques, "Une vie en équilibre" *Anciens combattants du monde*, May 1955, pp. 11–14 (with additions in Marie Marvingt's handwriting)

Border Watch [Mount Gambier, Australia], "Female Admirable Crichton," November 29, 1913, p. 6; *Brunswick and Coburg [Australia] Leader*, "Female Admirable Crichton," March 6, 1914, p. 4; *Camperdown [Australia] Chronicle*, "The Admirable Crichton," November 27, 1913, p. 5; *Dimboola [Australia] Banner and Wimmera and Mallee Advertiser*, "Female Admirable Crichton," March 10, 1914, p. 4; *The Gundagai [Australia] Independent and Pastoral, Agricultural and Mining Advocate*, "Female Admirable Crichton," April 8, 1914, p. 2; *The South Eastern Times [Millicent, Australia]*, "Female Admirable Crichton," August 1, 1922, p. 4; *West Gippsland [Australia] Gazette*, "Female Admiral Crichton," December 2, 1913, p. 6; *Zeehan and Dundas [Tasmania] Herald*, "Female Admirable Crichton," November 19, 1913, p. 3.

Boselli, Elisabeth, letter, October 28, 1982.

Boselli, Elisabeth, "Marie Marvingt 1875–1963 'la Fiancée du Danger,'" *Pionniers*, July 1984, pp. 18–22.

Boston Sunday Post, "Women Airbirds Eager to Serve in War," August 23, 1914, p. 47.

Botella, Anne, *Les Trésors du Musée de l'Air et de l'Espace* (Le Cherche Midi, 2013), pp. 152, 153.

Boucher, Hélène, "Mes Impressions," *Le Journal [Paris]*, August 12, 1934, p. 1.

Boulet, Jean, "De passage au Maroc: Mlle Marie Marvingt, la Fiancée du Danger, témoin ou vic-time de vingt catastrophes, elle est la femme la plus extraordinaire du monde," Moroccan newspaper, 1951, pp. 1, 4.

Bourcelot, Nicole, "Anniversaire Marie Marvingt (1875–1963), *L'Abeille*, December 19, 2003.

Bourdet, Maurice, "Les Amazones de l'Air, *Le Petit Parisien*, December 27, 1931, pp. 1, 4.

Bowen, Barry D., "Marie Marvingt: The Premier Sportswoman of France and Pioneer in the Development of Aeromedical Evacuation," special to AirAmbulanceServices.com, March 6, 2013.

Branchu, Marc, "Les cent vies de Marie Marvingt," *Magazine Air France*, no. 99, July 2005, pp. 42–44, 46.

Brandon [Manitoba] Weekly Sun, "The Aviatress," December 8, 1910, p. 8; *Sioux County [IA] Herald*, "The Aviatress," December 28, 1910, p. 3.

Brandon [Manitoba] Weekly Sun, "Fair Aviators," January 12, 1911, p. 23; January 19, 1911, p. 8;

Brion, Hélène, "Aviation," *L'Action féministe*, March-April, 1915, p. 3.

[Brisbane] Queensland Figaro, "Between Ourselves Entirely," December 16, 1909, p. 15.

The [Brisbane] Telegraph, "Marie Marvingt," April 10, 1911, p. 12; *The Western Champion and General Advertiser for the Central-Western [Australia] Districts*, "Here and There," April 8, 1911, p. 14.

Brun, Emmanuel, and Madame Brun, interview, June 22, 1982.

Brun, Mario, "Riviera Gazette," *Nice Matin*, November 22, 1956.

Bulletin de la Meurthe et Moselle, "Mlle Marvingt," March 11, 1917, p. 1.

Bulletin de la Meurthe et Moselle, "Mlle Marvingt," August 12, 1917, p. 2.

Bulletin de la Société de Géographie et d'Études Coloniales de Marseille, "50,000 kilomètres en Afrique du Nord, par Marie Marvingt, Vice-Présidente des Amis de l'Aviation Sanitaire, Déléguée de la Ligue Aéronatique," vol. 51, 1930, pp. 93–94.

Bulletin Mensuel de l'Administration des Postes, no. 138, February 1867, p. 50; *Bulletin Mensuel des Postes et Télégraphes*, March 1879, p. 103; *Bulletin Mensuel des Postes et Télégraphes*, novembre 1879, p. 680.

Le Bulletin Meusien, "Les propos d'une infirmière," November 26, 1938, p. 1.

Burlingham, H., "Les Grimpeuses de Cimes," *Femina*, no. 299, September 1, 1911, pp. 465–466, 470–471.

Burnley Express, "Perilous Balloon Journey," October 30, 1909, p. 8.

Les Cahiers de la Santé Publique, "Historique de l'Aviation Sanitaire," June 25, 1931, pp. 995, 1033.

Les Cahiers de la Santé Publique, "Journées de l'Aviation Sanitaire Coloniale: Réception à L'Aéro-Club de France," June 25, 1931, p. 1043.

Cairns [Australia] Post, "Lady Sky Pilot," December 1, 1910, p. 3; *[Adelaide, Australia] Observer*, "Frenchwoman's Ariel Flight," December 3, 1910, p. 33; *[Adelaide] Chronicle*, "Lady Aviator's Flight," December 3, 1910, p. 24; *The Argus [Melbourne, Australia]*, "Pilot of the Air: A Lady Qualifies," November 30, 1910, p. 13; *The [Australia] Gundagai Independent and Pastoral, Agricultural and Mining Advocate*, "Newsy Items," December 7, 1910, p. 4; *The [Brisbane] Telegraph*, "Lady Secures a Certificate," November 30, 1910, p. 4; *The [Brisbane] Week*, "Aeroplane Flight," December 9, 1910, p. 12; *Le Courrier Australien*, "Aviatrice française," December 2, 1910, p. 4; *Daily Mercury [Mackay, Australia]*, "A Sky Pilot," November 30, 1910, p. 4; *Daily Telegraph [Launceston, Tasmania]*, "Aviation: Lady Secures Certificate," November 30, 1910, p. 5; *Gympie [Australia] Times and Mary River Mining Gazette*, "Aviation," December 1, 1910, p. 3; *[Launceston, Tasmania] Examiner*, "News in Brief," November 30, 1910, p. 1; *[Launceston, Tasmania] Examiner*, "A Lady's Flight," November 30, 1910, p. 5; *[Lismore, Australia] Northern Star*, "A Lady Aviator," December 2, 1910, p. 3; *The [Melbourne] Australasian*, December 3, 1910, p. 42; *The [Melbourne] Age*, "Lady 'Air Pilot,'" November 30, 1910, p. 7; *[Perth] Sunday Times*, "Gist of the Cables," December 4, 1910, p. 8; *The [Perth] West Australian*, "A Lady Aeronaut," November 30, 1910, p. 7; *Queensland Times*, "A Lady Flyer," November 30, 1910, p. 5; *The Register [Adelaide, Australia]*, "Frenchwoman's Ariel Flight," December 1910, p. 22; *[Rockhampton, Australia] Morning Bulletin*, "Aerial Navigation," November 30, 1910, p. 5; *[Sydney, Australia] Evening News*, "Brevities," November 30, 1910, p. 1.

Cajelot, Maurice, "Marie Marvingt, 'la fiancée du diable,' 100 ans," *Républicain Lorrain*, February 13, 1975, p. 4.

The Canberra [Australia] Times, "Veteran Woman Air Ace Dies," December 17, 1963, p. 1.

The Capricornian, "Gossip," October 19, 1907.

Cartes Postales et Collections, "Marie Marvingt," November 15, 1987, no 118.

Casella, George, "Les Sports d'hiver," *Le Sport universel illustré*, February 14, 1909, p. 111.

Casella, Georges, "'Sportswomen' d'hiver à Chamonix," *La Culture Physique*, March 15, 1909, pp. 201–202.

The Catholic Press [Sydney, Australia], "Flying Women: Their Fearlessness and Nerve; A Frenchwoman Who Flies 40 Miles an Hour," May 26, 1910, p. 18.

Centre Psychothérapique de Nancy, "Constation de Décès," and medical notes, December 14, 1963.

C'est la vie!, "Marie Marvingt: femme universelle," April 28, 1950, no. 25, p. 3.

Chambe, René, *Histoire de l'aviation* (Flammarion, 1980), pp. 80, 198, 256.

Chancel, Jules, "A la conquête du brevet d'aviateur," *Lecture pour tous*, January 1911, pp. 355–366.

Charut, Gérard, "Les cent vies de Marie Marvingt," *L'Est Républicain*, March 4, 1990.

Charut, Gérard, "Marie Marvingt: savoir vouloir: interview posthume," *L'Est Républicain*, August 26, 2009.

Charvier, Annie, "Marie Marvingt, Première ambulancière de l'air (1875–1963)," *Cartes Postales et Collection*, November-December 1987, no 118, pp. 12–17.

Chevalier, Frédérique, "Marie Marvingt, la fiancée des Airs (1875–1963)," *Les Grandes Aventurières* (City Editions, 2007), pp. 85–95.

[Chicago] Daily Tribune, "Marie Marvingt, Noted French Flyer, Is Guest in City," October 19, 1935, p. 21.

Chicago Star Publications, "Breaks Airwomen's Record," February 2, 1911, p. 4; *Arlington Heights Cook County [IL] Herald*, January 27, 1911, p. 7; *Dakota Farmers' Leader [Canton, SD]*, "Breaks Airwomen's Record," January 27, 1911, p. 9; *Pocahontas County [Iowa] Sun*, "Breaks Airwomen's Record," January 26, 1911, p. 3; *The [St. Paul, MN] Appeal*, January 28, 1911, p. 3.

Chicago Star Publications, "Mistress of All Sports," 31 October 1913, p. 11; *Kempton [IN] Courier*, "Mistress of All Sports," November 7, 1913, p. 8; *New Oxford [PA] Item*, "Mistress of All Sports," November 13, 1913, p. 6.

The [Chicago] Sun, "Woman Aviator Hits Tree," August 7, 1911, p. 1.
The [Chicago] Sun, "Women Seek War Jobs: Want Positions As Ambulance Drivers in France," April 14, 1915, p. 2.
Chicago Tribune, "Marvingt, 88, Climber and Aviatrix, Dies," December 16, 1963, section 3, p. 8.
Clemitson, Suze, ed., *Ride the Revolution: The Inside Stories from Women in Cycling* (2015).
Clemitsen, Suze, and Mark Fairhurst, *P Is for Peloton: The A-Z of Cycling* (2015), p. 82.
Cline, Ernest, *Armada: A Novel* (2015), p. 175.
Codron, Charles, letter, circa 1905 (Lhérault collection).
The Colfax Chronicle, "Daring Mlle. Marvingt Makes Greatest Long-Distance Flight in a French Aeroplane," February 18, 1911, p. 7.
Colin, Marie-Thérèse, "On a retrouvé 'les ailes de la Colombe,' *L'Est Républicain*, May 9, 1993.
Combeau-Mari, Evelyne, et Valérie Boulain, "Transgression des normes sexuées et violence symbolique: Itinéraire de jeunesse d'une sportive, Marie Marvingt (1875–1963), *International Review on Sport and Violence* (2014), no. 8, pp. 48–61.
Comoedia, "Nos Aviatrices: Hommage à Mlle Marvingt," August 2, 1930, p. 4.
Comoedia, "L'oeuvre de Mlle Marvingt pour l'aviation sanitaire," November 18, 1931, p. 5.
Comoedia, "Les conférences scolaires de Mlle Marvingt sur l'activité de notre aviation," November 26, 1931, p. 4.
Comte, Marie-Gaëtane, "Marie Marvingt," *Oxygen*, February 2006, no. 4, p. 24.
"La conférence de Mlle Marvingt à Notre-Dame de Sion," newspaper article, January 25, 1952.
Congrès International d'Aéronautique à Turin 25–31 octobre 1911, *Procès Verbaux, Rapports, et Mémoires* (1912).
Conte, Arthur, *Le premier janvier 1920* (Pion, 1979), p. 277.
Cordier, Marcel, "Les Escales de Marie Marvingt," *Leurs Demeures en Lorraine: Tome II* (Éditions Pierron, 1983), pp. 162–177.
Cordier, Marcel, *Hommes et lieux de mémoire en Lorraine* (Éditions Pierron, 1991), p. 10, 93, 94.
Cordier, Marcel, "Marie Marvingt: une femme de Foi," *Chrétiens dans la ville*, no 73, June 1992, p. 6.
Cordier, Marcel, "Marie Marvingt: Gloire du sport," *La Dépêche Meusienne*, 1995.
Cordier, Marcel, "Marie Marvingt: Une Lorraine qui volait au secours des blessés," *La vie la santé*, July 1998, no. 6, pp. 28–29.
Cordier, Marcel, "Marie Marvingt (1875–1963)," *Revue Lorraine Populaire*, June 2004, p. 12.
Cordier, Marcel, *Lorraine: Secrète et insolite* (Éditions du Sapin d'or, 2011), pp. 27, 32, 83, 86, 156.
Cordier, Marcel, "La Fiancée du danger: Marie Marvingt (1875–1963), *La nouvelle revue lorraine*, no. 20, June-July 2013, p. 30.
Cordier, Marcel, "Marie Marvingt 1875–1964: "La fiancée du danger," *Revue semestrielle du Conseil Général de Moselle (Metz)*, no. 20, September 2013.
Cordier, Marcel, "Marie Marvingt," *L'Ami Hebdo*, December 15, 2013.
Cordier, Marcel, "Hommage à Marie Marvingt (1875–1963)," *L'Ami Hebdo*, January 1, 2014.
Cordier, Marcel, "Les combats de Marie Marvingt," *La nouvelle revue lorraine*, no. 27, August-September 2014, p. 20–23.
Cornishman, "Blown to Sea in a Balloon: A Woman's Adventurous Voyage Across Channel," November 4, 1909, p. 6.
Coulet, Jacques, and Madame Coulet, interview, June 24, 1982.
Le Courrier de Champagne, "L'oeuvre du Théâtre Chrétien," May 15, 1913, p. 2.
Le Courrier de Champagne, "Les obsèques du capitaine Féquant," September 12, 1913, p. 3.
Le Courrier de Champagne, "Après la coupe Gordon-Bennett," October 26, 1913, p. 3.
Le Courrier de Champagne, "An accident d'aéroplane à Machault," December 13, 1913, p. 3.
Le Courrier de Champagne,"Machault: Accident d'aéroplane," December 14, 1913, p. 4.
Le Courrier de Champagne, "Les Enfants de Lorraine," December 15, 1913, p. 2.
Le Courrier de Champagne, "L'accident de Mlle Marvingt," December 18, 1913, p. 3.
Le Courrier de Champagne, "Mlle Marvingt au Grand Prix de l'Aéro-Club," July 24, 1914, p. 3.
Le Courrier d'Éthiopie, "Pas de femmes!" May 11, 1928, p. 3.
Cousin, Daniel, "La Femme automobiliste," *La Presse*, January 10, 1914, p. 3.
Cousin, Daniel, "La Femme professionnelle," *La Presse*, December 10, 1925, p. 3.
Le Cri de Nancy, "La Séance solennelle de rentrée," December 5, 1908, p. 31.
Le Cri de Nancy, "Mlle M. Marvingt," no. 7, February 13, 1909, p. 167.
La Croix, "Le grand prix de l'Aéro-Club de France," no. 8363, June 28, 1910, p. 2.
La Croix, "Une élève de Latham," no. 8422, September 6, 1910, p. 6.
La Croix, "L'Aviation," no. 8422, December 23, 1910, p. 5.
La Croix, "Sports d'Hiver à Chamonix," no. 8542, January 25, 1911, p. 6.
La Croix, "Une aviatrice tombe dans un jeu de boules," no. 8708, August 8, 1911, p. 6.
La Croix, "Le meeting de Saint Etienne," no. 8714, August 15, 1911, p. 6.
La Croix, Mlle Marvingt conference en faveur de l'aviation militaire," no. 8972, June 16–17, 1912, p. 6.
La Croix, "Chronique Sportive," no. 8973, June 18, 1912, p. 6.
La Croix, "Mlle Marvingt est allée de Paris à Bruxelles en ballon," no. 9009, July 30, 1912, p. 6.
La Croix, "Une aviatrice emmène deux passagers à 8.000 mètres," no. 9052, September 19, 1912, p. 2.
La Croix, "Le record de la hauteur," no. 9372, October 3, 1913, p. 6.
La Croix, "L'aviatrice Marvingt," no. 9393, October 28, 1913, p. 5.
La Croix, "Chronique Sportive," no. 9402, November 8, 1913, p. 6.
La Croix, "Chute d'une aviatrice," no. 9432, December 13, 1913, p. 5.
La Croix, "Après le Congrès des Infirmières," August 18, 1933, p. 4.

La Croix, "Légion d'honneur," no. 15932, January 27, 1935, p. 5.
La Croix, "Evacuation aérienne d'un religieuse missionnaire," April 15, 1938, p. 5.
La Croix, "L'aviation sanitaire à Luxembourg," no. 26994, July 1, 1938, p. 6.
Crystal, Lyliane, "Mlle Marie Marvingt," *Annales africaines*, July 20, 1922, p. 975.
La Culture Physique, "Une belle école d'énergie pour la femme: La 'Stella,'" August 15,1911, p. 510.
Cuny, Jean-Marie, "Marie Marvingt (rue)," *Dictionnaire des rues de Nancy* (Edition Cuny, 2001).
Curral-Couttet, Gaby, "Marie Marvingt, la fiancée du danger," *Les Folles Années de Chamonix* (Éditions France-Empire, series "si 1900 m'était conté," 1984), pp. 75–78.
Custis, John Trevor, "Women Make Real Advent Into Aerial Navigation," *The Washington Post*, December 18, 1910, p. S4.
Daçay, Jean, "Le Tourisme Aérien, école de culture physique," *La Culture Physique*, January 15, 1911, pp. 60–61.
Dahlquist, Marina, ed., *Exporting Perilous Pauline: Pearl White and the Serial Film Craze* (2013), pp. 18, 72–75, 84, 92–93.
Daily Mirror, "Joan of Arc of the Clouds," May 20, 1915, p. 10.
Dalbanne, J., "Le Concours de Ski des Vosges," *Armée et marine*, March 20, 1909, pp. 95–96.
Dalbanne, J., "Le Ve Concours International de Ski du 'Club Alpin Français,'" *Armée et marine*, February 28, 1911, p. 246.
Dall'Asta, Monica, "Pearl, the Swift One, or the Extraordinary Adventures of Pearl White in France," in Marina Dahlquist, ed., *Exporting Perilous Pauline: Pearl White and the Serial Film Craze* (2013), pp. 72–75, 84, 92–93.
Daniel, Florence, "Une esquisse inédite de Friant ainsi dédicacée: 'Temps héroïque de l'aviation, Somer de passage à Nancy, à Marie Marvingt, hommage d'E. Friant,'" *La Gazette Lorraine*, no. 44, December 2005.
Darrow, Margaret H., *French Women and the First World War: War Stories of the Home Front* (2000), pp. 229, 240–241.
Daurat, Michel, "A quatre-vingts ans, Marie Marvingt demeure toujours la fiancée du danger," *Lorraine Magazine*, no. 9, February 1955, pp. 10–13.
Davis, Jeffrey R., Robert Johnson, Jan Stepanek, Jennifer A. Fogarty, eds., *Fundamentals of Aerospace Medicine*, 4th ed. (2008), pp. 9, 14.
de Beauregard, Marie-Josèphe, letter, April 12, 1984.
de Beauregard, Marie-Josèphe, *Femmes de l'air: Chronique d'une conquête* (1993), pp. 55–58, 66, 73, 75–76.
de Champclos, G. Davin, "Mlle Marie Marvingt et ses 56.000 kilomètres en Afrique du Nord: La première femme de sport du monde," *L'Éclaireur du soir*, January 3, 1928.
de Cilleuls, Jean, "Aperçu sur l'Aviation Sanitaire," *Société des lettres, sciences, et arts du Saumurois*, April 1937, pp. 17–40.
de Laroche, Raymonde, "Recordwomen," *Je sais tout*, vol. 6, no. 72, January 15. 1911, pp. 801–809.
de Mazières, M., "Mouvement touristique au Maroc en 1934," *Revue de géographie marocaine*, January 1935, p. 107.
Denning, Andrew, *Skiing Into Modernity: A Cultural and Environmental History* (2014), p. 46.
De Pauw, Linda Grant, *Battle Cries and Lullabies: Women in War From Prehistory to the Present* (2014), p. 224.
La Dépêche Meusienne, "Marie Marvingt (1875–1963)," November 15, 2003.
de Saint-Fégor, L., *Le royaume de l'air* (Société d'Édition et de Publications, 1910), pp. 338, 341.
Desmonceaux, G., "Nous avons une nouvelle aviatrice," *Le Sport universel illustré*, November 27, 1925, p. 768.
Desormière, Yves, "A 80 ans Marie Marvingt, la 'fiancée du danger,' est la seule femme au monde à être titulaire de quatre brevets de pilotage: ballon, avion, hydravion, secourisme de l'air," *La Croix*, March 4, 1955, pp. 1, 4.
Diehl, Nicole, letter, July 8, 1982.
Dille, J.R., "Women in Civil and Military Aviation: The First 125 Years (1804–1929)," *Aviation, Space, and Environmental Medicine*, vol. 71, September 2000, pp. 957–961.
Dinago, Georges, "Une aviatrice lorraine," *L'Éclair de l'Est*, January 7, 1911.
Dinan, R., "Les Enfants Prodiges: Mlle Marvingt lance un défi mondial," *L'Ordre*, July 2, 1948.
Le XIXème siècle, "La traversée de Paris à la nage," July 31, 1906, p. 2.
Le XIXème siècle, "Le Ski," January 31, 1909, p. 3.
Le XIXème siècle, "Aéronautique," October 30, 1909, p. 4.
Le XIXème siècle, "Le Grand Prix de l'Aéro-Club," June 30, 1910, p. 4.
Le XIXème siècle, "Chute d'une aviatrice," August 9, 1911, p. 4.
Le XIXème siècle, "Mlle Marvingt tombe encore," August 17, 1911, p. 3.
Le XIXème siècle, "Une aviatrice fait une conférence," June 14, 1912, p. 3.
Le XIXème siècle, "Aéronautique: Le Prix Salmagean," October 1, 1912, p. 3.
Le XIXème siècle, "Le Grand Prix des ballons, July 19, 1914, p. 3.
Dollet, Christophe, "Marie Marvingt sans visa," *L'Est Républicain*, July 11, 2011.
Dollfus, Charles, *The Orion Book of Balloons* (The Orion Press, 1960), p. 71.
Dollfus, Charles, and Henri Bouché, *L'Histoire de l'aéronautique* (Éditions de l'Illustration, 1938), pp. 233, 235.
Dollfus, Charles, Henry Beaubois, Camille Rougeron, *L'Homme, l'air et l'espace* (Éditions de l'Illustration, 1965), pp. 130, 144; Dollfus, Charles, Henry Beaubois, Camille Rougeron, *L'Aviation: son histoire, des origines à 1960* (Baschet et Cie., 1979), pp. 130, 144.
D'Orliac, J., "L'avion-ambulance," *Le Journal [Paris]*, June 23, 1914, p. 6.
Drigny, Georges, "L'Année de l'aviation" and "La Grande Semaine d'hiver du T.C.F.," *Le Sport universel illustré*, January 9, 1910, pp. 30, 139.
Drigny, Georges, "La fin de la saison," *Le Sport universel illustré*, December 4, 1910, p. 783.
Droz, Stanislas, *Les Femmes dans la Grande Guerre 1914–1918* (Vent d'Est, 2014), pp. 74, 75–78.

Dublin Daily Express, "Remarkable Flight," October 28, 1909, p. 5; *Aberdeen Journal*, "Balloon Trip From France to England," October 28, 1909, p. 5; *Exeter and Plymouth Gazette*, "Blown to England: A French Balloonist's Experience," October 28, 1909, p. 5; *Fife Free Press & Kirkcaldy Guardian*, October 30, 1909, pp. 3–4; *Grantham Journal*, "Blown to England," October 30, 1909, p. 7; *Manchester Courier and Lancashire General Advertiser*, "Balloonists' Adventure; Journey From France," October 28, 1909, p. 7; *The Scotsman*, "Balloonists' Peril, Blown from France to England," October 28, 1909, p. 7.

Dubreuil, Yvonne, "Marie Marvingt 'la fiancée du danger,' brochure from the Congrès de Nancy, October 1998, pp. 23–25.

Dundee Courier, "Other Balloonists Come to Grief," July 21, 1914, p. 5.

Dundee Evening Telegraph, "Girl Climber's Record," September 7, 1905, p. 3.

Dundee Evening Telegraph, "Women Conquerors of the Air," October 30, 1912, p. 6.

Dundee Evening Telegraph, "Aeroplane Ambulance," June 24, 1914, p. 4.

Dunkerque-Sports, "Natation: A Joinville le Pont," July 30, 1911, p. 2.

Duperey, Amy, *Almanach 2012 des terres de France* (Presses de la Cité, 2011), p. 14.

du Taillis, Jean, "Le cinquième salon d'aéronautique," *Les Annales politiques et littéraires*, December 7, 1913, p. 528.

du Taillis, Jean, "Sports d'hivers," *Les Annales politiques et littéraires*, February 1, 1914, p. 112.

Dutrieu, Hélène, "Au salon de l'Aéronautique: Le Vernissage," *Gil Blas*, December 17, 1911, pp. 1–2.

Duvernay, Dr. L., "Le Mont Revard," *Le Sport universel illustré*, March 10, 1912, pp. 158–159.

Dvorak, Petula, "A Bolt to the Blue," *The Washington Post*, April 10, 2015.

East Liverpool [OH] Review, "French Flyer Arrives," August 29, 1935, p. 14.

E.B., "La Fiancée du Danger: Une célèbre femme de l'air est à Limoges" *Notre Province [Limoges]*, June 1944, p. 160.

L'Echo d'Alger, "Mlle Marvingt emmène deux passagers à trois mille mètres de hauteur," September 19, 1912, p. 2.

L'Echo d'Alger, "L'aviatrice Marie Marvingt se prodigue au bénéfice des monuments commémoratifs," October 10, 1912, p. 3.

L'Echo d'Alger, "Une aviatrice capote et se blesse," December 14, 1913, p. 2.

L'Echo d'Alger, "Mlle Marvingt conférence: Elle veut réunir les fonds nécessaires à l'achat d'un avion-ambulance," May 23, 1914, p. 2.

L'Echo d'Alger, "Le Grand Prix de l'Aéro-Club de France," July 13, 1914, p. 2.

L'Echo d'Alger, "Le Grand Prix des sphériques," July 20, 1914, p. 2.

L'Echo d'Alger, "Quelques réservistes de marque," August 7, 1914, p. 2.

L'Echo d'Alger, "Conférence," July 8, 1922, p. 2.

L'Echo d'Alger, "Département d'Alger: Blida," July 17, 1922, p. 3.

L'Echo d'Alger, "Départment d'Alger: Ténès et Cherchell," August 8, 1922, p. 3.

L'Echo d'Alger, "Kolea: Conférence," August 19, 1922, p. 3.

L'Echo d'Alger, "Conférence," November 8, 1922, p. 2.

L'Echo d'Alger, "Aviation et aérostation," November 23, 1922, p. 4.

L'Echo d'Alger, "Département d'Alger: Blida," December 4, 1922, p. 4.

L'Echo d'Alger, "Département d'Alger: Mouzaiaville," December 5, 1922, p. 3.

L'Echo d'Alger, "Le vol à voile: Le meeting à Biskra commence aujourd'hui," January 26, 1923, p. 2.

L'Echo d'Alger, "Le vol à voile: Le meeting à Biskra," January 28, 1923, p. 2.

L'Echo d'Alger, "Conférence sur l'Aéronautique," March 11, 1923, p. 3.

L'Echo d'Alger, "Mlle Marvingt," June 21, 1924, p. 4.

L'Echo d'Alger, "Département de Constantine: Sétif," October 5, 1924, p. 3.

L'Echo d'Alger, "Au pays du merveilleux," April 30, 1925, p. 2.

L'Echo d'Alger, "Conférence," May 4, 1925, p. 3.

L'Echo d'Alger, "Initiation aux sciences psychiques," May 6, 1925, p. 2.

L'Echo d'Alger, "Conférence au Cercle Militaire," 9 mai 1925, p. 2

L'Echo d'Alger, "Huit cours et conférences à Alger de Mlle Marie Marvingt," May 15, 1925, p. 3.

L'Echo d'Alger, "Prochaines conférences de Mlle Marvingt," June 9, 1925, p. 4.

L'Echo d'Alger, "Prochaines conférences de Mlle Marvingt," June 11, 1925, p. 2.

L'Echo d'Alger, "Conférence à Orléansville," July 2, 1925, p. 3.

L'Echo d'Alger, "Département d'Alger: Teniet-el-Haad," July 2, 1925, p. 3.

L'Echo d'Alger, "Une conférence de Mlle Marvingt à Mascara," July 8, 1925, p. 6.

L'Echo d'Alger, "Conférence de Mlle Marvingt," July 22, 1925, p. 3.

L'Echo d'Alger, "Ceryville: Conférence," July 28, 1925, p. 4.

L'Echo d'Alger, "Nédroma: Conférence aéronautique," August 19, 1925, p. 4.

L'Echo d'Alger, "Le Congrès Afrique du Nord-Midi," August 4, 1927, p. 3.

L'Echo d'Alger, "Aux Beaux-Arts," March 9, 1929, p. 2.

L'Echo d'Alger, "Une conférence de Mlle Marvingt aux Beaux-Arts," March 11, 1929, p. 2.

L'Echo d'Alger, "Conférence touristique," March 12, 1929, p. 2.

L'Echo d'Alger, "La conférence de Mlle Marvingt aux Beaux-Arts," March 14, 1929, p. 2.

L'Echo d'Alger, "A travers notre Afrique de Nord avec Marie Marvingt," March 16, 1929, p. 2.

L'Echo d'Alger, "Médea: Conférence," 18 mars 1929, p. 6

L'Echo d'Alger, "Conférence de Mlle Marvingt, aviatrice," March 28, 1929, p. 6.

L'Echo d'Alger, "Un hommage grandiose de l'Algérie à Clemenceau et à tous nos poilus," January 2, 1930, p. 3.

L'Echo d'Alger, "Le banquet de la Betterave," January 20, 1930, p. 4.5

L'Echo d'Alger, "Les journées de l'aviation sanitaire à Orly," July 31, 1931, p. 4.
L'Echo d'Alger, "Nouvelles de Tunisie: Une conférence de Mademoiselle Marvingt," March 16, 1932, p. 2.
L'Echo d'Alger, "Constantine: Ain-M'lila," July 16, 1932, p. 7.
L'Echo d'Alger, "Le résident général Ponsot assiste à une prestigieuse revue des troupes qui ont participé à la pacification des confins sahariens du Maroc," March 20, 1934, p. 6.
L'Echo d'Alger, "Aviatrices," December 15, 1934, p. 2.
L'Echo d'Alger, "Armée de l'air," January 26, 1935, p. 2.
L'Echo de Bougie, "Mademoiselle Marie Marvingt à Bougie," December 31, 1922, p. 2.
L'Echo de Bougie, "Mademoiselle Marvingt à Bougie," October 18, 1924, p. 3.
L'Echo de Paris, "La traversée de Paris à la nage," July 30, 1906.
L'Echo de Sanflorain, "Une aviatrice tombe dans un café," August 12, 1911.
L'Echo Sportif du Centre et de l'Ouest, "Une Carrière féminine sportive," 1914.
L'Echo de Tiaret, "Conférences," July 11, 1925, p. 2.
L'Éclair, "Paris à la nage," July 30, 1906.
L'Éclair de l'Est, "Une journée de ski au Mont Revard," February 4, 1912.
Edinburgh Evening News, "Girl Climber's Record," September 7, 1905, p. 6.
Eiermann, J., grand chancelier, chef du bureau de la Légion d'Honneur, letter to Jacqueline Maire, December 31, 1982.
"Eine Sturmfahrt im Ballon über die Norsee wehrend der Nacht von Nancy nach Southwold," *Deutsche Zietschr. Luftsch*, September 21, 1910.
Emporia [KS] Gazette, "French Aviatrix Honored," October 12, 1935, p. 3.
En Lorraine, "Une héroïne Lorraine: Un exemple à suivre," no, 42, May 1998, pp. 32–33.
Enselme-Trichard, Renée, letters, July 3 and October 5, 1982.
En verre et contre Tout, "Marie Marvingt, la fiancée du danger," theater piece, August 24–25, 2013; www.enverreetcontretout.net.
L'Équipe, "Mlle Marvingt, détentrice des records féminins en aéroplane," October 1910.
Erbstein, Marguerite, "Marie Marvingt, 'la fiancée du danger,' entre à l'Olympe des plus grandes sportives," *L'Est Républicain*, September 8, 1987.
Erbstein, Roland, in collaboration with Jean Matt, "Cartes Postales d'hier et d'aujourd'hui: En 1909, 'la fiancée du danger' ouvrait aux femmes la voie des airs," *L'Est Républicain*," March 23, 1980.
Est Magazine, "D'hier à aujourd'hui la mongolfière," July 29, 2007, p. 15.
Est Magazine, "Il y a 100 ans dans l'Est: Mlle Marvingt tombe dans un café," August 28, 2011.
Est Magazine, "Il y a 100 ans dans l'Est: Une chute de Mlle Marvingt," December 15, 2013.
Estrada de Tourniel, Jérôme, "La fiancée du danger," *Passions Grand Est*, no. 26, December 2003, pp. 58–62.
Estrada de Tourniel, Jérôme, *Les Grands Événments de Meurthe-et-Moselle: De 1900 à Nos Jours* (De Borée, 2017), pp. 228–230.
L'Est Républicain, "Objets perdus ou trouvés," November 23, 1900.
L'Est Républicain, "Concours de tir," June 19, 1906, p. 2.
L'Est Républicain, "Concours de Tir," June 26, 1906, p. 2.
L'Est Républicain, "Concours de Tir," June 27, 1906, p. 2.
L'Est Républicain, "Concours de Tir," June 30, 1906, p. 2.
L'Est Républicain, "Concours de Tir: Journée de clôture," July 3, 1906, p. 2.
L'Est Républicain, "La traversée de Paris à la nage," July 30, 1906, p. 1.
L'Est Républicain, "La traversée de Paris à la nage," July 31, 1906, p. 1.
L'Est Républicain, "La fête de tir de Malzéville," August 5, 1906, p. 1.
L'Est Républicain, "Concours de tir à Malzéville," August 12, 1906, p. 1.
L'Est Républicain, "Société française de secours aux blessés," December 13, 1907, p. 2.
L'Est Républicain, "Au concours de ski de Chamonix," January 31, 1909, p. 2.
L'Est Républicain, "Le concours de skis à Gérardmer," February 23, 1909, p. 1.
L'Est Républicain, "Les fêtes de skis à Gérardmer," February 24, 1909, p. 2.
L'Est Républicain, "Ligue nationale aérienne et ligue de l'Est," March 27, 1909, p. 1.
L'Est Républicain, "Ligue aérienne de l'Est," April 27, 1909, p. 2.
L'Est Républicain, "Banquet en l'honneur de l'aviateur Sommer," September 10, 1909, p. 2.
L'Est Républicain, "A Jarville Aviation," 11 septembre 1909, p. 2
L'Est Républicain, "Pendant le Congrès d'aéronautique," September 21, 1909, p. 2.
L'Est Républicain, "Le Concours de ballons libres," September 25, 1909, p. 1.
L'Est Républicain, "Après le Concours de ballons libres: ce que dit Mlle Marvingt," September 26, 1909, p. 2.
L'Est Républicain, "Les deux ascensions de dimanche au parc aéronautique," October 10, 1909, p. 2.
L'Est Républicain, "Au parc d'aéronautique," October 11, 1909, p. 1.
L'Est Républicain, "L'Ascension de Mlle Marvingt," October 12, 1909, p. 2.
L'Est Républicain, "Ascension au ballon libre," October 19, 1909, p. 2.
L'Est Républicain, "Ascension à la Chiennerie," October 27, 1909, p. 3.
L'Est Républicain, "Une ascension mouvementée de Mlle Marvingt," October 28, 1909, p. 1.
L'Est Républicain, "Une ascension mouvementée de Mlle Marvingt (suite)," October 29, 1909, p. 2.
L'Est Républicain, "Mlle Marvingt—apprenti-pilote," December 1, 1909, p. 2.
L'Est Républicain, "Mlle Marvingt, aviatrice," December 28, 1909, p. 2.
L'Est Républicain, "Latham et Mlle Marvingt," January 9, 1910, p. 1.
L'Est Républicain, "La Grand Semaine de Gérardmer," February 15, 1910, p. 2.

L'Est Républicain, "Journée d'excursion," February 16, 1910, p. 3.
L'Est Républicain, "L'Aviation et les Etrangers au Camp de Châlons," April 17, 1910, p. 2.
L'Est Républicain, "Grande fête aéronautique du 15 mai," May 14, 1910, p. 2.
L'Est Républicain, "Ce nouvel accident ne décourage pas Mlle Marvingt," May 15, 1910, p. 1.
L'Est Républicain, "La fête aéronautique de dimanche," May 16–17, 1910, pp. 1–2.
L'Est Républicain, "Nouvelle ascension de Mlle Marvingt," May 18, 1910, p. 2.
L'Est Républicain, "Mlle Marvingt, pilote aéronaute," June 18, 1910, p. 2.
L'Est Républicain, "Prochaine ascension de Mlle Marvingt," June 21, 1910, p. 2.
L'Est Républicain, "Les ascensions de Mlle Marvingt," June 28, 1910, p, 2.
L'Est Républicain, "Mlle Marvingt passe brillamment son brevet de pilote," October 23, 1910, p. 2.
L'Est Républicain, "Mlle Marvingt aviatrice," November 28, 1910, p. 1.
L'Est Républicain, "Mlle Marvingt va s'attaquer au record de Mlle Dutrieu," December 9, 1910, p. 2.
L'Est Républicain, "Parmi les actualités de cette semaine," December 10, 1910, p. 3.
L'Est Républicain, "Mlle Marvingt," January 1–2, 1911, p. 1.
L'Est Républicain, "Pourquoi je n'ai pas pu reprendre la Coupe 'Femina,'" January 4, 1911, p. 2.
L'Est Républicain, "On patine," January 19, 1911, p. 2.
L'Est Républicain, "Aviation," January 22, 1911, p. 3.
L'Est Républicain, "Une course militaire internationale de ski," February 13, 1911, p. 1.
L'Est Républicain, "Roller Skating," February 26, 1911, p. 2.
L'Est Républicain, "Une aviatrice à Longwy," March 17, 1911, p. 3.
L'Est Républicain, "Un beau trait de camaraderie sportive," July 9, 1911, p. 1.
L'Est Républicain, "Mlle Marvingt blessée en automobile," August 4, 1911, p. 1.
L'Est Républicain, "Mlle Marvingt tombe dans un café," August 8, 1911, p. 2.
L'Est Républicain, "Incidents au meeting de Saint-Etienne," August 16, 1911, p. 2.
L'Est Républicain, "Avant le meeting d'aviation," April 5, 1912, p. 2.
L'Est Républicain, "Les Fêtes d'Aviation à Nancy: L'arrivée de Kimmerling; La Journée de Samedi" April 7, 1912, p. 2.
L'Est Républicain, "Fêtes d'aviation à Nancy," April 8–9, 1912, p. 2.
L'Est Républicain, "Les aviateurs militaires," April 13, 1912, p. 2.
L'Est Républicain, "Les Fétes de Toul: Ce fut une journée splendide: on applaudit de beaux vols," April 15, 1912, p. 1.
L'Est Républicain, "L'aéroplane-ambulance," June 14, 1912, p. 3.
L'Est Républicain, "Nancy: La journée," July 1, 1912, p. 2.
L'Est Républicain, "La Fin prévue," July 20, 1912, p. 4.
L'Est Républicain, "Les prouesses de Mlle Marvingt," July 29, 1912, p. 4.
L'Est Républicain, "Féte patriotique," November 7, 1912, p. 3.
L'Est Républicain, "Au Théâtre," November 12, 1912, p. 3.
L'Est Républicain, "Art et Aviation: La collaboration féminine," November 21, 1912, p. 2.
L'Est Républicain, "L'Aviation militaire à Nancy," December 19, 1912, p. 2.
L'Est Républicain, "Le monument de Caumont," May 17, 1913, p. 2.
L'Est Républicain, "Un vol de Mlle Marvingt," October 3, 1913, p. 2.
L'Est Républicain, "Une conference sur l'aviation," March 20, 1914, p. 3.
L'Est Républicain, "Pour l'avion-ambulance," May 8, 1914, p. 3.
L'Est Républicain, "Ministre de l'Instruction Publique est à Nancy," July 13, 1914, pp. 1–2.
L'Est Républicain, "Une nacelle se détache d'un ballon," July 20, 1914, p. 2.
L'Est Républicain, "Mlle Marvingt veut s'engager," May 21, 1915, p. 2.
L'Est Républicain, "Décès du 10 mars: Félix-Constant Marvingt," March 13, 1916, p. 2.
L'Est Républicain, "Mlle Marvingt à In-Saleh," February 21, 1923, p. 3.
L'Est Républicain, "Mademoiselle Marvingt," June 18, 1924, p. 5.
L'Est Républicain, "Le Rallye-Ballon automobile," September 13, 1924, p. 4.
L'Est Républicain, "Les Bals de Samedi: A l'Union des Femmes de France," November 28, 1926, p. 3.
L'Est Républicain, "L'Aviation et le Comité des Fêtes," September 26, 1927, p. 3.
L'Est Républicain, "Le concours international des avions legers," September 15, 1928, p. 1.
L'Est Republicain, "Au Congrès international de l'aviation sanitaire: Où l'on entend Mlle Marvingt," May 16, 1929, p. 1.
L'Est Républicain, "La conférence de Mlle Marvingt," June 24, 1929, p. 4.
L'Est Républicain, "Une nouvelle causerie de Mlle Marvingt: La jeunesse et l'aéronautique," July 5, 1929, p. 3.
L'Est Républicain, "Causerie scolaire sur l'aviation," July 24, 1929, p. 3.
L'Est Républicain, "Deux heures dans les airs," October 7, 1929, p. 2.
L'Est Républicain, "La conférence de Mlle Marvingt," October 10, 1929, p. 3.
L'Est Républicain, "Conférence de Mlle Marvingt," October 12, 1929, p. 3.
L'Est Républicain, "La conférence de Mlle Marvingt: Deux heures dans les airs," October 13, 1929, p. 3.
L'Est Républicain, "Conférence de Mlle Marvingt," October 16, 1929, p. 3.
L'Est Républicain, "Matinée dansante du Pelican Club," November 19, 1929, p. 3.
L'Est Républicain, "L'émulation parmi les groupes scolaires de la Ligue aéronautique," December 5, 1929, p. 3.
L'Est Républicain, "Chez les amis de l'Aviation sanitaire," November 30, 1930, p. 2.
L'Est Républicain, "La conférence de Mlle Marvingt," July 8, 1931, p. 3.

L'Est Républicain, "Mancieulles," July 12, 1931, p. 7.
L'Est Républicain, "La Cérémonie de Champenoux a revetu, hier, une particulière solennité," September 14, 1931, p. 2.
L'Est Républicain, "La remise du challenge de Mlle Marvingt," November 17, 1931, p. 2.
L'Est Républicain, "Une soirée sensationnelle mercredi 13 courant," January 6, 1932, p. 3.
L'Est Républicain, "Mlle Marvingt à la salle Poirel, mercredi prochain," January 10, 1932, p. 3.
L'Est Républicain, "La soirée exceptionnelle de demain mercredi," January 12, 1932, p. 3.
L'Est Républicain, "Aujourd'hui, Salle Poirel, la Société de Géographie reçoit Mlle Marvingt," January 13, 1932, p. 3.
L'Est Républicain, "Une conférence de Mlle Marvingt," January 14, 1932, p. 3.
L'Est Républicain, "Une bonne conference de Mlle Marvingt," January 15, 1932, p. 2.
L'Est Républicain, "Ligues aeronautique et d'aviation sanitaire," January 17, 1932, p. 3.
L'Est Républicain, "Conference," January 20, 1932, p. 3.
L'Est Républicain, "La soirée de la Faculté de Lettres," January 22, 1932, p. 3.
L'Est Républicain, "Les explorations de Mlle Marvingt," January 23, 1932, p. 3.
L'Est Républicain, "La vie musulmane en Afrique en Tunisie et aux colonies italiennes," January 25, 1932, p. 3.
L'Est Républicain, "Mlle Marvingt parle de la Tunisie," January 26, 1932, p. 3.
L'Est Républicain, "Ligue Aéronautique de France," January 31, 1932, p. 3.
L'Est Républicain, "Ligue Aéronautique de France," February 4, 1932, p. 3.
L'Est Républicain, "Ligue Aéronautique de France," February 7, 1932, p. 3.
L'Est Républicain, "A la Ligue Aéronautique de France," February 8, 1932, p. 2.
L'Est Républicain, "Le buste de Mlle Marvingt," July 9, 1932, p. 3.
L'Est Républicain, "Visions d'Orient," October 4, 1932, p. 3.
L'Est Républicain, "Visions d'Orient," October 6, 1932, p. 3.
L'Est Républicain, "Visions d'Orient," October 9, 1932, p. 3; "La journée à Nancy," p. 2.
L'Est Républicain, "Visions d'Orient," October 10, 1932, p. 3.
L'Est Républicain, "L'accueil de la Capitale lorraine au président de la République," November 6, 1932, p. 1.
L'Est Républicain, "Nancy a fait à son maire, M.K. Malval, des obsèques émouvantes empreintes de simplicité et de dignité," May 17, 1933, p. 1.
L'Est Républicain, "IIe Congrès International de l'Aviation Sanitaire," May 21, 1933, p. 1.
L'Est Républicain, "L'Est Illustré; Sommaire du no. 544, " May 27, 28, 29, 1933, p. 5.
L'Est Républicain, "Les Ailes qui sauvent," August 26, 1934, p. 3.
L'Est Républicain, "L'Aviatrice Nancéïenne Marie Marvingt Chevalier de la Légion d'honneur," January 21, 1935, p. 4.
L'Est Républicain, "Le Challenge Capitaine Echeman," June 9, 1935, p. 4.
L'Est Républicain, "Au pavillon de l'Aéronautique: Le Gala des ailes qui sauvent," July 23, 1936, p. 4.
L'Est Républicain, "A l'exposition de Nancy, Le succès de la Fête enfantine," July 24, 1936, p. 4.
L'Est Républicain, "Au stand de l'Aéronautique," July 26, 1936, p. 4.
L'Est Républicain, "Nancy, la ville de Stanislas, fait au général Rydz-Smigly une enthousiaste accueil," September 4, 1936, p. 3.
L'Est Républicain, September 25, 1936, p. 2.
L'Est Républicain, "La réception du Colonel Weiss," October 19, 1936, p. 4.
L'Est Républicain, "Le meeting au vélodrome du rassemblement national Lorrain," October 26, 1936, p. 4.
L'Est Républicain, "Ligue des Amis de l'aviation sanitaire," December 2, 1936, p. 4.
L'Est Républicain, "Henry de Monfreid chez les sauvages," December 7, 1936, p. 3.
L'Est Républicain, "Les obsèques du Professeur Gaston Michel," March 17, 1937, p. 4.
L'Est Républicain, "La séance recreative de l'A.M.C.," March 22, 1937, p. 3.
L'Est Républicain, "La conférence du docteur Benech," May 21, 1937, p. 5.
L'Est Républicain, "Gala de L'Aéro-Club de L'Est," June 5, 1937, p. 4.
L'Est Républicain, "Obsèques: Madame Humblot," September 9, 1937, p. 4.
L'Est Républicain, "Mlle Marie Marvingt, chevalier de la Santé Publique,"November 5, 1937.
L'Est Républicain, "Ligue des Amis de l'Aviation sanitaire," January 12, 1938, p. 4.
L'Est Républicain, "Le bal de la presse," March 2, 1938, p. 4.
L'Est Républicain, "L'Aéro-Club de l'Est a tenu son Assemblée générale annuelle," "Le cinquantenaire du comité de Nancy de l'Union des Femmes de France," "Mariage de Paul-Cavallier et de Suyrot," March 6, 1938, p. 4.
L'Est Républicain, "Une soeur blanche est evacuée en avion d'Alger à Paris," April 13, 1938, p. 5.
L'Est Républicain, "M. Jean Beziaud parle de l'aviation française," April 24, 1938, p. 4.
L'Est Républicain, "M. Jacques Mortane a parlé de Guynemer et de Mermoz," May 22, 1938, p. 4.
L'Est Républicain, "Le Salon de la Société Lorraine des Amis des Arts s'ouvre aujourd'hui," October 9, 1938, p. 4.
L'Est Républicain, "M. Robert Leurquin a raconté ce qu'il a vu en Chine," November 8, 1938, p. 5.
L'Est Républicain, "La soirée de l'Aviation populaire," November 11, 1938, p. 4.
L'Est Républicain, "Les fêtes du 11 novembre dans la capitale lorraine pavoisée," November 12, 1938, p. 4.
L'Est Républicain, "Necrologie: M. Kleber Saudry," October 12, 1939, p. 3.
L'Est Républicain, "Matinee de la Jeunesse de l'Empire Français," November 13, 1939, p. 3.
L'Est Républicain, "Le Gala de bienfaisance du 26e," December 25, 1939, p. 3.
L'Est Républicain, "La Matinee de bienfaisance de la Salle Poirel," January 23, 1940, p. 3.

L'Est Républicain, "La Matinee Artistique de la Salle Poirel," February 5, 1940, p. 3.
L'Est Républicain, "La Jeunesse de l'Empire Français a reçu les Gars du 170e," March 18, 1940, p. 3.
L'Est Républicain, "La matinee de la salle Poirel," April 22, 1940, p. 3.
L'Est Républicain, "Le Gala des Lions," May 3, 1940, p. 3.
L'Est Républicain, "Jeanne d'Arc, la libératrice et le cinquième anniversaire de la capitulation allemande ont été fêtés ensemble au cours de manifestations sobres, mais imposantes," May 15, 1950.
L'Est Républicain, November 18, 1952; June 6, 1958; November 8, 1958 (theses issues, too old to photocopy, were read into a tape recorder).
L'Est Républicain, "Avec les félicitations de Marie Marvingt," July 31, 1953.
L'Est Républicain, "La 'Victoire de Marie Marvingt," February 2, 1955.
L'Est Républicain, "'La fiancée du danger' fête ses 80 ans," February 22, 1955.
L'Est Républicain, "La 'fiancée' de Nancy," June 4, 1955.
L'Est Républicain, "Mlle Marie Marvingt reçoit deux nouveaux témoignages officiels de gratitude," February 21, 1958, p. 4.
L'Est Républicain, "La 'fiancée du danger' est morte, hier, à 88 ans," December 15, 1963.
L'Est Républicain, "Le suprême adieu de Nancy à MM," December 18, 1963.
L'Est Républicain, "A la mémoire de Marie Marvingt," December 1, 1983.
L'Est Républicain, "Il y a vingt ans mourait Marie Marvingt," December 14, 1983.
L'Est Républicain, "Marie Marvingt donnera son nom aux complexe sportif," December 28, 1983.
L'Est Républicain, "Tout est prêt pour l'inauguration du complèxe sportif Marie Marvingt," January 6, 1984.
L'Est Républicain, "La fiancée du danger Marie Marvingt donne son nom au gymnase de Maxéville," January 8, 1984.
L'Est Républicain, "La résidence Marie Marvingt a déjà une pierre," April 3, 1984.
L'Est Républicain, "Marie Marvingt aura sa plaque," May 4, 1984.
L'Est Républicain, "Nancy: en mémoire de Marie Marvingt, 'la fiancée du danger,'" June 4, 1984.
L'Est Républicain, "Samedi à Tomblaine: Baptême du LEP Marie Marvingt," May 21, 1987.
L'Est Républicain, "Le lycée professionnel dédié à la 'fiancée du danger,'" May 24, 1987.
L'Est Républicain, "Anniversaire de la mort de Marie Marvingt," December 10, 1988.
L'Est Républicain, "Le lycée Marie Marvingt de Tomblaine commémore la mort de 'la fiancée du danger, December 14, 1988.
L'Est Républicain, "L'éclat du neuf pour Zéphirine," February 4, 1990.
L'Est Républicain, "Les cent vies de Marie Marvingt," September 22, 1991.
L'Est Républicain, "Marie Marvingt à l'honneur," November 15, 1992.
L'Est Républicain, "Marie-Marvingt a des ailes," January 27, 1993.
L'Est Républicain, "Champ-le-Boeuf: Hommage à la fiancée du danger," April 19, 1993.
L'Est Républicain, "Marie Marvingt à l'affiche," April 30, 1993.
L'Est Républicain, "Marie Marvingt à l'honneur," May 28, 1993.
L'Est Républicain, "Tomblaine enclenche le turbo," May 30, 1993.
L'Est Républicain, "Marie Marvingt donne des ailes aux conférenciers," June 1, 1993.
L'Est Républicain, "Sous le signe de Marie Marvingt," June 28, 1993.
L'Est Républicain, "Bouxières-aux-Dames: Marie Marvingt, fiancée du danger," June 1994.
L'Est Républicain, "Hommage à Marie Marvingt," June 24, 1997.
L'Est Républicain, "La 'fiancée du danger' a désormais un visage," October 5, 1998.
L'Est Républicain, "Laxou: Honneur à Marie Marvingt," May 12, 2002.
L'Est Républicain, "Marie Marvingt dans le Cantal: L'aéro-club d'Aurillac s'est baptisé 'Marie Marvingt,'" September 11, 2002.
L'Est Républicain, "Voinémont: Marie Marvingt: la fiancée du danger," November 19, 2003.
L'Est Républicain, "Dommartemont: Marie Marvingt à l'honneur," March 30, 2012.
L'Est Républicain, "Marie Marvingt est décédée il y a 50 ans," December 15, 2013.
L'Européen, "Un film sur l'aviation," November 23, 1934, p. 5.
L'Européen, "Ce que fut le 3e Congrès International de l'Aviation Sanitaire," June 28, 1935, p. 5.
Evening Dispatch, "Women Pilots Fly Spitfires," December 20, 1940, p. 4.
Evening News [Sault Sainte Marie, MI], "Another Contender for Aviation Honors: Mme Marvingt," January 10, 1911, p. 6; *Waterloo [IA] Times Tribune*, "Another Contender for Aviation Honors," January 6, 1911, p. 8.
Examiner, "Mlle. Marie Marvingt," October 2, 1907.
Facts on File, "Obituaries," 1963, p. 497.
Faure, Odile, "Marie Marvingt, La fiancée du danger," *Massif Central Magazine*, March-April 1996, pp. 68–71.
Fédération des Internationaux du Sport Français (FISF), "Promotion 1995 des 'Gloires du Sport,'" 1995.
Femina, "Les Bavardages de Françoise," February 1, 1910, p. 72.
Femina, "Toutes au volant!" April 15, 1910, p. 208.
Femina, "Au pays de l'aviation," August 1, 1910, p. 402.
Femina, "La Coupe d'aviation 'Femina,'" December 15, 1910, p. 680.
Femina, "La Coupe Femina," January 1, 1911, pp. 14, 18.
Femina, "Mlle Hélène Dutrieu se prépare à disputer la 'Coupe Femina,'" January 15, 1911, p. 28.
Femina, "Nos aviatrices," May 1, 1911, p. 242.
Femina, "On court la Coupe Femina," September 1, 1912, p. 482.
La Femme, "Les aviatrices," November 1911, p. 173.
Féral, Roger, "Marie Marvingt (86 ans), 'la fiancée du danger,' a passé la Toussaint dans un hélicoptère à réaction," *Télé-Paris*, November 1960.

Le Figaro, "La traversée de Paris," July 30, 1906, p. 6.
Le Figaro, "Les promeneuses de l'air," September 21, 1907, p. 1.
Le Figaro, "Les Sports d'Hiver," January 9, 1908, p. 5.
Le Figaro, "Un voyage mouvementé," October 28, 1909, p. 1; "Echos et nouvelles," p. 3.
Le Figaro, "Sports d'Hiver," January 26, 1910, p. 7.
Le Figaro, "Grand Prix de l'Aéro-Club de France," June 23, 1910, p. 7.
Le Figaro, "Le Grand Prix de l'Aéro-Club de France," June 28, 1910, p. 7.
Le Figaro, "Le prix de l'héroïsme," October 21, 1910, p. 4.
Le Figaro, "Banquet d'Espérantistes," November 13, 1910, p. 3.
Le Figaro, "Aviation," November 16, 1910, p. 6.
Le Figaro, "Aviation," November 28, 1910, p. 6.
Le Figaro, "Coupe Femina," November 29, 1910, p. 6.
Le Figaro, "La Coupe Femina," December 2, 1910, p. 7.
Le Figaro, "Aviation," December 6, 1910, p. 6.
Le Figaro, "Aviation," December 31, 1910, p. 7.
Le Figaro, "Sports divers à Chamonix," January 24, 1911, p. 7.
Le Figaro, "Sports divers à Chamonix," January 25, 1911, p. 7.
Le Figaro, "Aviation," August 16, 1911, p. 6.
Le Figaro, "Les dames oiselles," October 13, 1911, p. 1.
Le Figaro, "Le Dîner de l'Académie des Sports," December 15, 1911, p. 3.
Le Figaro, "Une conférence de Mlle Marvingt," June 13, 1912, p. 7.
Le Figaro, "Aérostation," July 6, 1912, p. 7.
Le Figaro, "De Paris à Bruxelles, les ballons de l'Aéro-Club de France," July 31, 1912, p. 5.
Le Figaro, "Aviation," August 6, 1912, p. 5.
Le Figaro, "Aviation," August 19, 1912, p. 6.
Le Figaro, "Aviation," September 17, 1912, p. 5.
Le Figaro, "Aviation," October 1, 1912, p. 5.
Le Figaro, "Une conférence de Mlle Marvingt," November 12, 1912, p. 7.
Le Figaro, "Aviation," December 10, 1912, p. 7.
Le Figaro, "Les monoplans Deperdussin," August 9, 1913, p. 7.
Le Figaro, "Un vol de Mlle Marvingt," October 3, 1913, p. 8.
Le Figaro, "Chute de Mlle Marvingt," December 13, 1913, p. 7.
Le Figaro, "La fiancée du danger," January 3, 1914, p. 1.
Le Figaro, "Les femmes alpinistes," January 10, 1914, p. 7.
Le Figaro, "La femme et les sports," January 20, 1914, p. 7.
Le Figaro, "Le grand prix de l'Aéro-Club de France," July 19, 1914, p. 5.
Le Figaro, "Le grand prix des ballons: Un accident," July 20, 1914, p. 2.
Le Figaro, "Le grand prix de l'Aéro-Club de France," July 21, 1914, p. 11.
Le Figaro, "Une rectification," July 26, 1914, p. 12.
Le Figaro, "La saison à Vittel," July 28, 1914, p. 10.
Le Figaro, "Les femmes au Mont-Blanc 1808–1921," October 18, 1921, p. 4.
Le Figaro, "Le Congrès de l'Afrique du Nord," July 25, 1927, p. 2.
Le Figaro, "Marie-Marvingt parle et vole," July 31, 1930, p. 8.
Le Figaro, "Nouvelles aériennes," March 27, 1931, p. 4.
Le Figaro, "Nouvelles aériennes," October 14, 1931, p. 8.
Le Figaro, "Mort à 88 ans de Marie Marvingt, 'la fiancée du danger,'" December 15, 1963.
Le Figaro, "Le Musée de l'air fête ses héroïnes," March 7, 2006.
Figures libres, November 2009.
Fisher, Élise, *Le roman de la place Stanislas* (Éditions Place Stanislas, 2007), pp. 51–53.
Flobert, Laure-Paul, "La Femme et le costume masculin," *Le Vieux Papier* (Bulletin de la Société Archéologique, Historique et Artistique), July 1, 1911, pp. 349–366.
Floret, Robert, "Grand-Mère 'Casse-Cou,'" *Détective*, no. 751, November 18, 1960, pp. 10–11.
Foley, Michael, *Pioneers of Aerial Combat: Air Battles of the First World War* (2013).
Forest and Stream, "The World's Greatest Sportswoman," September 13, 1913, vol. 81, p. 323.
Forest City [SD] Press, "The Bride of Danger," April 17, 1914, p. 4.
Fradet, Camille, *L'Est Républicain*, "Aéroport régional: pourquoi pas Marie Marvingt?" September 23, 1989.
France, Geo, "Marie Marvingt espère fêter ses 90 ans ... dans la lune," *Ouest-France*, February 23, 1955.
France Aviation, "Nos lecteurs nous écrivent: Marie Marvingt," September 1957, p. C.
France Dimanche, "Les 'Vieilles Tiges' sont toujours vertes," January 2, 1954, p. 2.
François-Baillet, Brigitte, "La passion du ciel," *Est Magazine*, August 30, 2009, pp. 6–7.
Frémy, Dominique, et Michèle Frémy, *Quid* (Robert Laffont, 1988), p. 1166.
Gardner, Major Lester D., "Women Soon Took to Air: Their 'Firsts' in Various Fields of Avia-tion Are Recorded," *New York Times*, 21 July 1940.
Gateway Heritage: Quarterly Journal of the Missouri Historical Society, vols. 23–24, 2002, p. 66.
Le Gaulois, "Ascension aventureuse, de Nancy en Angleterre," October 28, 1909, p. 3.
Le Gaulois, "Nouveaux pilotes," November 13, 1910, p. 4.
Le Gaulois, "Petites nouvelles de la nuit," August 7, 1911, p. 33.
Le Gaulois, "L'Académie des Sports," November 11, 1911, p. 4.
Le Gaulois, "Chez Deperdussin à Courcy," July 28, 1912, p. 4.
Le Gaulois, "Petites nouvelles de la nuit," July 29, 1912, p. 3.
Le Gaulois, "Aux Ecole Deperdussin à Reims," July 30, 1912, p. 3.
Le Gaulois, "Les ballons de l'Aéro-Club," July 31, 1912, p. 6.
Le Gaulois, "Aux Ecoles Deperdussin à Bétheny," August 4, 1912, p. 4.
Le Gaulois, "Aux Ecoles Deperdussin," August 6, 1912, p. 3.
Le Gaulois, "Aux Ecoles Deperdussin—Aérodrome de la Champagne," August 8, 1912, p. 4.
Le Gaulois, "Aux Ecoles Deperdussin," August 13, 1912, p. 3.

Le Gaulois, "L'Ecole Deperdussin à Bétheny," August 22, 1912, p. 4.
Le Gaulois, "Aux Ecoles Deperdussin," September 8, 1912, p. 3.
Le Gaulois, "Aux Ecoles Deperdussin à Bétheny," September 17, 1912, p. 3.
Le Gaulois, "Aux Ecoles Deperdussin à Bétheny," September 21, 1912, p. 6.
Le Gaulois, "Aux Ecoles Deperdussin à Bétheny," September 23, 1912, p. 4.
Le Gaulois, "Aux Ecoles Deperdussin à Bétheny," September 26, 1912, p. 4.
Le Gaulois, "Aux Ecoles Deperdussin à Bétheny," September 28, 1912, p. 3.
Le Gaulois, "La Conquête de l'Air," September 30, 1912, p. 3.
Le Gaulois, "Aux Ecoles Deperdussin à Bétheny," October 2, 1912, p. 4.
Le Gaulois, "A Bétheny—Ecole Deperdussin militaire," October 11, 1912, p. 3.
Le Gaulois, "Aérodrome de la Champagne—Ecoles Deperdussin," October 31, 1912, p. 4.
Le Gaulois, "Chez Deperdussin," November 9, 1912, p. 4.
Le Gaulois, "Chez Deperdussin à Reims," November 23, 1912, p. 4.
Le Gaulois, "Aérodrome de la Champagne—Ecoles Deperdussin," December 6, 1912, p. 4.
Le Gaulois, "A l'Ecole Deperdussin, aérodrome de la Champagne," August 10, 1913, p. 4.
Le Gaulois, "Le 10e Grand Prix de l'Aéro-Club de France," July 19, 1914, p. 4.
Le Gaulois, "Le 10e Grand Prix de l'Aéro-Club de France," July 21, 1914, p. 5.
Le Gaulois, "La saison à Vittel," July 28, 1914, p. 5.
Le Gaulois, "Parmi ceux qui sont partis," August 7, 1914, p. 2.
Le Gaulois, "Le service du Duc de Chevreuse," February 5, 1918, p. 2.-
Le Gaulois, "Le meeting de Buc," October 9, 1920, pp. 1–2.
Le Gaulois, "Le baptême de l'air du cardinal Luçon," June 22, 1928, p. 4.
Le Gaulois, "Congrès international de l'aviation sanitaire," July 22, 1928, p. 5.
Gautier, Emile, ed., *Année scientifique et industrielle 1909* (Hachette, 1910), p. 92.
Gazette des Eaux, "Vittel: Aviation," January 6, 1912, p. 762.
La Gazette du Village, "Chute d'une aviatrice dans un café," August 13, 1911, p. 515.
Geelong [Australia] Advertiser, "Lady Aviator Runs Into Trouble," August 18, 1911, p. 4.
Géhin, Vincent, "Des frères Mongolfier à Marie Marvingt," *L'Est Républicain*, July 20, 2005.
Geoffroy, J.M., letter, March 29, 1983.
George, Jean-Claude, "La fiancée du danger," *Contes et légendes de la Meuse* (De Borée, 2009), pp. 407–411.
Georges, Jean-Luc, "Qui se souvient de Marie Marvingt?" *Ouest-France*, August 5, 1993.
Georges, Jean-Luc, "Marie Marvingt, le film," *L'Est Républicain*, June 18, 2005.
Gerbaud, A. Lorac, letter to Phillipe Bayart, June 28, 1988.
Gil Blas, "L'Eve moderne," August 4, 1906, p. 1.
Gil Blas, "Natation," July 12, 1907, p. 4.
Gil Blas, "Natation," July 17, 1907, p. 4.
Gil Blas, "La traversée de Toulouse," September 3, 1907, p. 4.
Gil Blas, "Patinage," January 30, 1909, p. 4.
Gil Blas, "Nouvelles de partout: Londres," October 28, 1909, p. 3.
Gil Blas, "Aviation," November 27, 1909, p. 4.
Gil Blas, "Les sports à Gérardmer," February 14, 1910, p. 3.
Gil Blas, "Aéronautique," June 21, 1910, p. 4.
Gil Blas, "Au meeting de Rouen," June 23, 1910, p. 4.
Gil Blas, "Aéronautique," June 25, 1910, p. 4.
Gil Blas, "Aéronautique," June 27, 1910, p. 3.
Gil Blas, "Le grand prix de l'Aéro-Club," June 29, 1910, p. 4.
Gil Blas, "Aéronautique," August 8, 1910, p. 3.
Gil Blas, "Aviation: Une aviatrice concourt," November 28, 1910, p. 3.
Gil Blas, "Mlle Marvingt vole 53 minutes," November 29, 1910, p. 3, 4.
Gil Blas, "Aviation," December 7, 1910, p. 4.
Gil Blas, "Aviation: A Mourmelon," December 9, 1910, pp. 3–4.
Gil Blas, "Aviation," December 29, 1910, p. 4.
Gil Blas, "Aviation: Au Camp de Châlons," December 31, 1910, p. 4.
Gil Blas, "Les sports d'hiver," February 16, 1911, p. 5.
Gil Blas, "Deux aviateurs blessés 'en automobile,'" August 4, 1911, p. 44.
Gil Blas, "Le meeting d'aviation à Saint-Etienne: Aviateurs victimes d'un accident d'automobile," August 5, 1911, p. 3.
Gil Blas, "L'aviation à Saint-Etienne," August 13, 1911, p. 2.
Gil Blas, "L'aviation en province," August 16, 1911, p. 4.
Gil Blas, "L'aviation," October 15, 1911, p. 4.
Gil Blas, "L'Académie des Sports," November 11, 1911, p. 4.
Gil Blas, "L'Académie des Sports," December 12, 1911, p. 5.
Gil Blas, "Le Dîner de l'Académie des Sports," December 16, 1911, p. 6.
Gil Blas, "Le Concours de ski du Club Alpin à Chamonix," February 8, 1912, p. 6.
Gil Blas, "Promenade d'aviatrice," September 19, 1912, p. 5.
Gil Blas, "Prouesse d'aviatrice," September 19, 1912, p. 5.
Gil Blas, "Aéronautique: le prix de distance," October 1, 1912, p. 5.
Gil Blas, "La Lorraine à Paris," December 9, 1912, p. 3.
Gil Blas, "Sports d'hiver: A Chamonix," January 21, 1913, p. 6.
Gil Blas, "Au meeting de Monaco," April 17, 1913, p. 6.
Gil Blas, "Mlle Marvingt se promène," October 26, 1913, p. 5.
Gil Blas, "Aéronautique et aviation: Les oiselles tombent," December 13, 1913, p. 5.
Gil Blas, "Le gala des aviatrices françaises," March 4, 1914, p. 4.
Gil Blas, "Le grand prix de l'Aéro-Club de France," July 10, 1914, p. 4.

Gil Blas, "Le grand prix de l'Aéro-Club de France," July 19, 1914, p. 5.
Gil Blas, "Le grand prix de l'Aéro-Club de France," July 20, 1914, p. 6.
Gil Blas, "Le grand prix de l'Aéro-Club de France," July 21, 1914, p. 5.
Giornale di Sicilia, "Una conferenza dell'aviatrice Marvingt al Circolo di Cultura," December 30, 1923.
Glenboro [Manitoba] Gazette, "A Plucky Woman," May 1, 1914, p. 5.
G.M., "Audacieux voyage en ballon: Interview de Mlle Marvingt," *L'Eclair de l'est*, October 30, 1909.
Godot, J.-H., "La vie de Marie Marvingt" et "Témoignage: A Nancy, sur les traces de Marie Marvingt: Rosalie, écrivain américain veut faire revivre 'la fiancée du danger,'" *L'Est Républicain*, June 25, 1982.
Godot, J.-H., interview, June 24, 1982.
Grahame-White, Claude, in collaboration with Harry Harper, *With the Airmen* (H. Frowde, Hodder and Stoughton, 1913), p. 47.
Grandidier, Gilbert, "Marie Marvingt (1875–1963), 'la fiancée du danger' ouvre aux femmes la voie des airs," newspaper article, May 1987.
"La grand-mère volante obtient, à quatre-vingt-six ans, le renouvellement de son brevet de pi-lote," newspaper article, October 1961.
Greenwell, Daisy, "The Top 10 Female Risk-Takers," *The London Times*, March 4, 2011, p. 40.
Grey, C.G., "The Utility of the Autogiro: A Debate," *The New York Times*, August 16, 1931.
Gygax, Georges, "La 'Fiancée du danger': La vie prodigieuse de Marie Marvingt," *L'Illustré*, April 24, 1958, pp. 34–36.
Gygax, Georges, "La Vie Prodigieuse de Marie Marvingt, la 'Fiancée du Danger,'" *L'Illustré*, May 1, 1958, pp. 40–42.
Hämmerle, Christa, Oswald Überegger, and Birgitta Bader-Zaar, *Gender and the First World War* (2014), p. 123.
Hammond Lake Country [IN] Times, "Fair French Women Rivals for Long Distance Aeroplane Flight Prize Offered by Paris Paper," December 27, 1910, p. 4.
Hammond Lake County [IN] Times, "Women to Fly in Chicago Meet," May 6, 1911, p. 3.
Hammond Lake County [IN] Times, "Lady-Birds Are Going to Take Wing at Squantum Aviation Meet Next Month; They'll Ask No Odds From Men Contestants," July 15, 1911, p. 4; *New Castle [PA] News*, "Lady-Birds Are Going to Take Wing at Squantum Aviation Meet Next Month; They'll Ask No Odds From Men Contestants," July 15, 1911, p. 7; *Sandusky [OH] Star Journal*, "Lady-Birds Are Going to Take Wing at Squantum Aviation Meet Next Month; They'll Ask No Odds From Men Contestants," July 15, 1911, p. 10.
Hartford Courant, "Famed Aviatrix Dies in France," December 16, 1963.
Hastings and St. Leonards Observer, "The Finest Sportswoman in the World," August 2, 1913, pp. 7–8
Hattiesburg [MS] American, "French Woman Flier Predicts Air Travel Will Predominate," November 16, 1928, p. 2; *Appleton [WI] Post Crescent*, "French Aviatrix Sees Big Flying Growth," December 8, 1928, p. 5; *Biloxi [MS] Daily Herald*, "Veteran French Aviatrix Sees Big Flying Growth," December 24, 1928, p. 14; *Borger Daily Herald [TX]*, "Veteran Aviatrix Sees Big Flying Growth," November 25, 1928, p. 7; *Cumberland [MD] Evening Times*, "Veteran French Aviatrix Sees Big Flying Growth," November 16, 1928, p. 11; *Davenport [IA] Democrat and Leader*, "French Aviatrix Sees Big Growth in Plane Usage," December 11, 1928, p. 10; *Decatur [IL] Review*, "Veteran Aviatrix Sees Flying Gain," November 16, 1928, p. 33; *Fairbanks [AL] Daily News Miner*, "Veteran French Aviatrix Sees Big Flying Growth," December 1, 1928, p. 5; *Gettysburg [PA] Times*, "Veteran French Aviatrix Claims Big Flying Growth," November 21, 1928, p. 2; *Hutchinson [KS] News*, "Veteran French Aviatrix Sees Big Flying Growth," December 22, 1928, p. 4; *Mason City [IA] Globe Gazette*, "Veteran French Aviatrix Sees Big Flying Growth," November 17, 1928, p. 6; *Ogden [UT] Standard Examiner*, "Vet Aviatrix Sees Big Flying Growth," November 25, 1928, p. 20; *Portsmouth [NH] Herald*, "Veteran French Aviatrix Sees Big Flying Growth," December 20, 1928, p. 11; *Wichita [TX] Daily Times*, "Veteran French Aviatrix Sees Big Flying Growth," November 23, 1928, p. 28.
Healy, Graham, *The Shattered Peloton: The Devastating Impact of World War I on the Tour de France* (2014), pp. 136–137.
Hébrard, Général, *L'Aviation des origines à nos jours* (Robert Laffont, 1954), p. 33.
Heinmuller, John P.V., *Man's Fight to Fly* (Aero Print Co., 1945), p. 276.
Hemardinquer, Didier, "Marie Marvingt crève l'écran," *L'Est Républicain*, March 20, 2005.
Hephaestus Books, *French Aviators, including: Roland Garros (aviator), Antoine de Saint-Exupéry, Henry Farman, Louis Blériot, Hélène Dutrieu, Marie Marvingt, Georges Guynemer, Jean-Pierre Blanchard, Didier Masson, Jean Mermoz, Raymonde de Laroche, Louis Charles Breguet* (2010), pp. 21–26.
Hephaestus Books, *French Balloonists, including: Montgolfier Brothers, Jacques Charles, Marie Marvingt, Henri Giffard, Camille Flammarion, Jean-Pierre Blanchard, Gaston Tissandier, Jean-François Pilâtre de Rozier, André-Jacques Garnerin, François Laurent d'Arlandes* (2010), pp. 11–16.
Hephaestus Books, *French Bobsledders, including: Marie Marvingt, Claude Brasseur, Bruno Mingeon, Emmanuel Hostache, Éric Le Chanony, Max Robert, Jean De Suarez d'Aulan, Jacques Bridou, Achille Fould (bobsleigh pilot), Henri Evrot, Robert Dumont, William Hirigoyen* (2010), pp. 1–6.
Hephaestus Books, *French Fencers, including: Eugène-Henri Gravelotte, Marie Marvingt, Jean-François Lamour, Brice Guyart, Laura Flessel-Colovic, Henri Callot, Henri Delaborde (fencer), Jean Maurice Perronet, Chevalier de Saint-George, Philippe Boisse, Régis Sénac* (2010), pp. 2–7'
Hephaestus Books, *French Mountain Climbers, including: Marie Marvingt, Gaston Rébuffat, Louis Lachenal, Maurice Herzog, Jean-Christophe Lafaille, Lionel Terray, Chantal Mauduit, Pierre Mazeaud, Alain Mesili, Marco Siffredi, Marcel Ichac, Benoît Chamoux, Jean Couzy* (2010), pp. 1–6.

Hephaestus Books, *French People Of World War I, including: Georges Clemenceau, Maurice Barrès, Théophile Delcassé, Aristide Briand, Marie Marvingt, Raymond Poincaré, Paul Painlevé, René Viviani, Hubert Lyautey, Madame Arno, Vincent de Moro-Giafferi, André Chevrillon* (2010), pp. 31–36.

Hephaestus Books, *French Sport Shooters, including: Marie Marvingt, Raphaël Poirée, Albin Lermusiaux, André De Schonen, Maurice Larrouy, Léon Moreaux, Eugène Balme, Roger de Barbarin, René Guyot, Justinien de Clary, Achille Paroche, Louis Duffoy, Maurice Lecog* (2010), pp. 1–6.

Hephaestus Books, *Women in War in France, including: Joan of Arc, Louise Labé, Susan Travers, Marie Marvingt, Isabella, Duchess of Lorraine, Geneviève de Galard, Louise Michel, Isabel of Conches, Pétroleuses, Renée Bordereau, Madame Arno, Pierronne, Caroline Aigle* (2010).

Historische, "Die französische Fliegerin Mlle. Marvingt," magazine clipping, 1910.

H.O., "Mlle Marvingt, la 'fiancée du danger,' nous révélera à Tunis les sources de son extraordi-naire dynamisme," *Le Petit Matin [Tunis]*, February 3, 1952.

The [Hobart, Tasmania] Mercury, "Gleanings from the World's Press," October 4, 1907 p. 2; *[Launceston, Tasmania] Examiner*, October 2, 1907, p. 3; *The Mercury [Australia]*, "Gleanings from the World's Press," October 4, 1907; *Riverine [Australia] Herald*, "Items of Interest," September 30, 1907. p. 2; *The [Rockhampton, Australia] Capricornian*, "Gossip," October 19, 1907, p. 5.

Holleran, Renee S., *ASTNA Patient Transport: Principles and Practice* (2009).

L'Homme libre, "Obsèques du commandant Félix," June 21, 1914, p. 2.

L'Homme libre, "Le grand prix de l'Aéro-Club: Mlle Marvingt atterrit sans encombre," July 21, 1914, p. 2.

L'Homme libre, "Premier congrès international de l'aviation sanitaire," October 11, 1928, p. 4.

L'Homme libre, "Les Amis de l'aviation sanitaire," July 10, 1929, p. 4.

L'Homme libre, "Le Challenge Capitaine-Eckeman," October 7, 1931, p. 4.

L'Homme libre, "Nouvelles aériennes: Mlle Marvingt à Argenteuil," January 10, 1933, p. 4.

L'Homme libre, "Rétrospective de l'Aviation féminine de 1910 à 1933," March 21, 1933, p. 2.

L'Homme libre, "Legion d'Honneur," January 26, 1935, p. 3.

L'Homme libre, "La Vie Sportive," November 5, 1937, p. 4.

L'Homme libre, "A l'Union des oeuvres de bienfaisance de l'Aéronautique," June 30, 1938, p. 4.

L'Homme libre, "Une intéressante conférence," December 19, 1938, p. 4.

Honolulu Star-Bulletin, "Activities of Women," June 30, 1915, p. 3; *The Broad Ax*, "Dames and Daughters, October 23, 1915, p. 2; *The Daily [OK] Ardmorite*, "Activities of Women," March 25, 1915, p. 7; *The Guthrie [OK] Daily Leader*, "Activities of Women," March 31, 1915, p. 3; *Hot Springs [SD] Weekly Star*, "Dames and Daughters," December 17, 1914, p. 7; *Milford [IA] Mail*, "Newsy Briefs About Woman's Activities," April 8, 1915, p. 3; *Racine [WI] Journal News*, "What Women Are Doing," July 17, 1915, p. 5.

Houard, Georges, "Marie Marvingt, pilote et propagandiste exemplaire," *Les Ailes*, January 1, 1964, p. 21.

Houart, Victor, and Edmond Petit, *Dictionnaire illustré de l'aviation* (Seghers, 1964), pp. 181–182.

Hoyt, Helen, "Helen Hoyt Has Instances of Daring of Women Aviators," *Gazette and Bulletin [Williamsport, PA]*, August 8, 1911, p. 7.

Huguenot, Vianney, "Notre Incroyable Amnésie," *L'Estrade*, no. 57, November 2015.

L'Humanité, "Paris à la nage," July 30, 1906, p. 2.

L'Humanité, "Natation," July 17, 1907, p. 4.

L'Humanité, "Les débuts d'une aviatrice à Mourmelon-le Grand," September 5, 1910, p. 4.

L'Humanité, "La grandmère (volante) fait (aussi) du vélo," no. 10, October 1961.

Huntingdon [PA] Daily News, "Aviation Is New Field for Women of Europe," May 3, 1926, p. 8.

Icare, "Les IPSA et les convoyeuses de l'air," no. 127, 1988, pp. 1–61.

The Illustrated London News, "Portraits and World's News," December 3, 1910, p. 862.

L'Illustration, "Les Sports d'Hiver," no. 3441, February 6, 1909, p. 95.

L'Illustration, "Une femme traverse la Manche en ballon," no. 3480, November 6, 1909, p. 335.

L'Illustration, "Une Ville d'Aviation," no. 3486, December 18, 1909, p. 460.

L'Illustration, "Les Dernières Journées d'Aviation en 1910," no. 3541, January 7, 1911, p. 4.

L'Illustration, "Un aéroplane dans un jeu de boules," no. 3572, August 12, 1911, p. 128.

L'Illustration "La Chute Mortelle du Capitaine Echeman," May 18, 1912, pp. 420–421.

L'Illustration, "Les Obsèques du Capitaine Echeman," no. 3613, May 25, 1912, p. 444.

L'Illustration, "Les Voyages Aeriens," February 15, 1919, no. 3963, p. 172.

L'Illustration, "Les Ailes Qui Sauvent," no. 4785, November 17, 1934, p. 33.

L'Illustration, "Course de dames en patin-ski a Chamonix," no. 4797, February 9, 1935, p. 170.

L'Illustration, "A Ski et à Chameau au coeur du Grand Erg Occidental," no. 4927, August 7, 1937, p. 456

L'Immeuble et la construction dans l'Est, "L'aviation et la future construction," January 30, 1910, p. 651.

Indépendant, "Mlle Marie Marvingt à Pau: Deux causeries littéraires et sportives," June 16, 1927.

Indianapolis Star, "Hopes to Win Prize for Sustained Flight," December 22, 1910, p. 2.

Indianapolis Star, "The Woman With War Wings," October 11, 1914, p. 55; *The Rocky Mountain News*, "The Woman With War Wings," October 11, 1914, p. 2.

L'Industrie vélocipédique, "Chute d'une aviatrice," December 20, 1913, p. 803.

L'Industrie vélocipédique, "L'accident de Mlle Marvingt," December 27, 1913, p. 816.

L'Intermédiaire des chercheurs et des curieux, "Femmes: Conquêtes des diplômes masculins," December 10, 1912, pp. 761.

International Women's Sports Hall of Fame, program, September 21, 1987.
Ireland, Norma Olin, *Index to Women of the World From Ancient to Modern Times: Biographies and Portraits* (F.W. Faxon, 1970).
Jacquemart, Claude, *Le Figaro*, "L'épopée des aventurières du ciel," December 26, 2003, p. 22.
Jansen, Marianne, "Marie Marvingt: 86 ans d'âge, 51 ans d'aviation," *Écho de la Mode*, no. 47, November 19, 1961, pp. 18–19.
J.B.L., "On peut fort bien 'être et avoir été,' démontre Marie Marvingt," *La Presse*, February 3, 1952.
J.G., "L'avion et le vélo: deux amours de Marie Marvingt, 'la fiancée du danger,'" *L'Est Républicain*, February 16, 1975, p. 3.
J.G., "Trois plaques pour la 'fiancée du danger': Marie Marvingt honorée dans sa ville," *L'Est Républicain*, June 8, 1984.
Le Journal [Paris], "La Fête de la Natation à la Sorbonne, December 10, 1906, p. 5.
Le Journal [Paris], "La Traversée de Paris à la Nage," July 29, 1906, p. 1.
Le Journal [Paris], "Paris à la Nage," July 30, 1906, p. 1.
Le Journal [Paris], "La Traversée de Paris à la Nage," July 7, 1907, p. 7.
Le Journal [Paris], "La Traversée de Paris à la Nage," July 12, 1907, p. 6.
Le Journal [Paris], "Un raid en ballon," October 28, 1909, p. 1.
Le Journal [Paris], "A propos d'une ascension," November 12, 1909.
Le Journal [Paris], "Les Aviateurs," November 23, 1909, p. 4.
Le Journal [Paris], "Seconde femme aviateur," June 18, 1910, p. 5.
Le Journal [Paris], "Concours de distance de l'Aéro-Club," June 22, 1910, p. 5.
Le Journal [Paris], "Les grandes manifestations sportives: Le concours de distance de l'Aéro-Club," June 26, 1910, p. 5.
Le Journal [Paris], "Le concours de l'Aéro-Club," June 27, 1910, p. 7.
Le Journal [Paris], "La course de distance de l'Aéro-Club," June 28, 1910, p. 7.
Le Journal [Paris], "Aviation," October 23, 1910, p. 6.
Le Journal [Paris], "Aviation," November 10, 1910, p. 5.
Le Journal [Paris], "Mlle Marvingt, détentrice de la Coupe d'Aviation Femina," November 29, 1910, p. 6.
Le Journal [Paris], "La Vie Sportive," December 6, 1910, p. 6.
Le Journal [Paris], "Records aériens," December 22, 1910. pp. 1–2.
Le Journal [Paris], "Pour les Coupes," December 31, 1910, p. 2.
Le Journal [Paris], "Dix Aviateurs ont, hier, tenté la conquête de records aériens," January 1, 1911, pp. 1–2.
Le Journal [Paris], "Sports d'Hiver," January 23, 1911, p. 6.
Le Journal [Paris], "Sports d'Hiver," January 26, 1911, p. 7.
Le Journal [Paris], "Sports d'Hiver," January 27, 1911, p. 7.
Le Journal [Paris], "La course militaire international de skis du Lioran," February 12, 1911, p. 4.
Le Journal [Paris], "Sports d'Hiver," February 23, 1911, p. 6.
Le Journal [Paris], "De l'aéroplane en ballon," March 4, 1911, p. 6.
Le Journal [Paris], "Aviateurs blessés en auto," August 3, 1911, p. 4.
Le Journal [Paris], "Dernières nouvelles sportives," August 7, 1911, p. 4.
Le Journal [Paris], "Aviation," August 14, 1911, p. 6.
Le Journal [Paris], "Un meeting d'Aviation fertile en incidents," August 15, 1911, p. 4.
Le Journal [Paris], "Nos aviateurs volent à Toul," April 15, 1912, p. 4.
Le Journal [Paris], "M. Raymond Poincaré inaugure le Champ d'Aviation de Bar-le-Duc," July 1, 1912, p. 4.
Le Journal [Paris], "Le concours d'atterrissage," July 6, 1912, p. 6.
Le Journal [Paris], "Les ballons de l'Aéro-Club," July 31, 1912, p. 5.
Le Journal [Paris], "Aviation," August 13, 1912, p. 7.
Le Journal [Paris], "De Limours à Beauville, September 19, 1912, p. 7.
Le Journal [Paris], "Le prix d'aéronautique," September 29, 1912, p. 5.
Le Journal [Paris], "Aviation," December 1, 1912, p. 8.
Le Journal [Paris], "Le patinage en Savoie," January 20, 1913, p. 4.
Le Journal [Paris], "A l'aérodrome de la Champagne," August 9, 1913, p. 5.
Le Journal [Paris], "Mlle Marvingt blessée par une chute d'aéroplane," December 13, 1913, p. 4.
Le Journal [Paris], "Des nouvelles de Mlle Marvingt," December 17, 1913, p. 8.
Le Journal [Paris], "Le grand-prix de l'Aéro-Club," July 8, 1914, p. 7.
Le Journal [Paris], "Vingt-quatre ballons partent aujourd'hui des Tuileries," July 19, 1914, p. 3.
Le Journal [Paris], "Le grand-prix de l'Aéro-Club," July 21, 1914, p. 9.
Le Journal [Paris], "La triomphale arrivée des autoschenilles qui ont traversé le Sahara," March 6, 1923, pp. 1–2.
Le Journal [Paris], "L'Arrivée à Biskra de la mission Haardt and Audouin," March 10, 1923, p. 3.
Le Journal [Paris], "La coupe de France de ski," January 29, 1928, p. 5.
Le Journal [Paris], "La mode pour aviatrices," September 19, 1928, p. 5.
Le Journal [Paris], "Le 1er Congrès de l'aviation sanitaire," May 16, 1929, p. 4.
Le Journal [Paris], "Comment s'habillent nos aviatrices," May 19, 1931, p. 4.
Le Journal [Paris], "Une conférence sur le rôle bienfaisant de l'avion sanitaire dans nos colonies," May 28, 1931, p. 4b.
Le Journal [Paris], "Le maréchal Lyautey à l'Aéro-Club de France, October 8, 1931, p. 5.
Le Journal [Paris], "M. J.-L. Dumesnil expose aux 'Vieilles Tiges' l'ensemble du programme du ministère de l'air," November 18, 1931, p. 3.
Le Journal [Paris], "A vol d'oiseau," June 7, 1932, p. 4.
Le Journal [Paris], "Le docteur Crochet parcourt

15.000 kilomètres pour participer à des congrès médicaux," July 3, 1932, p. 5.
Le Journal [Paris], "M. Pierre Cot est venu à La Baule accueillir les concurrents du Tour de France des avions de tourisme partis le matin de Biarritz," July 28, 1933, pp. 1, 4.
Le Journal [Paris], "L'aviation sanitaire au Grand-Palais," November 30, 1934, p. 3.
Le Journal [Paris], "Légion d'honneur," January 26, 1935, p. 4.
Le Journal [Paris], "Au Club du Faubourg," December 6, 1936, p. 7.
Le Journal [Paris], "Au Club du Faubourg," January 9, 1937, p. 7.
Le Journal [Paris], "Au Club du Faubourg," October 27, 1937, p. 7.
Le Journal [Paris], "A l'Élysée," January 12, 1939, p. 4.
Le Journal Amusant, "Le grand prix de l'Aéro-Club de France," July 2, 1910, p. 14.
Le Journal Amusant, "Commission d'aviation," December 3, 1910, p. 14.
Le Journal Amusant, "Pour la Coupe Femina," December 31, 1910, p. 14.
Le Journal Amusant, "Mlle Marvingt, aéronaute, va en sphérique de Paris à Bruxelles," August 10, 1912, p. 14.
Le Journal Amusant, "Aérostation," January 11, 1913, p. 14.
Journal de la Jeunesse, "Athlétisme," no. 1759, August 18, 1906.
Journal Officiel de la République Française, "Officiers d'Académie," November 19, 1911, p. 9207.
Journal des débats politiques et littéraires, "La traversée de Paris à la nage," July 30, 1906, p. 3.
Journal des débats politiques et littéraires, "Alpinisme," January 10, 1908, p. 3.
Journal des débats politiques et littéraires, "Alpinisme," January 24, 1909, p. 3.
Journal des débats politiques et littéraires, "Alpinisme," January 29, 1909, p. 3.
Journal des débats politiques et littéraires, "Aérostation," October 29, 1909, p. 3.
Journal des débats politiques et littéraires, November 24, 1909, p. 3.
Journal des débats politiques et littéraires, "Aviation," June 19, 1910, p. 3.
Journal des débats politiques et littéraires, "Aviation," October 24, 1910, p. 3.
Journal des débats politiques et littéraires, "Aviation," November 29, 1910, p. 3.
Journal des débats politiques et littéraires, "Une aviatrice tombe dans un jeu de boules," August 8, 1911, p. 3.
Journal des débats politiques et littéraires, "Sports divers," November 12, 1911, p. 3.
Journal des débats politiques et littéraires, "Manifestation patriotique," March 16, 1912, p. 3.
Journal des débats politiques et littéraires, "Conférences de Mlle Marvingt," June 13, 1912, p. 2.
Journal des débats politiques et littéraires, "Meeting pour aviation," July 3, 1912, p. 2.
Journal des débats politiques et littéraires, "L'Aviation au jour le jour," August 20, 1912, p. 2.
Journal des débats politiques et littéraires, "Prouesse d'aviatrice," September 19, 1912, p. 4.
Journal des débats politiques et littéraires, "L'Aviation," October 3, 1913, p. 4.
Journal des débats politiques et littéraires, "Chute de Mlle Marvingt," December 14, 1913, p. 2.
Journal des débats politiques et littéraires, "Obsèques d'Hubert Latham," January 18, 1914, pp. 2–3.
Journal des débats politiques et littéraires, "Les obsèques du commandant Félix," June 22, 1914, p. 4.
Journal des débats politiques et littéraires, "Le Grand Prix de l'Aéro Club," July 21, 1914, p. 4.
Journal des débats politiques et littéraires, "Le Grand Prix de l'Aéro Club," July 22, 1914, p. 5.
Journal des débats politiques et littéraires, "Les obsèques du capitaine Féquant," September 11, 1915, p. 3.
Journal des débats politiques et littéraires, "L'aviation sanitaire," July 7, 1929, p. 5.
Journal des débats politiques et littéraires, "A l'Exposition Coloniale: Les Journées d'aviation sanitaire," July 26, 1931, p. 4.
Journal des débats politiques et littéraires, "Le congrès d'aviation sanitaire coloniale," August 1, 1931, p. 2.
Journal des débats politiques et littéraires, "Légion-d'Honneur," January 27, 1935, p. 2.
Journal des débats politiques et littéraires, "La Vie Aérienne," June 9, 1935, p. 5.
Journal des débats politiques et littéraires, "A l'Elysée," January 13, 1939, p. 3.
Journal de la Meurthe et des Vosges, "Une conference sur l'aviation," March 20, 1914, p. 3.
Journal de la Meurthe et des Vosges, "Deux heures dans les airs," March 28, 1914, p. 3.
Journal de la Meurthe et des Vosges, "Ambulance de la place St-Georges," December 30, 1915, p. 2.
Journal de la Meurthe et des Vosges, "Etat-Civil quotidien de Nancy du 10 mars," March 12, 1916, p. 2.
Journal Officiel de la République Française, "Officiers d'Académie," November 19, 1911, p. 9207.
Journal Officiel de la République Française, "Ministère de l'Air: dons et legs., Officiers d'Académie," July 2, 1932, p. 7200.
Journal Officiel de la République Française, "Légion d'Honneur," January 26, 1935, p. 871.
J.P., "Marie Marvingt, pilote d'avion, est morte à 88 ans: Championne en tous genres," *Le Monde*, December 17, 1963, p. 13.
J.-P. G., "Le Portrait du Jour: Mlle Marie Marvingt," *L'Eclair de l'Est*, July 4, 1948.
Julliot, Ch.-L., "Une Ligue pour la propagande de l'Aviation Sanitaire," *Les Ailes*, February 13, 1930, p. 7.
Julliot, Ch.-L., "Les onze voeux de l'Aviation Sanitaire," *Les Ailes*, September 17, 1931, p. 7.
La Justice, "Nageuses," August 4, 1906, p. 1.
Kalgoorlie [Australia] Miner, "Two Hundred Flying Men; Growing Army of Aviators; 'Stars' and Their Pupils," May 5, 1910, p. 8; *Northam [Australia] Courier*, "Two Hundred Flying Men: Growing Army of Aviators," May 13, 1910, p. 4.
Kernel, Hélène, "Marie Marvingt, doyenne des aviatrices," *France Aviation*, August 1957, p. 8.
Khirouni, Chaynesse, Députée de Meurthe-et-Moselle, letter to the International Marie Marvingt Committee, December 17, 2013.

Kingston [Jamaica] Gleaner, "Aerial Flights," December 24, 1910, p. 6.

Knecht, Marcel, multiple communications and copies of Marie Marvingt's letters to him dated October 30, 1954, and January 14, 1962.

Koskenmaki, Rosalie, "Queen of the Air," *Jack and Jill*, December 1985, pp. 28–31

Koskenmaki, Rosalie, "The Fiancée of Danger," *Women Sports and Fitness*, February 1986, pp. 27–29, 54.

Koskenmaki, Rosalie, "An Amazing Sport," *Young American*, vol. 4, no. 18, September 1987, p. 18.

Krysaniac, C., *Le Cartophile de Meurthe-&-Moselle*, bulletin de liaison du Cercle Cartophile de Meurthe & Moselle, no. 1, February 1983.

Ladame, Albert, "Petite Histoire de l'Aviation Sanitaire," *Air ANSORAA*, no. 133, April-May-June 2013, pp. 15–18.

Ladame, Albert, "Mémoires de Lorraine: Marie Marvingt," *Air ANSORAA*, no. 137, April-May-June 2014.

Laget, Françoise et Serge, and Jean-Paul Mazot, with the collaboration of Elizabeth Foch, *Le Grand Livre du Sport Féminin* (FMT éditions, 1982), pp. 252, 269, 275, 296, 340, 369, 441, 462, 487, 495, 496.

Lahausse, Jean-Bernard, and Romain Sertelet, "Biographie du mois: Marie Marvingt, 'La Femme la Plus Décorée du Monde," March 2013, http://verdun-meuse.fr.

Lallier, Roger, "Par-Dessus les Moulins," *VU: La Journal de la Semaine*, no. 61, May 15, 1929, pp. 390–391.

Lam, David M., "Marie Marvingt and the Development of Aeromedical Evacuation," *Aviation, Space, and Environmental Medicine*, vol. 74, no. 8, August 2003, pp. 863–868.

Lancashire Evening Post, news item, November 29, 1910, p. 7.

Langeron, André, "La Femme et l'aviation," *Les Ailes*, November 10, 1932, p. 4.

La Lanterne, "Paris à la nage," July 31, 1906, p. 2.

La Lanterne, "Sports d'hiver: A Chamonix," January 30, 1909, p. 3.

La Lanterne, "Les aviateurs français," June 19, 1910, p. 4.

La Lanterne, "Le Grand Prix de l'Aéro-Club de France," June 30, 1910, p. 4.-

La Lanterne, "Une aviatrice vole 53 minutes," November 30, 1910, p. 4.

La Lanterne, "Aéronautique," December 8, 1910, p. 4.

La Lanterne, "Pour la Coupe Femina," January 1, 1911, pp. 1, 2.

La Lanterne, "De l'aéroplane au ballon sphérique," March 5, 1911, p. 4.

La Lanterne, "Incidents à un meeting d'aviation," August 17, 1911, p. 4.

La Lanterne, "Un nouveau champ d'aviation," July 2, 1912, p. 2.

La Lanterne, "Aéronautique: Paris-Bruxelles," July 30, 1912, p. 3.

La Lanterne, "Les voyages en sphérique," August 2, 1912, p. 3.

La Lanterne, "Les vols de Mlle Marvingt," December 23, 1912, p. 4.

La Lanterne, "Un vol de Mlle Marvingt," October 3, 1913, p. 4.

La Lanterne, "Chute de Mlle Marvingt," December 12, 1913, p. 3.

La Lanterne, "Un raid de l'aviateur Marvingt," October 26, 1913, p. 3.

La Lanterne, "La Course des ballons," July 20, 1914, p. 3.

Laprévote, Charles, "On a retrouvé Zéphirine," *L'Est Républicain*, February 1, 1990.

Lassalle, Émile J., *Les cent premiers aviateurs brevetés au monde et la naissance de l'aviation* (Nauticaero, 1960), p. 21.

Larue, Michèle, "Marie Marvingt: Reine du ciel et aventurière des sables/Queen of the skies and desert adventuress," *RAM (Royal Air Maroc) Magazine*, no. 136, March-April 2006, pp. 96–104.

Laurent, Michel-Yves, "Marie Marvingt, surnommée 'Marie Casse-Cou,' traversa au début du siècle, la mer du Nord en ballon; elle pilote aujourd'hui à 86 ans un hélicoptère à réac-tion," *L'Aurore* or *Paris Press*, no. 16, November 2, 1960.

Lauwick, Hervé, *Heroines of the Sky (Conquérantes du Ciel)*, trans. from French by James Cleugh (Frederick Muller, 1960), pp. 25, 28.

Lawson, Eric, and Jane Lawson, *The First Air Campaign: August 1914-November 1918* (2007), p. 57.

The Leader [Orange, Australia], "Balloonists in a Gale, 60 Miles an Hour," September 11, 1914, p. 1.

Leblond, J.-J. "Pour le 52e anniversaire de sa première ascension en ballon Marie Marvingt a piloté un avion sanitaire entre Bordeaux et Marseille," *Dauphiné*, November 1952, p. 3.

Lebow, Eileen F., *Before Amelia* (Brassey's, 2002), pp. 34–45, 283.

Leclerc, Jean, "Le vélo de Marie Marvingt," *CAIN*, March 10, 2016.

Lecornu, Joseph Louis, *La Navigation aérienne: histoire documentaire et anecdotique* (1913), p. 306.

Lederlé, Olivier, "Marie Marvingt, femme d'un siècle," *Républicain lorrain*, November 2, 1991.

Légion d'Honneur, La Grande Chancellerie de la Légion d'Honneur: documents supporting and confirming her grades of Chevalier and then Officer of the Légion d'Honneur, 1935 and 1950.

Lejeune, Dominique, *La France de la Belle Époque 1896–1914* (Armand Colin, 1991), p. 110.

Lexa-Chomard, Annette, *Lucien Cuénot, L'intuition naturaliste* (2004), p. 56–57.

La Liberté, "Marie Marvingt est morte à 88 ans," no. 6164, December 16, 1963, pp. 1 & 8.

Liégeois, Achille, "Les Nancéiennes loin de chez elles: Mlle Marie Marvingt rentre d'un long voyage: Elle nous raconte ses impressions," *L'Est Républicain*, August 23, 1926, p. 1.

Lienhart, M. le Professeur Robert, "Rapport: Prix Jeanne Goury 1961," January 14, 1962.

Ligue aéronautique de France, "Les Conférences de Mlle Marvingt," May 1914, p. 75.

Lincolnshire [England] Echo, "The First Sportswoman in the World," August 13, 1913, p. 4.

Logansport [IN] Pharos Reporter, "Dames and Daughters," October 12, 1914, p. 9.

London Daily News, "Lady's Thrilling Story," October 28, 1909, p. 5.

London Evening News, "Aerial Adventure," October 28, 1909, p. 16.
London Evening News, "Ladies' Flying Meeting: First International Contest Among Women at the Hendon Aerodrome," July 1, 1912, p. 7.
London Evening News, "Abroad," December 13, 1913, p. 1.
The London Monitor and New Era, "Lady Aviator's Record," December 10, 1910, p. 2.
London Standard, "Women and Aviation: Some Notable Achievements," January 12, 1912, p. 13.
London Standard, "Women Aviators' Meeting," June 29, 1912, p. 11.
London Standard, "Ladies in the Air: First International Contest for Women," July 2, 1912, p. 10.
Long Beach [CA] Independent, "Still Learning at 80," April 5, 1955, p. 7; *Anniston [AL) Daily Star]*, "Learning at 80," April 21, 1955, p. 32; *Bryan [TX] Daily Eagle*, "Learning at 80," April 25, 1955, p. 5; *Burlington [NC] Daily Times News*, "Aviation Pioneer Learns Something New," April 6, 1955, p. 5; *Carbondale Southern Illinoisan*, "Woman Takes Up Helicopter at 80," April 5, 1955, p. 16; *Chanute [KS] Tribune*, "80-Year-Old Woman Pilot," April 5, 1955, p. 10; *Chester [PA] Times*, "Learning at 80," April 4, 1955, p. 43; *Chillicothe [MO] Constitution Tribune*, "80-Year-Old Woman Tries 'Copter," April 12. 1955, p. 12; *Clearfield [PA] Progress*, "Learning at 80," April 27, 1955, p. 2; *Cumberland [MD] Evening Times*, "Never Too Old to Learn, She Says," April 4, 1955, p. 1; *Dixon [IL] Evening Telegraph*, "Learning at 80," April 18, 1955, p. 10; *Findlay [OH] Republican Courier*, "Learning at 80," April 28, 1955, p. 13; *Lubbock [TX] Morning Avalanche*, "Learning at 80," April 22, 1955, p. 2; *Mason City [IA] Globe Gazette*, "Learning at 80," May 18, 1955, p. 32; *Ottawa [KS] Herald*, "Marie Marvingt," April 5, 1955, p. 2; *San Rafael [CA] Daily Independent Journal*, "Learning at 80," June 16, 1955, p. 22; *Sikeston [MO] Daily Standard*, "80-Year-Old Woman Tries 'Copter," April 5, 1955, p. 1; *Snyder [TX]Texas) Daily News*, "Learning at 80," May 5, 1955, p. 6; *Thomasville [GA] Times Enterprise*, "Learning at 80," April 28, 1955, p. 8.
Le Lorrain, "Deux anciennes du pensionnat Sainte-Chrétienne: 'La fiancée du danger' and la soeur Blanche," May 11, 1950.
Los Angeles Herald, "Midwinter Aviators," December 26, 1910, p. 4.
Los Angeles Times, "Woman War Flyer in Los Angeles," October 1, 1935.
M.A., *Samedi Soir*, "Marie Marvingt, 'fiancée du danger,' attend encore que soit relevé son 'défi mondial,'" no. 505, March 3, 1955, pp. 1, 3.
Maggio, Rosalie, "Daredevil Marie," *Cricket*, August 1988, pp. 27–31; "Daredevil Marie, Part 2" *Cricket*, September 1988, pp. 50–55; "The Shooting Star," *Cricket*, October 1988, pp. 42–47.
Maggio, Rosalie, and Marcel Cordier, *Marie Marvingt: La femme d'un siècle* (Éditions Pierron, 1991).
Maguin, Frédéric, ed., *Femmes célèbres de Nancy* (Éditions Koidneuf, 2007), pp. 28–29.
Maire, Jacqueline, letters, July 5, 1982; July 23, 1983; July 29, 1982; August 5, 1982.
Marcacci, Philippe, "Marie Marvingt redécolle," *Est Magazine*, June 27, 2004, pp. 10–11.
Marchal, Mireille, "Du ciel à la Carrière pour la 'Fiancée du danger," *Republicain Lorrain*, June 8, 1984.
Marchand, Robert, *Le ciel n'a pas de toit* (Editions Berger Levrault, 1962), pp. 258–262.
Marchand, Robert, "Le ciel n'a pas de toit," *France Aviation*, July 1962, p. 8.
Marchand, Robert, "Les conquérantes du ciel," *Le petit echo de la mode Magazine*, no. 17, April 25, 1964, pp. 46–49.
Les Marches de l'Est, "Pour le Lieutenant Caumont," June 1913, p. 234.
Marck, Bernard, "Marie Marvingt," *Aéroports Magazine*, 1992.
Marck, Bernard, *Dictionnaire universel de l'Aviation* (Tallendier, 2005), pp. 688–689.
Marck, Bernard, *Elles ont conquis le ciel* (Arthaud, 2009), pp. 14–21/Marck, Bernard, *Women Aviators* (Flammarion, 2013), pp. 14–21, 27, 39, 40, 43.
Marck, Bernard, "Ces merveilleux fous volants," *Le Figaro Histoire*, no. 21, August-September 2015, pp. 42–51.
Margot, Olivier, "L'Indomptable: Marie Marvingt," *L'Équipe*, August 17, 2013, pp. 60–63.
Marie France, "La Fiancée du danger' lance un défi aux femmes du monde entier," June 22, 1948, p. 7.
"Marie Marvingt: 56.000 kilomètres en Afrique du Nord," North African newspaper, April 16, 1928.
"Marie Marvingt, surnommée 'la fiancée du danger,' parlera vendredi à 18 heures," Algerian newspaper, December 19, 1951.
"Marie Marvingt," circa 1955, French aviation publication, pp. 41–42.
"Marie Marvingt, Éleve-pilote à 80 ans," newspaper article, 1955.
"Marie Marvingt, inventeur de l'aviation sanitaire, est morte à Nancy: Une grande sportive, une grande Française," newspaper article, December 16, 1963.
"Marie Marvingt, 'la fiancée du danger,' est décédée à l'âge de 88 ans," newspaper article, De-cember 18, 1963.
"Marie Marvingt: quatre fois vingt ans!" newspaper article, February 25, 1975.
"Marie Marvingt, La fiancée du danger," newspaper article, circa 1988.
Martin-Auberdon, Astrid, "Marie Marvingt, une femme résistante célébrée," *Sud Ouest*, May 20, 2016.
Marty, le Général, homage to Marie Marvingt, *Ouest Éclair*, June 1, 1928.
Marvingt, Marie, "Naissance de Marie Félicie Élisabeth Marvingt," birth certificate from Auril-lac (Cantal) France.
Marvingt, Marie, "La mer du Nord traversée en ballon en pleine tempête au milieu de la nuit," *Le Petit Journal*, no. 17132, November 22, 1909, pp. 1–2.
Marvingt, Marie, "Winter Sports at Chamonix," *The Times [London]*, February 2. 1909, p. 17.
Marvingt, Marie, "The Alps in Winter, I," *The Times [London]*, December 14, 1909.
Marvingt, Marie, "The Alps in Winter, II," *The Times [London]*, December 28, 1909.
Marvingt, Marie, "Traversée nocturne de la Mer du Nord en Ballon," *L'Aérophile*, vol. 18, no. 1, January 1, 1910, pp. 17–20.

Marvingt, Marie, "Les sports d'hiver dans les Vosges," *Le Sport*, March 5, 1910, p. 1.
Marvingt, Marie, "Mon vol le plus émouvent," *La vie au grand air*, no. 638, December 10, 1910, p. 897.
Marvingt, Marie, "Les femmes et le ski," en Louis Magnus et R. de la Fregeolière, *Les Sports d'Hiver* (Éditions Pierre Lafitte et Cie, 1911; Éditions Slatkine, 1979), pp. 176–181.
Marvingt, Marie, "Dramatique traversée de la mer du Nord," *La Culture Physique*, January 1, 1911, pp. 7–12.
Marvingt, Marie, "Sky Women," *Collier's*, September 30, 1911, p. 15.
Marvingt, Marie, "Les obsèques du commandant du Plessis à Palerme," *La Dépêche Tunisienne*, January 1914.
Marvingt, Marie, "En ballon, de Paris à la mer d'Irlande: Mlle Marvingt raconte à *L'Éclair de l'Est* son intrépide voyage aérien," *L'Éclair de l'Est*, July 22, 1914.
Marvingt, Marie, "Dans Nancy bombardée," *Samedi Soir*, January 29, 1916.
Marvingt, Marie, "Un Défi Mondial!," copy with Marie's annotations on it, 1922.
Marvingt, Marie, "La 'Fiancée du danger" lance un défi aux femme du monde entier, newspaper article, 1922.
Marvingt, Marie, "A propos de la randonnée aérienne de S. Em. le cardinal Luçon," *La Croix*, no. 13897, June 22, 1928, p. 1.
Marvingt, Marie, "Sur les Alpes l'hiver: Au dessus de Mt Blanc en avion—La saison à Chamonix," *Lectures Pour Tous*, January 1929, pp. 46–53.
Marvingt, Marie, signed autograph statement, Paris, June 5, 1931.
Marvingt, Marie, "L'Aviation Sanitaire et les Femmes," *Revue aéronautique de France*, September-October 1931, pp. 4, 5–6.
Marvingt, Marie, "L'Aviation sanitaire coloniale aux derniers congrès d'aéronautique," *L'Aérophile*, vol. 26, no. 11, November 15, 1931, pp. 321, 345.
Marvingt, Marie, "Les ailes féminines au service de la charité," *La Croix*, no. 4956, November 28, 1931, pp. 1–2.
Marvingt, Marie, "Souvenirs de Chalcidique tirés de mon carnet de route: A propos du tremble-ment de terre de Grèce," *L'Éclair de l'Est*, October 1, 1932, p. 1.
Marvingt, Marie, "Patrons et patronnes des aviateurs et aviatrices," *La Croix*, no. 15552, November 1–2, 1933, p. 3.
Marvingt, Marie, "La Femme et l'aviation au service de la charité: les infirmières-convoyeuses de l'Aviation Sanitaire," *Les Ailes*, December 21, 1933, p. 11.
Marvingt, Marie, "Nos maréchaux de France vivants," *Petit Marocain*, September 26, 1934.
Marvingt, Marie, "Deux grands Lorrains réunis dans la mort," *Petit Marocain*, October 18, 1934.
Marvingt, Marie, "Leurs femmes," *La Vigie Marocaine*, October 22, 1934.
Marvingt, Marie, "Anniversaire de fondation de l'A.S.C.A.M.," *La Vigie Marocaine*, January 30, 1935.
Marvingt, Marie, "Comment j'ai crée l'A.S.C.A. (Aviation Sanitaire Civile au Maroc)," *L'Echo des Ailes*, June 12, 1935, p. 5.
Marvingt, Marie, "Un avion de la Swissair effectue au Havre une évacuation sanitaire," *Les Ailes*, August 8, 1935, p. 7.
Marvingt, Marie, "L'Aviation et la jeunesse des écoles en Amérique," *Revue aéronautique de France*, December 1936, pp. 2–3.
Marvingt, Marie, "Maryse Bastié, la grande championne des Ailes Françaises, *Revue aéro-nautique de France*, February-March 1937, pp. 2–3.
Marvingt, Marie, "L'hommage d'une aînée: Pauvre Amélia," seven handwritten pages apparent-ly never published, 1937, in the keeping of the International Marie Marvingt Committee.
Marvingt, Marie, "Trente minutes avec la trinité du banquet royal à Versailles," *La Croix*, no. 17016, July 28, 1938, p. 3.
Marvingt, Marie, "Mon filleul, le général Féquant," *Excelsior*, December 29, 1938.
Marvingt, Marie, "Comment j'ai conçu le premier avion sanitaire," *Revue de la Société Scientifique et Historique de Documentation Aérienne*, édition de guerre, no. 6, January-February 1940.
Marvingt, Marie, "L'Aviation sanitaire mondiale," report no. 1262m speech given at the 3rd Congrès National de L'Aviation Française, April 1–5, 1947.
Marvingt, Marie, "Sagesse d'une vieille messine: En marge de la croisade de l'amabilité," *Le Lorrain*, May 11, 1950.
Marvingt, Marie, "Trois morts mettent en deuil les ailes de Lorraine: Hommage d'une 'vieille tige,'" *Dimanche Éclair*, September 17, 1950, p. 3.
Marvingt, Marie, "Soeur Emile de la Doctrine Chrétienne: Une vie au service des orphelines," newspaper article, c. 1950.
Marvingt, Marie, handwritten postcards to: Marthe Roumier (April 27, 1908; February 12, 1909; August 10, 1914; May 11, 1915; December 23, 1916; April 29, 1922; July 7, 1922; Au-gust 26, 1922; August 20, 1923; July 10, 1925; August 14, 1926; July 13, 1934; April 27, 1936; December 29, 1937; June 28, 1941); Jeanne Roumier Lhérault and/or Louis Lhé-rault (February 10, 1922; February 13, 1934; April 18, 1935; August 25, 1960); Mesde-moiselles Winsbach (November 13, 1951; December 28, 1951; April 5, 1957; July 21, 1957); Dr. and Mrs. Thévenin (January 19, 1952; May 15, 1955); Marcel Knecht (October 30, 1954; January 14, 1962); Marie Thérèse and her husband (November 18, 1917); Madame Bouchet (November 1, 1941); Mademoiselle Haller (July 2, 1945); Claire (Jan-uary 19, 1956); Odile (January 19, 1956); Madame Coulet (June 23, 1958); Madame Goupit (1922); unidentified (January 24, 1912; July 21, 1919; March 21, 1940); cousins (December 28.1951; January 22, 1955; April 17, 1957; November 11, 1958); Renée Enselme-Trichard (undated); Marshal Graziani. There are a few of the postcards that have turned up over the years; most are in the keeping of the International Marie Marvingt Committee.
Marvingt, Marie, photocopy of the invitation handwritten by Marie, L'Imprimerie Berger-Levrault, Nancy and Paris.
Marvingt, Marie, Livres d'Or autograph books, whereabouts unknown; some autographs were seen by journalists and reported; a few pages

appear to have been photocopied; the ma-jority of these autographs are lost.

Marvingt, Marie, dossier of documents found in April 2012 and in the keeping of the Interna-tional Marie Marvingt Committee: September 1, 1908, laissez-passer in German for her as a journalist; July 18, 1917, request to see photos for her conferences in Milan; October 22, 1917, driver's license; March 1, 1921, the Council President invites her to come by "after 10:00."; August 24, 1921, safe conduct to German cities; December 5, 1926, safe conduct for several trips abroad; August 7, 1941, regulations stating that her conferences can contain nothing of a political nature; August 9, 1941, handwritten note from Colonel Blassell about gas rations; August 10, 1941, letter advising her to strictly avoid any issues in her conferences that might disoblige France's enemies; August 28, 1941, Pétain thanks her for a donation to a women's sports event; September 3, 1941, friendly note from Colonel Blasselle; November 15, 1941, the "censure" bureau of Périgueux authorizes her to hold conferences there; February 5, 1942, Nice similarly authorizes her; February 13, 1942, Aviation Ministry authorizes her to continue her conferences; February 14, 1942, press bureau in Nice also authorizes her conferences there; May 7, 1942, another note from press bureau in Nice; April 29, 1943, letter from the national director of education and sports telling her which cities have improved their swimming facilities; January 6, 1944, French and German authorities both authorize her to speak in the Angoulème area; May 5, 1944, special authorization given to travel and speak in prohibited zones; October 4, 1951, she is authorized to speak in Algeria; January 9, 1952, she is authorized to speak in Tunisia; January 18, further authorization to speak in Tunisia.

Marvingt, Marie, telegrams between the Prefect of Lyon and the prefect of Nancy, and letters related to her request for an Italian visa, October 26, 1917; in the keeping of the Interna-tional Marie Marvingt Committee.

Marvingt, Marie, death certificate, December 14, 1963.

Marvingt, Marie, Félicie, Elisabeth, extensive but unsourced chronology of her achievements and awards.

Masson, Lucien, "Programme des Spectacles: Marie Marvingt," May 29, 1993.

Master, Nancy Robinson, *Airplanes* (2008), p. 12.

Mathis, Suzy, "Aviatrices d'hier et d'aujourd'hui: Marie Marvingt," *Aviation Française*, May 8, 1946, p. 9.

Le Matin, "Paris à la nage no. 2," July 30, 1906, p. 4.

Le Matin, "Paris à la nage amateurs," July 17, 1907, p. 6.

Le Matin, "La vie sportive: Patinage," January 29, 1909, p. 5.

Le Matin, "Sous le ciel d'Egypte, un aviateur tombe de 45 mètres," February 2, 1910, p. 1.

Le Matin, "Dernières nouvelles sportives: Aéronautique," April 24, 1910.

Le Matin, "Une femme aviateur obtient son brevet," June 18, 1910, p. 5.

Le Matin, "Femmes aéronautes," June 19, 1910, p. 4.

Le Matin, "Le Grand Prix de l'Aéro-Club de France," June 27, 1910, p. 5.

Le Matin, "Chez les hommes oiseaux," July 29, 1910, p. 4.

Le Matin, "Mlle Marvingt, pilote brevetée," October 23, 1910, p. 5.

Le Matin, "Mlle Marvingt gagne un coupe," November 28, 1910, p. 5.

Le Matin, "La Coupe Femina," December 23, 1910, p. 4.

Le Matin, "Deux aviateurs se tuent," December 29, 1910, pp. 1–2.

Le Matin, "Mlle Marvingt, aviatrice, gagne une course de skis," January 24, 1911, p. 3.

Le Matin, "Course de luges à plat ventre," January 25, 1911, p. 3.

Le Matin, "L'aviatrice Mlle Marvingt a gagné la course de dames," February 12, 1911, p. 3.

Le Matin, "Deux aviateurs voulaient vaincre le vent; ils sont morts d'une mort affreuse," May 19, 1911, pp. 1–2.

Le Matin, "Les Ondines nagent dans la Marne, à Joinville," July 24, 1911, p. 5.

Le Matin, "Une aviatrice tombe dans un café," August 7, 1911, p. 1.

Le Matin, "Record d'aviation à deux," August 9, 1911, p. 1.

Le Matin, "L'aviatrice fait du tourisme," September 21, 1911, p. 5.

Le Matin, "Reims-Mourmelon-Reims en aéroplane avec six passagers," October 19, 1911, p. 2.

Le Matin, "La Fête de Nancy (7 et 8 avril)," March 23, 1912, p. 2.

Le Matin, "Les Fêtes d'aviation," April 8, 1912, p. 3.

Le Matin, "Les Fêtes d'aviations de Nancy," April 9, 1912, p. 3.

Le Matin, "Les Fêtes d'aviation," April 15, 1912, p. 3.

Le Matin, "Aéronautique," June 13, 1912, p. 5.

Le Matin, "Aéronautique," July 9, 1912, p. 4.

Le Matin, "A la recherche de l'aéroplane," July 17, 1912, p. 6.

Le Matin, "Voyage interrompu," August 5, 1912, p. 3.

Le Matin, "Conférencière," August 19, 1912, p. 4.

Le Matin, "L'aviatrice fait du ballon," September 19, 1912, p. 5.

Le Matin, "Aumont-Thiéville passe la Manche en ballon," September 30, 1912, p. 5.

Le Matin, "Une aviatrice très occupée," December 10, 1912, p. 6.

Le Matin, "Sports d'Hiver," January 26, 1913, p. 7.

Le Matin, "La conférencière prend le plus court," April 7, 1913, p. 5.

Le Matin, "Mlle Marvingt dans les nuages," June 14, 1913, p. 6.

Le Matin, "Etats généraux de tourisme: La dernière liste," September 2, 1913, p. 2.

Le Matin, "Tir: Mlle Marvingt se distingue au championnat d'Europe de dames," September 8, 1913, p. 5.

Le Matin, "Concours de hauteur," September 28, 1913, p. 2.

Le Matin, "Mlle Marvingt première femme sportive du monde va disputer la Coupe Femina," October 4, 1913, p. 4.

Le Matin, "Le tourisme aérien," October 17, 1913, p. 2.

Le Matin, "L'éclectique aviatrice s'entraîne," October 30, 1913, p. 5.
Le Matin, "Être blessé n'est rien," December 17, 1913, p. 5.
Le Matin, "Vingt-deux ballons sphériques s'élèvent des Tuileries pour le Grand Prix de distance de l'Aéro-Club," July 20, 1914, p. 5.
Le Matin, "Le Grand Prix de l'Aéro-Club," July 21, 1914, p. 6.
Le Matin, "L'Aviation sanitaire," October 24, 1921, p. 5.
Le Matin, "Echos sportifs," January 19, 1922, p. 4.
Le Matin, "La dernière course de chevaux pilotés par des femmes," September 23, 1928, p. 3.
Le Matin, "Le ministre de l'air chez les Vieilles Tiges de l'aviation," November 18, 1931, p. 3.
Le Matin, "La plus jeune aviatrice du monde," November 21, 1932, p. 4.
Le Matin, "L'assemblée générale des anciens des régiments d'aviation du Maroc et de la Syrie," March 25, 1933, p. 10.
Le Matin, "Légion d'honneur," January 26, 1935, p. 5.
Le Matin, "Légion d'honneur," January 27, 1935, p. 2.
Le Matin, "Constitution au Maroc d'une ligue de l'aviation sanitaire civile," April 11, 1935, p. 5.
Le Matin, "La première de ce soir," October 27, 1937, p. 10.
Le Matin, "La féte de l'air le 10 juillet à Villacoublay," June 30, 1938, p. 8.
Le Matin, "En souvenir d'Amelia Earhart, la célèbre aviatrice américaine," July 5, 1938, p. 2.
Mauchaussée, Jean, "Marie Marvingt, reine de l'air," *Prestige*, October 1, 1959, no. 3, p. 1.
Mazenod, Lucienne, et Ghislaine Schoeller, *Dictionnaire des Femmes Célèbres* (1992), p. 579.
McQueen, Elizabeth Lippincott (Mrs. Ulysses Grant), "Women's International Association of Aeronautics," letter held at the Smithsonian Archives, September 25, 1949.
Méchelle, Ginette, interview, June 23, 1982.
La Médecine internationale, "Les Amis de l'Aviation sanitaire," August 1929, p. lxi.
The [Melbourne] Age, "London Gossip," August 17, 1912, p. 4.
The [Melbourne] Australasian, "Australians Abroad," December 4, 1909, p. 39.
Mémoires de la Société d'agriculture, commerce, sciences et arts du département de la Marne, "Mourmelon Capitale de l'Aviation," 1960, p. 182.
Messidor, "La traversée de Paris à la nage," July 9, 11, 17, 18, 1907, p. 3.
Le Meusien, "Toujours pilote d'avion à 86 ans," October 20, 1961.
Micelli, Corinne, *Air Actualités*, "Marie Marvingt: Pionnière des ailes qui sauvent," no. 2006, April 2006, pp. 58–61.
Michel-Royer, Dr. Jean, recorded interview with Marie Marvingt, January 1963.
Micromegas, "Une Femme," *L'Echo d'Alger*, July 6, 1922, p. 2.
Micromegas, *L'Echo d'Alger*, "P.S.," July 9, 1922, p. 2.
Millward, Liz, *Women in British Imperial Airspace: 1922–1937* (2008), p. 32.
"Mlle Marvingt à Bône," Algerian newspaper, July 9, 1924.
Les Modes de la Femme de France, no. 463, March 23, 1924, p. 8, and no. 472, May 25, 1924, p. 28.
Mollier, Christian et Jean-Pierre Gallay, *La mémoire des Sports d'hiver au pied du Mont-Blanc* (Imprimerie Nouvelles Chamonix/Sallanches) 2000, pp. 7, 80–81, 88, 92, 154.
Le Monde, "Nouveaux Attentats," May 20, 1952.
Le Monde, "Marie Marvingt, pionnière de l'aviation," July 24, 2004.
La Montagne, "Conférences," 1906, vol. 2, p. 251.
La Montagne, "Nouvelles alpines," vol. 3, Septembre 1907, p. 418
La Montagne, "Vers la création d'un timbre 'Marie-Marvingt,'" April 8, 2002.
Moolman, Valerie, *Women Aloft* (Time/Life Books, 1981), p. 30 / Moolman, Valerie, *Les femmes aviateurs* (Time/Life Books, 1982), p. 30.
Moore, Alison, ed., *Sexing Political Culture in the History of France* (2012).
Morgane, Yves, "La fiancée du danger: Marie Marvingt fut l'une des premières femmes aviatri-ces égales des hommes dans la conquête de l'air," *Miroir de l'histoire*, no, 37, November-December 1978, pp. 80–84.
Mortane, Jacques, *La femme dans le sport et l'aviation* (Éditions J. Dupuis Fils et Cie, 1937), pp. 23, 28, 44.
Musée de l'Air, "Marvingt, Marie: Une aviatrice tombe dans un jeu de boules," photo (1911).
Musée de l'Air, Service Bibliothèque et Documentation, dossier on Marie Marvingt.
La Musette (Massif Central), "Nos Echos," March 1911, p. 15.
Myriel, "Impressions d'une Française sur Giullaume II à Urville," *L'Est Républicain*, May 22, 1905, p. 1.
Myriel, "Mes débuts d'alpiniste," *L'Éclair de l'Est*, December 6, 1908, p. 2.
Myriel, "Les sports d'hiver," *Éclair de l'Est*, March 8, 1909.
Myriel, "Le ski," *Éclair de l'Est*, March 21, 1909.
Myriel, "L'Aviation," *Éclair de l'Est*, May 4, 1909.
Myriel, "Le 'Ville de Nancy': Il est arrivé!" *Le Sport*, July 24, 1909, p. 2.
Myriel, "Un nouveau monoplan à Champigneulles," *L'Éclair de l'Est*, August 4, 1909.
Myriel, "La grande semaine d'aviation," *L'Éclair de l'Est*, August 27, 1909.
Myriel, "La grande semaine d'aviation," *L'Éclair de l'Est*, September 1, 1909, p. 2.
Myriel, "Ma première ascension comme pilote," *L'Éclair de l'Est*, September 29, 1909.
Myriel, "Comment j'ai gagné la Coupe Léon Auscher," *Le Sport*, March 5, 1910.
Myriel, "Clôture de la Grande Semaine des Vosges, organisée par le Touring-Club de France," *Le Sport*, supplement to no. 73, April 28, 1910.
Myriel, "Comment on devient Pilote-Aviateur," *Le Sport*, no. 102, November 24, 1910, p. 1.
Myriel, "Mon premier vol de durée," *Le Sport*, December 8, 1910.
Myriel, "Pourquoi je n'ai pas pu reprendre la coupe femina," *Courrier de l'Est*, January 4, 1911.
"Nageuse, écuyère, skieuse, aviatrice, etc., la femme la plus décorée du monde reste, à 74 ans, fidèle à la bicyclette," newspaper article, 1950.

Nancy-Ochey Flash, "Marie Marvingt, 'La fiancée du danger,' no. 171, July 2003, p. 17.
Nancy Sportif. "La Vie sportive de Mlle Marvingt," June 4 juin, 1914.
New Castle [PA] News, "Pilots Commended by French Aviatrix," January 9, 1936, p. 19.
New York Post, "France to England by Balloon," October 26, 1910.
The [New York] Sun, "Eight Women Aviators," May 8, 1910, p. 3.
The [New York] Sun, "Women As Aviators," January 18, 1911, p. 6.
The [New York] Sun, "Woman Aviator Hits Tree," August 7, 1911, p. 1.
The [New York] Sun, "Women Seek War Jobs," April 14, 1915, p. 2.
New York Times, "Winner of the Cup Offered by Femina," January 1, 1911, p. 7.
New York Times, "Early Aviator, 88," December 16, 1963, p. 33.
New York Times, "Sports People," September 11, 1987.
Nicolaou, Stéphane, et Élizabeth Mismes-Thomas, *Aviatrices: Un siècle d'aviation féminine française* (Altipresse: Musée de l'Air et de l'Espace, 2004), pp. 38–42.
Nicolas, Eric, "Sur les traces de Marie Marvingt," *L'Est Républicain*, March 8, 2009.
Nicolas, Eric, *L'Est Républicain*, "Marie Marvingt va revoler," April 4, 2010.
Nicolle, Jean-Loup, "Marie Marvingt racontée aux enfants," Prix Jeanne Goury, 1961, Musée des Beaux Arts, Nancy.
Nocher, Jean, *En direct avec vous*, émission de radio, January 1964.
Noël, Gérard, "Marie Marvingt: femme intrépide du début du siècle," *Des Livres*, October 22, 1991, pp 1–2.
Noetinger, Jacques, "Reflets du ciel: La première de toutes," *Vie et bonté: Revue officielle de la Croix Rouge Française*, no. 160, November 1965.
Nord-Touriste, "Le grand-prix de l'Aéro-Club de France," June 1910, p. 137.
Nord-Touriste, "La Coupe Femina," December 1910, p. 255.
Nord-Touriste, "Pour doter l'Armée d'un avion-ambulance," November 1912, p. 246.
Nord-Touriste, "Une conférence sur l'aérostation et l'aviation par Mlle Marvingt," November 1912, pp. 251–252.
Nord-Touriste, "La conférence de Mlle Marvingt sur l'aérostation et l'aviation," December 1912, pp. 269, 274–275.
Nord-Touriste, "Conférences Mensuelles," April 1913, p. 74.
Nord-Touriste, "Le Concours d'atterrissage," July 1913, p. 158.
Nord-Touriste, "Nos Aviateurs Militaires: lettre de Paul-Maurice Echeman à sa soeur," October 1912, pp. 231–232.
Normand, Suzanne, "Les Aviatrices," *Marianne*, March 4, 1936, p. 9.
The North West Post [Formby, Tasmania], "Woman Falls in Restaurant," August 17, 1911, p. 3; *[Adelaide] Daily Herald*, "Lost Her Nerve, Woman Comes to Grief," August 15, 1911 p. 5; *Daily Post [Hobart, Tasmania]*, "Woman Falls in Restaurant," August 15, 1911, p. 5; *[Wagga Wagga, Australia] Daily Advertiser*, "Lady Aviator's Fall," August 15, 1911, p. 2.
Norwalk Reflector Herald, "Three Frenchwomen Volunteer to Serve in Aviation Corps," September 11, 1914, p. 4; *Birtle Eye [Manitoba] Witness*, "Activities of Women," November 3, 1914, p. 3; *Bismarck [ND] Daily Tribune*, "Activities of Women," November 11, 1914, p. 2; *The Bourbon News [Paris, KY]*, "Activities of Women," October 9, 1914, p. 3; *Brian [Denton, TX] Daily Eagle and Pilot*, "Activities of Women," September 1, 1914, p. 4; *Chillicothe [MO] Constitution*, "Activities of Women," October 12, 1914, p. 3; *Cullman [AL] Times*, "Activities of Women," November 5, 1914, p. 8; *Indiana Weekly Messenger*, "Activities of Women," November 4, 1914, p. 2; *Jeffersonville [IN] Star*, "Activities of Women," October 3, 1914, p. 3; *The Ocala fl Evening Star*, "Activities of Women," December 28, 1914, p. 2; *Palo Alto [Emmetsburg, IA] Reporter*, "Activities of Women," October 8, 1914, p. 8; *Racine [WI] Journal News*, "What Women Are Doing," October 4, 1914, p. 6; *Racine [WI] Journal News*, "Concerning Women," November 28, 1914, p. 6; *Racine [WI] Journal News*, "What Women Are Doing," January 2, 1915, p. 6; *Sequoyah County [OK] Democrat*, "Activities of Women," September 25, 1914, p. 7.
Nottingham Evening Post, "Woman Flies for 53 Minutes," November 29, 1910, p. 4.
La Nouvelle République, "Marie Marvingt de Nancy fête ses 80 ans," no. 3178, February 23, 1955.
La Nouvelle République, "Marie Marvingt la 'fiancée du danger' est morte," no. 5855, December 16, 1963, pp. 1, C.
Oakland Tribune, "Daring Birdwomen Invade Cloudland," July 23, 1911, p. 12.
Oakland Tribune, "Flyerettes," August 7, 1911, p. 6.
Ogden [UT] Standard, "The Alpine Climbing Season," October 16, 1905, p. 1.
Olivaint, Maurice, "Causerie d'un Algérois," *Annales Africaines*, May 22, 1925, pp. 324–325.
Ottogalli-Mazzacavallo, Cécile, *Femmes et alpinisme (1874–1919): Un genre de compromis* (L'Harmattan, 2006), pp. 162–163.
Oudin, René, "Marie Marvingt succède à Gyp," *L'Est Républicain*, July 17, 1968.
L'Ouest Éclair [Caen] and *L'Ouest Éclair [Rennes]*, "Prouesse d'aviatrice: Elle s'élève à trois mille mètres avec deux passagers," September 19, 1912, p. 5.
L'Ouest Éclair [Caen] and *L'Ouest Éclair [Rennes]*, "Une aviatrice," October 3, 1913, p. 3.
L'Ouest Éclair [Caen], "Ligue Aéronautique de France," May 23, 1928, p. 6.
L'Ouest Éclair [Caen], "Le Général Marty félicite Mlle Marvingt," June 1, 1928, p. 5.
L'Ouest Éclair [Caen], "Anciens de l'Aéronautique," June 2, 1928, p. 6.
L'Ouest Éclair [Caen], "Pour l'Essor de l'aviation française," March 25, 1932, p. 6.
L'Ouest Éclair [Caen], "Le gala de l'Aéro-Club de Caen et du Calvados," March 10, 1933, p. 4.
L'Ouest Éclair [Caen], "Le gala de la Ligue Aéronautique de France," April 8, 1937, p. 4.
L'Ouest Éclair [Caen], "Le gala de la Ligue Aéronautique de France," April 16, 1939, p. 4.

L'Ouest Éclair [Nantes] and *L'Ouest Éclair [Rennes]*, "Une exploratrice vient de parcourir 56.114 kilomètres," June 16, 1926, p. 3.

L'Ouest Éclair [Rennes], "La première aviatrice brevetée," October 23, 1910, p. 1.

L'Ouest Éclair [Rennes], "Le grand prix de l'Aéro-Club," July 21, 1914, p. 3.

L'Oued-Sahel, "Conférence de Mlle Marie Marvingt," January 4, 1923, p. 2.

Pacific Stars and Stripes, "Never Too Old," April 4, 1955, p. 1.

Pacific Stars and Stripes, "Remembered," 17 June 1955, p. 19

Pacific Stars and Stripes, "Marie Marvingt Dies; Flight Pioneer," December 17, 1963, p. 5.

Paris-Esperanto, "Nos dîners," December 1910, p. 55.

Le Parisien, "Marie Marvingt, surnommée 'la fiancée du danger,' est décédée samedi à Nancy," December 16, 1963.

Paris-Match, "Mlle Marvingt: Depuis 60 ans 'Fiancée du Danger,'" no. 7, May 7, 1949, pp. 24–25.

Paris-Soir, "Sports d'hiver," January 30, 1928, p. 4.

Paris-Soir, "L'Automobile Club féminin au Bourget," May 20, 1928, p. 3.

Paris-Soir, "Avec les 'as,' en attendant les concurrents de la Coupe Deutsch de la Meurthe," May 30, 1933, p. 5.

Paris-Soir, "Aviatrices," December 5, 1934, p. 2.

Paris-Soir, "Légion d'Honneur," January 27, 1935, p. 3.

Le Pays Lorrain, "L'Entente cordiale et la Lorraine," no. 7, 1907, pp. 350–351.

Le Pays Lorrain, "A Metz," no. 4, 1914, p. 256.

Le Pays Lorrain, "Nouvelles Lorraines," no. 10, 1952, p. 142.

Pelletier, George E., "Wings for the Wounded," *Oakland Tribune*, 17 September 1939, p. 78; *Lincoln [NE] Sunday Journal and Star*, 17 September 1939, p. 33.

Pelot, Paul, "Marie Marvingt fut une des femmes les plus extraordinaires du siècle," *L'Équipe*, December 16, 1963.

The [Perth] Daily News, "Woman Aeronaut Blown Out to Sea, Travels From France to England and Has Narrow Escape, Leap from the Car," November 29, 1909, p. 2; *Bairnsdale Advertiser and Tambo and Omeo Chronicle*, "Woman Aeronaut Blown Out to Sea," January 6, 1910, p. 6.

The [Perth] Daily News, "Adventurous Voyages," December 19, 1910, p. 8; *[Melbourne] Punch*, "Prattle About People," December 8, 1910, p. 6; *The [Perth, Australia] Daily News*, "Mainly People," December 17, 1910, p. 6.

The [Perth] Daily News, "Woman's Chat," December 13, 1913, p. 3; *The Age [Melbourne]*, "Woman's World," October 29, 1913, p. 15.

Pessel, Jeanne-Georgette, letter, June 27, 1982.

Petit, Edmond, "Hommage à Marie Marvingt," *Forces Aériennes Françaises*, no. 200, February 1964, pp. 268–270.

Petit, Edmond, *Histoire mondiale de l'aviation* (Hachette, 1967), p. 73.

Petit, Edmond, and Patrick Facon *La vie quotidienne dans l'aviation en France au début du XXe siècle, 1900–1935* (Broché, 1977), p. 78.

Petit, Edmond, *Nouvelle histoire mondiale de l'aviation* (Hachette Réalités, 1973, 4th ed., 1978), p. 71.

Le Petit Journal, "Une femme a traversée la Manche en ballon," October 28, 1909, p. 1.

Le Petit Matin [Tunis], "La conférence de Mlle Marvingt au Cercle de l'Automobile-Club," March 10, 1932.

Le Petit Matin [Tunis], "Mlle Marvingt, la 'fiancée du danger,'" February 2, 1952.

Le Petit Matin [Tunis], "La fiancée du danger à l'Alliance Française," February 5, 1952.

Le Petit Parisien, "Les fêtes de Nancy," July 2, 1906, p. 2.

Le Petit Parisien, "La traversée de Toulouse à la nage," September 2, 1907, p. 5.

Le Petit Parisien, "La Grande Semaine du Touring-Club," January 29, 1909, p. 4.

Le Petit Parisien, "'Antoinette' à Mourmelon," June 12, 1910, p. 4.

Le Petit Parisien, "Petites Nouvelles Sportives," June 18, 1910, p. 4.

Le Petit Parisien, "Le Grand Prix de l'Aéro-Club," June 26, 1910, p. 4.

Le Petit Parisien, "Le Grand Prix de l'Aéro-Club," June 28, 1910, p. 3.

Le Petit Parisien, "A Mourmelon," August 27, 1910, p. 4.

Le Petit Parisien, "A Mourmelon," August 29, 1910, p. 5.

Le Petit Parisien, "A Mourmelon," September 4, 1910, p. 5.

Le Petit Parisien, "A Mourmelon," September 6, 1910, p. 4.

Le Petit Parisien, "A Mourmelon," September 24, 1910, p. 5.

Le Petit Parisien, "A Mourmelon," September 29, 1910, p. 5.

Le Petit Parisien, "Au camp de Châlons," October 3, 1910, p. 4.

Le Petit Parisien, "A Mourmelon," October 10, 1910, p. 4.

Le Petit Parisien, "A Mourmelon," October 18, 1910, p. 5.

Le Petit Parisien, "A Mourmelon," October 19, 1910, p. 4.

Le Petit Parisien, "Mlle Marvingt pilote-aviateur," October 23, 1910, p. 5.

Le Petit Parisien, "Exploit d'aviatrice: Mlle Marvingt vole 53 minutes," November 28, 1910, p. 2.

Le Petit Parisien, "Mlle Marvingt, détentrice de la coupe d'aviation femina," November 29, 1910.

Le Petit Parisien, "Aéronautique," December 5, 1910, p. 4.

Le Petit Parisien, "Exploit d'aviatrice," December 6, 1910, p. 3.

Le Petit Parisien, December 18, 1910, pp. 406, 408.

Le Petit Parisien, "A Mourmelon," December 22, 1910, p. 5.

Le Petit Parisien, "Contre les records," December 26, 1910, p. 4.

Le Petit Parisien, "Le Meeting de Saint Etienne," July 28, 1911, p. 4.

Le Petit Parisien, "Les Deuils de l'aviation," March 11, 1912, p. 1.

Le Petit Parisien, "La revue de printemps à Nancy," March 15, 1912, p. 5.
Le Petit Parisien, "Mlle Marvingt: Conférence en faveur de l'Aviation Militaire, June 13, 1912, p. 4.
Le Petit Parisien, "Mlle Marvingt, aéronaute, va en sphérique de Paris à Bruxelles, July 29, 1912, p. 5.
Le Petit Parisien, "Une aviatrice ammène deux passagers," September 19, 1912, p. 5.
Le Petit Parisien, "Le Voyage de Mlle Marvingt," December 18, 1912, p. 4.
Le Petit Parisien, "Les Voyages d'une Aviatrice," January 4, 1913, p. 5.
Le Petit Parisien, "Aéronautique," November 12, 1913, p. 4.
Le Petit Parisien, "Mlle Marvingt capote," December 13, 1913, p. 6.
Le Petit Parisien, "Le Grand-Prix de l'Aéro-Club," July 19, 1914, p. 5.
Le Petit Parisien, "M. Rumpelmayer est le grand vanqueur du Grand-Prix de l'Aéro-Club," July 22, 1914, p. 4.
Le Petit Parisien, "Le Miroir des sports," October 27, 1920, p. 3.
Le Petit Parisien, "Le meeting de vol à voile de Biskra," January 27, 1923, p. 3.
Le Petit Parisien, "Biskra la Blanche cité du vol à voile," February 4, 1923, p. 1.
Le Petit Parisien, "Un challenge pour l'Aviation Sanitaire," October 10, 1931, p. 6.
Le Petit Parisien, "Les principales émissions françaises," January 21, 1933, p. 10.
Peyrefitte, Alain, ed., "Les femmes pilotes affrontent le danger" (*Le Figaro*, 20 janvier 1914), *L'Aventure du XXe Siècle* (1989), p. 180.
Peyrègne, Bertrand (text) et Robert Rigot (illustrations), "On l'appelait 'La fiancée du danger,' *Jeunes*, no. 3, 1964.
Pfeiffer, Doug, "Musée Dauphinois: La Grande Histoire du Ski," *Skiing Heritage Journal*, March 2004, pp. 33–37.
Picot, Augusta, letter, June 30, 1982.
Pilâtre, Philippe Buron, *Lorraine, Fille de l'Air* (Éditions Serpenoise, 2010), pp. 52–58.
Pionniers: Revue Aéronautique, "Voici 120 ans ... Marie Marvingt," no. 124, April 1995, pp. 39–40.
Plancard, Frédéric, "La Fiancée du Danger," *L'Est Républicain*, September 6, 2013.
Planck, Charles E., *Women with Wings* (1942), p. 302.
Plauche-Gillon, Jéhanne, letter to the Lhérault family, November 5, 1963.
Poiré, Léopold, war photographs to 1917 dedicated to Marie Marvingt, Archives Départe-mentales de Meurthe-et-Moselle.
Poirier, Jean-Pierre, *La Véritable Jacqueline Auriol* (2005), pp. 51–52.
Poirier, Marcelle, "Danger's Sweetheart, at 80, still astonishes Paris," *Yorkshire Post and Leeds Intelligencer*, March 10, 1955, p. 5.
Polacco, Michel, "L'Avion qui sauve," *L'Aviation Autrefois* (Hoëbeke) 2007, p. 146.
Polyglotte (Journal du Comité Polyclinique de Gentilly), no 2, Spring 1984, pp. 2–3.
Popular Aviation, December 1935, vol 17, no. 6, p. 374.
Pound, Richard, ed., *Quotations for the Fast Lane* (2013), p. 199.
La Presse, "La Traversée de Paris à la Nage," July 30, 1906, p. 2.
La Presse, "Paris à la nage amateurs," July 17, 1907, p. 3.
La Presse, "La traversée de Toulouse à la nage," September 3, 1907, p. 3.
La Presse, "La Grande Semaine du T.C.F.," January 29, 1909, p. 1.
La Presse, "Aviation," June 18, 1910, p. 3.
La Presse, "Grand-Prix de l'Aéro-Club," June 23, 1910, p. 3.
La Presse, "Ca et Là," August 11, 1910, p. 3.
La Presse, "Mlle Marvingt continue," December 1, 1910, p. 3.
La Presse, "Au camp de Châlons: Une tentative de Mlle Marvingt," December 31, 1910, p. 11.
La Presse, "Le Championnat de bobsleighs," January 26, 1911, p. 3.
La Presse, "De l'aéroplane au ballon," March 5, 1911, p. 3.
La Presse, "Chute de Mlle Marvingt," August 8, 1911, p. 4.
La Presse, "Nouvelle chute de Mlle Marvingt," August 16, 1911, p. 3.
La Presse, "De plus en plus fort," October 20, 1911, p. 3.
La Presse, "Une conférence de Mlle Marvingt," June 14, 1912, p. 3.
La Presse, "Chez Deperdussin à Courcy," July 28, 1912, p. 3.
La Presse, "De Paris vers la Belgique," August 1, 1912, p. 3.
La Presse, "Aviation," August 7, 1912, p. 3.
La Presse, "A l'école Deperdussin," August 8, 1912, p. 3.
La Presse, "Série de beaux vols," August 14, 1912, p. 3.
La Presse, "Aérostation: De Paris à Worcester," October 1, 1912, p. 3.
La Presse, "Aviation," October 3, 1912, p. 3.
La Presse, "Les exploits de Mlle Marvingt," October 6, 1912, p. 3.
La Presse, "Pour aller faire une conférence," April 18, 1913, p. 3.
La Presse, "Les Grandes Emotions du Sport," August 12, 1913, p. 1.
La Presse, "L'Oevure de Mlle Marvingt: L'Avion-Ambulance," August 18, 1913, p. 3.
La Presse, "Le Grand Prix des Ballons," July 21, 1914, p. 4.
La Presse, "Le Grand Prix des Ballons," July 22, 1914, p. 3.
La Presse, "La fiancée du danger à l'Alliance Française," February 5, 1952.
La Presse Médicale, "L'Aviation sanitaire en Algérie, au Maroc et au Levant en 1929 and en France en 1930," January 17, 1931, pp. 85–86.
La Presse Médicale, "Journées d'aviation sanitaire coloniale," March 4, 1931, p. 335.
Prévost, Marie-Eudes Lauriot et Adélaïde de Clermont-Tonnerre, "Les chevalières du ciel: Marie Marvingt, la fiancée du danger," *Point de vue*, no. 3204, December 16, 2009, pp. 38–43.
Probst, Ernst, "Marie Marvingt, Die Mütter der Luftambulanz," *Königinnen der Lufte von A bis Z* (2010), pp. 249–252, 602.

Proctor, Tammy M., *Civilians in a World at War, 1914–1918* (2010), p. 97.
Procureur, Jean-Pierre, *La Grande Semaine d'Aviation de la Champagne: 22–29 août 1909* (Editions Dominique Fradet, 2009), p. 118.
Le Progrès, "Mlle Marvingt à Orléansville," July 2, 1925, p. 2.
The Queenslander, "A New Sport for Women," January 28, 1911, p. 6.
Le Radical, "Paris à la nage," July 30, 1906, p. 2.
Le Radical, "Informations," August 3, 1906, p. 3.
Le Radical, "De Lorraine en Angleterre par-dessus la Manche," October 30, 1909, p. 3.
Le Radical, "Une femme aéronaute," November 27, 1909, p. 3.
Le Radical, "Le grand prix de l'Aéro-Club," June 23, 1910, p. 5.
Le Radical, "Le grand prix de l'Aéro-Club," June 25, 1910, p. 5.
Le Radical, "Le grand prix de l'Aéro-Club," June 26, 1910, p. 7.
Le Radical, "Les dames aéronautes," August 10, 1910, p. 5.
Le Radical, "Une nouvelle aviatrice," September 5, 1910, p. 5.
Le Radical, "Mlle Dutrieu contre Mlle Marvingt," November 30, 1910, p. 5.
Le Radical, "La Coupe Femina pour aviation," December 6, 1910, p. 1.
Le Radical, "Pour la Coupe Femina," December 31, 1910, p. 1.
Le Radical, "Pour la Coupe Femina," January 1, 1911, p. 1.
Le Radical, "Le Championnat de France des bobsleighs," January 26, 1911, p. 5.
Le Radical, "Les sports: Natation," July 5, 1911, p. 7.
Le Radical, "Ministres en voyage," July 1, 1912, p. 2.
Le Radical, "Raid d'une aviatrice en ballon sphérique," July 29, 1912, p. 3.
Le Radical, "Aviation: Envolées diverses," July 30, 1912, p. 5.
Le Radical, "Où ont-ils atterri?" August 1, 1912, p. 6.
Le Radical, "Mlle Marvingt, avec deux passagers, s'élève à 3,000 mètres," September 19, 1912, p. 6.
Le Radical, "A l'aérodrome de la Champagne," December 7, 1912, p. 5.
Le Radical, "Mlle Marvingt tombe sans mal," December 13, 1913, p. 6.
Le Radical, "La grande fête des ballons," July 9, 1914, p. 5.
Le Radical, "Le grand prix des ballons," July 11, 1914, p. 4.
Le Radical, "Le concours international de ballons," July 19, 1914, p. 5.
Le Radical, "Le Grand Prix de l'Aéro Club—Les atterrissages," July 21, 1914, p. 5.
Le Radical, "Le grand prix de la Corse," April 22, 1921, p. 4.
Raffalovich, G.D., "Souvenirs des temps héroïques," *Le Monde Illustré*, November 17, 1934, pp. 964, 966.
Le Rappel, "Derrière la Toile," February 15, 1906, p. 3.
Le Rappel, "La traversée de Paris à la nage," July 31, 1906, p. 2.
Le Rappel, "La Vie Sportive: Le Ski," January 31, 1909, p. 3.
Le Rappel, "La Vie Sportive: Aéronautique," October 31, 1909, p. 4.
Le Rappel, "Second femme aviateur," June 21, 1910, p. 4,
Le Rappel, "Aéronautique: Le Grand-Prix de l'Aéro-Club," June 30, 1910, p. 4.
Le Rappel, "Chute d'une aviatrice," August 9, 1911, p. 4.
Le Rappel, "Mlle Marvingt tombe encore," August 17, 1911, p. 3.
Le Rappel, "Une aviatrice fait une conférence," June 14, 1912, p. 3.
Le Rappel, "MM. Poincaré et Lebrun à Bar-le-Duc," July 2, 1912, p. 3.
Le Rappel, "Aéronautique: Le prix Salmageane," October 1, 1912, p. 4.
Rastelli, E., "La Semaine d'Aviation de Turin," *L'Aérophile*," August 1, 1911, p. 357.
Ravailler, Jean-Pierre, "Marie Marvingt, aviatrice, une grande aventurière," *Elle sont passées par la Lorraine* (2011), pp. 39–40.
The Register [Adelaide, Australia], "Remarkable Adventure," December 23, 1909, p. 12; *[Adelaide] Evening Journal*, "Balloon Adventure: Occupants Drift Across the North Sea," December 18, 1909, p. 2; *The World's News [Sydney]*, "Balloon Adventure: Occupants Drift Across the North Sea," December 11, 1909, p. 9.
Reichardt, Gisèle, "'Marie Casse-Cou' un siècle d'exploits," *Républicain Lorrain*, April 26, 1992, p. 7.
Remy, Jean-Louis, *C'était hier ... Nancy et ses environs* (1978), no. 53, 54, 55.
Renac, Jean, "Les femmes et l'aéronautique," *France Aviation*, July-August 1982, p. 3.
Renoux-Barès, E., "Une héroïne parle à Pau de l'aviation, école d'énergie," *Revue Aéronautique de France*, October 1927, p. 11.
Républicain Lorrain, "A Nancy et à Metz, Marie Marvingt, la 'Fiancée du Danger' dira quels sont les buts humanitaires de sa mission," October 19, 1949, p. 2.
Républicain Lorrain, "Les anciennes élèves de Sainte-Chrétienne ont retouvé avec plaisir l'ambiance de leur établissement," May 14, 1955.
Républicain Lorrain, "Marie Marvingt, la 'fiancée du danger,' a fêté ses 85 ans," February 21, 1960, p. 3.
Républicain Lorrain, "Marie Marvingt, la 'fiancée du danger,' n'est plus," December 15, 1963, pp. 1, 4.
Républicain Lorrain, "Les obsèques de Mlle Marie Marvingt ont lieu ce matin à 10 h à Saint-Epvre," December 17, 1963.
Républicain Lorrain, "Les Ailes françaises ont rendu un ultime et émouvant hommage à Marie Marvingt, 'La colombe qui sauve,'" December 18, 1963, p. 2.
Républicain Lorrain, "Hommage à Marie Marvingt," March 2, 1990.
Républicain Lorrain, "Promotion 'Marie Marvingt.'" March 4, 1990.
Républicain Lorrain, "La fiancée du danger," June 23, 1996.
Républicain Lorrain, "Marie Marvingt dans le panthéon postal," July 1, 2004
Républicain Lorrain, "Sur les traces de Marie Marvingt," October 8, 2010.

Républicain Lorrain, "Marie Marvingt, une femme de tempérament et d'exception," February 20, 2014.
La Revue aérienne, "A Châlons," December 25, 1909, pp. 775–776.
La Revue aérienne, "A Reims," December 25 1909, pp. 776–777.
La Revue aérienne, "Les chutes du 4 janvier," January 10, 1910, p. 21.
La Revue aérienne, "Nos Prix," January 10, 1910, p. 28.
La Revue aérienne, "Nos Prix," July 10, 1910, p. 412.
La Revue aérienne, "Les aviatrices," September 10, 1910, p. 517.
La Revue aérienne, "La semaine d'aviation à Turin," July 10, 1911, p. 359.
La Revue aérienne, "Nouvelles sportives," September 25, 1911, p. 482.
La Revue aérienne, "Rapport du Comité directeur à l'Assemblée générale," October 10, 1911, p. 532.
La Revue aérienne, "Les Fêtes du 30 juin à Bar-le-Duc," July 25, 1912, pp. 398, 401.
La Revue aérienne, "Aéro Club de France grand prix des sphériques," July 25, 1914, p. 420.
Revue aéronautique de France, "L'Ecole des femmes ... pilotes," November-December 1924, p. 217.
Revue aéronautique de France, "Les distinctions de la Ligue Aéronautique de France," January 1926, p. 13.
Revue aéronautique de France, "Les Conférences du Comte de La Vaulx," April-May-June 1926, pp. 21–22.
Revue aéronautique de France, "Distinctions méritées," July 1926, p. 47.
Revue aéronautique de France, "La Ligue Aéronautique en Province," May 1927, pp. 8–9, 13.
Revue aéronautique de France, "La Ligue en Province: Muret," June 1927, p. 8.
Revue aéronautique de France, "Assemblée général statutaire du 27 juin 1927," July 1927, p. 11.
Revue aéronautique de France, "Mlle Marvingt," July 1927, pp. 15–16.
Revue aéronautique de France, "Partie Officielle," April 1928, p. 13.
Revue aéronautique de France, "Sur la Côte d'Azur: Conférences de Mlle Marvingt," May 1928, p. 12.
Revue aéronautique de France, "Assemblée générale statutaire de la Ligue Aéronautique," July 1928, p. 7.
Revue aéronautique de France, "Mlle Marvingt au Mans," July 1928, p. 10.
Revue aéronautique de France, "Une initiative de Mlle Marvingt," August 1928, p. 13.
Revue aéronautique de France, "La Ligue en Province," February 1929, p. 5.
Revue aéronautique de France, "Assemblée générale statutaire," July 1929, p. 14.
Revue aéronautique de France, "Mademoiselle Marvingt: Une grande propagandiste," October 1929, pp. 7, 15.
Revue aéronautique de France, "Une conférence de Mlle Marvingt à Nancy," November 1929, pp. 12–13.
Revue aéronautique de France, "Mademoiselle Marvingt," January 1931, pp. 11, 12; "Notre fête du 21 décembre," p. 14.
Revue aéronautique de France, "Partie Officielle," July-August 1931, p. 11.
Revue aéronautique de France, "La Ligue Aéronautique de France et La Jeunesse des Ecoles," March 1932, p. 14.
Revue aéronautique de France, "Partie Officielle," August-September 1932, pp. 3–4.
Revue aéronautique de France, "Une belle tournée de conférences," October 1932, p. 5.
Revue aéronautique de France, "Comité des Dames," January 1933, p. 8.
Revue aéronautique de France, "Partie Officielle," July 1933, pp. 10–11, 16.
Revue aéronautique de France, "Mlle Marvingt au Maroc," February-March 1934, p. 4.
Revue aéronautique de France, "La Ligue aéronautique de France et les écoles," February-March 1934, p. 8.
Revue aéronautique de France, "Avec Mademoiselle Marvingt: Présidente fondatrice de l'aviation sanitaire civile au Maroc," July 1934, pp. 2, 15.
Revue aéronautique de France, "Distinctions Honorifiques: Légion d'Honneur," February 1935, p. 2.
Revue aéronautique de France, "La Ligue aéronautique de France et les écoles," April 1935, p. 14.
Revue aéronautique de France, "Présentation au Bourget de la première promotion des infirmières de l'air," June 1935, pp. 5, 7.
Revue aéronautique de France, "Partie Officielle," November 1935, p. 11.
Revue aéronautique de France, "Grande Fête Annuelle Populaire d'Aviation," April 1936, pp. 10–11.
Revue aéronautique de France, "La Fête populaire annuelle de la Ligue Aéronautique de France," June-July 1936, pp. 3, 12, 14.
Revue aéronautique de France, "Sections Féminines," February-March 1937, p. 15.
Revue aéronautique de France, "Nouvelles," May-June, 1938, p. 10.
La revue de l'aviation, "Femmes volantes," January 7, 1911.
La revue du Mont-Blanc et de Chamonix, "Les prouesses d'une Française à Chamonix," August 1905.
Reynolds, Siân, *France Between the Wars: Gender and Politics* (Routledge, 1996), p. 67.
La revue du Touring Club de France, "Petites Annonces," no. 387, February 1927, p. 37.
Revue de l'Union des femmes de France, "Les journées d'Aviation sanitaire à l'Exposition Coloniale," October 1931, p. 402.
La Revue Lorraine Populaire, "Personnages de chez nous: Marie Marvingt," December 1, 2000, p. 45.
Revue Mensuelle du Touring-Club de France, "Liste des candidats," April 1896, p. 156.
Revue Mensuelle du Touring-Club de France, "Echanges," March 15, 1901, p. 130.
Revue Mensuelle du Touring-Club de France, November 1905, p. 516.
R.G., "Une femme qui pratique tous les sports: Mlle Marvingt," *Le Miroir des Sports*, October 28, 1920, pp. 260–261.
Ric et Rac, "Prouesses d'il y a vingt ans," December 6, 1930, p. 2.
Richet, Charles, "L'aviation sanitaire c'est le progrès sans larmes," *Le Matin*, July 6, 1935, p. 1.
Rio, Armand, "La fiancée du danger," *Lecture pour tous*, vol. 15, no. 7, April 1, 1913, pp. 63–71.
Riverain, Jean, *Dictionnaire des aéronautes célèbres* (Larousse, 1970), p. 106.

Rogez, Madame, letter, January 15, 1984.
Robaux, P. et D., *Les rues de Nancy*, 1984, p. 208.
Robertston, Patrick, *Robertson's Book of Firsts* (2011), p. 546.
Robertson, Priscilla, *An Experience of Women: Pattern and Change in Nineteenth-Century Europe* (Temple University Press, 1982), pp. 341–342.
The [Rockhampton, Australia] Capricornian, "Balloonist's Thrilling Voyage," January 1, 1910, p. 28; *[Launceston, Tasmania] Daily Telegraph*, "Blown Over North Sea, Balloonists' Thrilling Voyage at the Mercy of the Elements," December 31, 1909, p. 6; *[Hobart, Tasmania] Daily Post*, "Blown Over North Sea, Balloonists' Thrilling Voyage at the Mercy of the Elements," January 3, 1910, p. 6.
Rozet, George, "L'Avènement du sport féminin," *Lectures pour tous*, April 1, 1919, p. 1565.
R.P., "Mademoiselle Marvingt," *Le cri de Constantin*, June 28, 1924.
Russell, Jesse, and Ronald Cohn, *Marie Marvingt* (2012).
Saindizié, J.-M., "C'est Marie Marvingt, 85 ans, la plus extraordinaire femme du siècle," *Pilote*, no. 56, November 17, 1960, p. 10.
Saladin, Raymond, "Les aviatrices des temps héroïques," *Aviation Magazine*, no. 51, June 1, 1952, p. 13.
Saladin, Raymond, "Figures et événements: Marie Marvingt," *Aviation Magazine*, March 24, 1955, pp. 12–13.
Salisbury Times, "Balloon in a Gale," October 29, 1909, p. 3.
The Salt Lake [UT] Tribune, "Women Contending for Aerial Honors," December 11. 1910, p. 2.
San Antonio Express, "Activities of Women," January 12. 1936, p. 30; *Ackley [IA]) World Journal*, "In Women's Realm," January 16, 1936, p. 6.
San Antonio Light And Gazette, "Daring French "Aviatresses," January 1, 1911, p. 49.
Save, Colette, "L'âge n'est plus une question d'années," *Point de Vue Images du Monde*, March 3, 1955, p. 20.
Scott, R.J., *Albuquerque Journal*, "Scott's Scrap Book," July 10, 1940, p. 11.
La Sfaxienne, "Mademoiselle Marvingt à Sfax," April 15, 1924, p. 1.
Shayler, David J., and Ian A. Moule, *Women in Space—Following Valentina* (Springer/Praxis, 2006), pp. 11, 16.
Shipton, Elisabeth, *Female Tommies: The Frontline Women of the First World War* (The History Press, 2014), pp. 83–85, 128, 237.
Seigel, Jessica, "10 Greatest Female Athletes Ever," *Glamour*, August 2008, p. 118.
Sheffield Evening Telegraph, "Midnight Descent," October 28, 1909, p. 3.
Sheffield Evening Telegraph, "News of the World," England, November 28, 1910, p. 5.
Sheffield Evening Telegraph, "News in a Nutshell," December 13, 1913, p. 5.
"Ski (Débuts en France)," *La Grande Encyclopédie de La Montagne* (Éditions Atlas, 1978), pp. 2179–2180.
Smith, Elizabeth Simpson, *Breakthrough: Women in Aviation* (Walker, 1981), p. 145.
Smith, Sidonie, "Virtually Modern Amelia: Mobility, Flight, and the Discontents of Identity," in Mary Ann O'Farrell and Lynne Vallone, eds., *Virtual Gender: Fantasies of Subjectivity and Embodiment* (1999), pp. 11, 23.
Smith, Sidonie, "In the Air: Aerial Gender and the Familiarity of Flight," *Moving Lives: Twentieth Century Women's Travel Writing* (2001), pp. 73, 80, 81.
The [Spokane, WA] Spokesman-Review, "Women Aviators for the Allied Army?," November 8, 1914, p. 2.
Le Sport, photograph from issue October 2, 1909, at the Meurthe-et-Moselle Departmental Archives.
Le Sport, "La Grande Semaine à Gerardmer," February 19, 1910.
Le Sport, "La coupe femina: Mlle Marvingt (Myriel) va s'attaquer au record de Mlle Dutrieu," November 24, 1910.
Le Sport Universel Illustré, "La première Equipe Féminine de Hockey," November 13, 1910, p. 734.
Le Sport Universel Illustré, "Nos aviatrices se distingent," December 11, 1910, p. 799.
Le Sport Universel Illustré, "Choses et Autres," November 19, 1911, p. 752.
Le Sport Universel Illustré, "Le Mont Revard," January 12, 1912, pp. 157–158.
Sportif, "Les Ondines nagent dans la Marne, à Joinville," 1906.
St. Ani, "Marie Marvingt," January 1964, p. 13.
Stanley, Autumn, *Mothers and Daughters of Invention: Notes for a Revised History of Technology* (Scarecrow Press, 1993; Rutgers University Press, 1995), p. 201.
Stewart, Mary Lynn, *For Health and Beauty: Physical Culture for Frenchwoman 1880s-1930s* (The Johns Hopkins University Press, 2001), pp. 168–169, 170–171.
The Strand Magazine, "'The Bride of Danger,' An Interview with Mlle. Marie Marvingt," vol. 46, September 1913, pp. 187–194.
Suarez, Rebecca L.; Strickland, James; Kauffman, Glenn S., *Air Ambulance Services*, no date, no page number.
Sumner, Ian, *Kings of the Air: French Aces and Airmen of the Great War* (2015), pp. 23–24.
The [New York] Sun, "Mlle. Marvingt Has Record for Pluck: French Woman Aviation Showed Remarkable Courage in Recent Accident," February 1, 1914, p. 45.
Susset, Bruno, *L'Est Républicain*, "Un des livres de Marie Marvingt retrouvé à Nancy," July 1, 1993.
The Sydney Morning Herald, "Airwomen's Flying Ambulance Corps: The Work of the Pioneer Frenchwoman," November 6, 1939, p. 2; *Townsville Daily Bulletin*, "Flying Ambulance Airwoman's Corps: Origin in France," January 6, 1940, p. 12.
[Sydney] Sunday Times, "New 'Record' for an Airwoman," January 15, 1911, p. 19.
[Sydney] Sunday Times, "Women on Wings: Some Daring Dianas of the Air," June 11, 1911, p. 23.
[Sydney] Sunday Times, "World's Famous Aviators Write of Their Experiences in Learning to Fly," November 12, 1911, p. 27.
The Tacoma [WA] Times, "Four Little Bird Women, All in a Row, Soaring For Prizes, Hi-lee, Oh, Hi-low," January 3, 1911, p. 5; *Logansport [IN] Pharos Reporter*, "Four Little Bird Women, All in a Row, Soaring For Prizes, Hi-lee, Oh, Hi-low," December 30, 1910, p. 6.

Talbott, Earl G., "Marie Marvingt, 'Fiancée of Danger': Flirted With Death 88 Years," *New York Herald Tribune*, December 16, 1963.
Le Temps, "Paris à la Nage," July 30, 1906, p. 1.
Le Temps, "Le Congrès aéronautique international," September 25, 1909, p. 2.
Le Temps, "Une femme traverse la Manche en ballon," October 29, 1909, pp. 3–4.
Le Temps, "Le Grand-Prix de l'Aéro-Club de France," June 28, 1910, p. 3.
Le Temps, "Le Grand-Prix de l'Aéro Club de France," June 29, 1910, p. 4.
Le Temps, "Une aviatrice vole cinquante-trois minutes," November 29, 1910.
Le Temps, "Aéronautique," December 7, 1910, p. 3.
Le Temps, "Tentatives/Portateurs des Brevets d'Aviation," January 1, 1911, p. 3.
Le Temps, "De l'aéroplane au ballon sphérique," March 4, 1911, p. 3.
Le Temps, "Aviateurs blessés en automobile," August 4, 1911, p. 5.
Le Temps, "Chute d'une aviatrice," August 8, 1911, p. 5.
Le Temps, "L'Aviation en Province," August 14, 1911, p. 5.
Le Temps, "L'Aviation en Province," August 16, 1911, p. 6.
Le Temps, "L'Académie des Sports," November 7, 1911, p. 5.
Le Temps, "Les fêtes d'aviation de Nancy," April 10, 1912, p. 4.
Le Temps, "Conférences d'aviatrice," June 13, 1912, p. 5.
Le Temps, "Les Voyages en Sphériques," July 30, 1912, p. 5.
Le Temps, "Les Voyages en Sphériques," August 1, 1912, p. 5.
Le Temps, "Le Prix Salmagean," September 30, 1912, p. 5.
Le Temps, "Les Aviatrices," January 25, 1913, p. 5.
Le Temps, "Le Meeting d'Aviation de Reims," September 29, 1913, p. 4.
Le Temps, "Le Grand-Prix de l'Aéro-Club de France," July 21, 1914, pp. 5, 6.
Le Temps, "Congrès national de l'Aviation Sanitaire," July 23, 1928, p. 3.
Le Temps, "Au Congrès d'Aviation Sanitaire," June 1, 1933, p. 4.
Le Temps, "Légion d'Honneur," January 27, 1935, p. 4.
Tessier, Roland, "La belle histoire des femmes volantes," *Le Figaro*, November 5, 1941, p. 3.
Tessier, Roland, *Femmes de l'Air* (Flammarion, 1948), pp. 21–22, 26.
Thétard, Henri, "Le Général Estienne et M. Citroën sont arrivés à In-Salah," *Le Petit Parisien*, February 20, 1923, p. 1.
Thirion, M., "Hommage de l'Aéro-Club de l'Est à Marie Marvingt," *L'Est Républicain*, December 18, 1963.
Thompson, Christopher, "Un troisième sexe? Les bourgeoises et la bicyclette dans la France fin de siècle," *Le Mouvement social: bulletin trimestriel de l'Institut francais d'histoire social*, July-September 2000, p. 32.
Thompson, Christopher S., *Tour de France: A Cultural History* (2008), p. 129.
The Times of India, "Marie Marvingt Dead," December 16, 1963, p. 7.
The Times [London], "Aeronautics: Across the North Sea in a Balloon," October 28, 1909, p. 7.
The Times [London], "Aeronautics: Across the North Sea in a Balloon," letter to the editor, December 18, 1909, p. 12.
Le Tir national, "XIIIe Concours National et International de Tir," August 4, 1906, p. 365.
Le Tir national, "XIIIe Concours National et International de Tir," August 11, 1906, p. 376.
Le Tir national, "XIIIe Concours National et International de Tir," September 15, 1906, p. 425.
Le Tir national, "Académie des Sports," December 23, 1911, p. 752.
La Tribune Républicaine, "Débuts de Mlle Marvingt," August 4, 1911.
"Une délégation de Lorraine offre un object d'art au président," newspaper article, October 20, 1913.
L'Union, "'La fiancée du danger' est décédée à Nancy, samedi, à 88 ans," December 16, 1963.
Valode, Philippe, "Marie Marvingt, l'inégalable casse-cou," *Les Grandes Scandaleuses* (First Editions, 2015), pp. 227–235.
Van Nuys [CA] and The Van Nuys Call, "Women Serve as Air Scouts," August 14, 1914, p. 5.
Varlet, Charles, "La réalisation de l'avion sanitaire par Marie Marvingt" (1933), in Marcel Cor-dier, "Les Escales de Marie Marvingt," *Leurs Demeures en Lorraine: Tome II* (1983), p. 168.
Veillon, Jean, *Marie Marvingt* (2013), pp. 1–87.
Verviers, Pierre, "Une femme veut doter la France d'avions-ambulances," *La Vie Sportive du Nord et du Pas-de-Calais*, November 16, 1912, p. 5.
Vestkusten [Sweden], "Marie Marvingt," March 26, 1914, p. 1.
La Vie aérienne illustrée, "Mlle Adrienne Bollard a traversé la Manche," September 4, 1920, p. 77.
La Vie au Grand Air, "Par-dessus la Mer du Nord," November 6, 1909, p. 7.
La Vie au Grand Air, "Un Tour à Mourmelon Aviation," January 1, 1910, pp. 8–9.
La Vie au Grand Air, "La Grande Semaine des Vosges," no. 597, February 26, 1910, pp. 136, 142.
La Vie au Grand Air, "Mlle Marvingt Portée en Triomphe," December 3, 1910, cover.
La Vie au Grand Air. "La Coupe Femina," December 17, 1910.
La Vie au Grand Air, "Une Aviatrice tombe dans un jeu de boules," August 19, 1911, cover.
La Vie au Grand Air, "Les Aviateurs aux Sports d'Hiver," February 17, 1912, p. 3.
La Vie au Grand Air, "Ecole de ski," March 15, 1913, p. 187.
La Vie Sportive du Nord et du Pas-de-Calais, "Les Fêtes Aérostatiques de Lille," July 19, 1913, p. 5.
La Vie Sportive du Nord et du Pas-de-Calais, "La Fête Aérostatique de Lille," July 26, 1913, p. 2.
Les Vieilles Tiges: Annuaire, "Membres Titulaires," 1924, p. 52
Les Vieilles Tiges: Annuaire, "Membres Titulaires," 1929, p. 118.

Villard, Henry Serrano, *Contact! The Story of the Early Birds* (Crowell, 1969), p. 248.

Viry-Babel, Anne, "Les Audacieuses" (film), 8 mars 2015.

"Visages de nos temps: Marie Marvingt," unidentified publication, September 19, 1932.

Volot, Frédérique, "La fiancée du danger," *Les Mystères de Meurthe-et-Moselle* (De Borée, 2011), pp. 332–346.

The [Wagga, Australia] Worker, "About Women," April 13, 1911, p. 7.

Wanss, Annie, "La mode chez les aviatrices," *La revue aérienne*, no. 82, March 10, 1912, pp. 125–128.

[Washington, D.C.] Evening Star, "A Female Admirable Crichton," October 18, 1913, p. 5.

The Washington Herald, "Women Not Good Flyers," November 29, 1910, p. 7; *El Paso Herald*, "Flying Not the Sport for Women," December 6, 1910, p. 11; *Naugatuck [CT] Daily News*, "Not a Sport in Which Women Will Excel," November 29, 1910, p. 6; *Trenton Evening Times*, "Says Women Will Not Excel in Air," November 29, 1910, p. 7.

The Washington Post, "Chicago Girl Hero of the Alps," November 12, 1905, p. SM5.

The Washington Post, "Aviation Winning Many Devotees Among Daring Women," May 8, 1910, p. S4.

The Washington Post, "Alliance to Hear French Aviatrix," January 14, 1936, p. 10.

The Washington Post, "France's Pioneer Woman Flier Wins Acclaim in a Dozen Lines," January 15, 1936, p. 13.

The Washington Post, "Wings of the World," March 8, 1936, p. PY6.

Welch, Roseanne, *Encyclopedia of Women in Aviation and Space* (1998), pp. 61, 138.

Western Daily Press [South Yorkshire], "Aeronauts' Adventures," July 21, 1914, p. 10; *Dundalk [Ireland] Examiner and Louth Advertiser*"Aeronauts' Adventures," July 25, 1914, p. 3.

Western Gazette [Somerset], "Balloonists Cross Channel: Lady's Remarkable Adventure," October 29, 1909, p. 7.

Western Mail, "'The Bride of Danger': World's Greatest Lady Athlete," May 8, 1914, p. 45; *Burnley Express, [Lancashire, England]*, "The Bride of Danger," August 6, 1913, p. 5; *Cheltenham Looker-On [Gloucestershire]*, "Bride of Danger," August 9, 1913, p. 20; *Fife Free Press, & Kirkcaldy Guardian*, August 9, 1913, p. 4; *Leicester Chronicle*, "The Bride of Danger," August 9, 1913, p. 6. *Milford [IA] Mail*, "The Bride of Danger: World's Greatest Lady Athlete," April 16, 1914, p. 2; *Whitstable Times and Herne Bay Herald*, "The Bride of Danger," August 16, 1913, p. 3.

West Gippsland [Australia] Gazette, "Flights of Women: Steering Aeroplanes As Though They Were Perambulators," March 14, 1911, p. 6; *Zeehan and Dundas [Tasmania] Herald*, "Flights of Women: Steering Aeroplanes As Though They Were Perambulators," February 8, 1911, p. 1.

Willotte, Jeanne Pauline Henriette Mathieu, letter, August 6, 1982.

Winnipeg [Manitoba] Free Press, "Over the World With Women," January 7, 1911, p. 20.

Winnipeg [Manitoba] Free Press, "Daring French Woman Retires at Eighty," July 5, 1955, p. 14; *Medicine Hat [Alberta] News*, "Marie Marvingt Retires to Enjoy Serene Old Age," July 5, 1955, p. 8.

Women's Sports and Fitness, "1987 Hall of Fame: The Winners' Circle," October 1987, p. 58; *Cedar Rapids [IA] Gazette*, "Sports Briefly," September 11, 1987, p. 14; *Colorado Springs Gazette Telegraph*, "Carner Heads Hall Inductees," September 11, 1987, p. 24; *Cumberland [MD] Evening Times*, "Sports," September 11, 1987, p. 1; *Daily Herald Suburban Chicago*, "Women's Sports," September 10, 1987, p. 812; *Daily Herald Suburban Chicago*, "Carner Among Seven Gaining Hall of Fame," September 11, 1987, p. 91; *Helena [MT] Independent Record*, "Women's Sports Hall of Fame," September 11, 1987; *Joplin [MO] Globe*, "Carner, 6 others to women's shrine," September, 11, 1987, p. 7; *Syracuse [NY] Herald Journal*, "Hall Names Seven," September 11, 1987, p. 31; *Tyrone [PA] Daily Herald*, "Women Sports," 11 September 1987, p. 6.

Woodhouse, Henry, "Women of the Air," *Outdoor World & Recreation*, vol. 48, May 1913, pp. 305–309

The World's News [Sydney], "Girl Climber's Record: Remarkable Series of Alpine Ascents," October 21, 1905, p. 7.

X.B., "D'ultimes témoignages viennent encore préciser quelques traits attachants de Marie Marvingt, illustre figure lorraine," *Républicain lorrain*, January 24, 1964, p. 2.

Yelnick, Claude, "Adieu à Marie Marvingt," *Icare*, March 1964, no. 29, pp. 60–65.

Yorkshire Evening Post, "Lady's Adventurous Voyage: Journey of a Thousand Miles Ends in Suffolk," October 28, 1909, p. 3.

Young, Pam, Director of Collections, Mission Inn, Riverside, California, letter, November 24, 1982.

Zeehan and Dundas Herald [Australia], "Female Admiral Crichton," November 19, 1913.

Zeyons, Serge, "Marie Marvingt, Une femme d'exception," *Timbres Magazine*, no. 151, December 2013, pp. 80–82.

Zwang, Annie, *100 Femmes qui ont fait l'Histoire de France* (Ellipses, 2010), pp. 158–159.

Index

www.ingramcontent.com/pod-product-compliance
Ingram Content Group UK Ltd.
Pitfield, Milton Keynes, MK11 3LW, UK
UKHW060612180726
13836UKWH00012B/2513

9 781476 675503